Theoretical Approaches to Biological Control

Biological control is the suppression of pest populations using predators, parasitoids and pathogens. Historically, biological control has usually been on a trial-and-error basis, and has failed more often than it has succeeded. However, by developing theories based upon fundamental population principles and the biological characteristics of the pest and agent, we can gain a much better understanding of when and how to use biological control. This book gathers together recent theoretical developments and provides a balanced guide to the important issues that need to be considered in applying theory to biological control. It will be a source of productive and stimulating thought for all those interested in pest management, theoretical ecology and population biology.

Bradford Hawkins has been at the University of California, Irvine since 1994, and has been an Associate Professor since 1997. He has held positions in a number of universities in the United States of America and the United Kingdom. He currently serves on the editorial board of the journals *Ecological Entomology* and *Ecology Letters*. Professor Hawkins has also co-edited *Parasitoid Community Ecology* (with William Sheehan) in 1994, and has written *Pattern and Process in Host-Parasitoid Interactions* (1994).

Howard Cornell has been a faculty member of the University of Delaware since 1975, and has been Professor of Biology since 1990. He was on the editorial board of *Ecology* and *Ecological Monographs* between 1979–1982. He has been a regular academic visitor to the Centre for Population Biology at Imperial College, Silwood Park, UK, and at the National Center for Ecological Analysis and Synthesis in Santa Barbara, California.

T0215442

Theoretical Approaches to Biological Control

EDITED BY

Bradford A. Hawkins
and Howard V. Cornell

CAMBRIDGE
UNIVERSITY PRESS

CAMBRIDGE UNIVERSITY PRESS
Cambridge, New York, Melbourne, Madrid, Cape Town, Singapore, São Paulo, Delhi

Cambridge University Press
The Edinburgh Building, Cambridge CB2 8RU, UK

Published in the United States of America by Cambridge University Press, New York

www.cambridge.org
Information on this title: www.cambridge.org/9780521572835

First published 1999
This digitally printed version 2008

A catalogue record for this publication is available from the British Library

Library of Congress Cataloguing in Publication data

Theoretical approaches to biological control / edited by Bradford A.
Hawkins and Howard V. Cornell.
      p.   cm.
Includes index.
ISBN 0 521 57283 5 (hb)
1. Pests – Biological control.   2. Insect pests – Biological
control.   3. Pests – Biological control – Mathematical models.
4. Insect pests – Biological control – Mathematical models.
5. Biological pest control agents.   I. Hawkins, Bradford A., 1952–   .
II. Cornell, Howard Vernon, 1947–   .
SB933.3.T48   1999
632′.96–dc21   98-12057   CIP

ISBN 978-0-521-57283-5 hardback
ISBN 978-0-521-08287-7 paperback

Contents

Contributors

Michael Barfield, Institut d'Ecologie, Université Pierre et Marie Curie, 7 quai St. Bernard, cc 237, 75232 Paris Cedex 05, France

Nigel D. Barlow, Biological Control Group, AgResearch, P.O. Box 60, Lincoln, New Zealand

Michael Begon, Population Biology Research Group, School of Biological Sciences, Nicholson Building, The University of Liverpool, Liverpool L69 3BX, UK

Alan A. Berryman, Department of Entomology, Washington State University, Pullman, WA 99164, USA

Cheryl J. Briggs, Department of Integrative Biology, University of California, Berkeley, CA 94720, USA

Jan Bruin, Institute of Systematics and Population Biology, University of Amsterdam, Kruislaan 320, 1098 SM Amsterdam, The Netherlands

Liebe F. Cavalieri, Division of Natural Sciences, Purchase College, State University of New York, 735 Anderson Hill Road, Purchase, NY 10577, USA

Gary C. Chang, Department of Zoology, P. O. Box 351800, University of Washington, Seattle, WA 98195–1800, USA

Martijn Egas, Institute of Systematics and Population Biology, University of Amsterdam, Kruislaan 320, 1098 SM Amsterdam, The Netherlands

H. Charles J. Godfray, Department of Biology & NERC Centre for Population Biology, Imperial College at Silwood Park, Ascot, Berkshire SL5 7PY, UK

Andrew P. Gutierrez, Department of Environmental Science Policy and Management, University of California, Berkeley, CA 94720, USA

Alan Hastings, Department of Environmental Science and Policy, University of California, One Shields Avenue, Davis, CA 95616, USA

Michael E. Hochberg, UPMC-UMR 7625, Fonctionnement et évolution des systèmes écologiques, Université Pierre et Marie Curie, 7 quai St. Bernard, cc 237, 75232 Paris Cedex 05, France

Robert D. Holt, Natural History Museum, Department of Ecology and Systematics, University of Kansas, Lawrence, KS 66045, USA

Martha S. Hunter, Department of Entomology, 410 Forbes Building, University of Arizona, Tucson, AZ 85721, USA

Vincent A. A. Jansen, NERC Centre for Population Biology, Imperial College at Silwood Park, Ascot, Berks SL5 7PY, UK. *Present address*: Department of Zoology, University of Oxford, South Parks Road, Oxford, OX1 3PS, UK

Arne Janssen, Institute of Systematics and Population Biology, University of Amsterdam, Kruislaan 320, 1098 SM Amsterdam, The Netherlands

Mark A. Jervis, School of Biosciences, University of Wales, P.O. Box 915, Cardiff CF1 3TL, UK

Kenneth B. Johnson, Department of Botany and Plant Pathology, Oregon State University, Corvallis, OR 97331-2902, USA

Peter Kareiva, Department of Zoology, P. O. Box 351800, University of Washington, Seattle, WA 98195-1800, USA

Neil A. C. Kidd, School of Biosciences, University of Wales, P.O. Box 915, Cardiff CF1 3TL, UK

Huseyin Koçak, Department of Mathematics and Computer Science, Univerity of Miami, Coral Gables, FL 33124, USA

Edward A. Kornkven, Department of Mathematics and Computer Science, South Dakota School of Mines and Technology, Rapid City, SD 57701, USA

Andreas Kruess, Fachgebeit Agraökologie, Georg-August-Universität, Waldweg 26, D-37073 Göttingen, Germany

Robert F. Luck, Department of Entomology,University of California, Riverside, CA 92521, USA

Nick J. Mills, Department of Environmental Science Policy and Management, University of California, Berkeley, CA 94720, USA

William W. Murdoch, Department of Ecology, Evolution and Marine Biology, University of California, Santa Barbara, CA 93106, USA

Gösta Nachman, Institute of Population Biology, University of Copenhagen, 15 Universitetsparken, DK-2100, Copenhagen Ø, Denmark

Roger M. Nisbet, Department of Ecology, Evolution and Marine Biology, University of California, Santa Barbara, CA 93106, USA

Leonard Nunney, Department of Biology,University of California, Riverside, CA 92521, USA

David W. Onstad, Department of Natural Resources and Environmental Sciences, University of Illinois, Urbana, IL 61801, USA

Bas Pels, Institute of Systematics and Population Biology, University of Amsterdam, Kruislaan 320, 1098 SM Amsterdam, The Netherlands

Maurice W. Sabelis, Institute of Systematics and Population Biology, University of Amsterdam, Kruislaan 320, 1098 SM Amsterdam, The Netherlands

Steven M. Sait, Population Biology Research Group, School of Biological Sciences, Nicholson Building, The University of Liverpool, Liverpool L69 3BX, UK

Veronica Solorzano, The Mathematical Institute, University of St. Andrews, North Haugh, St. Andrews, Fife KY16 9SS, UK

Matthew B. Thomas, Leverhulme Unit for Population Biology and Biological Control, NERC Centre for Population Biology, Imperial College, Silwood Park, Ascot, Berkshire SL5 7PY, UK

David J. Thompson, Population Biology Research Group, School of Biological Sciences, Nicholson Building, The University of Liverpool, Liverpool L69 3BX, UK

Teja Tscharntke, Fachgebeit Agraökologie, Georg-August-Universität, Waldweg 26, D-37073 Göttingen, Germany

Minus van Baalen, Institute of Systematics and Population Biology, University of Amsterdam, Kruislaan 320, 1098 SM Amsterdam, The Netherlands

Joop C. van Lenteren, Departments of Entomology and Theoretical Production Ecology, C.T. de Wit Graduate School Production Ecology, Wageningen Agricultural University, P.O.Box 8031, 6700 EH Wageningen, The Netherlands

Herman J. W. van Roermund, Departments of Entomology and Theoretical Production Ecology, C.T. de Wit Graduate School Production Ecology, Wageningen Agricultural University, P.O.Box 8031, 6700 EH Wageningen, The Netherlands

Sandra J. Walde, Department of Biology, Dalhousie University, Halifax, Nova Scotia, Canada B3H 4J1

Simon N. Wood, The Mathematical Institute, University of St. Andrews, North Haugh, St. Andrews, Fife KY16 9SS, UK

Preface

The scientific basis for successful biological control has been a relentlessly pursued but elusive goal. Nevertheless there has been progress. Some of this has come from retrospective approaches which examine the average behavior of many biological control systems to look for consistent patterns. Thus, we arrive at empirical rules such as 'Importation of parasitoids that have maximum parasitism rates of less than 30% in the home region are doomed to failure or at best partial success.' Such rules, when true, are useful, but it is important to better characterize successful introductions in order to refine our ability to make accurate predictions. In other words, what are the specific attributes of biological control agents and the systems in which they operate that are likely to enhance success? The tools of the theoretician can be used to explore the effects of such attributes on the dynamic properties of biological control systems and the ability of agents to suppress pest populations. Biological control systems are after all simply systems of interacting populations that should follow the same rules as natural systems, at least to some extent. Although it has been argued that traditional population dynamic theory has been of little practical use in biocontrol, modern theoretical developments attending to the complexities of real populations and their contexts are more promising. The application of recent theoretical advances to the problem of biological control is the concern of this book.

The general thrust of the book is to apply modern theory to the problem of biological control and to identify biological details that will help us to formulate better (i.e., more useful) theory. Mathematical models are central to such an enterprise, but not all chapters are strictly mathematical. Rather, we wanted the book to be a blend of biology and math in order to illustrate the heuristic power which lies at their interface. The chapter topics thus range from purely theoretical considerations of multiple attractors in three-species food chains, and the effects of spatial structure on the dynamic stability of populations, to such biological considerations as parasitoid sex ratio, mating behavior and nutrition and their impact on effective control. Some chapters present specific, detailed models of one system which provide a high degree of prediction of their dynamics. Other chapters might use a specific system as a starting point, but then go on to make general predictions about the links between biology and dynamics. We have also included chapters on genetics and evolution since it is becoming increasingly obvious that evolution occurs over time scales relevant to medium and long-term system management. While the application of theory to biological control issues is certainly not a new development, this book represents the first attempt as far as we know to gather a wide diversity of current theoretical positions into one place. We wanted to organize theoretical approaches in a systematic way so that we could see what are the most fruitful directions for further development.

It will become immediately obvious that we ourselves have not contributed chapters to this volume. Neither of us is a theoretician, but both of us have been involved in biological control issues, and we thought that we could develop a useful book for non-theoreticians by taking an 'outsiders' approach. To be useful, theory has to be comprehensible to non-specialists, and our editorial philosophy was to insist that authors write chapters that we could easily understand with the hope that that would be the case for other empirically oriented biologists. Most theoreticians in 1997 seem to understand this need.

Because of the diversity of viewpoints, it was a bit more difficult deciding how the book should be organized, and for any failings in that regard we take full responsibility. We have organized the book into five parts. Part 1. Biological Control Theory: Past and Present contains 2 chapters which provide a broad overview of the history and application of theory to biological control, and a systematic review of mathematical models applied to biological control systems. Part 2. Ecological Considerations contains a diversity of chapters with an ecological slant that range from pure modeling exercises to a focus on the details of parasitoid nutrition. Part 3. Spatial Considerations applies the relatively new techniques of spatial modeling to show that space definitely matters to biological control systems as well as to natural systems. This section also contains a discussion of habitat fragmentation, which has normally been studied in the context of conservation biology, but which must have effects on biological control systems. Part 4. Genetic/Evolutionary Considerations focuses on the interface between evolution and population dynamics to explore the consequences when genetic changes occur at time scales relevant to the regulation of populations in agricultural systems. Finally, Part 5. Microbes and Pathogens treats the role of microorganisms in biological control programs, both as agents and as targets.

If our book is successful in meeting its goals, it will be due in no small part to the efforts of many individuals. The authors of individual chapters were the foundation of this enterprise, and they have been uniformly creative and supportive. They have also played a major editing role by reviewing chapters in their respective areas of expertise. Dr Tracey Sanderson, our editor at Cambridge University Press, has offered helpful guidance and encouragement during all stages of this project.

It is our sincere desire that this book provides a balanced presentation of the important issues in the application of theory to biological control, and that it will be the source of productive and stimulating thought on the problem of how to make biological control predictable.

Bradford A. Hawkins, Irvine
Howard V. Cornell, Newark

PART I

Biological control theory: past and present

The need to provide a *conceptual* foundation for biological control to support its *empirical* basis was realized by early workers. Indeed, the spectacular successes achieved by some early biological control projects directly stimulated the development of host–parasitoid theory in the 1930s. This legacy continues to the present, and population dynamic theory generated 60 years ago continues to form much of the theoretical basis of biological control. In Chapter 1, Berryman provides a cogent review of the historical development of ecological theory as it relates to biological control, focusing on discrete-time (difference) models that best describe systems in which the insects reproduce seasonally. Since biological control at its best is permanent and stable (in the sense that fluctuations in pest densities are heavily constrained over large geographic areas), the notion of 'regulation' is deeply ingrained in biological control theory. Berryman discusses regulation in the context of control theory and ecological engineering, engineering principles that share a large number of attributes with standard population dynamic theory. Although some biological control workers may be uncomfortable with the application of mechanical constructs to biological systems, all will readily recognize the overlap between the underlying principles of control theory and ecological theory. Perhaps fewer will be familiar with the state of predator–prey theory. An increasingly intense debate has been developing in recent years concerning whether predator–prey dynamics is best described by the long-standing 'prey-dependent' models or by 'ratio-dependent' models. Berryman outlines the historical development of both types of models and describes the basic differences between the approaches. It is beyond the scope of this book to attempt to resolve the debate on the issues involved. As far as many biological control workers are concerned, the best approach will largely depend on which methodology best explains the success or failure of their projects. Having a range of theoretical tools at our disposal should not be considered a liability.

Many insects reproduce year-round; the dynamics of these are best described by continuous-time (differential) models. In Chapter 2, Briggs *et al.* focus on such models. Furthermore, they discuss the development of models that incorporate some of the more obvious biological attributes of parasitoids and their hosts; namely, that host-insects have distinct life stages, some of which are not susceptible to attack by parasitoids, that the fecundity of parasitoids is influenced by adult females feeding on potential hosts, and that few parasitoids really search randomly for hosts. Although stage-structured models have a relatively short history, they have come a long way quickly, and Briggs *et al.* provide examples illustrating how adding a little biological detail to the general models can provide insight to specific biological control systems. Finally, Briggs *et al.* address the problem of stability and biological control. A major stimulus for theoretical innovations is that the early, simple models of host–parasitoid population dynamics are unstable, which appears at odds with empirical experience. Briggs *et al.* discuss two potential solutions to the problem:

(i) metapopulation dynamics may engender global stability into systems characterized by local instability, and (ii) generalist parasitoids can persist in otherwise unstable systems. As usual, theory is developing faster than the field data needed to test it, so both of these solutions, albeit sensible, are not well documented. Even so, this represents an example of where theory has identified research areas that field workers would be remiss not to consider.

Undoubtedly, the most common complaint about theory is that it is too general and vague to be applicable to real biological control systems. Even when theory is sufficiently specific and applicable, it is scattered in the literature and can be difficult to locate. Barlow (Chapter 3) compiles and briefly describes 50 models that have been applied to specific biological control problems. He also classifies models by types in order to initiate a synthesis of what the specific models can teach us. One clear message is that the application of theory to solve biological control problems is accelerating, with no cases before 1980, 18 in the 1980s, but 32 in the 1990s. The claim that biological control is entirely *ad hoc* is rapidly becoming dated.

1

The theoretical foundations of biological control

ALAN A. BERRYMAN

Introduction

'Theory without practice is fantasy, practice without theory is chaos' (abridged from a forgotton author).

In this chapter I outline the fundamental theoretical concepts that I believe must underlie the practice of biological control, for without a theoretical foundation it is unlikely that we will succeed in any practical endeavor: it would be like trying to put a man on the moon without knowledge of the theories of gravity and planetary motion (Berryman, 1991). The question is, what is the theoretical foundation upon which the practice of biological control should rest?

In the context of this chapter, the term biological control is used to describe the control or regulation of pest populations at inocuous densities by their natural enemies. Hence, the theory of biological control contains elements of (i) *control theory*, which is concerned with the regulation of dynamic variables at, or near, set points or equilibria by negative feedback mechanisms, and (ii) *predator–prey theory*, which is concerned with the dynamics of interactions between predators and their prey.

Control theory

'Negative feedback is the necessary technique by which long-term stability in complex systems can be obtained' (Milsum, 1966).

Control theory was originally developed by engineers for the purpose of regulating the dynamics of mechanical systems. However, the ideas of control theory are also part of the more general theories of dynamic systems and non-linear dynamics (or chaos theory). Control theory is concerned with time-dependent changes (*dynamics*) of complex systems and how these changes are controlled or regulated by *negative feedback*. Some ecologists and pest managers may argue that these ideas are not useful because biological systems are uniquely different from mechanical systems. Yet there are similarities, for both mechanical and biological systems change with time and can, therefore, be considered within the broad class of *dynamic systems*. In addition, they are both dissipative in the sense that they require energy to run and, in the process of running, dissipate energy. Finally, they both have finite life expectancies (i.e., they 'die'). The only thing truly unique about biological systems is that they reproduce themselves, but it is not too difficult to conceive of machines doing the same in the future. For these reasons, there is no reason to expect biological systems to be governed by different rules than other dynamical systems (Berryman, 1989, 1995).

Although the literature on control theory can be intimidating for a biologist, once one has broken the jargon barrier, the basic ideas are quite simple and intuitive. Consider a *variable* of interest, say the density of a pest population. A measure of this variable (its value) at a particular time can be used to identify the condition, or state, of the system at that point in time. The variable is then called a *state variable*. Changes in the value of a state variable can be brought about by either *exogenous* or *endogenous* processes; exogenous processes affecting a change in the state variable but themselves being unaffected by current or past values of the state variable, and endogenous processes affecting and being affected by changes in the current or past values of the state variable.

Exogenous processes result from the interaction between a state variable and one or more *exogenous factors or variables* as, for instance, severe winter weather may cause a decline in a pest population (weather is an exogenous variable because it affects but is not affected by pest densities). The equations describing an exogenous process are called *forcing functions* if the process is applied in a predictable time-dependent fashion (e.g., seasonal effects on insect survival), or *random functions* if the process is applied in an unpredictable manner (e.g., weather variations from the climatic norm). Exogenous forcing functions cause changes in the average values of state variables, while random functions cause them to fluctuate around their average values. (Note that exogenous variables are the same as *density-independent* variables).

Endogenous processes, on the other hand, result from the action of *feedback loops*. For example, suppose that warm spring weather (an exogenous variable) causes a pest population to produce more offspring than normal in a particular year (e.g., an increased birth rate). This large pest population provides more food for predators, resulting in increased predator reproduction. More predators eat more pests, and this may cause the pest population to move back towards its original density. Both predator and prey populations are linked together by a circular causal pathway or feedback loop (Hutchinson, 1948). In this case it is a *negative feedback* loop because the original change in state (a population increase) was opposed by the action of the endogenous process (the predator–prey interaction). When variables are linked together in feedback loops they are called *endogenous variables* and the number of variables involved in a feedback loop determines the *dimension* or *order* of the generating process. Hence, a system composed of a mutually interacting population of predators and prey has a dimension of two and can be considered to be a second order dynamic process (Royama, 1977). (Note that negative feedback acting on population density is conceptually identical to *density-dependence*, and that the negative feedback between predators and their prey is synonomous with *delayed density dependence*).

Of course it is also possible to have endogenous processes that only involve one state variable as, for example, when pests of the same species compete with each other for food or space but do not cause quantitative changes in the abundance of food or space (equivalent to *direct or immediate density dependence*). In this case the quantity of food or space acts as an exogenous variable, and the only endogenous variable is the density of the pest population. Hence, the dimension of this dynamic system will be one and the generating process is first order. As we will see later, natural enemies can also generate first order feedback on the pest population. Of course, it is possible to have third or higher order feedback processes, involving three or more interdependent variables, in ecological systems. For example, the destruction of food plants by herbivores may remove hiding places and escape routes, making them more vulnerable to predation, and giving rise to a feedback process involving plant, herbivore and predator populations (Krebs *et al.*, 1995).

Because negative feedbacks cause state variables to return towards their original values, they act as stabilizing forces in dynamic systems. Hence, negative feedbacks can lead, with time, to *equilibrium* or balance in mechanical and ecological systems; i.e., they *regulate* or *control* the state variables. However, although negative feedback is a *necessary condition* for stability of dynamic systems, it is not sufficient to ensure stability. To guarantee stability, the negative feedback must act *rapidly* and *gently*, otherwise the state variables may oscillate to varying degrees around their equilibrium points.

The speed of action of a negative feedback loop depends, to a large degree, on its dimension, or the number of variables involved in the feedback loop. This is because each variable in the loop needs time to change quantitatively in response to the value of the preceeding variable, as a predator population requires time to produce more offspring after being confronted with a superabundance of prey. In general, first order dynamic processes act faster, and give greater stability, than second or higher order processes. Gentleness, on the other hand, has more to do with the rate at which state variables approach equilibrium, which is to a large degree dependent on the rates of change of the variables involved in the feedback loop. A useful analogy is an automobile approaching a stop sign: one approaching slowly can be stopped by gentle brake pressure while one approaching rapidly will require vigorous braking and may go out of control. Hence, populations growing slowly and regulated by first order negative feedbacks will normally show a greater degree of stability than those growing rapidly and regulated by higher order interactions (e.g., predator–prey interactions). In this way, elephants with their low reproductive rates are expected to have more stable populations than insects, and insects regulated by competition for food are usually expected to have more stable populations than those controlled by parasitoids.

It is important to realize that endogenous processes can also give rise to *positive feedback* which, in contrast to stabilizing negative feedback, induces instability in dynamic systems by accentuating or amplifying changes in state. Positive feedback is the force behind population explosions, inflation spirals, arms races, and organic evolution. As we will see later, positive feedback can also result from interactions between populations of predators and their prey, and that this can create unstable breakpoints or thresholds in dynamic systems. With this brief introduction to the basic ideas of control theory, we now proceed to the theory of predator–prey dynamics.

Predator–prey theory

'we constantly see a great fluctuation in numbers, the parasite rapidly increasing immediately after the increase of the host species, overtaking it numerically, and reducing it to the bottom of another ascending period of development.' (Howard, 1897)

Predator–prey theory concerns the dynamic interaction between populations of predators, or more generally consumers, and their prey, or more generally resources. As we saw earlier, this interaction can create a negative feedback loop, the necessary condition for regulation and stability of dynamic systems. This fundamental property of predator–prey interactions forms the basis for the theory of biological control of pest organisms.

There are two basic philosophies in modeling predator–prey interactions. One emerges from the venerable *logistic* equation (Verhulst, 1838), which describes the growth of a consumer population living in a finite environment, the other from the Lotka–Volterra equations (Lotka, 1925; Volterra, 1928), which describe the physics of interacting predator and prey populations. Because different conclusions about the practice of biological control can be drawn from each approach, I will present both in this chapter.

There are also two ways in which predator–prey models are usually formulated mathematically. Both logistic and Lotka–Volterra models were originally presented as continuous time *differential* equations, and this approach has been carried forward by ecologists working with organisms that reproduce continuously (e.g., aphids and scales). Others working with organisms that reproduce seasonally (salmon and many insects with discrete generations) tend to employ *difference* equations written in discrete time (e.g., Hassell, 1978). In this chapter I take the latter approach because most of the insects I have worked with have discrete annual generations. However, with the proviso that difference equations are generally less stable than analogous differential equations, the conclusions reached about biological control are not usually affected by the type of equation employed.

Finally, because this chapter attempts to lay down a basic theoretical framework upon which the practice of biological control should rest, it is mainly concerned with broad generalities rather than specific questions or problems. My aim is to show how a general understanding of predator–prey theory can act as a guide to successful biological control. Others in this volume show how more detailed models can be used to answer more specific questions (see Chapter 2) and to describe more specific situations (see Chapter 7).

Logistic predator–prey theory

The first mathematical model for the growth of a population of consumers was Verhulst's (1838) logistic equation, which he derived from Malthus' (1789) concept of a 'struggle for existence', a struggle that inevitably ensues when populations grow in a finite

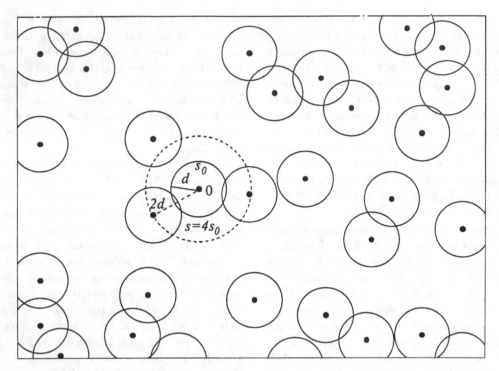

Figure 1.1. Royama's (1992) geometric model for competition between randomly distributed organisms living in a fixed density of resources. Each organism is surrounded by a circle of area s_0 and diameter d, which it needs to meet its resource demands and competes with other organisms that are closer than $2d$ to it. It is easy to see that competition will be less intense as resource density rises (because s_0 will become smaller) and will intensify as population density rises (because the number of circles will increase).

environment (notice that Darwin also employed this idea in his theory of evolution). The logistic was rediscovered as an empirical model for population growth by Pearl & Reed (1920) and then by Lotka (1925) as a limiting case of his 'fundamental equations of kinetics of evolving systems'. Perhaps the most convincing derivation of the logistic equation was Royama's (1992) geometric model of consumers drawing resources from a fixed area (Fig. 1.1).

Following some mathematical argument and simplification, the following equation emerges for the growth of a consumer population (see also Berryman *et al.*, 1995a):

$$N_t = N_{t-1} e^{a-(N_{t-1}/cD)} = N_{t-1} e^{a-bN_{t-1}}; \; b = 1/cD \quad (1.1)$$

where N_t is the density of consumers at time t, a is the maximum rate of change of an average consumer in a

given environment ($a = \ln(1 + \text{births}/N/t - \text{deaths}/N/t)$, c is the value of the resource to the consumer, D is the fixed density of resources available to the consuming population, and b is a coefficient of intraspecific competition for resources. This equation is a discrete form of the original logistic equation (Cook, 1965) and is also identical to the Ricker (1954) model familiar to fisheries ecologists (Berryman, 1978). For those more familiar with the differential form of the logistic, note the following correspondence between difference and differential forms:

$$\frac{1}{N}\frac{dN}{dt} = \ln\left(\frac{N_t}{N_{t-1}}\right) = a - bN \quad (1.2)$$

Leslie (1948) seems to be the first to have suggested that the logistic equation can be converted into a model for a predator population feeding on a depletable prey resource by letting D be a dynamic variable. For example, if $N_{i,t}$ is now the density of a population in the ith trophic level of a food chain at time t, then we can rewrite eqn (1.1):

$$N_{i,t} = N_{i,t-1}e^{a_i[N_{i,t-1}/(c_iN_{i-1,t-1})]} \qquad (1.3)$$

Mortality from higher trophic levels is easily incorporated into this model. For example, Thompson (1924) proposed a simple model for the interaction between populations of insect parasitoids and their insect hosts based on the assumption that parasitoids deposit eggs randomly among the host population. Under these conditions, the probability of an individual prey surviving parasitism is given by the zero term of the Poisson distribution

$$S_i = e^{-(d_iN_{i+1}/N_i)} \qquad (1.4)$$

where S_i identifies the survival of prey species i from the attack of N_{i+1} parasitoids, and d_i is the average fecundity of female parasitoids. Berryman (1992) suggests using an expression like this to expand Leslie's model to any population in a food chain of arbitrary length:

$$N_{i,t} = N_i e^{a_i - (N_i/c_iN_{i-1})}S_i = N_i e^{a_i - (N_i/c_iN_{i-1}) - (d_iN_{i+1}/N_i)} \qquad (1.5)$$

Notice that the time subscripts have been omitted from the right hand side of the equation for purposes of clarity. From now on, a population variable appearing on the right hand side without a time subscript should be assumed to represent the density of the population at time $t-1$.

Equation (1.5) provides the basic structure for the logistic food chain model whose biological meaning becomes clear if we rewrite it as follows:

$$N_{i,t} = N_i \, e^{a_i}e^{-(N_i/c_iN_{i-1})}e^{-(d_iN_{i+1}/N_i)} \qquad (1.6)$$

The first term, $N_i e^{a_i}$, describes the geometric growth of a consumer population when food is unlimited ($N_{i-1} \rightarrow \infty$) and predators are absent from the environment ($N_{i+1} \rightarrow 0$). The second term, $e^{-(N_i/c_iN_{i-1})}$, describes the probability of an individual consumer surviving the struggle with its cohorts for a limited

supply of food as a function of the consumer/resource *ratio*. The last term, $e^{-(d_iN_{i+1}/N_i)}$, describes the probability of an individual consumer surviving attack by its enemies, again as a function of the consumer/resource ratio. Because the dynamics of all populations in the food chain are driven by the *ratio* of consumers to their resources, logistic equations are sometimes classified as *ratio-dependent* or **RD** models (other **RD** models will be encountered later).

It is fairly straightforward to expand this equation to general predators by incorporating alternate food into the denominator of the ratios (Berryman, 1992):

$$N_{i,t} = N_i e^{a_i - [N_i/c_i(F_{i-1}+N_{i-1})] - [d_iN_{i+1}/(F_i+N_i)]} \qquad (1.7)$$

where F_i is the availability of alternative food in the ith trophic level (i.e., F_i is the cumulative density of alternative food species, times a parameter that specifies the preference for and value of those food species relative to the species in question). Of course, this model can also be extended to *food webs* of arbitrary complexity by expanding the alternative food constant to a system of dynamic food species (Berryman *et al.*, 1995b). However, because we are only concerned with two-species food chains in this chapter we will not complicate the model further.

Although eqn (1.7) suffices for any species in a food chain be it a predator, prey, or both, it is useful to write the equation for each species in a predator–prey interaction because this is how many of the models discussed below are usually presented. To generate the prey equation let N be the density of prey, $D = F_{i-1} + N_{i-1}$ is the constant density of resources for the prey, and $P = N_{i+1}$ is the density of predators. Under these conditions, eqn (1.6) becomes:

$$N_t = Ne^{a_n - b_nN - [d_nP/(F+N)]} \quad \text{Prey} \qquad (1.8a)$$

with the coefficient of competition being defined by $b_n = 1/c_nD$). Similarly, for the predator equation let $P = N_i$ and $N_{i+1} = 0$, so that:

$$P_t = Pe^{a_p - \{P/[c_p(F+N)]\}} \qquad \text{Predator} \qquad (1.8b)$$

These equations form our basic model for logistic (**RD**) predator–prey interactions.

Lotka–Volterra predator–prey theory

In the mid 1920s Lotka (1925) and Volterra (1928) independently proposed a model for predator–prey interactions that has become the mainstay of classical predator–prey and food web theory. Although the LV model was originally formulated as differential equations, I transform them here into difference equations (following Royama, 1971, see Appendix):

$$N_t = Ne^{a_n - d_n P} \qquad \text{Prey} \qquad (1.9a)$$

$$P_t = Pe^{-d_p + a_p N} \qquad \text{Predator} \qquad (1.9b)$$

Ten years later, Nicholson & Bailey (1935) derived another model for the interaction between an insect parasitoid and its insect host which, like Thompson's, evolved from the assumption of random (Poisson) search. In the NB model, however, the probability of prey surviving predation:

$$S_n = e^{-d_n P} \qquad (1.10)$$

is only dependent on predator density rather than on the predator/prey ratio. Given an exponentially growing prey population, and multiplying by the NB survival function, yields a prey model:

$$N_t = Ne^{a_n} e^{-d_n P} = Ne^{a_n - d_n P} \qquad (1.11a)$$

identical to the LV model eqn (1.9a). In fact, Royama (1971: p. 28) showed that the NB model is really a special case of the LV model. However, the NB parasitoid model is usually formulated by calculating the number of parasitized hosts $N(1 - S_n)$ (e.g., Hassell, 1978) and, if more than one parasitoid can be produced per parasitized host, multiplying by the number of offspring produced per host, m_p:

$$P_t = m_p N (1 - S_n) = m_p N (1 - e^{-d_n P}) \qquad (1.11b)$$

Notice that the NB parasitoid model has a different structure from the LV predator equation (1.9b).

As originally formulated, the LV and NB models have a number of problems, one of which is the infinite growth of prey in the absence of predation. To correct this, prey growth is often assumed to be logistic and so the term $-b_n N$ is added to the exponent of eqn (1.9a). Another problem is that the reproductive rate of individual predators, the term $e^{a_p N}$ in the LV predator equation eqn (1.9b), is unbounded (i.e., there is no

limit to the appetite nor the fecundity of predators). To correct this requires a new concept, that of the predator's feeding response to the density of its prey.

Predator responses to prey density

Solomon (1949) originally separated the response of predators into functional and numerical components, with the former describing how the feeding rate of individual predators changes with respect to prey density and the latter how the reproductive rate responds to changes in prey density. However, I prefer to call the former a *behavioral response* because it really describes the behavior of a predator searching a fixed space in a fixed interval of time for different densities of prey. In contrast, the numerical response describes changes in predator *numbers* in response to changes in prey density and is, therefore, a true *population response*.

Holling (1959) identified three basic kinds of behavioral responses:

Type I (linear or straight) response, where the attack rate of the individual predator increases linearly with prey density but then suddenly reaches a constant value at the attack rate that satiates the predator (Fig. 1.2(a)).

Type II (cyrtoid or C-shaped) response, where the attack rate of the predator increases at a decreasing rate with prey density until it reaches a constant (Fig. 1.2(b)). Cyrtoid responses are typical of many arthropod parasitoids and predators, particularly those with fairly narrow host preferences (specialists).

Type III (sigmoid or S-shaped) response, where the attack rate accelerates at first and then decelerates towards its constant satiation rate (Fig. 1.2(c)). Sigmoid responses are typical of predators that readily switch from one host to another (i.e., generalists) and that concentrate their feeding in areas where hosts are abundant (i.e., exhibit spatial density dependence).

A number of models have been proposed for the behavioral response but the most general one comes from the Michaelis–Menten equation of enzyme kinetics (Real, 1977):

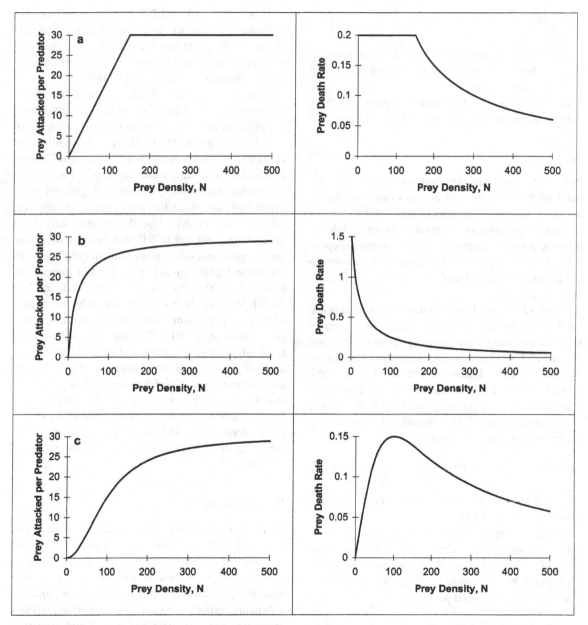

Figure 1.2. Behavioral responses of predators to prey density (left) and resulting *per capita* death rate of the prey (right): (a) Type I or linear response, (b) Type II or cyrtoid response, (c) Type III or sigmoid response.

$$N_a = \frac{d_n N^x}{h + N_x} \qquad (1.12)$$

where h is a constant and x is a coefficient of non-linearity: when $x = 1$ the response is cyrtoid and when $x > 1$ it becomes sigmoid.

If we divide the behavioral response by prey density, we obtain the *per capita* prey death rate:

$$\frac{N_a}{N} = \frac{d_n N^{x-1}}{h + N^x} \qquad (1.13)$$

Notice that the death rate declines with prey density in all behavioral responses, at least when prey density is high (Fig. 1.2). This means that the dynamics of the prey population are under the influence of destabilizing positive feedback (inverse density dependence). The exception is the sigmoid response which can produce negative feedback in sparse prey populations (Fig. 1.2(c)).

Much of the modern literature and conventional wisdom on the dynamics of predator–prey interactions comes from incorporating cyrtoid behavioral responses ($x = 1$) and logistic limitation into **LV** or **NB** models (e.g., Rosenzweig & MacArthur, 1963; Hassell, 1978). To do this we replace the prey death rate constant, d_n, in the prey eqn (1.9a) by the cyrtoid death rate function, $d_n/h + N$, and add a logistic self-limitation term to yield the model:

$$N_t = N e^{a_n - b_n N - [d_n P/(h+N)]} \quad \text{Prey} \qquad (1.14a)$$

Notice that when $h = F$ this model is identical to the logistic prey eqn (1.8a).

In the corresponding predator equation, the reproduction term can be calculated from the number of prey killed per predator (the behavioral response) multiplied by the number of predator offspring produced per consumed prey:

$$P_t = P e^{-d_p + [m_p N/(h+N)]} \quad \text{Predator} \qquad (1.14b)$$

Note that this equation is quite different from the logistic predator model eqn (1.8b). If we use the **NB** parasitoid model eqn (1.11b), however, then the consumer population is given by:

$$P_t = m_p N \left(1 - e^{-[d_n P/(h+N)]}\right) \text{Parasitoid} \qquad (1.14c)$$

Although the **LV** and **NB** consumer models have different structure, the dynamics are quite similar. It is also worth noting that logistic models capture the concept of the cyrtoid response in the predator/prey ratio (compare eqns (1.8a) and (1.14a)).

Arditi & Ginzburg (1989) have argued that predators should respond behaviorally to the ratio of prey to predators rather than to prey density alone as in eqn (1.12). Similarly, Gutierrez and colleagues developed physiologically based models of energy acquisition and allocation in which the energy acquisition functions (behavioral responses) were dependent on predator/prey ratios (Gutierrez, 1992). These models, like the logistic, have been called *ratio-dependent* or **RD** and, because they have similar properties and make similar predictions to the basic logistic model, they are not considered seperately in this chapter. In contrast, **LV** and **NB** models and their derivatives are driven by the density of the prey population alone, and are therefore called *prey-dependent*, or **PD**. **RD** and **PD** models arise from different philosophies and make different predictions about the dynamic behavior of populations and communities. The following citations are provided for those who wish to explore the heated debate that has arisen over the relative merits of these two philosophies (see Abrams, 1994; Gleeson, 1994; Sarnelle, 1994; Akcakaya *et al.*, 1995; Berryman *et al.*, 1995a).

Population dynamics

In this section I analyze and compare the dynamics of **PD** and **RD** predator–prey models using the methods of phase space analysis. Phase space is defined as an n-dimensional coordinate system describing all possible states of a system composed of n dynamic variables. As only two dynamic variables are considered, the density of prey N and predators P, we will only need to work in two-dimensional (N, P) phase space. Phase space can be divided into zones of increase or decrease for each species by the equilibrium *isoclines* of the species (the set of points in phase space on which each species remains unchanged or in equilibrium). The isoclines are calculated by setting the population to equilibrium,

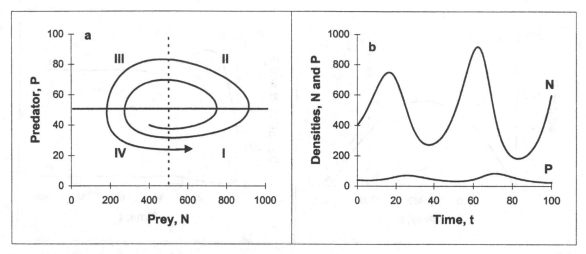

Figure 1.3. (a) Equilibrium isoclines for prey (continuous line) and predator (dashed line) from the discrete LV model (eqns (1.9a and b)) and (b) the corresponding time series trajectories for prey, N, and predator, P.

$P_t = P_{t-1}$ and $N_t = N_{t-1}$, and solving the equations for either N^* or P^*. For example, converting the original LV predator eqn (1.9b) to logarithms and doing some algebra we find the predator isocline:

$$\ln P_t = \ln P_{t-1} - d_p + a_p N_{t-1}$$
$$\ln P_t - \ln P_{t-1} = 0 = -d_p + a_p N^* \qquad (1.15)$$
$$N^* = d_p / a_p$$

This isocline is a vertical line parallel to the predator axis and intercepting the prey axis at d_p / a_p (Fig. 1.3(a), dashed line). Similarly, the prey isocline is also a straight line parallel to the prey axis and intercepting the predator axis at a_n / d_n (Fig. 1.3(a), continuous line). Where the two isoclines intersect, both species are in equilibrium, so we have a community equilibrium. Phase space is divided into four quadrats; in quadrat I both predator and prey populations grow, in quadrat II predators increase and prey decrease, in quadrat III both populations decrease, and in quadrat IV the prey increase and predators decrease. Beginning at an arbitrary location (N_0, P_0), the time evolution of the coordinate (N_t, P_t) forms a circular anticlockwise orbit in phase space (Fig. 1.3(a)). Notice that the trajectory crosses the prey isocline vertically, because only the predator population is changing at this point, and the predator isocline horizantally, because only the prey popula-

tion is changing. Notice also that the dynamics of the discrete LV (also the NB) model are unstable because the orbit gets wider and wider as time progresses until the prey and/or predator go extinct. If we plot predator and prey densities against time, we see that the two populations oscillate in a cyclical manner with the predator population lagging behind that of its prey (Fig. 1.3(b)). This is an identical pattern to that described by Howard (1897: quoted above on p. 5) from observations of real predator–prey interactions, and is the main insight arising from the original LV model. (Note that the solution of the differential LV equations is a stable cycle with amplitude dependent on the initial condition = neutrally stable limit cycle.)

PD models containing cyrtoid behavioral responses and logistic prey limitation (eqn 1.14) have more complex, parabolic or 'humped', prey isoclines, but the predator isoclines remain the same (Fig. 1.4(a)). The ascending left arm of the isocline is caused by the positive feedback due to predator satiation and the descending right arm to logistic prey limitation. Much of the conventional wisdom on predator–prey interactions has arisen from the analysis of such models (see e.g., Rosenzweig & MacArthur, 1963). These models are characterized by stable dynamics as the community equilibrium

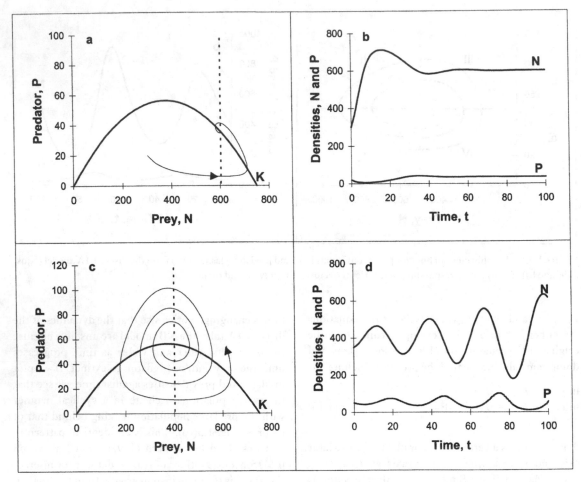

Figure 1.4. (a) and (c) equilibrium isoclines for prey (continuous line) and predator (dashed line) from the discrete **LV** model with cyrtoid behavioral response and logistic limitation (eqns (1.14a and b)), and (b) and (d) time series trajectories for prey, N, and predator, P.

approaches the prey carrying capacity, $K = a_n/b_n$, the place where the prey isocline meets the prey axis (Figs 1.4(a) and (b)), and unstable cycles as the community equilibrium moves to the left (Figs 1.4(c) and (d)). It is this property that creates the paradoxes associated with conventional predator–prey theory: the *paradox of enrichment* states that, because the peak of the prey isocline moves to the right as the prey's carrying capacity increases, say when nutrients are added to the environment, while the predator isocline is unaffected, the community equilibrium will be destabilized as a result of enrichment

(Rosenzweig 1971). The *paradox of biological control* asks: how is it possible to control pests at low densities, far to the left of the prey 'hump', when the interaction between pest and predators becomes highly unstable under these conditions? (Luck, 1990; Arditi & Berryman, 1991).

RD models (eqn (1.8)) have parabolic prey isoclines similar to **PD** models with cyrtoid behavioral responses and logistic limitation, but their predator isoclines are quite different (Fig. 1.5(a)). This right-sloping predator isocline confers greater stability on **RD** predator–prey interactions (Fig. 1.5) and also

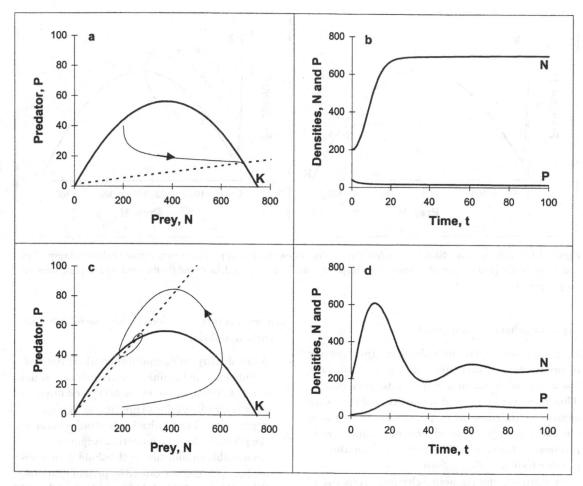

Figure 1.5. (a) and (c) equilibrium isoclines for prey (continuous line) and predator (dashed line) from the discrete logistic **RD** model (eqns (1.8a and b)), and (b) and (d) time series trajectories for prey, N, and predator, P.

solves the paradoxes of enrichment and biological control: enrichment now has no effect on stability because the position of the community equilibrium does not change in relation to the hump of the prey isocline when the prey carrying capacity is elevated. The paradox of biological control is also solved because it is now possible to have a stable community equilibrium with the prey population regulated at very sparse densities (Figs 1.5(c) and (d)). In fact, it is possible for the predator–prey system to persist in a variable environment even when the community equilibrium is zero! (Arditi & Berryman, 1991). For example, if the predator is very efficient or has a very

high *per capita* rate of change, relative to its prey, then its isocline may be so steep that it does not intercept the prey isocline at all. Under these conditions, the community equilibrium can be zero for both species and, in a constant environment, both species will go extinct. In a more natural environment, however, minor disturbances such as the immigration of a few prey into the system can bring the trajectory into quadrat I where both populations begin another growth cycle. Hence, the predator–prey system can persist in a *stochastic equilibrium* as is observed in many real situations (Murdoch *et al.*, 1985; Luck, 1990; Arditi & Berryman, 1991).

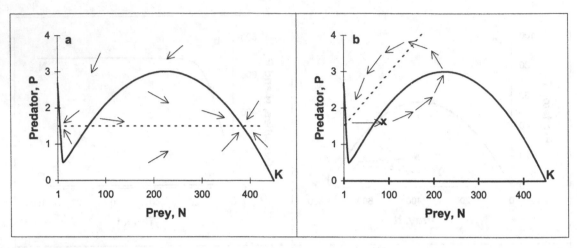

Figure 1.6. Equilibrium isoclines for prey (continuous line) when attacked by predators with sigmoid behavioral responses and where (a) the predators (dashed line) do not respond numerically to prey density and (b) the predators respond numerically to prey density.

Sigmoid behavioral responses

Up to this point I have only discussed the dynamics of predator–prey interactions in which the predator has a cyrtoid behavioral response to prey density. This response by itself does not contribute to the control or stabilization of pest (prey) populations. On the contrary, as noted previously, the cyrtoid response actually destabilizes the system through positive feedback (Fig. 1.2(b)).

In contrast, the sigmoid behavioral response of predators can act as a stabilizing influence on very sparse pest populations because it can then give rise to negative feedback (Fig. 1.2(c): Holling, 1966). For example, if we set $x = 2$ in eqn (1.13), then the prey death rate is $d_n N/(h + N^2)$ and we see that deaths rise at first with prey density but later fall as N becomes very large. This property gives rise to prey isoclines which are n-shaped (Fig. 1.6(a)). Sigmoid behavioral responses usually result from the action of generalist predators that switch to and/or aggregate on dense prey populations (Holling, 1966; Murdoch, 1969; Gould *et al.*, 1990). Because generalist predators and parasitoids often do not respond numerically to the density of their prey, their isoclines will run parallel to the prey axis (Fig. 1.6(a)). Under these conditions the isoclines

can cross at three locations giving rise to three community equilibria:

1. A low-density stable equilibrium where the pest population is maintained at extremely sparse densities (4–5 pests in our example) by first order negative feedback due to predator switching/aggregation. The feedback is first order because the predators have no numerical response.
2. An unstable equilibrium or threshold at an intermediate pest density caused by predator satiation. When the population is below this point (about 50 pests) densities are drawn towards the low-density community equilibrium, and when it is above this point an outbreak erupts.
3. A high-density stabilizing equilibrium where the pest population is regulated by shortage of food or other resources (around 400 pests in Fig. 1.6(a)).

Pest populations regulated by predators with sigmoid behavioral responses are said to be *metastable* because their stable equilibria are bounded by unstable thresholds (Lotka, 1925; Berryman *et al.*, 1984). Under these conditions the pests can be very rare but, if their numbers rise above the unstable equilibrium, an outbreak may occur. In addition, because pests migrating from outbreak regions can raise local

Table 1.1. *Key for identifying types of pest outbreaks (after Berryman, 1987)*

Space–time dynamics	Outbreak class
1. Restricted outbreaks that do not spread out from local epicenters.	**Gradient** outbreak
No positive feedback due to escape from regulating factors	
(a) High frequency or 'saw-toothed' oscillations due to rapid (first order) feedback	
(i) Outbreaks permanently associated with particular localities	*Sustained* gradient
(ii) Outbreaks follow changing environmental conditions	*Pulse* gradient
(b) Low frequency or 'cyclical' oscillations due to delayed (second or higher order) feedback	
(i) Outbreaks permanently associated with particular localities	*Cyclical* gradient
(ii) Outbreaks follow changing environmental conditions	*Pulse* gradient
2. Expanding outbreaks that spread out from local epicenters.	**Eruptive** outbreak
Positive feedback due to escape from regulating factors	
(a) High frequency or 'saw-toothed' oscillations at sparse densities	
(i) High frequency oscillations at high density	*Sustained* eruption
(ii) Cyclical oscillations at high density	*Pulse* eruption
(b) Low frequency or 'cyclical' oscillations at sparse density	
(i) High frequency oscillations at high density	*Permanent* eruption
(ii) Cyclical oscillations at high density	*Cyclical* eruption

densities in non-outbreak areas above the *escape threshold*, the outbreak can spread over large areas. For this reason, the expanding populations of pests with metastable isoclines are classified as *eruptive outbreaks* (Table 1.1).

It is easy to see from Fig. 1.6(a) that the equilibrium structure of this system is sensitive to the abundance of predators. If the predator density is very low ($P < 0.5$), then only the high-density equilibrium will remain, while if predators are more dense ($P > 3$), then the pest will be driven to extinction or very sparse densities without the possibility of outbreaks. It is also apparent that, when predator populations respond numerically to the density of the pest, the slope of the predator isocline will be greater than zero and, if the numerical response is sufficient, we will be left with a single low-density community equilibrium (Fig. 1.6(b)). However, the transient dynamics of the system are now much more complicated. For example, if an immigration of pests causes a local population to increase above the rising arm of the prey isocline, a second order predator–prey cycle will be initiated causing a short-term outbreak pulse (follow the arrows from the point *x* in Fig. 1.6(b)). In addition, the emigration of pests into adjacent non-outbreak regions could lead to outbreaks rippling through space, much like the outspreading ripples caused by a pebble thrown into a pond. This type of spreading outbreak wave is classified as a *pulse eruption* (Table 1.1).

Control hierarchies and feedback dominance

Pest populations can be influenced by a large number of negative feedbacks, most of which are due to predator–prey, or more broadly speaking, consumer–resource interactions. Some of these interactions, such as the feeding responses of generalist predators with sigmoid behavioral responses, can create stabilizing first order negative feedbacks. Others, such as interactions with specialist parasitoids, can create second order negative feedbacks that are generally less stable and often give rise to cyclical dynamics. If all these feedbacks operated simultaneously and synchronously on the dynamics of the pest population, it would be difficult to predict and manage them in any logical way. In reality, however, different feedback processes tend to operate seperately and sequentially on the dynamics of populations (Berryman, 1993). To illustrate this process, consider an insect pest population that has been drastically reduced by a catastrophic event. Soon afterwards, the population will grow back towards its previous density. As it grows, however, it may come to the attention of generalist predators, which then concentrate their feeding on this increasingly abundant food supply. If the negative impact of predation is sufficient to overcome the positive effect of pest reproduction, the population could be regulated at a very sparse density by first order negative feedback with these predators (Fig. 1.6(a)). On the other hand, if the effect of predation is insufficient, the numerical responses of specialist predators or parasitoids may then take over and eventually suppress the pest population *via* a second order feedback cycle. At the end of this cycle of pest increase and collapse, generalist predators may again be able to reassume dominance and control the population at a very sparse density (Fig. 1.6(b)). Even if specialists fail to suppress the pest population, other negative feedback processes such as disease epizootics or food depletion will eventually come into play. In this way, populations of pests are naturally subjected to a *hierarchy* of feedback processes that *dominate* the dynamics over different density ranges, and which may regulate the pest at or around several different equilibria. Within this dominance hierarchy may be one or more unstable equilibria (caused by saturating behavioral responses), which would allow the population to escape from low-density regulation following minor environmental disturbances or immigrations. This escape in time and space may result in a population eruption that spreads over vast areas but, in the end, will be terminated by the second order responses of specialist parasitoids and predators, pathogens and the destruction of food supplies (Berryman, 1987; Berryman *et al.*, 1987).

Ecological engineering

'The techniques of designing and operating the economy with nature . . . so that humans become partners with their environment' (Odum, 1989).

I have made the point that the practice of biological control should rest upon two basic theoretical foundations – control theory and predator–prey theory. Knowledge of the rules of feedback and control help us to understand the causes of change in ecological systems and how to control ecological variables at desirable levels (desirable from the point of view of human beings and their environments, or from the more general viewpoint of the global ecosystem). Activities such as this fall under the general heading of *ecological engineering* (Mitsch & Jorgensen, 1989; Berryman *et al.*, 1992). As we will see, biological control can be considered an important part of the practice of ecological engineering.

There are three general ways in which the feedback structure of ecological systems can be engineered to produce stable, selfsustaining dynamics. These are listed below.

1. *Construct new negative feedback loops.* For instance, new negative feedback loops are created when pest organisms are introduced into new environments because negative feedback is automatically created by the pest–host interaction. Likewise, the purpose behind the introduction of natural enemies in classical biological control programs is to build a dominant negative feedback between enemies and pest. Feedback loops can also be built by directing human actions in response to changing levels of a pest population. For example, if farmers respond to increasing pest damage by spraying a pesticide whenever the damage exceeds a predefined level, often called the economic damage level or EDL, then a negative feedback loop is created. The integration of EDLs, natural enemies, pesticide sprays and other human actions into pest control operations has become known as *integrated pest management*.

The integration of human feedbacks into ecosystems, however, must be based on an intimate understanding of predator–prey theory. Let me try and illustrate this by example: suppose we have an effective parasitoid, or guild of parasitoids, that regulates a pest insect around a stable low-density equilibrium; i.e., the stable community equilibrium is on the ascending arm of the pest isocline (Fig. 1.5(c)). In a variable environment, the pest population will tend to cycle around the community equilibrium and may occasionally exceed the EDL (Fig. 1.7). The job of

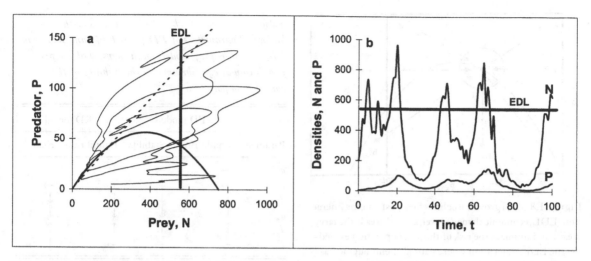

Figure 1.7. Equilibrium isoclines for a stable **RD** predator–prey system (same as Fig. 1.5(c)) showing prey fluctuations in a variable environment that occasionally exceed the economic damage level (EDL).

the pest manager is to minimize the probability of the pest population exceeding the EDL.

The first thing to notice is that the further the system is away from the community equilibrium, the wider will be the cyclical orbits, and the greater the likelihood of the pest population exceeding the EDL as it goes through its natural predator–prey cycle. Hence, the pest manager should try to keep the system as close to the equilibrium point as possible, say within the circular target in Fig. 1.8. To do this requires tactics that change with the particular location of the pest/parasitoid density coordinate (the points in Fig. 1.8) in the predator–prey cycle. For example, if the populations are at location **a** (moderate pest and low parasitoid density), then the only tactic that will bring the system into the target area is to increase the parasitoid population in some way. This is usually called natural enemy *augmentation*. At location **b**, on the other hand, we need to reduce the pest population as well as increase parasitoids. This could be done with a insecticide that only kills the pest, a *selective* pesticide, plus enemy augmentation. At location **c** a selective pesticide alone would be the best tactic while at **d** we need to kill both pest and parasitoid so the use of a non-selective pesticide is justified, or even essential. Finally, if the system is at positions **e** or **f** an augmentative release of the *pest* is

required while the use of a selective pesticide against the *parasitoid* at location **e** would also help to minimize the probability of future outbreaks! These seemingly counterintuitive tactics may be difficult for the manager to swallow but the effectiveness of the former has been verified empirically (Maksimovic *et al.*, 1970). The purpose of these treatments, of course, is to prevent the parasitoid population from overexploiting its food supply and dying out from starvation, or even going extinct locally.

2. Manipulate the biological parameters. There are two important considerations for successful biological control; first, the density of the pest population at equilibrium should be lower than the EDL, and second, the community equilibrium should be as stable as possible (Hassell, 1978; but see Murdoch *et al.*, 1985). How can we manipulate the predator–prey interactions to achieve these ends? First we can investigate how the parameters of the predator–prey equations affect these two conditions (Table 1.2) and then we can try and imagine how to change the values of these parameters by various management actions.

Notice that the equilibrium pest density can be lowered by decreasing d_p, and increasing d_n and a_p, in both models, and by decreasing the prey parameters

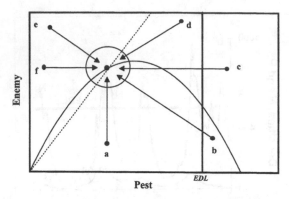

Figure 1.8. Zero growth isoclines for a pest–enemy interaction: EDL, economic damage level; circled area is the target area that minimizes the risk of the pest population exceeding the EDL; **a**, location where natural enemy augmentation is the best tactic; **b**, location where a combination of selective pesticide against the pest and natural enemy augmentation are the best tactics; **c**, location where the use of a selective pesticide against the pest is the best tactic; **d**, location where a non-selective pesticide that kills both species should be used; **e**, location where a selective pesticide that kills the natural enemies and augmentation of the pest population are the best tactics; **f**, location where augmentation of the pest population should be employed. Notice that pesticides should be used to effect optimal kill to bring the system into the target area. Tactics aimed at obtaining maximum kill can reduce the populations below the target area, resulting in future pest outbreaks or what has been called 'the boomerang effect', 'pest resurgence' or 'bounceback' (after Berryman, 1995).

a_n and $K = a_n/b_n$ in **RD** models. Furthermore, notice that the same changes also cause undesirable reductions in stability, with the exception of K, which has no influence on stability in **RD** models and has an inverse effect on stability in **PD** models (the paradox of enrichment). Alternative food apparently has no effect on equilibrium pest densities in either model, but does have a direct effect on stability in **RD** models.

Parameters of the predator–prey equations can be manipulated by either genetic or environmental engineering. For example, the *per capita* growth rate parameters (a_n and a_p) reflect the maximum birth rates minus the minimum death rates of prey and predators living in a particular habitat, given that the

Table 1.2. *Effects of changing the values of the biological parameters of* **PD** *(eqns 1.14) and* **RD** *(eqns 1.8) predator–prey models on the density of the pest population at equilibrium and the stability of the equilibrium point*

	PD models		RD models	
Parameter	Equilibrium	Stability	Equilibrium	Stability
a_n	→	→	↑	↑
K	→	↓	↑	→
d_n	↓	↓	↓	↓
a_p	↓	↓	↓	↓
d_p	↑	↑	↑	↑
F	→	→	→	↑

Notes:
Symbols: →, ↑, and ↓ indicate no effect, a direct effect, and an inverse effect of the parameter on the equilibrium properties, respectively.

effects of density-induced negative feedback are absent or minimal. Obviously, these properties are dependent on both genetic and environmental factors, and so the value of the parameters can be altered by manipulating the genetics of the organisms or changing the environment in which they live. For example, the *per capita* rate of change of the pest population, a_n, may be reduced by growing crops in regions less favorable for the reproduction and/or survival of the pest, such as mixed-species intercropped agroecosystems. Likewise, the predator rate of change, a_p, can be raised by growing crops in regions more favorable for predator reproduction and/or survival. One could also lower the carrying capacity of the prey, K, and increase the availability of alternative food for the predator, F, by mixed-species intercropping and other manipulations that reduced the availability of food for the prey and increased it for the predator. If predator or prey populations are limited by factors other than food supplies, such as nesting or hibernation sites, then manipulating these factors could have similar effects. Finally, we have the parameter d_n, which measures the impact of individual predators on individual prey, including such things as searching efficiency and appetite, and d_p, which measures the influence of an individual prey on the reproduction and survival of an individual predator. These parameters may be much more difficult to alter through environmental manipulations. However, there may be possibilities to do so through breeding programs or genetic engineering. For example, we could weaken the impact of a pest on its food plant (in this case d_p because the pest is now the predator) by breeding or engineering resistant plants. Of course, if we obtain a completely resistant strain, then the feedback between plant and pest is broken, and the pest population will go extinct on that strain of plant.

3. *Manipulate feedback structures.* Finally we can employ the principles of control hierarchies and feedback dominance to manipulate the feedback structure to our advantage. For example, if we can elevate the density of generalist predators so that their sigmoid behavioral responses dominate the feedback structure, then the pest equilibrium can probably be reduced significantly and at the same time stabilized by the first order negative feedback (Fig. 1.6). This might be accomplished by providing nesting places, alternative foods, and/or cover for generalist predators. Of course, feedback structures can also be altered by weakening or removing dominant feedbacks. For example, we may wish to weaken the negative feedback between a native herbivore and its generalist natural enemy and thereby release an eruptive outbreak to control a native weed (Berryman *et al.*, 1992). Although such manipulations for the biological control of *native* pests with outbreaks of *native* natural enemies (native × native interaction) is theoretically plausible, it has yet to be demonstrated in practice.

There are, of course, innumerable ways in which new feedbacks can be built into ecosystems, or by which the feedback structure of ecosystems can be altered. The only limitation is imagination and ingenuity. What is essential is that we clearly understand the rules of the game – the laws of dynamic systems and the principles of population dynamics and predator–prey interactions. In other words, we should know how feedbacks affect the dynamic behavior of complex systems, how trophic and other ecological interactions affect ecological feedback structures, and how the manipulation of genes and environment can change the level of equilibrium and its stability. With these rules in mind, it may be possible for ecologists to engineer stable, sustainable ecosystems that can support humans and the creatures on which they depend indefinately. On the other hand, without an intimate understanding of these rules, humans will continue to make decisions that disrupt feedback loops, or their stability, resulting in unpredictable and catastrophic events.

References

Abrams, P. A. (1994) The fallacies of ratio-dependent predation. *Ecology* 75: 1842–50.

Akcakaya, H. R., Arditi, R. & Ginzburg, L. R. (1995) Ratio-dependent predation: an abstraction that works. *Ecology* 76: 995–1004.

Arditi, R. & Berryman, A. A. (1991) The biological control paradox. *Trends in Ecology and Evolution* 6: 32.

Arditi, R. & Ginzburg, L. R. (1989) Coupling in predator–prey dynamics: ratio-dependence. *Journal of Theoretical Biology* 139: 311–26.

Berryman, A. A. (1978) Population cycles of the Douglas-fir tussock moth (Lepidoptera: Lymantriidae): the time-delay hypotheses. *Canadian Entomologist* 110: 513–18.

Berryman, A. A. (1987) The theory and classification of outbreaks. In *Insect Outbreaks*, ed. P. Barbosa & J. C. Schultz, pp. 3–30. New York: Academic Press.

Berryman, A. A. (1989) The conceptual foundations of ecological dynamics. *Bulletin of the Ecological Society of America* 70: 234–40.

Berryman, A. A. (1991) Population theory: an essential ingredient in pest prediction, management, and policy-making. *American Entomologist* 37: 138–42.

Berryman, A. A. (1992) The origins and evolution of predator–prey theory. *Ecology* 73: 1530–5.

Berryman, A. A. (1993) Food web connectance and feedback dominance, or does everything really depend on everything else? *Oikos* 68:183–5.

Berryman, A. A. (1995) Biological control. In *Encyclopedia of Environmental Biology*, vol. 1, pp. 291–8. New York: Academic Press.

Berryman, A. A., Stenseth, N. C. & Wollkind, D. J. (1984) Metastability of forest ecosystems infested by bark beetles. *Researches on Population Ecology* **26**: 13–29.

Berryman, A. A., Stenseth, N. C. & Isaev, A. S. (1987) Natural regulation of herbivorous forest insect populations. *Oecologia* **71**: 174–84.

Berryman, A. A., Valenti, M. A., Harris, M. J. & Fulton, D. C. (1992) Ecological engineering: an idea whose time has come? *Trends in Ecology and Evolution* **7**: 268–70.

Berryman, A. A., Guiterrez, A. P. & Arditi, R. (1995a) Credible, parsimonious and useful predator prey models – a reply to Abrams, Gleeson and Sarnelle. *Ecology* **76**: 1980–5.

Berryman, A. A., Michalski, J., Gutierrez, A. P. & Artiditi, R. (1995b) Logistic theory of food web dynamics. *Ecology* **76**: 336–43.

Cook, L. M. (1965) Oscillations in the simple logistic growth model. *Nature* **207**: 316.

Gleeson, S. K. (1994) Density dependence is better than ratio dependence. *Ecology* **75**: 1834–5.

Gould, J. R., Elkinton, J. S. & Wallner, W. E. (1990) Density-dependent suppression of experimentally created gypsy moth, *Lymantria dispar* (Lepidoptera: Lymantriidae), populations by natural enemies. *Journal of Animal Ecology* **59**: 213–33.

Gutierrez, A. P. (1992) Physiological basis of ratio-dependent predator–prey theory: the metabolic pool model as a paradigm. *Ecology* **73**: 1552–63.

Hassell, M. P. (1978) *The Dynamics of Arthropod predator–prey Systems*. Princeton: Princeton University Press.

Holling, C. S. (1959) The components of predation as revealed by a study of small mammal predation of the European pine sawfly. *Canadian Entomologist* **91**: 293–320.

Holling, C. S. (1966) The functional response of invertebrate predators to prey density and its role in mimicry and population regulation. *Memoirs of the Entomological Society of Canada* **48**: 1–86.

Howard, L. O. (1897) A study in insect parasitism: a consideration of parasitism of the white-marked tussock moth, with an account of their habits and interrelations and with descriptions of new species. *United States Department of Agriculture, Technical Series* **5**: 5–27.

Hutchinson, G. E. (1948) Circular causal systems in ecology. *Annals of the New York Academy of Science* **50**: 221–46.

Krebs, C. J., Boutin, S., Boonstra, R., Sinclair, A. R. E., Smith, J. N. M., Dale, M. R. T. & Turkington, T. (1995) Impact of food and predation on the snowshoe hare cycle. *Science* **269**: 1112–15.

Leslie, P. H. (1948) Some further notes on the use of matrices in population mathematics. *Biometrica* **35**: 213–45.

Lotka, A. J. (1925) *Elements of Physical Biology*. Baltimore: Williams and Wilkins.

Luck, R. F. (1990) Evaluation of natural enemies for biological control: a behavioral approach. *Trends in Ecology and Evolution* **5**: 196–9.

Maksimovic, M., Bjegovic, P. L. & Vasiljevic, L. (1970) Maintaining the density of the gypsy moth enemies as a method of biological control. *Zastita Bilja* **1107**: 3–15.

Malthus, T. R. (1789) *An Essay on the Principle of Population*. London: Johnson. (Reprinted as *Malthus-Population: The First Essay*. Ann Arbor Paperbacks: University of Michigan.)

Milsum, J. H. (1966) *Biological Control Systems Analysis*. New York: McGraw-Hill Book Company.

Mitsch, W. J. & Jorgensen, S. E. (eds) (1989) *Ecological Engineering: An Introduction to Ecotechnology*. New York: John Wiley and Sons.

Murdoch, W. W. (1969) Switching in general predators: experiments on predator specificity and stability of prey populations. *Ecological Monographs* **39**: 335–54.

Murdoch, W. W., Chesson, J. & Chesson, P. L. (1985) Biological control in theory and practice. *American Naturalist* **125**: 344–66.

Nicholson, A. J. & Bailey, V. A. (1935) The balance of animal populations. Part 1. *Proceedings of the Zoological Society of London* **3**: 551–98.

Odum, H. T. (1989) Ecological engineering and self-organization. In *Ecological Engineering: An Introduction to Ecotechnology*, ed. W. J. Mitsch & S. E. Jorgensen, pp. 79–101. New York: John Wiley and Sons.

Pearl, R. & Reed, L. J. (1920) On the rate of growth of the population of the United States since 1790 and its mathemetical representation. *Proceedings of the National Academy of Science of the USA* **6**: 275–88.

Real, L. (1977) The kinetics of functional response. *American Naturalist* **111**: 289–300.

Ricker, W. E. (1954) Stock and recruitment. *Journal of the Fisheries Research Board of Canada* **11**: 559–623.

Rosenzweig, M. P. (1971) Paradox of enrichment: destabilization of exploitation ecosystems in ecological time. *Science* **171**: 385–7.

Rosenzweig, M. P. & MacArthur, R. H. (1963) Graphic representation and stability conditions of predator–prey interaction. *American Naturalist* **97**: 209–23.

Royama, T. (1971) A comparative study of models for predation and parasitism. *Researches on Population Ecology*, **Supplement no. 1**: 1–91.

Royama, T. (1977) Population persistence and density dependence. *Ecological Monographs* **47**: 1–35.

Royama, T. (1992) *Analytical Population Dynamics*. London: Chapman and Hall.

Sarnelle, O. (1994) Inferring process from pattern: trophic level abundances and imbedded interactions. *Ecology* **75**: 1835–41.

Solomon, M. E. (1949) The natural control of animal populations. *Journal of Animal Ecology* **18**: 1–35.

Thompson, W. R. (1924) La théorie mathematiques de l'action des parasites entomophages et le facteur du hasard. *Annales de la Faculté des Sciences de Marseille* **2**: 69–89.

Verhulst, J.-P. (1837) Notice sur la loi que la population suit dans son accroissement. *Correspondance Mathématiques et Physique* **10**: 113–21.

Volterra, V. (1928) Variations and fluctuations of the number of individuals in animal species living together. Translated from the original paper and republished in *Animal Ecology*, ed. R.N. Chapman (1931), pp. 409–48. New York: McGraw-Hill. (First published in 1928 in Italian).

Appendix: Derivation of the discrete Lotka–Volterra equations

Start with the original Lotka–Volterra differential equations:

$$\frac{dN}{dt} = a_n N - d_n NP$$

$$\frac{dP}{dt} = -d_p P + a_p NP$$

where N and P are the density of prey and predators, and a_n and a_p are the increase and d_n and d_p the decrease rates, respectively. This equation cannot be solved analytically, so we have to arrive at the discrete formalism indirectly (following Royama, 1971). Because births occur at the end of each generation we assume that no predators are born during the generation so that the predator equation becomes (Royama, 1971):

$$\frac{dP}{dt} = -d_p P$$

which can be integrated to yield:

$$P_t = P_0 e^{-d_p t}$$

Substituting in the prey equation we get:

$$\frac{dN}{dt} = a_n N - d_n N P_0 e^{-d_p t}$$

which can then be integrated:

$$N_t = N_0 e^{a_n t - d_n P_0 (1-e^{-d_p})/d_p}$$

Setting the time-step to unity:

$$N_t = N_{t-1} e^{a_n - d_n P_{t-1}(1-e^{-d_p})/d_p}$$

and if we let the death rates due to factors other than predation be very small (as most models do), then $(1 - e^{-d_p})/d_p \to 1$ and:

$$N_t = N_{t-1} e^{a_n - d_n P_{t-1}} \qquad \text{(discrete prey equation)}$$

which we consider to be a discrete form for the **LV** prey equation.

The corresponding predator equation should then be:

$$P_t = P_{t-1} e^{-d_p + a_p N_{t-1}} \qquad \text{(discrete predator equation)}$$

2

Recent developments in theory for biological control of insect pests by parasitoids

CHERYL J. BRIGGS, WILLIAM W. MURDOCH AND ROGER M. NISBET

Introduction

Historically, most models of insect–parasitoid interactions have been framed in terms of systems of difference equations of the Nicholson–Bailey type. These have been extensively reviewed by Hassell (1978), May & Hassell (1988), and Jones *et al.* (1994). Berryman (Chapter 1) explores the structure of discrete-time models of various types. Although the discrete-time framework of these models is consistent with the discrete generation natural history of a number of temperate insect systems, they are less appropriate for many pest species that can have a number of overlapping generations within a year. They are further limited by the fact that only the net effect, summed up over the year, of all of the processes affecting the species within the year can be included in the equations. This results in rather abstract models in which it is difficult to include explicitly specific behavioral or physiological mechanisms.

In this chapter we will concentrate on continuous-time models of host–parasitoid interactions that incorporate host stage- and parasitoid physiological state-dependent behaviors. We will discuss the utility of these types of models in comparing the predicted efficacy of different parasitoid species in biological control programs. We will discuss how models that incorporate within-season dynamics and parasitoid behaviors into discrete-time non-overlapping generations produce predictions that are not possible with standard discrete-time models. In much of the chapter we concentrate on the stability of, or types of dynamics produced by, the host–parasitoid interaction, because that was the focus of many of the models reviewed. However, in the section on Local stability and biological control, we consider approaches that question whether a stable local pest equilibrium is needed for successful biological control. We have not included the vast number of recent spatially explicit models of host–parasitoid and predator–prey interactions (c.f., Hassell *et al.*, 1991; Comins *et al.*, 1992; McCauley *et al.*, 1993; Rand & Wilson, 1995; Wilson *et al.*, 1995; Rohani & Miramontes, 1995; Ruxton & Rohani, 1996), as that would take an entire review in itself.

Stage-structured models

In the last 10–15 years, a wide array of individual-based modeling approaches have been introduced to population ecology. These range from large simulations of collections of individuals to stage-structured models that treat all individuals as identical within a developmental stage, but allow for differences in demographic rates between stages. It is this latter approach we concentrate on here. Insect populations are natural candidates for the stage-structured modeling framework. Their life history is conveniently divided into a number of distinct stages (eggs, larvae of various instars, pupae, and adults), which can have very different biological properties. In some cases, however, our 'stages' will refer to size or age classes rather than distinct developmental stages.

Parasitoids can behave very differently towards the different stages of a host insect. Many parasitoids attack only a certain stage or stages of the host (e.g., egg parasitoids, larval parasitoids, etc), and the parasitoid may further discriminate among host sizes. A parasitoid may also use the different-sized hosts for different purposes. In particular, larger host individuals represent a larger packet of nutrients for the parasitoids, and many parasitoid species will lay larger clutches of eggs on larger hosts. Synovigenic species (those that continually mature new eggs from the nutrients gained from feeding on the tissues of some host individuals) typically feed off smaller, and oviposit in larger, hosts (Walde *et al.*, 1989; Kidd & Jervis, 1991a,b; and see Chapter 8). These oviposition decisions can strongly affect population dynamics, the level of suppression of pest populations, and competition between potential biological control agents. The stage-structured modeling approach allows incorporation of such features of real systems into models, and provides a link between the wealth of studies done on the behavior and evolutionary ecology of parasitoids and the effects of these properties of individuals on the dynamics of populations.

In this section we review stage-structured models of host–parasitoid systems. The goal of many of these models was to look separately at different aspects of the stage-specific interactions between parasitoids and hosts.

Invulnerable adult stage

The first continuous-time stage-structured host–parasitoid model (Murdoch *et al.* 1987) was motivated by the observation that most parasitoid species attack only the juvenile stage(s) of their host, while the adult stage is invulnerable to attack. As with many of the other models written by Murdoch and co-workers, this one was inspired by the extremely successful biological control of California red scale (*Aonidiella aurantii*) by the parasitoid wasp *Aphytis melinus*, in an apparently stable interaction.

The model is illustrated in Fig. 2.1(a). It makes all of the assumptions of the Lotka–Volterra consumer–resource model: density-independent host growth and parasitoid death, a linear (Type I) functional response, and a constant conversion between attacks on hosts and births of parasitoids. The only exception is that in the stage-structured model, both the host and parasitoid populations are divided into a juvenile and an adult stage. Only adult hosts produce new juvenile hosts, only adult parasitoids search for hosts, and only juvenile hosts can be successfully parasitized.

The main result from this model was that a long invulnerable adult stage can stabilize the host–parasitoid interaction (see also Smith & Mead, 1974). While the basic continuous-time Lotka–Volterra model without time lags has a neutrally stable equilibrium (the discrete-time form is unstable: see Chapter 1), adding a time lag such as that between parasitoid oviposition and development of a new searching adult parasitoid destabilizes the equilibrium. However, if the host adult stage is relatively long, then a stable equilibrium can result, even when the parasitoid has a non-zero development time (Fig. 2.2(a)). With a short adult stage, multi-generation stable limit cycles with a period related to that of the Lotka–Volterra model occur. The remarkable stability of red scale populations may in part be due to the relatively long-lived adult stage. However, stability in the model is bought at the cost of an increased density of the pest (specifically, an increase in the density of the adult stage), because a fraction of the host population is protected from parasitism.

Figure 2.2(a) provides a way of classifying combinations of parasitoid and host species according to major features of their life cycles. Thus classes of hosts with short-lived adults, such as moths, butterflies, midges, mayflies etc cluster close to the *y*-axis, whereas grasshoppers, beetles, aphids, scale insects etc, which have relatively long-lived adults, would be spread to the right of the diagram. The *y*-axis quantifies the number of generations that the parasitoid can pass through within a host generation, with values close to the *x*-axis representing many parasitoid generations per host generation. The next section will highlight the dynamical importance of this ratio of development times.

Relative development times of host and parasitoids and parasitoid density-dependent attack

Godfray & Hassell (1987, 1989) studied a model with much the same framework as Murdoch *et al.* (1987), except that they allowed the parasitoids to attack during only a restricted portion of the juvenile host developmental period (illustrated in Fig. 2.1(b)), and they included a density-dependent parasitoid attack rate (i.e., the death rate of hosts due to parasitoids, which is a linear function of both hosts and parasitoids

(a) Murdoch *et al.*(1987)

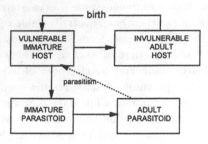

(b) Godfray & Hassell (1989)

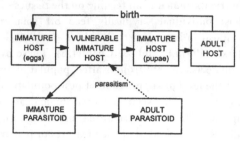

(c) Gordon *et al.*(1991)

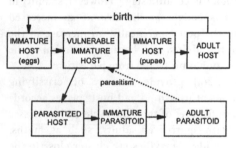

(d) Reeve *et al.*(1994a,b)

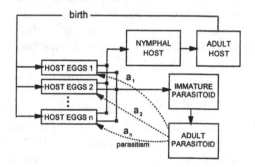

(e) Murdoch *et al.*(1992)

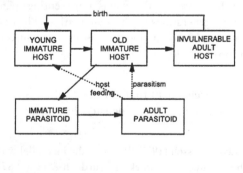

(f) Murdoch *et al.*(1996a)

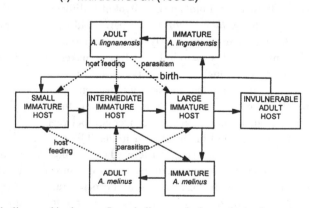

Figure 2.1. Diagram of the stage–structure of the models discussed in the text. In each diagram the boxes depict the stages of the host and parasitoid(s). The continuous arrows represent transitions between the various stages, and the broken arrows represent attack by the parasitoid(s).

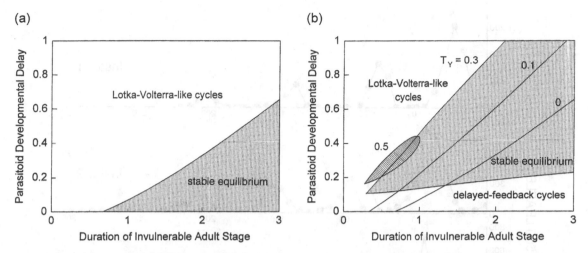

Figure 2.2. Stability boundaries showing the regions of local stable or unstable equilibria in terms of the durations of the invulnerable adult stage and the parasitoid developmental delay, for (a) the Murdoch *et al.* (1987) model, and (b) the Murdoch *et al.* (1992) model. In each case, the equilibria are locally stable in the shaded region or to the right of the boundary. Lotka–Volterra-like stable limit cycles occur to the left of the boundary. All stage durations are relative to the total host development time, $T_J = 1$ ($= T_Y + T_O$). In (b) the boundaries are shown for three different values of the duration of the young immature host stage (T_Y) that is attacked by the parasitoid but not used for the production of female parasitoid offspring. Both the duration of the adult host stage and the parasitoid developmental delay are relative to the duration of the immature host stage. The regions below the boundaries for $T_Y = 0$ and $T_Y = 0.1$ are stable. The region below the shaded area for $T_Y = 0.3$ shows delayed feedback stable limit cycles, and such cycles also occur in a region to the right and below the shaded region when $T_Y = 0.5$.

in the Lotka–Volterra and Murdoch *et al.* (1987) models, is now a decelerating function of the number of parasitoids present). Their model was motivated by tropical plantation pest species (mostly Lepidoptera) that have a relatively short adult lifespan, so they investigated the dynamics that can occur when the adults are short-lived.

The major insight gained from this model was that the interaction between hosts and parasitoids can lead to a sustained pattern of apparently discrete generations in systems without strong seasonal forcing, even though continuous reproduction is possible. The ratio of host:parasitoid development times is one of the key parameters. If the parasitoid development time is approximately 0.5 (or 1.5) times the host development time, then these 'generation cycles' are most likely. This lack of synchrony between parasitoids and hosts reinforces generation cycles because parasitism during a peak in host density will lead to new adult parasitoids half a host

generation later, instead of during the next host generation peak, thereby decreasing the host density between the generational peaks. This mechanism was raised as a potential explanation for discrete generation pulses observed in a number of tropical plantation pests (e.g., Fig. 2.3: and see Notely, 1948, 1956; Bess, 1964; Wood, 1968; Bigger & Tapely, 1969; Bigger, 1973, 1976). These generation cycles can occur only if the invulnerable adult stage is relatively short. Long-lived adults would distribute their offspring over a long period of time, smearing out any generational structure.

Mathematically, roots with a period of a single host generation are also present in the Murdoch *et al.* (1987) model if the adult stage is very short-lived (for parameters very close to the *y*-axis in Fig. 2.2(a)). However, in that region they are obscured by unstable Lotka–Volterra-like cycles. The density-dependence in the parasitoid attack rate in the Godfray & Hassell model has a stabilizing effect on

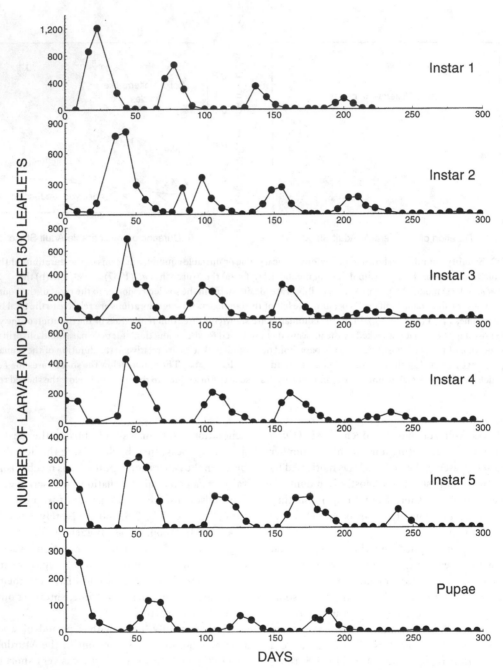

Figure 2.3. An example of generation cycles observed in a tropical plantation pest (the coconut caterpillar, *Opisina arenosella* Walker), illustrating the progression of dominant cohorts through the host age classes (figure reproduced from Perera *et al.*, 1988, with permission from CAB International). The host development time is 46 days and adult moths live for about 7 days. The period of the cycles is between 50 and 60 days.

the dynamics. In order for the generation cycles to be observed, this density-dependence needs to be strong enough to stabilize the Lotka–Volterra cycles, but not so strong as to damp out the generation cycles.

Gordon *et al.* (1991) presented a modification of this model appropriate for koinobiotic parasitoids (these species typically attack a wide range of host ages but all emerge from hosts of the same age, so that parasitoids take longer to develop in younger hosts, see Fig. 2.1(c)). They found that generation cycles are also possible with koinobiotic parasitoids, but that the 0.5 ratio of host:parasitoid development time is not a fixed rule. In their model, motivated by the stored-product moth, *Cadra cautella* (Walker), attacked by the parasitoid *Venturia canescens* (Gravenhorst), generation cycles were most likely when the host:parasitoid development ratio was around 0.2–0.3 (or 1.2–1.3) instead. This is because there is a stage after the host has been parasitized before the immature parasitoid begins to develop (labeled 'parasitized host' in Fig. 2.1(c)) that effectively increases the parasitoid developmental delay.

Single-generation cycles can be exposed in the Murdoch *et al.* (1987) model, when the adult host is short-lived, simply by making the *per capita* parasitoid attack rate decrease with parasitoid density. Again, the density-dependence needs to be strong enough to suppress the Lotka–Volterra-like limit cycles, but not so strong as to suppress the single generation cycles. The potential for single-generation cycles thus appears to be a basic property of stage-structured host–parasitoid interactions.

Differential vulnerability to parasitism

In the Godfray & Hassell (1987, 1989) model, the density-dependent parasitoid attack rate that suppressed the Lotka–Volterra-like cycles was included as a generalized function intended to represent non-random searching behavior by parasitoids. Motivated by the observation of generation cycles in populations of the salt-marsh planthopper, *Prokelisia marginata*, Reeve *et al.* (1994a,b) proposed a *mechanistic* model that exhibits single generation cycles. They hypothesized that in the planthopper system, the mechanism leading to a density-dependent attack rate was differential vulnerability among hosts to parasitism. In the salt-marsh system, hosts located at different tidal heights were under water, and therefore protected from parasitism, for different lengths of time. They included this in a model by dividing the juvenile host population into a number of different types on each of which the parasitoids had different attack rates (illustrated in Fig. 2.1(d)). The adult host population develops from all juvenile hosts of all types that have escaped parasitism, and a fraction of all offspring recruit to each of the patch types. There is a single adult parasitoid population that searches on all types of hosts. They found from the model that this type of differential vulnerability to parasitism does in fact lead to density-dependence in the parasitoid attack rate. When incorporated into a model with parameter values appropriate for the planthopper system, generation cycles are a potential outcome.

Two features of the Reeve *et al.* (1994a) model allow for the stabilizing effect of differential vulnerability of hosts to parasitism. First, the hosts are assigned their vulnerability at birth. This cannot change in response to parasitoid or live host density. The second is that parasitoids spend the same amount of effort searching for hosts of all types, even though their successful attack rate is much higher on some host types than others. This results in a lower realized parasitoid attack rate at high parasitoid densities than at low parasitoid densities (i.e., density-dependence in the parasitoid attack rate), because at high parasitoid densities the density of hosts of all types is suppressed more than at low parasitoid densities, but the density of the most vulnerable host types are suppressed to the greatest degree. Therefore, the effective parasitoid attack rate decreases as parasitoid density increases (see also discussion of Rohani *et al.*, 1994, below).

Insecticides and single-stage outbreaks

The finding that single generation cycles (such as those in Fig. 2.3) are one possible outcome of host–parasitoid interactions is of relevance to biological control because generation cycles appear as outbreaks of a single stage of the pest insect. In the past, the prevalent view among biological control workers who have observed repeated single-stage outbreaks (that form effectively discrete generations in populations with continuous reproduction) in tropical plantation pests is that they are caused by the action of insecticides. The hypothesis is as follows: insecticides kill all stages of the hosts and parasitoids, except those that are in a protected stage (such as the pupa). This synchronizes the pest population into a single age cohort. If the parasitoid is shorter lived than the pest, then it will not be able to persist between generations on the synchronized pest population. Thus, the pest population, escaping parasitism, will be able to reach outbreak densities and retain its discrete-generation age structure.

Godfray & Chan (1990) used a model to investigate the plausibility of this hypothesis. For illustrative purposes they assumed a less common, but simpler, situation in which it is the adult stage of the host that is attacked by the parasitoid and the juvenile host stage is invulnerable. Models in which the adult is the stage attacked were also investigated by Hastings (1984) and Murdoch *et al.* (1987). The Godfray & Chan (1990) model added a density-dependent host birth rate in addition to a parasitoid attack rate dependent on parasitoid density. This model produces the same range of dynamics as those observed when the juvenile host is attacked, and generation cycles are possible when the ratio of host:parasitoid development times is ≈ 0.5, and the adult host stage is relatively short-lived.

Godfray & Chan (1990) simulated a catastrophic synchronization event in the model by removing all host individuals not in a short protected (pupal) stage, and killing off all parasitoids. They found that, in the absence of the parasitoids, the single-stage quality of the outbreaks was quickly lost, and a mixed age-structure resulted. This finding is consistent with standard theory of age-structured populations (Caswell, 1989 and references therein). They showed that some factor, such as the parasitoid, is needed to continuously reinforce the synchronized age-structure in order for the discrete-generation, single-stage outbreaks to persist for very long. Insecticides can however trigger transient single-stage outbreaks even

if the host–parasitoid system eventually returns to a stable equilibrium.

Host stage-dependent parasitoid oviposition behaviors

Murdoch *et al.* (1992) developed a model to look at size-selective sex allocation and host-feeding in parasitoids. In parasitoids, the size of the host frequently determines the size of the developing parasitoid, which can influence its fitness as an adult. Sex allocation theory predicts that, if hosts are limited, a female parasitoid should use the larger-sized hosts she encounters for the production of offspring of the sex that will benefit most from being large (Charnov *et al.*, 1981). In parasitoids, females generally benefit more than males from being large, because female fecundity is more closely related to body size (larger females can produce more eggs) than is male mating success (although this is difficult to measure in the field). Many experimental studies have shown that parasitoids do indeed allocate female eggs to larger host individuals and male eggs to smaller hosts (e.g., Jones, 1982; King, 1989). Similarly, in species that destructively host-feed, it is often the case that smaller host individuals get fed upon, and larger hosts are parasitized (Walde *et al.*, 1989; Kidd & Jervis, 1991a,b; and see Chapter 8).

To look at both of these parasitoid oviposition behaviors, Murdoch *et al.* (1992) used the model illustrated in Fig. 2.1(e). As in most of the previous models, only juvenile hosts can be attacked by the parasitoid and the adult stage is invulnerable. However, in this case the juvenile host stage is divided into two classes. Younger juvenile hosts are used either for host-feeding or for the production of male parasitoid offspring, and older juvenile hosts receive a female parasitoid egg. The rest of the assumptions are the same as in the Murdoch *et al.* (1987) model. The new model follows only female parasitoids, and there is no gain to the female population from either host-feeding or production of males. Neither of these assumptions is strictly valid: if males become limiting, production of more males will contribute to the female population, and females do gain nutrients from host-feeding for the production of new eggs. However, the qualitative results are the same even if these assumptions are relaxed (our unpublished results).

The main result from this paper was that through attacking and killing young juvenile hosts without receiving any gain, the parasitoids are reducing the density of hosts that achieve the size from which they can successfully produce female offspring. This effectively creates a type of delayed pseudo-density-dependence in the parasitoid attack rate: high parasitoid densities lead to high levels of attack on young hosts, which leads to few hosts surviving to the older juvenile stages, and therefore few potential new adult female parasitoids. This delayed pseudo-density-dependence has two effects on the dynamics (see Fig. 2.2(b)): first, it tends to stabilize the multi-generational Lotka–Volterra cycles (i.e., a stable equilibrium is possible with a shorter invulnerable adult lifespan than in the Murdoch *et al.* (1987) model: the Lotka–Volterra boundary moves to the left). Second, it leads to a new class of delayed feedback cycles, that usually have a period between 1 and 2 times the total host development time. (An added complication is that parts of the stable region may contain multiple attractors, i.e. a stable equilibrium or stable limit cycles, depending on the initial conditions.)

This type of dynamical effect of size-dependent sex allocation and host-feeding was shown to be more general by Murdoch *et al.* (1997) and our unpublished results. Whenever parasitoids attack younger hosts, and the gain to the searching female parasitoid population is less than that from attacks on large hosts, delayed pseudo-density-dependence ensues. Increasing the pseudo-density-dependence (either through increasing the attack rate on young hosts, or increasing the potential gain from large hosts), leads to a decrease in the parameter space in which Lotka–Volterra cycles occur, and an increase in the parameter space in which delayed-feedback cycles occur. (This can be seen by following the changes in the stability boundaries shown in Fig. 2.2(b) as the duration of the younger immature host stage, T_Y, increases.) Such mechanisms include: increasing parasitoid clutch size, increasing juvenile parasitoid survival, or increasing the fecundity of the resulting parasitoid adult, with host age. The same result is

found when host-feeding or sex-allocation depends not only on host size but on parasitoid egg load (see below). Finally, the same dynamical result also arises in koinobiont parasitoids, but in this case *via* altering the time delays. We have found however, that the effects of egg limitation and limited numbers of males can weaken this dynamical effect, by causing the small host to make a larger contribution to the searching female parasitoid population (our unpublished results). These analyses thus show that apparently different responses by parasitoids have effectively the same dynamic consequences and can be viewed as variants of a single, broader, phenomenon.

Use of the comparative approach with stage-structured models

One of the benefits of stage-structured models is that most of the parameters are easily interpreted biologically and can be measured directly. They include such things as stage durations and through-stage survival probabilities. This increases the utility of these models to address questions in real biological control systems. Waage (1990) has argued that models might be most useful to biological control programs in the future when they are used in a comparative way to screen competing candidate biological control agents. In the past, theory has been used to suggest characteristics of parasitoids that might make them good biological control agents. But in actuality, not all combinations of desirable attributes occur in any real parasitoid species; only some limited combinations of these characteristics occur in the species that actually attack the target species. The comparative approach could be used to predict the relative success of two or more proposed real species. We next present some examples, and then make some comments on insights obtained from general theory.

The mango mealybug and two candidate parasitoids

Godfray & Waage (1991) used the comparative approach to suggest which of two parasitoid species would be more effective at reducing the density of the mango mealybug, *Rastrococcus invadens*, although in this case the modeling exercise was done after the fact. The model suggested that one parasitoid species, *Gyranusoidea tebygi*, would lead to considerably lower pest density. The main difference between the two species was that *G. tebygi* attacked an earlier stage of the mealybug than did the other parasitoid species (*Anagyrus* sp.). In reality, *G. tebygi* was the species released in the biological control program, but not because of the predictions of the model: it was the easier species to raise in culture, and has proven to be remarkably successful at reducing densities of the mealybug.

California red scale *and* Aphytis melinus *and* A. lingnanensis

As mentioned above, the parasitoid *A. melinus* has been extremely successful at controlling California red scale, a pest of citrus. However, other parasitoids were tried, including the congener, *A. lingnanensis*, before success was achieved with *A. melinus*. *Aphytis lingnanensis* was moderately successful at controlling red scale, but it was rapidly displaced (within only a few years) from most of southern California following the addition of *A. melinus*, when the density of red scale was also further suppressed. This is one of the most widely cited examples of competitive displacement of one species by another.

Luck & Podoler (1985) suggested that *A. melinus* might gain a competitive advantage over *A. lingnanensis* because a female offspring could successfully develop on slightly smaller host individuals (see also Chapter 16). The plausibility of this hypothesis was confirmed by the stage-structured model of Murdoch *et al.* (1996a), illustrated in Fig. 2.1(f). Both *Aphytis* species will attack and host-feed or lay a male egg on virtually all stages of juvenile hosts. However *A. melinus* can lay female eggs in scale individuals that are as small as 0.39 mm^2, whereas *A. lingnanensis* can only use individuals greater than 0.55 mm^2 for a female offspring. The model suggests that this small difference in life history is sufficient to give *A. melinus* enough of an advantage to displace *A. lingnanensis* within a year or so and to reduce scale density substantially.

Cassava mealybug *and* Epidinocarsis lopezi *and* E. diversicornis

Similarly, the stage-structured model of Gutierrez *et al.* (1993) indicates that one parasitoid species (*Epidinocarsis lopezi*) gains a competitive advantage over its congener *E. diversicornis* through its ability to lay a female egg in smaller individuals of the cassava mealybug, *Manihot esculenta*. *Epidinocarsis lopezi* has been found to be the more successful of the two parasitoid species in the field. The modeling approach of these authors is discussed in Chapter 5.

Theory of stage-structured parasitoid competition

Briggs (1993) used general models (not based on a particular biological control system) to compare the predicted biocontrol success of parasitoids that attack early or late developmental stages of their host. This model emphasized that in such stage-structured interactions, it is important to distinguish which stage(s) of the host causes the most economic damage, and is therefore the main target of biological control.

A parasitoid gains a competitive advantage, and can exclude a competitor, by its ability to attack an earlier developmental stage of the host. In the Briggs (1993) model, one parasitoid can displace another when it can suppress the density of the *stage* of the host *used by its competitor* to a density below that at which the competitor's population can increase from low density (a stage-structured extension of classic competition theory; MacArthur & Levins, 1967; Holt *et al.*, 1994). Therefore, the parasitoid that wins in competition will drive to a lower density both the stage that it attacks and the stage that its competitor attacks. However, if the adult stage is not attacked by either species, and is the major economic pest, then it is possible for a parasitoid that attacks an early host stage (e.g., the egg stage), at a relatively low attack rate, to exclude a competitor with a higher attack rate that parasitizes a later host stage (e.g., the larval stage). This can lead to poorer control of the adult pest density by the early attacking parasitoid species that wins in competition (see Fig. 2.4). This can happen because the early attacking parasitoid leads to

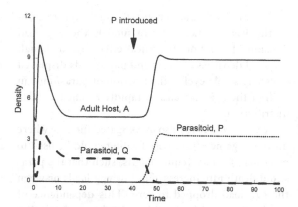

Figure 2.4. A simulation of the Briggs (1993) model in which parasitoid species *P* attacks host eggs and parasitoid species *Q* attacks host larvae. Parasitoid *Q* has a higher attack rate than parasitoid *P*, but *P* wins in competition, excluding *Q*, because it is able to attack an earlier stage of the host. The introduction of *P* and exclusion of *Q* leads to an increase in the density of adult hosts, *A*.

a lower *total* density of the host stage that the later parasitoid attacks, however through attacks during the later developmental stage, the later species can lead to a lower density of hosts that survive to the age at which they mature into the adult stage, and therefore a lower adult host density.

However, given *equal* attack and death rates, a parasitoid that attacks earlier in the juvenile host stage will generally lead to the same or better suppression of the adult host density than a parasitoid that attacks late (Murdoch & Briggs, 1996). Two species that have the same attack and death rates can be expected to lead to exactly the same adult host density only under the restrictive conditions that the host is vulnerable to both species for equal lengths of time, and the host suffers mortality from no sources other than parasitism. However, if there is non-parasitoid-induced juvenile host mortality, then the parasitoid species that attacks earlier will suppress the host to the lower density.

Students of biological control have often tried to specify the characteristics that distinguish successful biological control agents, emphasizing for example a high *per capita* attack rate. The stage-structured modeling framework in Murdoch & Briggs (1996)

emphasizes, by contrast, that success depends on the ensemble of traits that constitute the parasitoid's life history. The species that suppresses the density of a host stage to the greatest degree is the one that requires the presence of the fewest pest individuals of that stage in order for the searching adult female population to replace itself in the next generation or time interval, and this outcome depends on many attributes. A non-exhaustive list includes: the minimum size of host from which a female parasitoid can develop; the number of female progeny gained from a host of a given size; the number of host-meals gained from a host of a given size; the fraction of juvenile parasitoids that survive to adulthood; and the longevity of the searching adult female parasitoids. In addition, for a given set of parasitoid attributes, the pest's life history and demography also affect the outcome, for example the duration and relative mortality of the different host stages. The modeling framework integrates and weighs all of these attributes in evaluating each potential natural enemy species.

Parasitoid state-dependent behaviors

The behavior of a parasitoid when it encounters a host can depend not only on the size and developmental stage of the host, but on the physiological state of the searching parasitoid itself. One important component of physiological state is egg load (the number of mature eggs that it is currently carrying). A number of parasitoid behaviors and properties have been found to depend on egg load (see Minkenberg *et al.*, 1992 for a review), including search intensity (Jones, 1977; Collins & Dixon, 1986; Trudeau & Gordon, 1989; Odendaal, 1989; Odendaal & Rausher, 1990; Rosenheim & Rosen, 1991), host-feeding rate (Rosenheim & Rosen, 1992; Chan & Godfray, 1993; Collier *et al.*, 1994), and clutch size (Courtney *et al.*, 1989; Courtney & Hard, 1990; Odendaal & Rausher, 1990; Rosenheim & Rosen, 1991).

Synovigenic parasitoids are those species in which the female emerges as an adult with fewer than the total number of eggs that she might lay in her lifetime. Females of most of these species need to obtain nutrients from host fluids or tissues in order to mature extra eggs. In the process of host-feeding the parasitoid usually kills the host and makes it unsuitable for subsequent parasitism. A high fraction of hosts killed by some parasitoid species are killed through host-feeding, and ecologists have long suggested that host-feeding may have important consequences for population dynamics (DeBach, 1943; Flanders, 1953). We have discussed the dynamical outcome of one aspect of host-feeding in the Murdoch *et al.* (1992) model of size-selective host-feeding, in which parasitoids host-feed on small hosts and oviposit on large hosts. Oviposition decisions that depend on the egg load of the searching parasitoid might be expected to have a dynamical effect as well, because current egg load represents past host-meals, and therefore past densities of hosts. Such decisions represent the trade-off between current and future reproduction for the parasitoid.

Host-feeding decisions dependent on egg load

Briggs *et al.* (1995) investigated several versions of a model in which the decision of the parasitoid, upon encountering a host, to use that host for a host-meal or for oviposition, depends on her current egg load. In the models, the probability of host-feeding (rather than parasitizing) is a decreasing function of egg load. This is based on the empirical relationship measured for the parasitoid *A. melinus* on California red scale (Collier *et al.*, 1994), and is qualitatively consistent with the predictions of dynamic optimization models by Chan & Godfray (1993), Collier *et al.* (1994) and Collier (1995).

No age-structure is included in either the host or parasitoid populations in the models. The parasitoid population is structured according to egg load (i.e., the model keeps track of the number of individuals carrying 0, 1, 2, ... eggs), and transitions between the egg load classes are made through oviposition and host-feeding (Fig. 2.5(a)).

Perhaps surprisingly, the decision to host-feed or oviposit based on egg load was found to have no effect on stability, but it can strongly affect the equilibrium densities (this result was also found from the simulation model of Kidd & Jervis (1989), and see Chapter

8). The model produced neutrally stable dynamics, as in the basic Lotka–Volterra model, the egg load dynamics having no dynamical effect. Even though the total densities of hosts and parasitoids displayed neutrally stable cycles, the fraction of parasitoids in each of the egg load classes rapidly reached a stable distribution.

Briggs *et al.* (1995) also investigated the case where nutrients gained from host-feeding could be used to meet maintenance requirements, such that the parasitoid death rate increased at low egg loads or when the egg load dropped to zero. This dependence of death rate on egg load was found to have a destabilizing effect. However, a gain of nutrients from an outside source that can be used for egg production, such as a gain of protein from pollen, had a stabilizing effect.

Host-feeding decisions dependent on host size and parasitoid state

Murdoch *et al.* (1997) showed that egg-load-dependent host-feeding does have a dynamical effect when the decision also depends on host size. In *Aphytis*, the probability an encountered host is host-fed decreases with egg load, but does so faster when the host is older, i.e., larger (Fig. 2.5(b)). The effect, however, is best interpreted as another manifestation of host-size-dependence, and leads to dynamics similar to those shown in Fig. 2.2.

Suppose, for illustration, that the young immature host stage comprises 30% of the immature stage duration, so $T_Y = 0.3$. When there is no difference in the rates at which the probability of host-feeding decline with increasing egg load on the two host stages (i.e., the two juvenile host stages are indistinguishable), we effectively recover the Murdoch *et al.* (1987) model (Fig. 2.1(a)) whose dynamics are shown in Fig. 2.2(a). When the difference is effectively infinite (the slope for the older immature host stage is, say, 1000 times larger than that for the younger immature host stage) we regain the model of size-dependent host-feeding and sex allocation (Murdoch *et al.*, 1992; Fig. 2.1(e)), and the dynamics are described by Fig. 2.2(b) for the case where $T_Y = 0.3$. The stable region is the shaded triangle

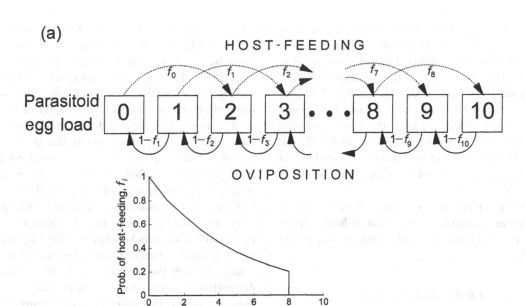

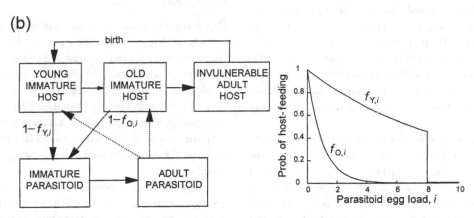

Figure 2.5. Diagram of the parasitoid state-dependent models discussed in the text. In (a) the parasitoid egg load transitions in the Briggs *et al.* (1995) model are shown. f_i is the probability that a parasitoid in egg-load class i will host-feed on a host that they encounter, with $1-f_i$ being the probability that they oviposit instead. f_i is a decreasing function of host density. Parasitoids gain 2 eggs from a host-feeding, and lose 1 egg when they oviposit in a host. (b) Depicts the Murdoch *et al.* (1997) model in which the host-feeding/oviposition decisions depend on both parasitoid egg load and host stage. Both young and old hosts can be used for either host-feeding or oviposition, but the probability of host-feeding on large hosts ($f_{O,i}$) decreased more rapidly with egg load than does the probability of host-feeding on young hosts ($f_{Y,i}$).

and both Lotka–Volterra-like cycles and delayed feedback cycles exist. When the ratio lies between these extremes the stability diagram is intermediate between Figs. 2.2(a) and 2.2(b). For example, if T_Y is still equal to 0.3, and if the slope in Fig. 2.5(b) for older hosts is 10 times greater than that for younger hosts, then we recapture approximately the stability boundary corresponding to $T_Y = 0.1$ in Fig. 2.2(b). Thus, the crucial effect here is not the dependence of host-feeding on egg load but rather its dependence on host size.

The rather surprising message regarding state-dependent decisions by parasitoids, at least so far, is that it has no independent effect on population stability.

Host-feeding and biological control

Historically, the conventional wisdom of biological control workers was that parasitoids that destructively host-feed make better biological control agents than those that do not, because of the extra mortality to the host caused by host-feeding (see e.g., Greathead, 1986). However models such as those of Yamamura & Yano (1988), Kidd & Jervis (1989), Murdoch *et al.* (1992), and Briggs *et al.* (1995) predict that the equilibrium host density is increased if a parasitoid host-feeds, because the parasitoid now needs more individual hosts for each new parasitoid produced.

Jervis *et al.* (1996) recently investigated the apparent disparity between these two views. From the BIOCAT database (Greathead & Greathead 1992) of all recorded biocontrol introductions, Jervis *et al.* (1996) compared the establishment and success rates of parasitoids that did and did not host-feed, and found that host-feeders did better on both accounts. From this they concluded that, contrary to theory, destructive host-feeders are probably better biocontrol agents, or at least no worse, than those that do not host-feed.

However, the theory requires careful interpretation. What the models say is that, if you compare two parasitoid species that are alike in all attributes, such as attack rate, maximum fecundity, longevity, etc, except that one of the species host-feeds and the other does not, then the species that host-feeds will lead to the higher host density of the two (because more hosts are needed to sustain the parasitoid population). But it is certainly the case that all else is not equal between the species. For example, if a parasitoid of a given size did not host-feed, then it would need to store more nutrients as a juvenile to have the same adult fecundity and longevity that is achieved by the host-feeder. This suggests it would need to avoid parasitizing a class of smaller hosts, which in turn would probably reduce the parasitoid's efficacy (e.g. see Use of the comparative approach with stage-structured models section, above). There are thus unavoidable trade-offs between host-feeding to gain new eggs and having a larger initial egg complement. Thus, although the observation of the success of destructive host-feeders in biocontrol is important, it does not disprove the theory. We therefore need to be cautious in interpreting theory to gain useful insights for biological control.

Egg limitation

One feature that was not included in the host-feeding models of Briggs *et al.* (1995) was any limitation to the number of eggs that a parasitoid could carry, or to the rate at which parasitoids could convert host-meals into new parasitoid eggs. Shea *et al.* (1996) looked at the effect of egg limitation (defined as any situation in which a parasitoid has an opportunity to oviposit on a host, but cannot due to lack of eggs) in models of both pro-ovigenic and synovigenic parasitoid species.

In pro-ovigenic species, the female parasitoid emerges as an adult with all of the eggs that she will ever lay. In this case, a parasitoid becomes egg-limited when she runs out of eggs. The Shea *et al.* (1996) model structured the parasitoid population according to egg number. Parasitoids are born into their maximum egg number class and egg number is decremented as they encounter and parasitize hosts. Egg limitation in this situation has no effect on stability. However, this is probably due to the fact that the model includes no parasitoid development time. Eggs laid in hosts instantly become new searching parasitoids, so there is no loss in the efficiency of the parasitoid population when a parasitoid individual runs out of eggs. She is instantaneously replaced by her offspring emerging with a full egg complement. Egg limitation is likely to be destabilizing in this context if the parasitoids have a finite development time.

Modeling egg limitation in synovigenic species poses a greater technical problem. Two types of physiological limitation can act in synovigenic parasitoids. First, as in pro-ovigenic species, there can be an absolute upper limit to the number of eggs that a parasitoid can hold, or to the number of meals that can be held in the gut or nutrients in the body. Second, finite amounts of time are needed for the material ingested from a host-meal to pass through the gut and be processed into new eggs. Parasitoids are able to host-feed after they have run out of eggs, and they can oviposit after their gut is full. The Shea *et al.* (1996) model for synovigenic parasitoids structures the parasitoid not only into egg-load classes, but also into gut-load classes (full or empty). The model assumes that after host-feeding there is a latent period during which the parasitoid cannot host-feed again, but during which she can parasitize if she has any eggs.

As in the pro-ovigenic case, the maximum number of eggs that a parasitoid can carry has no effect on stability. However, the second aspect of egg limitation, namely the latent period during which host-meals are processed into new parasitoid eggs, does have an effect. Increasing this latent period has a destabilizing effect on the equilibrium. Egg limitation, as expected, thus has a somewhat analogous dynamical effect to predator satiation, though through a different mechanism.

Hybrid models: discrete reproduction, but within-season dynamics

All of the models discussed so far have been continuous-time models, which are most appropriate for systems without strong seasonal forcing (such as tropical systems) in which reproduction can occur continuously and all life stages of the organism potentially can be present throughout the year. (Such models can be adapted to seasonal environments by including a periodic forcing function.) We have discussed a range of realistic behaviors that can conveniently be included in these models. Developing models to look at the dynamical effect of such behaviors in temperate systems, in which reproduction often occurs in discrete pulses, however, poses a problem. Furthermore, discrete-time difference equation models allow for only the net effect of the behavior over the year to be included. Recently, a number of authors have taken a hybrid modeling approach to this problem.

In the hybrid approach, the model is divided into two parts: a between-generation (i.e., between-season) phase and a within-generation phase. The between-generation phase is a discrete-time map, mapping the density at the end of one season onto the density that starts the beginning of the next season. The discrete pulse of reproduction is included in this phase. The within-generation phase consists of continuous-time (possibly delayed-) differential equations that describe any processes that the author wants to investigate in detail within a season. This part includes the processes of parasitism, mortality, movement, etc. This approach is technically cumbersome. Analytical results are possible only for certain limiting cases, so most of the results are obtained through numerous simulations. The hybrid approach can also be used in cases where there are many generations per season and when there is some dormant period (diapause or quiescence) (Driessen & Hemerik, 1990).

Parasitoid aggregation and within-season redistribution

Rohani *et al.* (1994) used the hybrid approach to look at aggregation of parasitoids to patches of high host density, and especially to look at the effect of redistribution of parasitoids within a season. Discrete-time models, in which the parasitoids aggregate to host patches solely on the basis of the density at the beginning of the season, have invariably concluded that strong aggregation can have a stabilizing effect on the host–parasitoid interaction (Hassell & May 1973, 1974; Pacala *et al.*, 1990). Strong aggregation also leads to large increases in host density because the parasitoid keeps revisiting already-parasitized hosts. In these models, the parasitoids remain for the entire season on the patch that they chose at the start of the season. However, continuous-time models in which the parasitoids can redistribute instantaneously among patches in response to changes in the density of unparasitized hosts have found that strong aggregation is more likely to have a destabilizing effect on the dynamics – though it does lead to greater reduction in prey density (Murdoch & Stewart-Oaten, 1989).

For the most part, the Rohani *et al.* (1994) model has resolved this controversy. They found that even a small amount of redistribution of the parasitoids within the season, in response to changes in host density, destroyed the stabilizing effect of parasitoid aggregation. This is because in the discrete-time models, parasitoids are assumed to remain in the patch that they chose at the start of the season, even after the host density has been depleted, thus causing the parasitoids to be inefficient. This inefficiency increases as the density of searching parasitoids increases, leading to an effective parasitoid attack rate that decreases with parasitoid density. Allowing parasitoid redistribution in the Rohani *et al.* (1994) model reduces this inefficiency of the parasitoids, and thereby reduces the stabilizing effect.

Rohani *et al.* (1994) also examined parasitoid aggregation to patches independent of the host density in the patch. Discrete-time models without parasitoid redistribution within the season have shown that parasitoid aggregation independent of host density can have an even greater stabilizing effect than aggregation to patches of high host density (Chesson & Murdoch, 1986; Pacala & Hassell, 1991). However, Murdoch & Stewart-Oaten (1989) showed that in continuous-time models aggregation independent of host density has no effect on stability.

In the hybrid model of Rohani *et al.* (1994), host-density-independent parasitoid aggregation has a stabilizing effect. This result hinges on two features of their model: (i) hosts remain in the patch in which they were born, and (ii) for a given patch, the same constant fraction of parasitoids visits that patch for the entire season, even after that patch has been depleted. Thus, the key feature for stability is once again the inefficiency caused by parasitoids revisiting

patches even after the host density has been depleted. The Rohani *et al.* (1994) model of host density-independent aggregation, and the Reeve *et al.* (1994a) model described earlier, are both cases in which the individual host carries a fixed 'relative risk of attack' throughout its life (Chesson & Murdoch, 1986). Heterogeneity within the host population in this relative risk of attack, such that a small fraction of the population receives most of the risk, can have a stabilizing effect in such models.

Phenological asynchrony between hosts and parasitoids

Heterogeneity in the risk of attack is also possible if there is variability in the timing of host emergence into the vulnerable stage relative to when the parasitoids are present. The Bailey *et al.* (1962) discrete-time model included heterogeneity of risk among the hosts that was meant to represent phenological asynchrony between the hosts and parasitoids, and Griffiths (1969) introduced the negative binomial model as a phenomenological description of this situation. Godfray *et al.* (1994) have recently used the hybrid approach to model this process explicitly. In their model, there is a single generation of both hosts and parasitoids each year. However, there can be variability in the timing within the year at which the hosts enter the vulnerable stage and at which the parasitoids emerge as searching adults.

They found that if the parasitoids emerged after the hosts, then hosts that emerged early in the season could partially or completely escape parasitism. Thus, the phenological asynchrony created a temporal refuge that had a stabilizing effect on the dynamics. However, because some hosts are escaping parasitism, this stabilizing effect is again bought at the cost of increased host density. For biological control programs, synchrony between the emergence times of hosts and parasitoids is probably preferable.

Multiple natural enemy generations per host generation

The hybrid modeling framework can be applied also to situations where the host has a single generation per year, but the parasitoid can go through multiple generations within a host generation (see Beddington *et al.*, 1978, for a model of this sort in the classical Nicholson–Bailey framework). There are no published models of this sort for parasitoids; however, Briggs & Godfray (1996) presented a model of insect baculoviruses that can be interpreted as an insect–parasitoid model (see Chapter 17). Hosts infected (read parasitized) early in the season die after a fixed period of time (parasitoid development time) and release new infectious units (searching parasitoids) that can produce additional rounds of infection (parasitize more hosts). They found that the system was most stable and the host densities were reduced to the greatest degree when the development time of the natural enemy was the shortest, so that multiple enemy generations could occur per host generation. This obviously reduces the lag in the numerical response of the natural enemy to changes in host density. This result is mirrored in the continuous-time stage-structured framework, discussed above, where a shorter parasitoid development time also reduces pest density and enhances stability (Murdoch, 1990).

This modeling framework could also be applied to the seasonal interaction between different parasitoid species. Biological control programs involving more than one parasitoid species sometimes may be successful because the species in some way complement one other (e.g., Huffaker & Kennett, 1966; DeBach *et al.*, 1971; Murdoch, 1990; Takagi & Hirose, 1994). One species may do better in the winter and the other in the summer. The hybrid approach would allow for explicit modeling of these interactions.

Although somewhat cumbersome, hybrid models have bridged the gap between discrete- and continuous-time models and have confirmed insights (e.g., concerning aggregation of parasitoid attacks, Rohani *et al.* (1994)) suggested from a comparison of predictions from the two different approaches.

Local stability and biological control

Most of the models discussed in this chapter investigate factors that induce stability in two-species, spatially homogenous, host–parasitoid systems. This focus on stability among theoreticians can be attributed in part to the inherent instability of Nicholson–Bailey and Lotka–Volterra models. But, while the factors leading to the persistence and stability of populations are of great theoretical interest for ecologists, an emphasis on local stability may be misleading for biological control practitioners in some circumstances (Murdoch *et al.*, 1985; Murdoch, 1990). As we noted above, stability in models is often bought at the expense of increased pest density. This is perhaps most striking in models of aggregated attacks. For example, the continuous-time Murdoch & Stewart-Oaten (1989) model suggests prey density is strongly suppressed by parasitoids that focus attacks where hosts are most abundant, but this also tends to be destabilizing.

Waage (1990) has argued that lack of stability has never been the known cause of the failure of a biological control program. When biological control programs fail, it is generally because the initial establishment of the natural enemy fails or the agent does not sufficiently reduce the pest density.

Nevertheless, even if we move our focus away from local population stability and towards low mean pest density, it is still important to understand how the natural enemy is able to keep the temporal variability of pest density low enough to avoid outbreaks above an economically acceptable threshold. Thus, regulation of the pest population, perhaps in the sense of boundedness below some economic threshold (Murdoch, 1990), remains an important issue for biological control. Here we briefly consider two conditions under which a (spatially) local stable pest equilibrium is not needed.

First, the models discussed have concentrated only on mechanisms that act at a local spatial scale (i.e., the populations are well mixed), and expanding our horizons to include spatially distributed metapopulations may yield different insights. In that context, one desirable attribute of a control agent is a higher dispersal rate than that of the pest (Reeve, 1988; Murdoch, 1990). Recent reviews however, have failed to find convincing evidence for metapopulation dynamics either in long-lived crop systems or elsewhere (Taylor, 1991; Harrison & Taylor, 1996;

see also Murdoch *et al.*, 1996b for an experimental test), and this remains an area greatly in need of empirical evidence to match the burgeoning theory.

Second, we concentrated on two-species interactions in which the parasitoid is implicitly specialized on a single host species. Once we leave this framework it is again possible to have biological control without a stable pest equilibrium. For example, Murdoch *et al.* (1985) introduced a model of a generalist parasitoid or predator that attacks the pest but also has other, unspecified, hosts or prey. In this model, even when the pest is absent from the interaction, the population of the parasitoid is always above some minimum density, P_{min}, sustained by other host species. This model has an equilibrium, $(0, P_{min})$, in which the pest is extinct and the parasitoid is maintained by the other hosts, and this equilibrium will be stable if infusions of the pest are driven to extinction. Whether or not sufficiently large infusions of the pest allow it to escape, or to escape long enough to exceed an economic threshold, will depend on the details of the parasitism functions.

Schoener & Spiller (1996) describe a situation that is probably explained by a model of this sort. On a group of small Caribbean islands, those with lizards have fewer species of spiders than do those lacking lizards. When lizards were introduced into the latter islands the 'extra' spider species were driven extinct, and the difference persisted for 7 years. The islands are near large sources of immigrants, so presumably spiders immigrate and the lizards, which are generalists, continually drive them extinct but are able to persist on alternative prey species.

Discussion

Theory for the dynamics of interacting parasitoid and host populations has become increasingly general and coherent over the last decade. The addition of stage- and state-structure to models has allowed the incorporation of a substantial amount of biological realism into biocontrol theory. Many apparently different life-history and behavioral details can be seen to be manifestations of broader phenomena that have common dynamical effects. In some cases, structured models simply reinforce results from classical theory (e.g., a high parasitoid attack rate is always desirable for biocontrol), however the more realistic models provide a framework for evaluating the different ensembles of life-history and demographic properties of potential biological control agents, and point out new attributes that determine the agents' success (such as the minimum host size from which a female parasitoid can develop). Furthermore, apparently conflicting results from different modeling approaches have been resolved. While there are certainly more phenomena to be investigated, the major gap at present is a lack of tests of these theoretical predictions.

The above is true for the classical approach, mainly followed here, that emphasizes the dynamical properties of the interaction between one host and one parasitoid species in a spatially homogeneous environment. Theoretical insights are also being developed for spatially subdivided and heterogeneous environments, including those based on the properties of individual organisms (e.g., Wilson *et al.*, 1995). As might also be expected in this difficult area, however, experimental investigations lag behind the theory. Finally, there is little theory in other areas that relax the assumption that the pest needs to be maintained at a local stable equilibrium, for example where the natural enemy is polyphagous. Such situations are likely to be central for insect pests that live on short-lived crops. Theory for biological control of such pests is virtually non-existent and might prove very useful.

Acknowledgements

This work was funded by National Science Foundation grant DEB 94–20286.

References

Bailey, V. A., Nicholson, A. J. & Williams, E. (1962) Interaction between hosts and parasites when some host individuals are more difficult to find than others. *Journal of Theoretical Biology* 3: 1–18.

Beddington, J. R., Free, C. A. & Lawton, J. H. (1978) Characteristics of successful enemies in models of biological control of insect pests. *Nature* 273: 513–19.

Bess, H. A. (1964) Populations of the leaf miner *Leucoptera meyricki* Ghesq. and its parasites in sprayed and unsprayed coffee in Kenya. *Bulletin of Entomological Research* **55**: 59–82.

Bigger, M. (1973) An investigation of Fourier analysis into the interaction between coffee leaf-miners and their parasites. *Journal of Animal Ecology* **42**: 417–34.

Bigger, M. (1976) Oscillations in tropical insect populations. *Nature* **259**: 207–9.

Bigger, M. & Tapley, R. G. (1969) Prediction of outbreaks of coffee leaf-miners on Kilimanjaro. *Bulletin of Entomological Research* **58**: 601–17.

Briggs, C. J. (1993) Competition among parasitoid species on a stage-structured host, and its effect on host suppression. *American Naturalist* **141**: 372–97.

Briggs, C. J. & Godfray, H. C. J. (1996) The dynamics of insect-pathogen interactions in seasonal environments. *Theoretical Population Biology* **50**: 149–77.

Briggs, C. J., Nisbet, R. M., Murdoch, W. W., Collier, T. R. & Metz, J. A. J. (1995) Dynamical effects of host-feeding in parasitoids. *Journal of Animal Ecology* **64**: 403–16.

Caswell, H. (1989) *Matrix Population Models.* Sunderland: Sinauer Associates.

Chan, M. S. & Godfray, H. C. J. (1993) Host-feeding strategies of parasitoid wasps. *Evolutionary Ecology* **7**: 593–604.

Charnov, E. L., Los-den Hartogh, R. L., Jones, W. T. & van den Assem, J. (1981) Sex ratio evolution in a variable environment. *Nature* **289**: 27–33.

Chesson, P. L. & Murdoch, W. W. (1986) Aggregation of risk: relationships among host-parasitoid models. *American Naturalist* **127**: 696–715.

Collier, T R. (1995) Adding physiological realism to dynamic state variable models of parasitoid host feeding. *Evolutionary Ecology* **9**: 217–35.

Collier, T. R., Murdoch, W. W. & Nisbet, R. M. (1994) Egg load and the decision to host-feed in the parasitoid, *Aphytis melinus. Journal of Animal Ecology* **63**: 299–306.

Collins, M. D. & Dixon, A. F. G. (1986) The effect of egg depletion on the foraging behavior of an aphid parasitoid. *Journal of Applied Entomology* **102**: 342–52.

Comins, H. N., Hassell, M. P. & May, R. M. (1992) The spatial dynamics of host-parasitoid systems. *Journal of Animal Ecology* **61**: 735–48.

Courtney, S. P. & Hard, J. J. (1990) Host acceptance and life-history traits in *Drosophila buskii*: tests of the hierarchy-threshold model. *Heredity* **64**: 371–5.

Courtney, S. P., Chen, G. K. & Gardner, A. (1989) A general model for individual host selection. *Oikos* **55**: 55–65.

DeBach, P. (1943) The importance of host-feeding by adult parasites in the reduction of host populations. *Journal of Economic Entomology* **36**: 647–58.

DeBach, P., Rosen, D. & Kennett, C. E. (1971) Biological control of coccids by introduced natural enemies. In *Biological Control*, ed. C. B. Huffaker, pp. 165–94. New York: Plenum.

Driessen, G. & Hemerik, L. (1990) Studies on Larval Parasitoids of *Drosophila*: From individuals to populations. PhD thesis, University of Leiden, The Netherlands.

Flanders, S. E. (1953) Predation by the adult hymenopterous parasite and its role in biological control. *Journal of Economic Entomology* **46**: 541–4.

Godfray, H. C. J. & Chan, M. S. (1990) How insecticides trigger single-stage outbreaks in tropical pests. *Functional Ecology* **4**: 329–38.

Godfray, H. C. J. & Hassell, M. P. (1987) Natural enemies can cause discrete generations in tropical insects. *Nature* **327**: 144–7.

Godfray, H. C. J. & Hassell, M. P. (1989) Discrete and continuous insect populations in tropical environments. *Journal of Animal Ecology* **58**: 153–74.

Godfray, H. C. J. & Waage, J. K. (1991) Predictive modelling in biological control: the mango mealy bug (*Rastrococcus invadens*) and its parasitoids. *Journal of Applied Ecology* **28**: 434–53.

Godfray, H. C. J., Hassell, M. P. & Holt, R. D. (1994) The population dynamic consequences of phenological asynchrony between parasitoids and their hosts. *Journal of Animal Ecology* **63**: 1–10.

Gordon, D. M., Nisbet, R. M., De Roos, A., Gurney, W. S. C. & Stewart, R. K. (1991) Discrete generations in host-parasitoid models with contrasting life cycles. *Journal of Animal Ecology* **60**: 295–308.

Greathead, D. J. (1986) Parasitoids in biological control. In *Insect Parasitoids*, ed. J. Waage & D. Greathead, pp. 289–318. London: Academic Press.

Greathead, D. J. & Greathead, A. H. (1992) Biological control of insect pests by insect parasitoids and predators: the BIOCAT database. *Biocontrol News and Information* **13**: 61N-8N.

Griffiths, K. J. (1969) The importance of coincidence in the functional and numerical responses of two parasites of the European pine sawfly, *Neodiprion sertifer. Canadian Entomologist* **101**: 673–713.

Gutierrez, A. P., Neuwenschwander, P. & van Alphen, J. J. M. (1993) Factors affecting biological control of cassava mealybug by exotic parasitoids: a ratio-dependent supply-demand driven model. *Journal of Applied Ecology* **30**: 706–21.

Harrison, S. & Taylor, A. D. (1996) Empirical evidence for metapopulation dynamics. In *Metapopulation Biology: Ecology, Genetics and Evolution*, ed. I. Hanski & M. E. Gilpin, pp. 27–42. New York: Academic Press.

Hassell, M. P. (1978) *The Dynamics of Arthropod Predator–Prey Systems*. Princeton: Princeton University Press.

Hassell, M. P. & May, R. M. (1973) Stability in insect host-parasite models. *Journal of Animal Ecology* **42**: 693–736.

Hassell, M. P. & May, R. M. (1974) Aggregation in predators and insect parasites and its effect on stability. *Journal of Animal Ecology* **43**: 567–94.

Hassell, M. P., Comins, H. N. & May, R M. (1991) Spatial structure and chaos in insect population dynamics. *Nature* **353**: 255–8.

Hastings, A. (1984) Delays in recruitment at different trophic levels: effects on stability. *Journal of Mathematical Biology* **21**: 33–44.

Holt, R. D., Grover, J. & Tilman, D. (1994) Simple rules for interspecific dominance in systems with exploitative and apparent competition. *American Naturalist* **144**: 741–71.

Huffaker, C. B. & Kennett, C. E. (1966) Studies of two parasites of olive scale, *Parlatoria oleae* (Colvee). IV. Biological control of *Parlatoria oleae* (Colvee) through the compensatory action of two introduced parasites. *Hilgardia* **37**: 283–335.

Jervis, M. A., Hawkins, B. A. & Kidd, N. A. C. (1996) The usefulness of destructive host feeding parasitoids in classical biological control: theory and observation conflict. *Ecological Entomology* **21**: 41–6.

Jones, R. E. (1977) Movement patterns and egg distribution in cabbage butterflies. *Journal of Animal Ecology* **46**: 195–212.

Jones, T. H., Hassell, M. P. & Godfray, H. C. J. (1994) Population dynamics of host–parasitoid interactions. In *Parasitoid Community Ecology*, ed. B. A. Hawkins & W. Sheehan, pp. 371–94. Oxford: Oxford University Press.

Jones, W. T. (1982) Sex ratio and host size in a parasitoid wasp. *Behavioural Ecology and Sociobiology* **10**: 207–10.

Kidd, N. A. C. & Jervis, M. A. (1989) The effects of host-feeding behaviour on the dynamics of parasitoid–host interactions, and the implications for biological control. *Researches on Population Ecology* **31**: 235–74.

Kidd, N. A. C. & Jervis, M. A. (1991a) Host-feeding and oviposition strategies of parasitoids in relation to host stage. *Researches on Population Ecology* **33**: 13–28.

Kidd, N. A. C. & Jervis, M. A. (1991b) Host-feeding and oviposition by parasitoids in relation to host stage: consequences for parasitoid–host population dynamics. *Researches on Population Ecology* **33**: 87–100.

King, B. H. (1989) Host-size-dependent sex ratios among parasitoid wasps: does host growth matter? *Oecologia* **78**: 420–6.

Luck, R. F. & Podoler, H. (1985) Competitive exclusion of *Aphytis lingnanensis* by *Aphytis melinus*: potential role of host size. *Ecology* **66**: 904–13.

MacArthur, R. H. & Levins, R. (1967) The limiting similarity, convergence, and divergence of coexisting species. *American Naturalist* **101**: 377–85.

May, R. M. & Hassell, M. P. (1988) Population dynamics and biological control. *Philosophical Transactions of the Royal Society of London, Series B* **318**: 129–170.

McCauley, E., Wilson, W. G. & de Roos, A. M. (1993) Dynamics of age-structured and spatially structured predator–prey interactions individual-based models and population-level formulations. *American Naturalist* **142**: 412–42.

Minkenberg, O. P. J. M., Tatar, M. & Rosenheim, J. A. (1992) Egg load as a major source of variability in insect foraging and oviposition behavior. *Oikos* **65**: 134–42.

Murdoch, W. W. (1990) The relevance of pest–enemy models to biological control. In *Critical Issues in Biological Control*, ed. M. Mackauer, L. E. Ehler & J. Roland, pp.1–24. Andover: Intercept.

Murdoch, W. W. & Briggs, C. J. (1996) Theory for biological control. *Ecology* **77**: 2001–13.

Murdoch, W. W. & Stewart-Oaten, A. (1989) Aggregation by parasitoids and predators: effects on equilibrium and stability. *American Naturalist* **134**: 288–310.

Murdoch, W. W., Chesson, J. & Chesson, P. L. (1985) Biological control in theory and practice. *American Naturalist* **125**: 344–66.

Murdoch, W. W., Briggs, C. J. & Nisbet, R. M. (1996a) Competitive displacement and biological control in parasitoids: a model. *American Naturalist* **148**: 807–26.

Murdoch, W. W., Briggs, C. J. & Nisbet, R. M. (1997) Dynamical effects of parasitoid attacks that depend on host size and parasitoid state. *Journal of Animal Ecology* **66**: 542–56.

Murdoch, W. W., Nisbet, R. M., Blythe, S. P. & Gurney, W. S. C. (1987) An invulnerable age class and stability in delay-differential parasitoid–host models. *American Naturalist* **129**: 263–82.

Murdoch, W. W., Nisbet, R. M., Luck, R. F., Godfray, H. C. J. & Gurney, W. S. C. (1992) Size-selective sex-allocation and host feeding in a parasitoid–host model. *Journal of Animal Ecology* **61**: 533–41.

Murdoch, W. W., Swarbrick, S. L., Luck, R. F., Walde, S. & Yu, D. S. (1996b) Refuge dynamics and meta-population dynamics: an experimental test. *American Naturalist* **147**: 424–44.

Notely, F. B. (1948) The *Leucoptera* leaf miners on coffee in Kilimanjaro: *Leucoptera coffeella* Guer. *Bulletin of Entomological Research* **39**: 399–416.

Notely, F. B. (1956) The *Leucoptera* leaf miners on coffee in Kilimanjaro: *Leucoptera caffeina* Washbn. *Bulletin of Entomological Research* **46**: 899–912.

Odendaal, F. J. (1989) Mature egg number influences the behavior of female *Battus philenor* butterflies. *Journal of Insect Behavior* **2**: 15–26.

Odendaal, F. J. & Rausher, M. D. (1990) Egg load influences search intensity, host selectivity, and clutch size in *Battus philenor* butterflies. *Journal of Insect Behavior* **3**: 183–94.

Pacala, S. W. & Hassell, M. P. (1991) The persistence of host–parasitoid associations in patchy environments. II. Evaluation of field data. *American Naturalist* **138**: 584–605.

Pacala, S. W., Hassell, M. P. & May, R. M. (1990) Host–parasitoid-associations in patchy environments. *Nature* **344**: 150–3.

Perera, P. A. C. R., Hassell, M. P. & Godfray, H. C. J. (1988) Population dynamics of the coconut caterpillar, *Opisina arenosella* Walker (Lepidoptera: Xyloryctidae), in Sri Lanka. *Bulletin of Entomological Research* **78**: 479–92.

Rand, D. A. & Wilson, H. B. (1995) Using spatio-temporal chaos and intermediate-scale determinism to quantify spatially extended ecosystems. *Proceedings of the Royal Society of London, Series B* **259**: 111–17.

Reeve, J. D. (1988) Environmental variability, migration, and persistence in host–parasitoid systems. *American Naturalist* **132**: 810–36.

Reeve, J. D., Cronin, J. T. & Strong, D. R. (1994a) Parasitism and generation cycles in a salt-marsh planthopper. *Journal of Animal Ecology* **63**: 912–20.

Reeve, J. D., Cronin, J. T. & Strong, D. R. (1994b) Parasitoid aggregation and the stabilization of a salt marsh host–parasitoid system. *Ecology* **75**: 288–95.

Rohani, P. & Miramontes, O. (1995) Host–parasitoid metapopulations: the consequences of parasitoid aggregation on spatial dynamics and searching efficiency. *Proceedings of the Royal Society of London, Series B* **260**: 335–42.

Rohani, P., Godfray, H. C. J. & Hassell, M. P. (1994) Aggregation and the dynamics of host–parasitoid systems: a discrete-generation model with within-generation redistribution. *American Naturalist* **144**: 491–509.

Rosenheim, J. A. & Rosen, D. (1991) Foraging and oviposition decisions in the parasitoid *Aphytis lingnanensis*: distinguishing the influences of egg load and experience. *Journal of Animal Ecology* **60**: 873–94.

Rosenheim, J. A. & Rosen, D. (1992) Influence of egg load and host size on host-feeding behaviour of the parasitoid *Aphytis lingnanensis*. *Ecological Entomology* **17**: 263–72.

Ruxton, G. D. & Rohani, P. (1996) The consequences of stochasticity for self-organized spatial dynamics, persistence and coexistence in spatially extended host-parasitoid communities. *Proceedings of the Royal Society of London, Series B* **263**: 625–31.

Schoener, T. W. & Spiller, D. A. (1996) Devastation of prey diversity by experimentally introduced predators in the field. *Nature* **381**: 691–94.

Shea, K., Nisbet, R. M., Murdoch, W. W. & Yoo, H. J. S. (1996) The effect of egg limitation in insect host–parasitoid population models. *Journal of Animal Ecology* **65**: 743–55.

Smith, R. H. & Mead, R. (1974) Age structure and stability in models of predator–prey systems. *Theoretical Population Biology* **6**: 308–22.

Takagi, M. & Hirose, Y. (1994) Building parasitoid communities: the complementary role of two introduced parasitoid species in a case of successful biological control. In *Parasitoid Community Ecology*, ed. B. A. Hawkins & W. Sheehan, pp. 437–448. Oxford: Oxford University Press.

Taylor, A. D. (1991) Studying metapopulation effects in predator–prey systems. *Biological Journal of the Linnean Society* **42**: 305–23.

Trudeau, D. & Gordon, D. M. (1989) Factors determining the functional response of the parasitoid *Venturia canescens*. *Entomologia Experimentalis et Applicata* **50**: 3–6.

Waage, J. K. (1990) Ecological theory and the selection of biological control agents. In *Critical Issues in Biological Control*, ed. M. Mackauer, L. E. Ehler & J. Roland, pp. 135–57. Andover: Intercept.

Walde, S. J., Luck, R. F., Yu, D. S. & Murdoch, W. W. (1989) A refuge for red scale: the role of size-selectivity by a parasitoid wasp. *Ecology* **70**: 1700–6.

Wilson, W. G., McCauley, E. & de Roos, A. M. (1995) Effect of dimensionality on Lotka–Volterra predator–prey dynamics: individual based simulation results. *Bulletin of Mathematical Biology* **57**: 507–26.

Wood, B. J. (1968) *Pests of Oil Palm in Malaysia and their Control*. Kuala Lumpar: Incorporated Society of Planters.

Yamamura, N. & Yano, E. (1988) A simple model of host–parasitoid interaction with host-feeding. *Researches on Population Ecology* **30**: 353–69.

3

Models in biological control: a field guide

NIGEL D. BARLOW

Introduction

This chapter considers the use of specific, as opposed to general, models in biological control. General models are considered by Berryman in Chapter 1 and Briggs *et al.* in Chapter 2. The term specific is used in preference to 'tactical' because 'tactical' carries connotations of real-time use. Here I consider models applied to particular real-world systems or to very specific questions about such systems in general. Having outlined the potential benefits of such models for the practice of biological control, 50 biological control models from the literature are tabulated and briefly discussed. These cover a range of different biological control systems, and critical comment is made where appropriate on aspects of interest or concern relating to the specific case histories or to the particular approaches adopted. Finally, some conclusions are drawn from the survey, in terms of the types of approaches adopted in the past, types of systems addressed and, in particular, how the modeling has matched its theoretical potential to aid in practical biological control. This potential embodies the following benefits:

(1) Predicting the outcome and success of a specific introduction.
(2) Predicting the impact of introduced agents on ecosystems and non-target species.
(3) Aiding in the selection of the most appropriate agent(s).
(4) Predicting optimum release and management strategies.
(5) Aiding in the identification and interpretation of critical field data.
(6) Increasing understanding of the processes involved and the reasons for success or failure.

A survey of specific biological control models

This review is necessarily selective, but has covered a reasonable range of systems and models from 1980 onwards, which, in practice, embraces the vast majority of biological control models. Some bounding was necessary and, somewhat arbitrarily, I have focused on models applied to all forms of biological control in which an agent is released, excluding the use of *Bacillus thuringiensis* which, in terms of models, has more in common with pesticide application than with other forms of biological control. This restriction excludes a vast literature on models for naturally occurring predator–prey, parasitoid–host and pathogen–host systems. The taxonomy of specific models is relatively simple: they can be categorized either by the type of system or by the type of model. Here the focus is first on the type of biological control system and second on the different types of model, the distinction between which becomes increasingly blurred the more one attempts to define it.

Table 3.1 lists the models by type of biological control system. The definitions of Waage & Greathead (1988) are adopted for the types of biological control, which are broadly as follows. 'Classical' refers to a single release giving sustained control; 'augmentative' to releases that supplement an existing introduced or naturally occurring agent; 'inoculative' to releases that give control sustained for more than one pest generation but not indefinitely; and 'inundative' to releases that give immediate control only, confined to one pest generation. Many models, particularly for types of biological control other than classical, focus on one part of the pest–enemy life system, such as the larval stage of an insect pest

Table 3.1. *Biological control models for specific systems or questions*

Authors	System	Control	Model type	Complexity	Model use
Insect/pathogen					
1. Valentine & Podgwaite (1982)	Gypsy moth/virus	Classical	ODE	Intermediate	Understanding
2. Brain & Glen (1989)	Codling moth/virus (p)	Inundative	Discrete	Intermediate	Prediction (m)
3. Hochberg & Waage (1991)	Rhinoceros beetle/virus	Classical	ODE	Intermediate	Prediction (o,m)
4. Barlow & Jackson (1992)	Grass grub/bacteria	Augmentative	Discrete	Simple	Understanding
5. Barlow & Jackson (1993)	Grass grub/bacteria	Augmentative	Markov	Simple	Prediction (o,m)
6. Hajek *et al.* (1993)	Gypsy moth/fungus (p)	Inundative	Simulation	Complex	Understanding
7. Weseloh *et al.* (1993)	Gypsy moth/fungus (p)	Inundative	Simulation	Intermediate	Understanding
8. Dwyer & Elkinton (1993)	Gypsy moth/virus (p)	Classical	DDE	Simple	Understanding
9. Dwyer & Elkinton (1995)	Gypsy moth/virus (p)	Classical	Diffusion	Intermediate	Understanding
10. Elkinton *et al.* (1995)	Gypsy moth/virus (p)	Classical	DDE	Intermediate	Understanding
11. Thomas *et al.* (1995)	Locust/fungus	Inundative	ODE	Intermediate	Prediction (o)
12. Larkin *et al.* (1995)	Mosquito/microsporidian	Augmentative	Simulation	Complex	Prediction (o)
Weed/pathogen					
13. De Jong *et al.* (1990a,b, 1991)	*Prunus*/mycoherbicide (p)	Augmentative	Discrete	Simple	Prediction (ont)
14. Smith & Holt (1996)	*Rottboellia*/head smut	Classical	ODE	Simple	Prediction (o,m)
Vertebrate/pathogen					
15. McCallum & Singleton (1989); Singleton & McCallum (1990)	House mouse/nematode	Augmentative	ODE/DDE	Intermediate	Prediction (o)
16. Dwyer *et al.* (1990)	Rabbit/myxomatosis	Classical	Simulation	Complex	Understanding
17. Barlow (1994, 1997)		Classical	ODE	Simple	Prediction (o,m)
Plant pathogen/microbial antagonist					
18. Knudsen & Hudler (1987)	Fungus/*Pseudomonas*	Inoculative	Simulation	Complex	Prediction (s)
Weed/herbivore					
19. Caughley & Lawton (1981)	Prickly pear/*Cactoblastis*	Classical	ODE	Simple	Understanding
20. Martin & Carnahan (1983)	Noogoora burr/cerambycid (p)	Classical	Simulation	Intermediate	Prediction (s)
21. Smith *et al.* (1984)	Musk thistle/*Rhinocyllus* (p)	Classical	Simulation	Complex	Understanding
22. Cloutier & Watson (1989)	Knapweed/seed-head flies (p)	Classical	Simulation	Intermediate	Prediction (s)
23. Powell (1989)	Knapweed/seed-head flies, beetle (p)	Classical	Discrete	Simple	Prediction (s)
24. Hoffmann (1990)	*Sesbania*/three weevils (p)	Classical	Simulation	Intermediate	Prediction (o,s)
25. Santha *et al.* (1991)	Aquatic vegetation/grass carp	Inoculative	Simulation	Intermediate	Prediction (m)
26. Akbay *et al.* (1991)	Waterhyacinth/two weevils	Classical	Simulation	Complex	Understanding
27. Smith *et al.* (1993)	*Striga*/gall-forming weevil (p)	Classical	Simulation	Intermediate	Prediction (o)
28. Lonsdale *et al.* (1995)	*Sida*/chrysomelid (p)	Classical	Simulation	Intermediate	Prediction (s)
29. Hayes *et al.* (1996)	Gorse/spider mite (p)	Classical	Simulation	Intermediate	Understanding
30. Grevstad (1996)	Purple loosestrife/two chrysomelids	Classical	Simulation	Intermediate	Prediction (m)

Table 3.1. (*cont.*)

Authors	System	Control	Model type	Complexity	Model use
Insect/parasitoid					
31. Hassell (1980); Roland (1988)	Winter moth/*Cyzenis*	Classical	Discrete	Simple	Understanding
32. van Hamburg & Hassell (1984)	Stalk borer/ *Trichogramma* (p)	Augmentative	Discrete	Simple	Prediction (o)
33. Weidhaas (1986)	House fly/*Spalangia* + predator	Augmentative	Simulation	Intermediate	Prediction (m)
34. Murdoch *et al.* (1987)	Red scale/*Aphytis*	Classical	DDE	Intermediate	Understanding
35. Hill (1988)	Armyworm/*Apanteles* (p)	Classical	Discrete	Simple	Understanding
36. Gutierrez *et al.* (1988, 1993)	Cassava mealybug/ *Epidinocarsis*	Classical	Simulation	Complex	Understanding
37. Liebregts *et al.* (1989)	White peach scale/ *Encarsia*	Classical	Simulation	Simple	Prediction (o)
38. Yano (1989a,b)	Whitefly/*Encarsia*	Inoculative	Simulation	Intermediate	Prediction (m)
39. Smith & You (1990); You & Smith (1990)	Spruce budworm/ *Trichogramma*	Inundative	Simulation	Complex	Prediction (m)
40. Godfray & Waage (1991)	Mango mealybug/two encyrtids	Classical	DDE	Intermediate	Prediction (s)
41. Boot *et al.* (1992)	Leafminer/*Diglyphus*	Augmentative	Simulation	Intermediate	Understanding
42. Hopper & Roush (1993)	General host/parasitoid	Classical	diffusion	Simple	Prediction (m)
43. Heinz *et al.* (1993)	Leafminer/*Diglyphus*	Augmentative	Simulation	Complex	Prediction (m)
44. Barlow & Goldson (1993)	*Sitona* weevil/*Microctonus*	Classical	Discrete	Simple	Understanding
45. Barlow & Goldson (1994)	Argentine stem weevil/ *Microctonus*	Classical	Simulation	Intermediate	Prediction (o,m)
46. DeGrandi-Hoffman *et al.* (1994)	*Lygus*/two parasitoids	Augmentative	Simulation	Complex	Prediction (o,s)
47. Hearne *et al.* (1994)	Stalk borer/*Goniozus*	Augmentative	ODE	Intermediate	Prediction (o,m)
48. Flinn & Hagstrum (1995)	Rusty grain beetle/ *Cephalonomia*	Inoculative	Simulation	Intermediate	Prediction (m)
49. van Roermund & van Lenteren (1995)	Whitefly/*Encarsia*	Inoculative	Simulation	Intermediate	Understanding
50. Barlow *et al.* (1996)	Common wasp/ *Sphecophaga*	Classical	Discrete	Intermediate	prediction (o,s)

Notes:
The types of biological control system are defined in the text (as described in Waage & Greathead, 1988), and are qualified by (p) if the model addresses only part of the control agent and target pest's life system. The types of model are: ODE ordinary differential equations; DDE, delayed differential equations; diffusion, partial differential equations for reaction/diffusion; discrete, discrete-time difference equations; Markov, stochastic with discrete states; simulations, where the model's equations are not listed in full and the only solutions presented are from simulation. Models are classified as simple, intermediate, or complex, reflecting the number of parameters and/or assumptions but with some consideration of the scope of the system addressed. Use of the model is considered under the categories of understanding and prediction. Prediction implies that, in addition to increasing understanding of the system, the model makes at least one specific prediction regarding management (m), the type of agents that should be selected (s), or the outcome of the biological control in terms of the target (o) or non-target species (ont).

within one season or one generation, or the dynamics of a weed population with a herbivore represented simply as an arbitrary mortality. Such partial (p) life system models are indicated as such in the Table.

Categorizing the models was rather more difficult, since distinctions between analytical/mathematical and simulation, or simple and complex tend to blur with closer investigation. For example, most simulation models can be written as sets of difference or differential equations, and simulation and mathematical models converge when the mathematical equations have to be solved numerically: given a suitable time-step, simulation is simply differential equations solved by the Euler method. So the distinction adopted here has more to do with how the authors presented their models and the number of equations involved, the term 'simulation' being used for models in which the equations were not listed in full and the only solutions presented were obtained by simulation. Most other kinds of models gave analytical as well as numerical/simulation results, and the ordinary differential equation ones were generally of the Anderson–May (A–M) type, mainly applied to disease–host systems, while the discrete-time, difference equation models were generally variants of the Nicholson–Bailey (N–B) type, applied most commonly to parasitoid–host systems. The complexity of the models was also hard to categorize in many cases. Number of parameters was one obvious criterion, but some systems are larger than others and necessarily demand larger models, other things being equal. Consequently, the complexity category used in Table 3.1 also embodies an element of relativity: whether a model is complex in relation to the system it describes and the question it asks. A slightly more detailed consideration of the models follows.

Insect/pathogen

1 Valentine & Podgwaite (1982) – Gypsy moth (*Lymantria dispar*)/nuclear polyhedrosis virus (NPV)

A differential equation model for the population dynamics of gypsy moth, including interaction with foliage, virus infection and transmission of disease

separately from free-living particles on foliage and in soil. The model has eight differential equations for the larval stage then a discrete relationship between final infected larvae, uninfected larvae and free-living virus in one year and initial infected and uninfected larvae in the next. The disease transmission rate is proportional to $(C/LC_{50})^{0.77}/[1 + (C/LC_{50})^{0.77}]$, where C is the concentration of virus particles and LC_{50} the concentration required to give 50% mortality. Unfortunately only a model description is given and no results are presented.

2 Brain & Glen (1989) – Codling moth (*Cydia pomonella*)/granulosis virus (CpGV)

Analytical models for (i) cumulative larval mortality as a function of virus dose and time in the laboratory, and (ii) as above for the field, together with associated estimates of reduction in fruit damage. Larval mortality was shown to vary with 1/10 power of virus concentration, and a relationship was obtained between percent reduction in fruit damage and virus concentration. Damage for a given virus concentration increased if larvae enter fruit from the side rather than near the calyx. The actual application of the model was not described.

3 Hochberg & Waage (1991) – Rhinoceros beetle (*Oryctes rhinocerus*)/baculovirus

A differential equation Anderson–May (A–M) disease–host model, with direct transmission but a long-lived reservoir in adult beetles. The model includes susceptible and infected classes in each of three life stages (juveniles, feeding adults, and breeding adults) and density-dependent mortality among juveniles to represent population limitation by availability of breeding sites. The effect of the fungus *Metarhizium anisopliae* was mimicked as an additional insecticide-like mortality. Transmission coefficients were estimated from a comparison of model and actual infection levels at equilibrium. The model predicted a more intense initial epidemic than in the few field data sets available, with a higher prevalence of virus and greater host suppression, and also a greater equilibrium suppression of around 95% rather than the 50–85% usually observed, although infection levels were realistic. Discrepancies were attributed to possible spatial effects, ignored in the model. The latter also predicted a far higher ratio of breeders to juveniles than observed, after release of the virus. Sensitivity analysis suggested that feeder-to-feeder transmission was the most important pathway. Adding *Metarhizium* has no effect on numbers of damaging adults but causes instability and may eliminate the virus – at best it replaces free virus with expensive biopesticide. Although not stated, it may be that these conclusions depend on details of the model, which currently gives overoptimistic estimates of the effectiveness of the virus. The model predicts that control by the virus should be maintained in the long term, even if storm damage increases dead wood breeding sites and leads to an outbreak of beetles in the short term. This contrasts with empirical predictions that the virus is ineffective above a particular density of dead palms.

4 Barlow & Jackson (1992) – Grass grub (*Costelytra zealandica*)/bacteria (*Serratia* spp.)

A very simple discrete-generation two-equation model similar to the Nicholson/Bailey but including host density-dependence and applied to a host–pathogen rather than host–parasitoid interaction, on the grounds that the host was largely univoltine and the disease appeared to have one main cycle of infection each year. The host rate of increase was temperature-dependent. The model was used to predict the effect of augmentation of naturally occurring *Serratia* spp. in New Zealand grass grub, using the commercial biocide, Invade®. It was also applied to naturally occurring protozoan diseases in grubs (*Nosema* and *Mattesia* spp.), and suggested that grub populations were maintained at around 50% of their carrying capacity by naturally occurring disease, and that augmentative use of Invade® could prevent an outbreak.

5 Barlow & Jackson (1993) – Grass grub (*Costelytra zealandica*)/bacteria (*Serratia* spp.)

A Markov transition matrix, dividing grub densities and proportions diseased into arbitrary classes, then using the extensive raw data to estimate transition probabilities from each density/disease class to every other in the following year. This allowed stochastic

projection of population trends with no inbuilt assumptions about underlying relationships, the only assumption being in the definition of the class intervals. Addition of Invade® was simulated as an increase in the proportion of grubs diseased, based on field trials. Contrary to expectations, the model showed that outbreaks frequently did not occur when grubs and disease were re-set to low levels by the ploughing of pastures. If an outbreak could be predicted, Invade® applied to a low, pre-outbreak population gave excellent economic returns, but without this predictive ability its use could result in a significant loss. It was also profitable to apply it to a high population in order to hasten its decline, in this case with more surety of a favorable result.

6 Hajek *et al.* (1993) – Gypsy moth (*Lymantria dispar*)/fungus (*Entomophaga maimaiga*)

An object-orientated, weather-dependent simulation model for within-generation epidemiology linked with experiments on transmission, development, and mortality. The model could be used to optimize or assess the likely impact of augmentative biological control using *Entomophaga*, but this was not discussed. The transmission coefficient was estimated. Secondary infection was found to be more significant than primary infection and virtually independent of timing of the latter. Including behaviorally-based differences in transmission between instars gave more realistic epizootics but no comparisons were presented of model results and actual field data.

7 Weseloh *et al.* (1993) – Gypsy moth (*Lymantria dispar*)/fungus (*Entomophaga maimaiga*)

A phenological, weather-based simulation model of fungus development in larvae, to account for the presence and timing of epizootics. The model embodied two transmission mechanisms for large larvae: the encountering of germinating resting spores in leaf litter; and sporulation from infected cadavers. As in the previous paper, there was no discussion of the model's possible use in augmentative biological control. Three 'waves' of infection were predicted. The interaction between temperature and pattern as well as amount of rainfall was shown to be important. Comparison with field data suggests that infection depends on the 3 day rainfall average (= dampness of the forest floor).

8 Dwyer & Elkinton (1993) – Gypsy moth (*Lymantria dispar*)/nuclear polyhedrosis virus (NPV)

A modified A–M model using delayed differential equations with a fixed delay for incubation time, a free-living infectious stage and no host reproduction, applied to infection and mortality of larvae within a generation. The model was linked with experiments to estimate the transmission coefficient. Although this and related gypsy moth–virus models have considerable relevance and potential application to augmentative biological control using NPV, this is not discussed in the papers. The model showed that the timing and severity of epidemics is determined by initial larval densities and time from infection to death. Transmission was underestimated at low densities, and may be increased by density-related changes in larval behavior.

9 Dwyer & Elkinton (1995) – Gypsy moth (*Lymantria dispar*)/nuclear polyhedrosis virus (NPV)

A spatial version of the simple A–M model (Dwyer & Elkinton, 1993), with diffusion but no time delay for incubation, linked to an experiment on dispersal of larvae and virus. Ballooning of first instars and subsequent crawling of larvae was insufficient to account for the observed virus spread. Possibilities are vectoring by parasitoids (e.g., tachinids), or greater than predicted transmission at low larval densities (i.e., on fringes of dispersal range).

10 Elkinton *et al.* (1995) – Gypsy moth (*Lymantria dispar*)/nuclear polyhedrosis virus (NPV)

Three models were compared, in terms of within-generation larval infection and mortality: (i) a delayed differential equation A–M model with a free-living infectious stage (Dwyer & Elkinton, 1993); (ii) a detailed gypsy moth life system model (Valentine & Podgwaite, 1982), with complex non-linear transmission; (iii) a simplified version of the Valentine & Podgwaite model with A–M transmission as described in model (i). All three were tested against eight real larval populations with different initial densities and infection rates. There were two cycles of infection during larval stage. Neither of the two models with simple transmission gave correct results for low-density larval populations, so the basic A–M model with time delays (Dwyer & Elkinton, 1993) is inappropriate and non-linear transmission is probably required. The value of building complex and simple models simultaneously is stressed.

11 Thomas *et al.* (1995) – Grasshoppers and locusts/fungus (*Metarhizium anisopliae*)

A modified A–M host–disease model with measured values for the horizontal transmission coefficient. The model showed the importance of secondary cycling after a single *Metarhizium* spray, which provides a biological substitute for the now-prohibited persistent chemicals (see also Chapter 20).

12 Larkin *et al.* (1995) – Mosquito (*Culex annulirostris*)/microsporidian (*Amblyospora dyxenoides*)

A complex, object-orientated simulation model using a package (HERMES) for interactive model construction and evaluation, for the effect of the microsporidian on mosquito populations. Reasonable agreement was obtained between observed and predicted mosquito and pathogen dynamics, and the model demonstrated the effects of augmentative pathogen releases.

Weed/pathogen

13 De Jong *et al.* (1990a,b, 1991) – Black cherry (*Prunus serotina*)/mycoherbicide (*Chondrostereum purpureum*)

Simple mathematical models for risk assessment to non-target fruit crops from biological control of *Prunus* in nearby forests. Risk was assessed using a Gaussian plume model for the spread of spores, given their release rate from the controlled area, considered as a point source (De Jong *et al.*, 1990a,b). Spore release from the controlled area was estimated from their rate of production, area affected and escape fractions derived from a multilayer turbulent diffusion model, solved at steady state (De Jong *et al.*

1991). The model indicated that the risk to non-target trees was high 500 m from the controlled area but negligible at 5 km downwind. Maps were produced of relative risk over the whole of the Netherlands (additional infection/background infection), based on the model results and proximity of target and non-target crops. Over most of the country the relative risk was close to 1, indicating little risk from biological control.

14 Smith & Holt (1996) – *Rottboellia cochinchinensis*/head smut (*Sporisorium ophiuri*)

An elegant treatment, using a simple, analytical, differential equation model for the full life system of the weed and pathogen. Inevitably this involved some simplifications, such as assuming that the dynamics were continuous in time whereas crops are actually continuously replanted. This was justified by considering the model to apply on a large spatial scale. Also, a saprophytic stage for the pathogen was ignored in the model. Both these complications are being addressed in subsequent work, but the basic model allowed a number of questions to be answered regarding the impact of the sterilizing rust and its interaction with other controls. For example, simple field observations of seed- and spore-production rates on isolated infected plants may allow post-release equilibria to be estimated, without the need for estimates of intra-specific competition and density-dependence. Best results correspond to certain smut characteristics, which could lead to selection of particular strains. Many non-biological controls appear to be quite ineffective, and the model indicates which of these interact synergistically with biological control. Attainment of equilibrium from a small inoculum can take many years and may involve oscillatory behavior.

Vertebrate/pathogen

15 McCallum & Singleton (1989); Singleton & McCallum (1990) – House mouse (*Mus musculus*)/nematode (*Capillaria hepatica*)

A simple A–M differential equation model and an age-structured differential equation model with time delays, simulating the use of a liver parasite to control outbreaks in Australia. The simple model showed that the host–pathogen system was unstable, partly because nematode eggs are released and transmission occurs only when the host dies (an uncommon feature for macroparasites but widespread in micro-parasitic diseases of insects). This does not preclude the use of *Capillaria* as a biocontrol agent, but means that the parasite life cycle operates on a time-scale similar to that of the host. The more detailed model suggests that the parasite would need to be re-introduced for each outbreak, as early as possible and at doses sufficient to give a significant initial knock-down of the populations. Thereafter the disease can persist at host densities lower than the threshold for initial establishment.

16 Dwyer *et al.* (1990) – Rabbit (*Oryctolagus cuniculus*)/myxomatosis

An extension of an A–M model with seasonality (in host reproductive rate but not mortality, which should also be seasonal), age-structure, and infectivity as a function of time since infection. The effect of vectors (fleas and mosquitoes) is subsumed within the linear transmission term. No host density-dependence was included. Six myxomatosis strains were considered, with virulence, recovery rate, and transmission coefficient being different for each. The model was used to investigate the dynamics of the system and the evolution of virulence by simulation. Interaction of natural periodicity due to maturation, and seasonal forcing, can lead to complex and chaotic behavior. The most competitive virus has the highest basic reproductive rate (R_0), corresponding to Grade IV (intermediate virulence) but reducing to Grade III in the presence of host resistance. (This is relative to the resistant hosts, but the Grades are defined relative to naive ones, and recent field surveys suggest that myxomatosis has retained maximum virulence on this basis (Grade I, A. Robinson, pers. comm.)). If resistance continues to evolve at the present rate, control should last 15–20 years more. The long-term steady-state is unpredictable without more data, and the model showed that dynamics depend on host background mortality rate, so density-dependence (ignored in the model) may be significant.

17 Barlow (1994, 1997) – Possums (*Trichosurus vulpecula*)/immunocontraceptive virus

A very simple A–M disease–host model predicting the possibility of biological control of brush-tailed possums in New Zealand, using viral-vectored immunocontraception. The model showed that significant suppression of possums was possible, but identified extremely stringent targets that would be required from the biotechnology and virology to achieve this. The ideal vector would be only mildly pathogenic and sexually transmitted, and at least 80% of infected (not infectious) animals would need to be successfully sterilized by the immunocontraceptive.

Plant pathogen/microbial antagonist

18 Knudsen & Hudler (1987) – Fungal pathogen (*Gremmeniella abietina*)/bacterium (*Pseudomonas fluorescens*)

A simulation model for the growth of applied bacterium and its inhibition of the fungal pathogen on the foliage of red pine (*Pinus resinosa*). The model assumes logistic growth of the *Pseudomonas*, with a constant upper asymptote but a rate depending on leaf wetness, temperature, and nutrient availability, and with a rather unusual complication that, in the absence of water, the bacteria in the model first decline, then resume logistic growth, then decline again over set periods of time. This is supposed to mimic the retention of moisture within the needle fascicles for longer than on the leaf surface, but the representation is somewhat obscure. The equations also appear to be written incorrectly. An age-class model is used for the pathogen's conidia, incorporating germination as a function of nutrients, temperature and wetness, aging, and mortality. Mortality from the control agent is a function of bacterial concentration and conidial age. The model appeared to perform reasonably well in controlled environment validation experiments, but the authors noted a number of constructive deficiencies that are likely to lead to improved understanding and models for this and similar systems. Three of these related to the logistic model: it did not include an initial lag in growth, characteristic of this and other bacterial

colonies; it assumed a fixed upper asymptote whereas bacterial populations sometimes reached a peak then declined (suggesting an interaction with limiting nutrients or accumulated waste products); and the upper asymptote is also likely to depend in practice on nutrient levels. In spite of these, possibly minor, deficiencies, the model showed clearly that the ability to survive dry periods may be a critical attribute for foliar biocontrol agents, and that only agents that are inhibitory at relatively low concentrations are likely to be successful; very large increases in bacterial numbers were necessary in the model to give significant disease reduction, essentially because inhibition was proportional to the logarithm of bacterial numbers. It is suggested, with reason, that the model could serve as a useful screening device for new agents, complementing and partly replacing extensive experimental trials.

Weed/insect herbivore

19 Caughley & Lawton (1981) – Prickly pear (*Opuntia* spp.)/moth (*Cactoblastis cactorum*)

This model, for the successful control of prickly pear in Australia by the moth *Cactoblastis*, represents another worthy attempt to fit an off-the-shelf theoretical model of two equations to a real-world system. Together with the host–parasitoid model of Hassell (1980) and the host–pathogen model of Barlow & Jackson (1992), it is probably the simplest in this review. In this case the model is an 'interferential' (Caughley & Lawton, 1981) plant–herbivore one in continuous time. The key to the interferential model is ratio-dependence in the herbivore equation (but not in the plant one): the herbivore rate of increase is related logistically to the ratio of herbivores to plants. This is viewed as herbivores interfering with each others' feeding, hence the term 'interferential'. In this case the interference results from the highly aggregated distribution of *Cactoblastis* eggs between plants, which results in the destruction of some plants but the survival of others. Given appropriate values for the six parameters, four of which were derived more or less independently, the model accurately mimics the large-scale dynamics of the system, predicting a highly stable postcontrol equilibrium of 11 plants/acre, each carrying an average of five egg sticks. Clearly it subsumes the effect of herbivore aggregation and interference, which, along with a low, stable host-plant equilibrium, is one hypothesis explaining the success. However, another hypothesis, referred to in Caughley & Lawton (1981), is a hide-and-seek process of local extinctions and reinvasion. Since the model gives the right answer, could it be that the equations are equally applicable to this metapopulation process as well? It would be interesting to test the behavior and parameter values of an interferential model against an explicit metapopulation model for this particular system.

20 Martin & Carnahan (1983) – Noogoora burr (*Xanthium occidentale*)/cerambycid beetle (*Nupserha vexator*)

A population dynamics simulation model for the weed, with age-structure (seed reservoir, seedlings, mature plants, and seeds produced), and asymptotic density-dependence from seedlings to mature plants, and from mature plant density to numbers of seeds produced. There was significant seed predation by a number of granivores. No results were presented comparing the model's predictions with real data, possibly because much of the available data had been used to parameterize the relationships. Attack by the beetle was expressed in terms of its impact on seeds produced relative to seedlings emerged in one year, but the long-term dynamics could not be predicted because of a lack of data on cerambycid mortality in the field. However, the model indicated the necessary attributes for a successful agent: tolerance of climate variability; ability to survive on isolated plants for several seasons; and ability to rapidly attain an equilibrium in which the plant density is significantly reduced. Some of these are sufficiently obvious, given knowledge of the plant's biology, that a model is unnecessary to highlight them. However, the model would allow testing of the impact of any specific agent found with these general attributes.

21 Smith *et al.* (1984) – Musk thistle (*Carduus thoermeri*)/weevil (*Rhinocyllus conicus*)

A physiological time, within-season simulation model for the development of musk thistles and the oviposition and development of the thistle-head-feeding weevil at a local scale (1.6 ha). Overall, quite a complex and detailed model for the limited system modeled and results presented, which are essentially phenological. The model for the plant includes size classes of rosettes, number of buds/heads related to plant size, subclasses of buds to allow for differential susceptibility to weevil oviposition, and distributed developmental delays for seed maturation. The model simulates terminal and subsequent bud initiation, stem elongation, and subsequent maturation of buds into seeds. The model for the weevil includes distributed temperature-dependent development times with a skewed distribution and allocates the daily oviposition among susceptible (early) buds according to an observed exponential distribution. Heads with sufficient eggs abort. The model overestimated early-season eggs/head at two of the three sites used, but this inaccuracy was removed by a 1 °C increase in inputted temperatures, demonstrating a sensitivity to temperature and a likely inaccuracy in temperature measurements. The main finding from the study was that the effect of the weevil on the thistle, and a possible reason for the slow build-up to levels giving successful control, is limited by variable, temperature-related asynchrony between weevil oviposition and thistle bud development. The authors argue for the use of phenological models preceding biological control introductions, as a test for likely agent–target synchrony and effectiveness of the former.

22 Cloutier & Watson (1989) – Knapweed (*Centaurea diffusa* and *C. maculosa*)/seed-head flies (*Urophora quadrifasciata* and *U. affinis*)

Leslie matrix simulation models of two knapweed species, with density-dependence included through specified maximum densities of seeds, seedlings, rosettes, and vernalized shoots. The basis for these maximum figures is not given, and the model's predictions for equilibrium seed numbers, under different intensities of seed removal by control agents, are likely to be highly sensitive to them. The model predicts that more than 99.5% of the seeds must be destroyed to reduce populations. Given the 94% reduction observed for seed-head flies, an additional agent attacking seedlings would have to kill 90% of these as well.

23 Powell (1989) – Knapweed (*Centaurea diffusa*)/two seed-head flies (*Urophora affinis* and *U. quadrifasciata*) and a buprestid root-feeding beetle (*Sphenoptera jugoslavica*).

A simple difference equation model for the weed, with most component population processes being density-dependent, giving an asymmetric, concave-up *per capita* growth rate/rosette density curve. Thus the plant is extremely resistant to reductions in density by biological control agents as the density becomes low, and the model suggests that the new equilibrium in the presence of biological control is around 70 plants/m^2. This is thought to be above the damage threshold for forage production, so to obtain further reductions in density, given the shape of the growth curve revealed by the model, it is suggested that insects should be selected that attack the larger, more prostrate and faster growing plants common at lower densities.

24 Hoffmann (1990) – *Sesbania punicea*/three weevil species (*Trichapion lativentre*, *Rhyssomatus marginatus* and *Neodiplorammus quadrivittatus*)

This is a particularly clear and well-presented case study, involving a simulation model for age-specific, logistic growth of *Sesbania*, with seed bank and density-dependent reductions in seed-set and mortality of seedlings as functions of mature plant density. Mortalities are included for weevils that destroy flower buds (*Trichapion lativentre*) and seed (*Rhyssomatus marginatus*), the effects of these being combined in the model), and that bore stems (*Neodiplorammus quadrivittatus*). It is shown that very high mortalities of buds and seeds (99%) are needed to suppress plant density to the extent required (approx. 10% of 'normal'). This could be achieved eventually by the stem borer but only after significant transient increases in plant density due to a reduction in mature plants and a huge increase in recruitment. But combining all three weevils gives both an immediate and substantial decline because the bud/seed feeders eliminate the compensation possible through recruitment. The model predicted

that the age-structure of *Sesbania* stands, analysed once, 4–5 years after introduction of control agents, can be used as a test of its predictions (populations should be dominated by old plants whereas young plants would dominate in the absence of control).

25 Santha *et al.* (1991) – Aquatic vegetation (mainly *Hidrilla verticillata*)/grass carp (*Ctenopharyngodon idella*)

A simulation model, using a modeling software package (STELLA), for inoculative biological control of vegetation biomass using variable stocking rates of non-reproducing grass carp. The model includes temperature- and daylength-dependent vegetation biomass dynamics, grass carp bioenergetics and growth, and management operations. The model accurately reproduced biomass dynamics in Texas ponds under different carp densities, and grass carp growth rates, and suggested that complete removal of the vegetation (disadvantageous for recreational fishing) could be obviated by sequential harvesting and re-stocking.

26 Akbay *et al.* (1991) – Waterhyacinth (*Eichhornia crassipes*)/two weevil species (*Neochetina eichhorniae* and *N. bruchi*)

A detailed, photosynthesis-based simulation model for the growth of waterhyacinth, and submodels for temperature-dependent development, mortality, reproduction, and emigration of two weevil species. Limitations on weevil population growth take several forms: larvae are removed in proportion to the amount of dead matter relative to total plant biomass; a proportion of newly-recruited adults is removed whenever adult density exceeds 225/m^2, presumed to represent emigration; and at a critical lower limit of standing plant biomass there is an additional loss rate from all weevil stages. The weevil's impact on the plant is through the feeding of large larvae, which remove leaf biomass and meristem tissue. The model gave excellent agreement between observed and simulated adult weevil densities, but consistent discrepancies between observed and predicted plant biomass. Agreement was claimed for some data sets in which the confidence limits for the data were so wide as to accommodate virtually any model's output, but this was not the case for all the validation data sets. It is not clear why the evaluation of the plant's biological control required such a detailed, photosynthesis-based model of plant growth.

27   Smith *et al.* (1993) – *Striga hermonthica*/gall-forming weevil (*Smicronyx umbrinus*)

A simulation model of the dynamics of a parasitic weed in Mali millet crops, plus weevil effects on seed production (galled seed capsules produce no seed), for long-term biological control through erosion of the seed bank. Seeds live 14 years so the seed bank is important. The model also considers integrated control using biological and cultural methods. Cultural controls include crop rotation (no seed production and 50% weed mortality each year that the host is absent), trap/catch crops (some seeds stimulated and killed between millet crops), and weeding (density of plants decreased but proportion with galls remains the same). The model showed that 80% of seeds must be removed each year to have any effect on emergent plant density and 95% of seeds must be destroyed annually to reduce plant density by 50%. The result is robust – other key parameters relating to survival must be reduced by more than 80% to significantly change the outcome. Weeding has the same effect as the biocontrol agent and trap crops are more effective than simple rotations. Long rotations (6 years of other crops or 5 of trap crops) cause extinction of the weed without biological control. Biological control is unlikely to be effective used alone but may be of value where the seed bank has already been reduced. This depends on the host-finding abilities of the weevil (e.g., in crop rotations where their locations change yearly), which is unknown. An acknowledged deficiency in the model is the non-dynamic treatment of the weevil. The whole plant model could probably be simplified and made analytical: e.g., two classes of plant (seeds and emerged) hence two equations.

28   Lonsdale *et al.* (1995) – *Sida acuta*/chrysomelid beetle (*Calligrapha pantherina*)

A stage-structured simulation model for annual plants, used to predict *Sida* density in northern Australia in the year following a measured 10-fold

reduction in seed output due to the beetle, and long-term, assuming that the same level of defoliation is maintained. The model does not include population dynamics and grazing behavior of the beetle. It predicted a 44% reduction in flowering plants, not significantly different from the actual reduction of 34%. If the defoliation level was sustained in subsequent years, the final plant density could be very significantly reduced, depending on the searching efficiency of the beetle. The model showed that seed carry-over (generational dormancy) lowers plant density, but reduces the rate of extinction when herbivory is patchy.

29   Hayes *et al.* (1996) – Gorse (*Ulex europaeus*)/spider mite (*Tetronychus lintearius*)

A simulation model with an hourly time-step for relative, temperature-dependent, population growth of the UK mite strain in different parts of New Zealand. Temperature affected phenology, mortality, and fecundity in the model, which predicted more generations and more rapid build-up of populations in Auckland (North island) than in Lincoln (South Island). But actual establishment was more successful in Lincoln, indicating that factors not included in the model affect establishment success. One possibility suggested was that incomplete diapause in warmer areas may subject the overwintering, non-diapausing mites to greater mortality.

30   Grevstad (1996) – Purple loosestrife (*Lythrum salicaria*)/two chrysomelid beetle species (*Galerucella calmariensis* and *G. pusilla*)

A stochastic simulation model for the design of biocontrol release strategies. The model predicts establishment success as a function of demographic stochasticity, environmental stochasticity, and Allee effects. For the system considered, demographic stochasticity had virtually no effect on establishment success, for any size of release. Environmental stochasticity had a significant effect that could not be over-ridden by increased release size, above about 50 individuals. In contrast, an Allee effect made establishment success critically dependent on release size, with a strong threshold effect. The Allee effect was represented by an asymptotic relationship between the probability of a female mating and the number of males in the population, and the threshold population size for establishment was just over that at which half the females were mated (assuming a 50:50 sex ratio). Given a fixed number of agents available for release, the model suggested that environmental stochasticity dictates many small releases, whereas Allee effects require few large ones to optimize establishment. Consequently, prerelease knowledge of the likely importance of these effects could greatly enhance establishment success.

Insect/parasitoid

31   Hassell (1980); Roland (1988) – Winter moth (*Operophtera brumata*)/tachinid (*Cyzenis albicans*)

Hassell's (1980) paper represents one of very few attempts to apply a simple theoretical host–parasitoid model to a real situation, namely the winter moth–*Cyzenis* interaction in Wytham Wood, UK, and Nova Scotia. The model represents an interesting comparison of approaches with that of Gutierrez *et al.* (1988, 1993). For, while all components of the winter moth–*Cyzenis* life systems were included, these were successfully summarized in a simple 2–equation, discrete-generation model for parasitoid and host. This was based on a negative binomial model, with the addition of host density-dependence at Wytham, through predation of moth pupae in the soil. The negative binomial parameter for parasitoid attacks was related positively to host density. The model showed that parasitism was unimportant at Wytham because the host was regulated at low densities by the density-dependent pupal mortality in the soil. In Nova Scotia this mortality was absent, the host reached high levels prior to introduction of the parasitoid, but the latter then successfully regulated the moth at low densities because of its high potential for increase and the parasitoid's negative binomial response. A significant modification to the model by Roland (1988) incorporated an interaction between parasitism and density-dependent predation of winter moth pupae, through parasitized pupae being more susceptible to predation. This gave the same realistic decline in winter moth abundance as in Hassell's (1980) model, but much better agreement with observed parasitism rates and pupal predation rates, which were respectively overestimated and underestimated by Hassell's (1980) model.

32   van Hamburg & Hassell (1984) – Stalk borer (*Chilo partellus*)/trichogrammatid wasps (*Trichogramma spp.*)

A simple analytical model for the overall survival of stalk borers given density-dependent larval survival preceded by different egg mortalities due to augmentative egg parasitism by *Trichogramma*. The model showed that variable success in the field may be due to host density-dependence rather than the size of larval mortality *per se*. Best results from parasitism are at the lowest initial host density, assuming that the level of parasitism can be sustained at this low density.

33   Weidhaas (1986) – House fly (*Musca domestica*)/pteromalid wasp (*Spalangia endius*) and predator (*Carcinops pumilio*)

A simulation model for augmentative predation and parasitism of house flies. Few data were used (e.g., no functional response but rather a constant number eaten or parasitized per predator or parasitoid per day), but the model was fitted to data from experimental releases of both predator and parasitoid. The model required an arbitrary change in fly growth rate halfway through the experiment, and was fitted by varying three unknown but critical biological parameters. It helped in understanding the interactions and requirements for control; e.g., two agents at densities that were ineffective for each alone would together give good control, depending on timing. It also emphasized the need for field as opposed to laboratory estimates of crucial parameters such as fly growth rate.

34   Murdoch *et al.* (1987) – Red scale insect (*Aonidiella aurantii*)/aphelinid wasp (*Aphytis melinus*)

Simple, delayed differential equation model applied to the biological control of red scale in southern California. The model was used in a strategic sense to determine simply whether the existence of overlapping generations and an invulnerable stage in the host could allow apparently stable control at a low pest density. No attempt was made to mimic the real system in detail, the model omitting features such as host-feeding by the parasitoid (which is suggested as a reason for the degree of control exerted at an average level of parasitism of around 20%), variable parasitoid sex ratios, and defense of the scale by ants. There was also no host density-dependence in the model, which showed that an invulnerable age class tends to stabilize unstable host–parasitoid models with overlapping generations, and this is particularly so if it is the adult rather than the juvenile stage that is invulnerable. As the authors point out, it is more common for parasitoids to attack juvenile than adult insects in practice. These features, an invulnerable class and overlapping generations, are quite likely to contribute to the stability of the red scale–*Aphytis* system, compared with the olive scale–*Aphytis* interaction in northern California, which is subject to local extinction and has discrete generations with no class invulnerable to the two parasitoid species attacking it. More details on this model and its variants are given by Briggs *et al.* in Chapter 2.

35 Hill (1988) – Armyworm (*Mythimna separata*)/braconid wasp (*Apanteles ruficrus*)

A simple density-dependence model to assess the impact of a new (Pakistani) strain of *Apanteles* that attacks armyworm larvae, in interaction with existing density-dependent parasitism of the following pupal stage. The model takes account of the effects of parasitoids on the feeding rates of the host and expresses results in terms of the host densities necessary to cause significant defoliation, given parasitism levels before and after introduction of the *Apanteles* strain. It predicts that the observed success of *A. ruficrus* is due as much to its role in suppressing feeding by the host as to its effect on host density, and that the latter may be positive when followed by over-compensating pupal density-dependence, as occurs at one study site.

36 Gutierrez *et al.* (1988, 1993) – Cassava mealybug (*Phenacoccus manihoti*)/encyrtid wasp (*Epidinocarsis lopezi*)

Two of a series of papers describing a detailed analysis of a weather–plant–pest–predator–parasitoid system, deliberately modeled with a consistent level of detail across all components, and therefore involving a substantial model with many parameters. These papers describe the model for the mealybug and coccinellid predators (*Hyperaspis* spp. and *Exochomus* spp.) and the introduced parasitoids *Epidinocarsis lopezi* and *Ep. diversicornis*, including mealybug–plant interactions and weather (temperature and rainfall). The model uses physiological time and a consistent framework of differential equations with distributed time delays to track changes in age, numbers and mass of all species and subunits (e.g., embryos), though in the case of the parasitoid only numbers were modeled. Trophic interactions also used a common framework based on a functional response equation involving the ratio of resource supply to demand (see Chapter 5). This model is therefore unique among those described here for biological control, in that it incorporates considerably more biological detail. Not only does this include simulation of the plant–pest interaction and effects of weather on both, but the model also uses a mechanistic physiological basis for the ecological processes. The advantage is that this yields a general framework for inter- and intra-trophic level interactions; the disadvantage is that it requires many more assumptions about the mechanisms involved (e.g., a Type II predation equation for phloem removal by the mealybug) and many more parameters. The model confirmed field observations that the introduced parasitoid *Ep. lopezi* is the most important factor controlling mealybug populations in the dry season, and rainfall, directly or through fungal diseases, during the rainy season. The parasitoid regulates the mealybug in Nigeria, despite the disruptive effects of rain- and drought-induced mortality and predation by native coccinellids. The large variance in parasitoid development time may help it bridge periods of low host abundance, while low rates of parasitoid immigration contribute significantly to regulation of the mealybug. *Epidinocarsis diversicornis* failed to establish because *Ep. lopezi* has a greater searching efficiency, differential sex allocation according to host size favors *Ep. lopezi*, and *Ep. lopezi* is superior at within-host competition in cases of multiple parasitism (see also Chapter 2).

37 Liebregts *et al.* (1989) – White peach scale (*Pseudaulacapsis pentagona*)/aphelinid wasp (*Encarsia* spp.)

A simple discrete-time simulation model for scale and the parasitoids *Encarsia berlesi* and *E. diaspidicola* on passion fruit vines in Western Samoa. The model updates numbers of parasitized (*P*) and unparasitized (*S*) adult female scale insects daily and was fitted to 1.5 years' data on female numbers and percent parasitism. There are only three parameters (daily proportional increase in unparasitized female scales, daily egg-laying rate of parasitoids given abundant hosts (*b*), and daily mortality of adult parasitoids), but multiple generations of host and parasitoid, so the model could effectively be considered as a continuous-time one. New numbers parasitized per day are calculated as $bPS/(P+S)$. The model allowed estimation of the three population parameters and suggested that real levels of parasitism were about twice those estimated by dissection of adults in the field, thereby accounting for successful biological control of the pest.

38 Yano (1989a,b) – Greenhouse whitefly (*Trialeurodes vaporariorum*)/aphelinid wasp (*Encarsia formosa*)

A simulation model for a host–parasitoid system with host development being temperature-dependent but simplified to constant development times and survival rates characteristic of temperatures during the two main tomato-growing seasons. The parasitoid attack equation is a modified Type II functional response equation, applied both to host-feeding and parasitism (with different parameters), for one parasitoid attacking multiple host stages, and with handling time and searching efficiencies related to parasitoid density. The model was parameterized from experiments. It confirmed other results for the effects of functional response parameters on stability,

and that non-concurrent host-feeding by parasitoids enhanced stability. It demonstrated an optimum time for parasitoid release following release of whiteflies, and showed that the same number of parasitoids introduced on multiple occasions spread over time was superior to fewer, more concentrated introductions. There is also an optimum number of parasitoids per release, given the initial number of whiteflies released, and an overall optimum number of both parasitoids (eight per release) and hosts (four) released. In general, results appear best for early releases of relatively low numbers of both hosts and parasitoids.

39 Smith & You (1990); You & Smith (1990) – Spruce budworm (*Choristoneura fumiferana*)/trichogrammatid wasp (*Trichogramma minutum*)

An age- and stage-class simulation model for the spruce budworm, designed to assess the effects of, and to optimize inundative releases of, a widespread but normally ineffective egg parasitoid. The structure is rather cumbersome, with a large matrix of age by stage classes in calendar time, and temperature-dependent development making transitions between classes rather awkward. A neater alternative, more easily represented mathematically, would have been a single vector of age-classes in physiological time, divided into stages according to the accumulated development represented by the progressive classes. Budworm development, oviposition, and mortality were based on early life table studies, and *Trichogramma* release represented as a reduction in budworm egg survival. A further model was then used to generate year-to-year changes in egg density and realistic outbreaks, based on a stochastic, density-dependent autoregressive relationship between log change in egg densities from one generation to the next and densities in the current and previous generations. This allowed simulation of *Trichogramma* releases over time (years), varying the intensity of each release so as to achieve specified threshold densities in specified years of an outbreak, given the population trend indices predicted over the 12 years of the study. Frequency, timing, and rate of application within a year were varied in the daily simulation model, and the optimum regime, giving a predicted 49% suppression, was a single release spread over 5 days, beginning 14 days after appearance of the first budworm eggs. The year-to-year autoregression model predicted that populations could be maintained at moderate but not low levels, using *Trichogramma* releases towards the end of the inclining phase of an outbreak.

40 Godfray & Waage (1991) – mango mealybug (*Rastrococcus invadens*)/encyrtid wasps (*Gyranusoidea tebygi* and *Anagyrus* sp.)

A delayed differential equation analytical 'model of intermediate complexity' for the mango mealybug and two encyrtid parasitoids, with age-structure, no host density-dependence, and an instantaneous risk of parasitism described by aP^β. The model showed that *Gyranusoidea* (already introduced) was likely to be a more effective parasitoid than *Anagyrus* (not yet introduced), because it acts earlier in the host's life cycle, but that stability was greater with *Anagyrus*. These findings were robust to the exact parameters associated with parasitism (and largely unknown), which gives the model, and the approach, particular value in predictive biological control. The outcome of combining both parasitoids was less clear and depended more on the details of the parameters. For example, if *Anagyrus* were highly efficient its introduction could adversely affect the greater level of suppression obtained from *Gyranusoidea*, and lead to a reduced overall level of biological control. Thus the model does not offer support for a generalized policy of multiple parasitoid releases (see also Chapter 2).

41 Boot *et al.* (1992) – Leafminer (*Liriomyza bryoniae*)/braconid wasp (*Diglyphus isaea*)

A simulation model for population growth of the leafminer on tomato plants, with leaf nitrogen and temperature as driving variables, and for seasonal inoculative releases of the parasitoid. The paper emphasized model validation against experimental data from a glasshouse. The model's performance was variable, but it correctly simulated near 100% mortality from parasitism in the second generation after parasitoid introduction.

42 Hopper & Roush (1993) – General host/parasitoid

A general reaction–diffusion sex-specific model for the establishment and spread of introduced parasitoids (chalcids, ichneumonids, and tachinids), and an equivalent model incorporating demographic stochasticity. Parasitoid increase at any point was density-independent but depended on the availability of mates: virgin females produced all males or no males. The equations were solved numerically to determine the critical number of females initially released for a parasitoid to establish, as a function of the net reproductive rate, mate detection distance, mean-square displacement per generation (km^2), and distance travelled per generation (km). The paper includes a retrospective analysis of factors influencing establishment success in practice. The model showed that an Allee effect may explain the failure of parasitoid establishment: a threshold number of initial females was necessary for establishment, smaller for a fast-growing population with high mate-finding ability and *vice versa*. At least 1000 females should be released at each location. Parasitoid populations may remain below the detection threshold for a long time after release because of spread and exponential growth from a small base. Experiments were advocated on parasitoid establishment *versus* numbers released.

43 Heinz *et al.* (1993) – Serpentine leafminer (*Liriomyza trifolii*)/braconid wasp (*Diglyphus begini*)

A tactical, real-time simulation model to predict the need for augmentative releases of a parasitoid against a leafminer in greenhouse chrysanthemums, assuming a constant, homogenous environment but with a detailed treatment of leafminer and parasitoid age-structure and parasitoid age-specific attack behavior, including host-feeding. Rather unnecessarily, an original unrealistic model was presented in addition to a 'reconstructed' model that included temperature effects and adjusted leafminer fecundity. The model showed that the timing of parasitoid releases was more important than intensity, that release rates required to give the necessary suppression of the pest were not linearly related to pest densities, and hence that the correct release strategy could not be expressed in terms of a parasitoid–pest ratio.

44 Barlow & Goldson (1993) – Weevil (*Sitona discoideus*)/braconid wasp (*Microctonus aethiopoides*)

Three different models were used to account retrospectively for the observed impact of an introduced braconid parasitoid of adult weevils in New Zealand lucerne, and to predict the long-term future effect of the control. Initially an attempt was made to fit simple continuous-time models or discrete, single-equation analogs to the system, but its dynamics proved too complex and necessitated a hybrid approach, with different models targeted at different questions. These complexities included summer aestivation and migration of the weevil, continuous reproduction over large parts of the year, sympathetic diapause of a proportion of developing parasitoids in summer weevils, and multiple, overlapping parasitoid generations. The models followed an extensive ecological analysis, both of the weevil and the parasitoid. The simplest of these models was a discrete three–equation one for the single host and two main parasitoid generations. The models together accounted for the observed decline of *Sitona* densities and larval damage, including a reduction of around 50% before the parasitoid was introduced. They predicted sustained control in the future with around 50% suppression of the weevil, and showed why control succeeded in New Zealand but not in Australia, largely because of the atypical development of the small proportion of parasitoids in summer and the consequent reduction in autumn egg-laying by the weevil.

45 Barlow & Goldson (1994) – Argentine stem weevil (*Listronotus bonariensis*)/braconid wasp (*Microctonus hyperodae*)

This simulation model predicted phenology of the parasitoid in New Zealand prior to its release from quarantine, based on temperature-dependent parasitoid development rates measured in the laboratory and observed seasonal patterns of the susceptible (adult) host stage. The model indicated good synchronization of parasitoid adults and weevil adults,

with the exception of one possible 'bottleneck' of low host abundance in spring, and that the parasitoid would probably have two generations in the South Island but three generations and therefore a more rapid initial rate of increase in the North Island. Diapause observed in some parasitoid ecotypes was unlikely to significantly affect its phenology in either site. Of importance for predicting and modeling likely parasitoid impact on host density, the model showed the timing of the main periods of parasitism during the year, relative to the times of reproduction and density-dependence in the host.

46 DeGrandi-Hoffman *et al.* (1994) – Myrid (*Lygus hesperus*)/mymarid and braconid wasps (*Anaphes iole* and *Leiophron uniformis*)

This is the only model in the present review that purports to be a general one, applicable to a range of parasitoid–pest–plant systems. Going by the name BIOCONTROL-PARASITE, it is a detailed simulation model with menus for entry of initial conditions, parameter values and functional forms, and is applied to the augmentative release of an egg parasitoid (*Anaphes*) and a nymphal one (*Leiophron*), against *Lygus hesperus* in seed alfalfa. As a general model it occupies a similar niche to POPULATION EXPLORER, currently under development in Australia (G. A. Norton & R. Sutherst, pers. comm.). Because it includes the population dynamics, phenology, and behavior of parasitoids and pest, and growth and herbivory of the plant, all at an equivalent level of detail, the model is complex and requires the input of many parameters. It does offer the facility for making predictions in real-time and correcting these as further field samples are obtained. In spite of claims in the paper, the model appeared to perform poorly in comparison with field data, particularly in terms of predicted *Lygus* egg densities in relation to adult densities. This highlights the general point that the only real test of a model is subjective: a comparison of observed and predicted patterns in a figure. Given such a figure, one can either accept or reject the model for one's purpose, or for the authors'. Correlations or percent of occasions on which the model is within the confidence limits of the data reveal little, as they can completely obscure consistent or significant departures in trends over time (see Discussion, below). In this case the discrepancies led to a re-examination of the assumptions and an increase in understanding, which is one of the main purposes of a model. With this caveat about its accuracy, the model offered the useful feature of being able to assess the impact of biological control on crop yield. It suggested that augmentative releases, at the rates used in the model, would have little effect on feeding damage or seed yield, largely because of immigration of the pest. The main reservation one would have about this and other supposedly general models, is that they could never cope with all the possible biology of all the possible species at the level of detail they seek to mimic. As only one example, the capacity of this model to represent density-dependence in the pest through processes other than food limitation appears limited.

47 Hearne *et al.* (1994) – Sugar cane stalk-borer (*Eldana saccharina*)/bethylid wasp (*Goniozus natalensis*)

A differential equation model for the interaction between the stalk borer in South Africa and its parasitoid, designed to aid in analysing the parasitoid's efficiency. The model related reduction in crop damage to magnitude and frequency of parasitoid releases, and showed that a change in farming practices could allow sustained biological control with 60% damage reduction.

48 Flinn & Hagstrum (1995) – Rusty grain beetle (*Cryptolestes ferrugineus*)/bethylid wasp (*Cephalonomia waterstoni*)

A simulation model for temperature-dependent development and population growth of the beetle, subject to inoculative releases of the parasitoid in stored wheat. The model accurately mimicked timing and magnitude of peak parasitoid density and predicted the optimum time for parasitoid release.

49 van Roermund & van Lenteren (1995) – Greenhouse whitefly (*Trialeurodes vaporariorum*)/aphelinid wasp (*Encarsia formosa*)

A detailed, behaviourally-based, temperature-dependent simulation model of the effect of the parasitoid on the whitefly on glasshouse tomatoes. Output agreed well with observed data and showed the sensitivity of biological control success to behavioral parameters for the parasitoid, such as walking speed and giving-up time (see Chapter 7 for further details).

50 Barlow *et al.* (1996) – Common wasp (*Vespula vulgaris*)/ichneumonid wasp (*Sphecophaga vesparum vesparum*)

The aim of this model was to account for the observed impact of the introduced parasitoid on wasps in New Zealand southern beech forests (*Nothofagus* spp.), and to predict the likely final result. Wasps here reach the highest densities in the world, because of the food supply offered by honey-dew-secreting coccids on the beech trees. The model treats wasps in terms of colonies rather than individuals (although a within-nest model is being developed (D.M.Leathwick, unpublished results), and comprises a single, Ricker equation for density-dependent population change from year to year. The parasitoid model required several equations because of the complexity of its life cycle, including an unknown number of multiple, overlapping generations within a year, and emergence of over-wintering cocoons spread over a period of several years. The model showed that the parasitism level achieved – less than 10% in 8 years – was explicable entirely in terms of R_0, the parasitoid's net reproductive rate, and that the reason this is so low is largely the delay in overwintering pupal emergence. Allowing 50% of cocoons to emerge after 1 year rather than 2 converted the parasitoid from an ineffective control agent to a spectacularly successful one in the model. The latter also indicated that the ideal control agent was a pathogen that impaired queen fertility without affecting her competitive ability.

Discussion

If this review is remotely representative, it suggests that models have been most commonly applied to host–parasitoid systems, followed by both weed–herbivore and insect–pathogen biological control systems. Twenty-nine out of the 50 models were applied to classical biological control systems, including all but one of the weed–herbivore ones, 11 were used in augmentative systems, five in inoculative and five in inundative ones. In terms of the models, about half (27) were simulations and half (23) were at least partially analytical, with 14 using continuous-time and 11 discrete-time. Analytical models were most commonly used in insect–pathogen systems, though several of these came from the same authors. The majority of models (27) were considered to be of intermediate complexity, 10 were complex and 13 simple, though it was not always possible to assess a model's complexity relative to that of the system addressed. It was a little easier when considering interactive models applied to whole life systems of both trophic levels, and the proportions remained similar: 16 out of 33 were of intermediate complexity, eight were complex and nine simple. The simplest interactive, whole-system models for real systems were those of Hassell (1980), Caughley & Lawton (1981), Barlow & Jackson (1992), Barlow & Goldson (1993), Barlow (1994, 1997), and Smith & Holt (1996). Models involving fungal pathogens tended to be relatively complex because of the pathogens' sensitivity to precise environmental conditions, often of short duration. However, this poses an interesting challenge: is it possible to develop simpler models of these systems based on critical selection of fewer key driving variables? Most weed–herbivore models were inherently more simple than those for insect targets because they concentrated on the plant's dynamics and treated the herbivore as a constant additional mortality of differing intensity. They posed the question 'what degree of control is necessary to obtain the suppression required?', a similar approach to the vertebrate immunocontraception model of Barlow (1994, 1997), but only half the problem. The other half is whether a given insect herbivore can provide that degree of control, and sustain it. None of the weed models considered this.

How have these 50 models contributed to the practice of biological control, given that most of them (32) made some prediction about management or the outcome of biological control?

Predicting the outcome of specific introductions

Predicting the effect of a biological control agent before introduction is still a largely futuristic goal, particularly in terms of an input to any decision. It demands both adequate models for the target pest and more estimates for natural enemy attack behavior and rate of increase in the field. The latter involves measurements of searching, spatial spread, and dispersive behavior, which cannot be undertaken in the laboratory or under quarantine. They could potentially be made in the place of origin of the agent, yet there appear to be no examples of such prerelease ecological studies of natural enemies.

No models appear to have predicted the effect of a classical control agent before its release. However, Barlow & Goldson (1994: Model 45) took a step in this direction by predicting the phenology of an introduced parasitoid before release, including the significance of its diapause behavior, its likely synchrony with the pest, and the likely order of parasitism, reproduction and density-dependent mortality in the pest's life cycle. The latter is an important requirement for developing simple models of parasitoid impact.

Predicting the ultimate impact of a classical biological control agent after its establishment, once data are available on its population behavior, is a more straightforward prospect than predicting prior to release. Moreover, such predictions are potentially useful because they enable the process of trial-and-error, with sequential release of different natural enemies, to be short-circuited. Given that ecological analysis and modeling reveals the mechanisms responsible, then it is only necessary to wait until a lack of success can be predicted rather than demonstrated before considering the next candidate agent for introduction. Eight of the models for classical biological control (Models 3, 14, 17, 24, 27, 36, 44, 50) predicted final outcomes after release, and those of Hoffman (1990: Model 24) and Barlow *et al.* (1996: Model 50) specifically referred to their use in deciding whether more or different agents should be released.

Four of the models (5, 15, 32, 46) predicted the outcomes of augmentative control, one (Model 25) that of inoculative control, and one (Model 11) the result of inundative control, again using data from previous releases to forecast the results of future ones.

Predicting impacts on non-target species

At present there is relatively little demand for such predictions, but the future is likely to see this change. Potentially, modeling can be used in the same ways for non-target species as for target ones, although in reality there is likely to be far less information available on the dynamics of threatened species than on those of pests. In this review, only the models of de Jong *et al.* (1990a,b,1991: Model 13) for mycoherbicide use addressed the issue of risk. General theory may help at another, more strategic level. For example, all host–parasitoid models predict the existence of a critical threshold pest density, below which the parasitoid cannot establish. If non-target species occur in habitats geographically isolated from those of the target pests, then they may be at no risk simply because of their generally lower densities (Barlow & Wratten, 1996). It may even be possible to establish the thresholds for parasitoid persistence through limited but targeted field experiments, and in future these may provide fertile areas for university PhD programs. If the ranges of the pest and non-target species overlap, the situation is different. In this case the pest will remain as a constant source of parasitoid attack on the non-target species, irrespective of the latter's density; there will be no diminution, therefore, of parasitoid pressure as the threatened species declines in abundance.

Selection of agents

Pre-introduction selection based on prospective modeling of natural enemy impacts remains a goal for the future, because of the lack of current knowledge upon which to base such predictions. However, the difficulty can be averted in two ways and predictions made that can usefully affect selection decisions. The first option is to compare the likely impacts of alternative agents under a range of scenarios covering the unknown parameters associated with their essential behavior. An excellent example is the study comparing two parasitoids of the mango mealybug in Africa (Godfray & Waage, 1991: Model 40). Using a relatively simple model, these authors showed that the searching efficiency of the parasitoid was the main unknown in predicting its impact, and that whatever value this took, *Gyranusoidea tebygi* had a greater impact on mealybug populations than did *Anagyrus* sp. This approach demands realistic but simple models, so that the number of unknown parameters is minimized.

The second option is to compare agents on the basis of the pest stage attacked and models for the pest's dynamics. The key to the difference in performance of the mealybug parasitoids was the different pest stages attacked by each, and a particularly important aspect is the presence and position of pest density-dependence in its life cycle. The model of Barlow *et al.* (1996: Model 50) exemplifies this approach, suggesting that the ideal biocontrol agent for wasps would be one that attacked early nests, reduced their output of workers or reduced queen fertility without impairing her combat ability. Agents that reduce queen production from late-season nests, or that kill queens, are unlikely to be effective because they act after the stage of concern (autumn nests) but prior to the spring over-compensating density-dependence caused by queen combat and nest usurpation.

In total, 10 of the models in the review (17, 18, 20, 22, 23, 24, 28, 40, 46, 50) provided conclusions about the types of agents most likely to give effective biological control against specific targets. However, there was no evidence for adoption of any of these conclusions in practical decision-making.

Predicting optimum release and management strategies

Two complementary models (30 and 42) considered the design of optimal release strategies for classical control agents. That of Grevstad (1996: Model 30) was stochastic, focused on weed biocontrol agents and the influence of Allee effects and environmental stochasticity. In contrast, Hopper & Roush (1993: Model 42) used both deterministic diffusion equations, and a demographically stochastic equivalent model, to investigate the influence of Allee effects and dispersal on establishment success and the optimum release strategies. Optimum release strategies in terms of timing, intensity, and frequency, were considered by a number of models for augmentative, inoculative or inundative agents (Models 2, 5, 15, 33, 38, 39). Where models other than those of Grevstad (1996) and Hopper & Roush (1993) addressed the management of classical biocontrol agents, it was generally in the context of their integration with other forms of control (Models 3, 5, 14, 17, 25). However, Barlow & Goldson (1994: Model 45) used a phenological model for host and parasitoid in a tactical, real-time way to estimate the likely need for further parasitoid releases, given concerns about the effect of a particularly cool spring immediately after release. The model indicated that synchrony would be maintained between host and parasitoid, meaning that emerging parasitoid adults would have adult weevils available to parasitize, so that a second major release at a more favorable time would be unnecessary. This was found to be the case (Barlow, 1993; Barlow & Goldson, 1994).

None of the models in the review considered the interaction between classical or other forms of biological control and the use of resistant plants. Yet there is growing worldwide interest in the extent to which semi-resistant crop varieties and semi-effective biological control agents, both rejected in the past for their indifferent performance, may nevertheless interact to give highly effective pest suppression (J. K. Waage, pers. comm.). Modeling offers considerable potential to help in these evaluations, which would otherwise be costly and complex.

Aiding in the identification and interpretation of critical field data

Ideally, model development should accompany or precede any biological control programme. If it does, it will identify data needs for building the model, which will also be the data required for understanding and predicting the outcome of the biological control program. For example, following the successful development of a prospective phenological model for *Microctonus hyperodae* attacking Argentine stem weevil in New Zealand (Barlow & Goldson, 1994: Model 45), the next step was a population model that would allow early prediction of the long-term success or failure of control, in interaction with the use of resistant grasses containing fungal endophyte toxins. Not surprisingly, a key requirement for such a model was an experiment to estimate the attack behavior of the parasitoid under near-field conditions, which was subsequently undertaken. In addition to providing necessary information for the model, this yielded the incidental but reassuring result that the parasitoid was a highly efficient searcher, so giving additional confidence in the program at an early stage. Hoffmann (1990: Model 24) identified critical field data of a different kind. His model suggested that the age-structure of the weed population under biological control would be diagnostic of the success of control, before this was obvious through a significant change in weed density. In similar vein, the model of Smith & Holt (1996: Model 14) for smuts on the weed *Rottboellia*, indicated that observations on isolated plants may allow the postrelease equilibrium to be estimated in advance.

Increasing understanding of the processes involved

All of the models reviewed made contributions to the understanding of the specific biological control programms involved. In this way, whether or not they were one of the 32 that additionally made specific predictions related to management, they potentially contribute to a general improvement in biological control. For such improvement cannot occur without an improvement in understanding, and modeling is one of the most effective routes to this goal. What is required, though, is first the accumulation of sufficient numbers of well-analysed case studies, and second, an attempt to extract some general rules from their results. This review indicates a growing number of model case studies, but probably still too few to extract meaningful generalities. The only obvious one that does emerge is the significance of density-dependence in weed populations and the consequent difficulty in obtaining suppression of populations without enormous mortality from biological control. It seems to be a phenomenon that has been widely appreciated only relatively recently, coinciding, possibly coincidentally, with the developing use of models in weed biological control.

Another key point in developing an understanding of biological control systems is the choice of modeling approach. The models must be neither too simple to capture reality, nor so complex that qualitative insights are obscured by unnecessary detail. The optimum approach is likely to be with 'models of intermediate complexity' (Godfray & Waage, 1991; Barlow, 1993) where the systems are complex, and simple models where the systems permit this. For example, there seems little point in developing photosynthetic-based models for plant growth when the question at issue is the impact of a beetle on plant abundance; why not model plant biomass growth at a higher level? Given the appropriate level of complexity, the actual methodology used is probably unimportant, though it would help the cause of many simulation models to present their equations explicitly.

The most important requirement, however, is that the models must fit the reality, not only statistically, since this can obscure consistent deviations in patterns that would be diagnostic of model failure, but as graphical output compared with that observed. The reason for graphical, as opposed to statistical, model validation can be illustrated by a hypothetical but realistic example. Consider a model of an aphid outbreak, which tracks the increase in density well but predicts a decline due to natural enemy action 2 weeks later than that observed. Subsequently the model faithfully tracks the sustained, low, post-outbreak densities. A statistical fit of observed to predicted trends could give a high correlation coefficient, or equivalent measure of goodness-of-fit, because most of the points represent good agreement at low densities and during the increase phase of the outbreak. Yet this satisfyingly good fit obscures the important biological fact, obvious from a graphical inspection, that the model did not predict the population crash correctly, and that this may have been due to its assumption of the wrong mechanism; natural enemies as opposed to density-related dis-

persal. The model could have enhanced understanding by showing that natural enemies could never have provided the early decline observed, had its hypothetical authors been more self-critical about its fit to the data. A number of models in this review gave very poor fits to observed data, yet were subsequently used for predictions or evaluations before the obvious problem had been addressed – that the model resembled nothing found in nature.

Conclusions

Of the 50 models reviewed here, 32 contained predictions for management or for the outcome of biological control, yet there was little evidence that any of them had actually been used in practical decision-making and pest management. The only documented examples were a phenological model that helped a tactical decision not to carry out a second release of a parasitoid (Barlow, 1993; Barlow & Goldson, 1994), and of a model for another host–parasitoid system that contributed to a decision to abandon further releases of an ineffective parasitoid and concentrate on a search for other agents (Barlow *et al.*, 1996). Why have the other 30 model predictions not been used? One answer is that they may have been adopted but that there is no documented evidence to this effect, in which case it would greatly assist the development of biological control theory if such documentation were to be obtained. Certainly, earlier reservations about biological control theory, namely that it has addressed questions different to those asked by practitioners (Kareiva, 1990), or has been pitched at too strategic a level to be useful in management (Barlow, 1993), do not apply to these case-specific models. This review does suggest that the use of specific models as components of biological control programmes *ab initio* is growing, and given the fact that they appear to be generating useful predictions it would seem only a matter of time before these are more widely implemented. In addition to their specific predictions, a great benefit of models lies in their contribution to the understanding of biological control case studies. A growing library of well-analysed case studies and natural enemy characteristics will provide the necessary basis for more general predictions and empirically derived theory, to complement that obtained directly from the analysis of more strategic analytical models. It may even be that empirically derived theory ultimately has more to offer, and there is certainly room for cross-fertilization between specific and general models. Thus, Murdoch (1990) considers the relevance of theory to biological control and considers features such as possible temporal and spatial refuges interfering with pest suppression in specific case studies. These arguments would be stronger if there were specific models for these systems that showed whether or not the hypothesized features were present and as important as the general theory suggests. In other words, it may be more fruitful to test general theory against specific models rather than against specific case studies in a vaguer and more general way.

If general theory is to come from specific models, can anything be learnt so far from the findings discussed in this review? The number of specific biological control models is still small, so it would be unreasonable to expect too much in the way of synthesis as yet. Nevertheless, it is worth considering some of the perennial questions about biological control in the light of this brief review, some of which have already been addressed in the above discussion. What are the characteristics of successful biocontrol agents, for example? In the case of insect parasitoids, predators or pathogens, specific models bear out earlier general model findings, that successful agents need to act after density-dependence in the life cycle but before the damaging stage (e.g. model 50). For weed biocontrol agents, similar arguments and similar practical findings suggest that successful agents are likely to attack plant stages after the strong density-dependence that characterizes early seedling establishment. Alternatively, they must inflict extremely high mortality on the seed stages. Additional features in weed biocontrol models, which appear to be quite specific in many cases, include tolerance of climatic variability and the ability to persist at low plant densities (Model 20), ability to attack the larger growth forms of plants characteristic of low densities (Model 23), and high searching ability (Model 28).

Another key question is, what factors allow good predictions of biological control outcomes? For specific weed biocontrol agents, some suggestions were studies of vital rates of host and pathogen on individual plants in the field, without needing to assess density-dependence (Model 14) and analysis of population age-structures postrelease (Model 24). For insect parasitoids, estimates of searching efficiency in the field, in the country of origin or immediately after release (Model 45; and see Discussion, above), and knowledge of the extent and timing of density-dependence in the host (Model 50) may allow early predictions of ultimate impact. What is needed to improve these predictions? For host–parasitoid systems, phenology models suggesting the order of biological control, reproduction and density-dependence in the pest life cycle, and measurements of searching efficiency in the country of origin should all help in making better predictions. Although not yet obvious, there is likely to be a future need for models of tritrophic interactions including the pest's resource (e.g., Model 36, or simpler versions thereof) and models for non-target species at risk, Above all, there should be adequate models for all significant target pests. Apart from models that specifically considered spatial spread of biocontrol agents (Models 9, 13, and 42), there appear to be no explicitly spatial ones for particular biological control systems, and it is not yet clear whether these are necessary for realistic predictions in the field. For weed biocontrol systems there is certainly a need for models that include the dynamics of the biocontrol agent as well as the plant.

In terms of modeling approaches, models of intermediate complexity appear to offer the most promise, be they analytical or simple simulations. Overall, the outlook for specific models in biological control is positive. Their use is increasing, they are offering concrete predictions at a useful level, and this review suggests that they have outgrown the initial temptation offered by computer simulation technology, to develop unnecessary and confusing complexity. This is one of the ever-present dangers for specific models, just as that for strategic models is to forget that the goal is to understand, not the behavior of models, but the behavior of nature.

References

Akbay, K. S., Howell, F. G. & Wooten, J. W. (1991) A computer simulation model of waterhyacinth and weevil interactions. *Journal of Aquatic Plant Management* 29: 15–20.

Barlow, N. D. (1993) The role of models in an analytical approach to biological control. In *Proceedings of the 6th Australasian Conference on Grassland Invertebrate Ecology*, ed. R. A. Prestidge, pp. 318–25. Hamilton, New Zealand: AgResearch, Ruakura Agriculture Centre.

Barlow, N. D. (1994) Predicting the impact of a novel vertebrate biocontrol agent: a model for viral-vectored immunocontraception of New Zealand possums. *Journal of Applied Ecology* 31: 454–62.

Barlow, N. D. (1997) Modelling immunocontraception in disseminating systems. *Reproduction, Fertility and Development* 9: 51–60.

Barlow, N. D. & Goldson, S. L. (1993) A modelling analysis of the successful biological control of *Sitona discoideus* (Coleoptera: Curculionidae) by *Microctonus aethiopoides* (Hymenoptera: Braconidae) in New Zealand. *Journal of Applied Ecology* 30: 165–78.

Barlow, N. D. & Goldson, S. L. (1994) A prospective model for the phenology of *Microctonus hyperodae* (Hymenoptera: Braconidae), a potential biological control agent of Argentine stem weevil in New Zealand. *Biocontrol Science and Technology* 4: 375–86.

Barlow, N. D. & Jackson, T. A. (1992) Modelling the impact of diseases on grass grub populations. In *Use of Pathogens in Scarab Pest Management*, ed. T. A. Jackson & T. R. Glare, pp. 127–140. Andover: Intercept.

Barlow, N. D. & Jackson, T. A. (1993) A probabilistic model for augmentative biological control of grass grub (*Costelytra zealandica*) in Canterbury. In *Proceedings of the 6th Australasian Conference on Grassland Invertebrate Ecology*, ed. R. A. Prestidge, pp. 290–6. Hamilton, New Zealand: AgResearch, Ruakura Agriculture Centre.

Barlow, N. D. & Wratten, S. D. (1996) The ecology of predator–prey and parasitoid–host systems: progress since Nicholson. In *Frontiers of Population Ecology*, ed. R. B. Floyd, A. W. Sheppard & P. J. De Barro, pp. 217–44. Collingwood, Australia: CSIRO Publishing.

Barlow, N. D., Moller, H. & Beggs, J. R. (1996) A model for the impact of *Sphecophaga vesparum vesparum* as a biological control agent of the common wasp in New Zealand. *Journal of Applied Ecology* 33: 31–44.

Boot, W. J., Minkenberg, O. P. J. M., Rabbinge, R. & Moed, G. H. de (1992) Biological control of the leafminer *Liriomyza bryoniae* by seasonal inoculative releases of *Diglyphus isaea*: simulation of a parasitoid–host system. *Netherlands Journal of Plant Pathology* 98: 203–12.

Brain, P. & Glen, D. M. (1989) A model of the effect of codling moth granulosis virus on *Cydia pomonella*. *Annals of Applied Biology* 115: 129–40.

Caughley, G. & Lawton, J. H. (1981) Plant–herbivore systems. In *Theoretical Ecology*, 2nd edn, ed. R. M. May, pp. 132–66. Oxford: Blackwell Scientific.

Cloutier, D. & Watson, A. K. (1989) Application of modelling to biological weed control. In *Proceedings of the VIIth International Symposium on the Biological Control of Weeds*, ed. E. S. Delfosse, pp. 5–16. Melbourne, Australia: CSIRO Publishing.

DeGrandi-Hoffman, G., Diehl, J., Donghui, Li, Flexner, L., Jackson, G., Jones, W. & Debolt, J. (1994) BIOCONTROL-PARASITE: parasitoid–host and crop loss assessment simulation model. *Environmental Entomology* 23: 1045–60.

de Jong, M. D., Scheepens, P. C. & Zadoks, J. C. (1990a) Risk analysis applied to biological control of a forest weed. *Grana* 29: 139–45.

de Jong, M. D., Scheepens, P. C. & Zadoks, J. C. (1990b) Risk analysis for biological control: a Dutch case study in biocontrol of *Prunus serotina* by the fungus *Chondrostereum purpureum*. *Plant Disease* 74: 189–94.

de Jong, M. D., Wagenmakers, P. S. & Goudriaan, J. (1991) Modelling the escape of *Chondrostereum purpureum* spores from a larch forest with biological control of *Prunus serotina*. *Netherlands Journal of Plant Pathology* 97: 55–61

Dwyer, G. & Elkinton, J. S. (1993) Using simple models to predict virus epizootics in gypsy moth populations. *Journal of Animal Ecology* 62: 1–11.

Dwyer, G. & Elkinton, J. S. (1995) Host dispersal and the spatial spread of insect pathogens. *American Naturalist* 145: 546–62.

Dwyer, G., Levin, S. A. & Buttel, L. (1990) A simulation model of the population dynamics and evolution of myxomatosis. *Ecological Monographs* 60: 423–47.

Elkinton, J. S., Dwyer, G. & Sharov, A. (1995) Modelling the epizootiology of gypsy moth nuclear polyhedrosis virus. *Computers and Electronics in Agriculture* **13**: 91–102.

Flinn, P. W. & Hagstrum, D. W. (1995) Simulation model of *Cephalonomia waterstoni* (Hymenoptera: Bethylidae) parasitising the rusty grain beetle (Coleoptera: Cucujidae). *Environmental Entomology* **24**: 1608–15.

Godfray, H. C. J. & Waage, J. K. (1991) Predictive modelling in biological control: the mango mealybug (*Rastrococcus invadens*) and its parasitoids. *Journal of Applied Ecology* **28**: 434–53.

Grevstad, F. S. (1996) Establishment of weed control agents under the influences of demographic stochasticity, environmental variability and Allee effects. In *Proceedings of the IXth International Symposium on Biological Control of Weeds*, ed. V. C. Moran & J. H. Hoffmann, pp. 261–7. Stellenbosch, South Africa: University of Cape Town.

Gutierrez, A. P., Neuenschwander, P., Schulthess, F., Herren, H. R., Baumgaertner, J. U., Wermelinger, B., Lohr, B. & Ellis, C. K. (1988) Analysis of biological control of cassava pests in Africa. II. Cassava mealybug *Phenacoccus manihoti*. *Journal of Applied Ecology* **25**: 921–40.

Gutierrez, A. P., Nevenschwander, P. & van Alphen, J. J. M. (1993) Factors affecting biological control of cassava mealybug by exotic parasitoids: a ratio-dependent supply–demand driven model. *Journal of Applied Ecology* **30**: 706–21.

Hajek, A. E., Larkin, T. S., Carruthers, R. I. & Soper, R. (1993) Modeling the dynamics of *Entomophaga maimaiga* (Zygomycetes: Entomophthorales) epizootics in gypsy moth (Lepidoptera: Lymantriidae) populations. *Environmental Entomology* **22**: 1173–87.

Hassell, M. P. (1980) Foraging strategies, population models and biological control: a case study. *Journal of Animal Ecology* **49**: 603–28.

Hayes, A. J., Gourlay, A. H. & Hill, R. L. (1996) Population dynamics of an introduced biological control agent for gorse in New Zealand: a simulation study. In *Proceedings of the IXth International Symposium on Biological Control of Weeds*, ed. V. C. Moran & J. H. Hoffmann, pp. 357–63. Stellenbosch, South Africa: University of Cape Town.

Hearne, J. W., van Coller, L. M. & Conlong, D. E. (1994) Determining strategies for the biological control of a sugarcane stalk borer. *Ecological Modelling* **73**:117–33.

Heinz, K. M., Nunney, L., & Parrella, M. P. (1993) Toward predictable biological control of *Liriomyza trifolii* (Diptera: Agromyzidae) infesting greenhouse cut chrysanthemums. *Environmental Entomology* **22**: 1217–33.

Hill, M. G. (1988) Analysis of the biological control of *Mythimna separata* (Lepidoptera: Noctuidae) by *Apanteles ruficrus* (Braconidae: Hymenoptera) in New Zealand. *Journal of Applied Ecology* **25**: 197–208.

Hochberg, M. E. & Waage, J. K. (1991) A model for the biological control of *Oryctes rhinoceros* (Coleoptera: Scarabaeidae) by means of pathogens. *Journal of Applied Ecology* **28**: 514–31.

Hoffmann, J. H. (1990) Interactions between three weevil species in the biocontrol of *Sesbania punicea* (Fabaceae): the role of simulation models in evaluation. *Agriculture, Ecosystems and Environment* **32**: 77–87.

Hopper, K. R. & Roush, R. T. (1993) Mate finding, dispersal, number released, and the success of biological control introductions. *Ecological Entomology* **18**: 321–31.

Kareiva, P. (1990) Establishing a foothold for theory in biological control practice: using models to guide experimental design and release protocols. In *New Directions in Biological Control*, ed. R. R. Baker & P. E. Dunn, pp. 65–81. New York: Alan R. Liss.

Knudsen, G. R. and Hudler, G. W. (1987) Use of a computer simulation model to evaluate a plant disease biocontrol agent. *Ecological Modelling* **35**: 45–62.

Larkin, T. S., Sweeney, A. W. & Carruthers, R. I. (1995) Simulation of the dynamics of a microsporidian pathogen of mosquitoes. *Ecological Modelling* **77**: 143–65.

Liebregts, W. J. M. M., Sands, D. P. A. & Bourne, A. S. (1989) Population studies and biological control of *Pseudaulacapsis pentagona* (Targioni-Tozzetti) (Hemiptera: Diaspididae) on passion fruit in Western Samoa. *Bulletin of Entomological Research* **79**: 163–71.

Lonsdale, W. M., Farrell, G. & Wilson, C. G. (1995) Biological control of a tropical weed: a population model and experiment for *Sida acuta*. *Journal of Applied Ecology* **32**: 391–9.

McCallum, H. I. & Singleton, G. R. (1989) Models to assess the potential of *Capillaria hepatica* to control population outbreaks of house mice. *Parasitology* **98**: 425–37.

Martin, R. J. & Carnahan, J. A. (1983) A population model for Noogoora burr (*Xanthium occidentale*). *Australian Rangeland Journal* 5: 54–62.

Murdoch, W. W. (1990) The relevance of pest–enemy models to biological control. In *Critical Issues in Biological Control*, ed. M. Mackauer, L. E. Ehler & J. Roland, pp. 1–24. Andover: Intercept.

Murdoch, W. W., Nisbet, R. M., Blythe, S. P., Gurney, W. S. C. & Reeve, J. D. (1987) An invulnerable age class and stability in delay-differential parasitoid–host models. *American Naturalist* 129: 263–82.

Powell, R. D. (1989) The functional forms of density-dependent birth and death rates in diffuse knapweed (*Centaurea diffusa*) explain why it has not been controlled by *Urophora affinis*, *U. quadrifasciata* and *Sphenoptera jugoslavica*. In *Proceedings of the VIIth International Symposium on the Biological Control of Weeds*, ed. E. S. Delfosse, pp. 195–202. Melbourne, Australia: CSIRO Publishing.

Roland, J. (1988) Decline in winter moth populations in North America: direct versus indirect effect of introduced parasites. *Journal of Animal Ecology* 57: 523–31.

Santha, C. R., Grant, W. E., Neill, W. H. & Strawn, R. K. (1991) Biological control of aquatic vegetation using grass carp: simulation of alternative strategies. *Ecological Modelling* 59: 229–45.

Singleton, G. R. & McCallum, H. I. (1990) The potential of *Capillaria hepatica* to control mouse plagues. *Parasitology Today* 6: 190–3.

Smith, L. M. II, Ravlin, F. W., Kok, L. T. & Mays, W. T. (1984) Seasonal model of the interaction between *Rhinocyllus conicus* (Coleoptera: Curculionidae) and its weed host, *Carduus thoermeri* (Campanulatae: Asteraceae). *Environmental Entomology* 13: 1417–26.

Smith, M. C. & Holt, J. (1996) Theoretical models for biological control: *Rottboellia cochinchinensis* infection with sterilising fungi. In *Proceedings of the IXth International Symposium on Biological Control of Weeds*, ed. V. C. Moran & J. H. Hoffmann, pp. 319–325. Stellenbosch, South Africa: University of Cape Town.

Smith, M. C., Holt, J. & Webb, M. (1993) A population model for the parasitic weed *Striga hermonthica* (Scrophulariaceae) to investigate the potential of *Smicronyx umbrinus* (Coleoptera: Curculionidae) for biological control in Mali. *Crop Protection* 12: 470–6.

Smith, S. M. & You, M. (1990) A life system simulation model for improving inundative releases of the egg parasite, *Trichogramma minutum*, against the spruce budworm. *Ecological Modelling* 51: 123–42.

Thomas, M. B., Wood, S. N. & Lomer, C. J. (1995) Biological control of locusts and grasshoppers using a fungal pathogen: the importance of secondary cycling. *Proceedings of the Royal Society of London, Series B* 259: 265–70.

Valentine, H. T. & Podgwaite, J. D. (1982) Modeling the role of NPV in gypsy moth population dynamics. In *Proceedings of the 3rd International Colloquium on Invertebrate Pathology*, pp. 353–6. Brighton: University of Sussex.

van Hamburg, H. & Hassell, M. P. (1984) Density-dependence and the augmentative release of egg parasitoids against graminaceous stalkborers. *Ecological Entomology* 9: 101–8.

van Roermund, H. J. W. & van Lenteren, J. C. (1995) Simulation of biological control of greenhouse whitefly with the parasitoid *Encarsia formosa* on tomato. *Proceedings of the Section of Experimental and Applied Entomology of the Netherlands Entomological Society, Amsterdam* 6, 153–60.

Waage, J. K. & Greathead, D. J. (1988) Biological control: challenges and opportunities. *Philosophical Transactions of the Royal Society of London, Series B* 318: 111–28.

Weidhaas, D. E. (1986) Models and computer simulations for biological control of flies. *Miscellaneous Publications of the Entomological Society of America* 61: 57–68.

Weseloh, R. M., Andreadis, T. G. & Onstad, D. W. (1993) Modeling the influence of rainfall and temperature on the phenology of infection of gypsy moth, *Lymantria dispar*, larvae by the fungus *Entomophaga maimaiga*. *Biological Control* 3: 311–18.

Yano, E. (1989a) A simulation study of population interaction between the greenhouse whitefly, *Trialeurodes vaporariorum* Westwood (Homoptera: Aleyrodidae), and the parasitoid *Encarsia formosa* Gahan (Hymenoptera: Aphelinidae). I. Description of the model. *Researches on Population Ecology* 31: 73–88.

Yano, E. (1989b) A simulation study of population interaction between the greenhouse whitefly, *Trialeurodes vaporariorum* Westwood (Homoptera: Aleyrodidae), and the parasitoid *Encarsia formosa*

Gahan (Hymenoptera: Aphelinidae). II. Simulation analysis of population dynamics and strategy of biological control. *Researches on Population Ecology* **31**: 73–88.

You, M. & Smith, S. M. (1990) Simulated management of an historical spruce budworm population using inundative parasite release. *Canadian Entomologist* **122**: 1167–76.

PART II

Ecological considerations

In this section we move from the general development and application of theory to more specific treatments of the ecological, behavioral and physiological factors identified by theory as being important for the implementation of biological control programs. The range of topics is varied, reflecting the complexities that arise when workers want to maximize and sustain population-level impacts of natural enemies of pests.

As previously mentioned, a major goal of theoreticians has been to reconcile the instability of simple host–parasitoid models with the permanence that characterizes successful biological control. This has traditionally been accomplished by adding heterogeneity to the models, whether environmental stochasticity, variability in parasitoid behavior, or differential susceptibility of potential hosts to attack. In Chapter 4, Hochberg & Holt discuss differential susceptibility in the context of host refuges from parasitism. The existence of host refuges is well documented in the empirical literature, and theoreticians have considered their role since the initial development of host–parasitoid theory. Hochberg & Holt use a modified Nicholson–Bailey discrete-time model to illustrate the effect of host refuges, coupled with the host's infinite rate of increase, on the ability of parasitoids to depress host populations.

Mills & Gutierrez (Chapter 5) apply one of the ratio-dependent models introduced by Berryman in Chapter 1. To those raised intellectually on traditional prey-dependent models, this approach will appear much more complex than is usual for a general model. But what matters to biological control workers is that it provides a useful framework for understanding why biological control works. Mills & Gutierrez apply the model to the control of whiteflies in which autoparasitoids are included in the control regime. Furthermore, they highlight a theme that is repeated in several other chapters in the book. For simplicity, biological control theory mostly models the interaction between the second and third trophic levels. But in reality, most biological control is attempted on phytophagous pest species, and there is a strong potential interaction among plants, enemies and pests. Whereas plant-interactions may be included implicitly in ditrophic level models, Mills & Gutierrez's model is tritrophic, and plant effects are modeled explicitly.

A central tenet of biological control is that specialist enemies are to be preferred over generalist enemies. Chang & Kareiva (Chapter 6) challenge this, and develop a simple continuous-time model incorporating a trade-off between the higher killing efficiency of specialists *versus* the better buffered population densities of generalists. They use the model to define conditions where generalists may be better biological control agents. Given the concern over the potential impact of generalist enemies on non-target prey/hosts, they are careful to point out that their argument is not relevant to the 'classical' method of introducing exotic enemies, but instead applies to native generalist enemies already present in the system.

To this point the emphasis has been on general models, the basic principles of which can be applied to specific systems. In Chapter 7, van Lenteren & van Roermund illustrate the alternative approach, in which a model was developed with a specific system in mind, and the emphasis was on integrating empirical experimental observations and theory. They present an individual-based tomato–whitefly–*Encarsia* model that incorporates spatial structure, parasitoid foraging behavior, parasitoid and host developmental rates, and leaf production. Their goal is to provide a more scientifically based framework for selecting potential biological control agents. The system-specific approach is sometimes criticized as not being sufficiently general, but if it helps solve a problem, it meets the most fundamental criterion for applying theory to biological control.

For most predators, the relationship between adult nutrition and their population dynamics is obvious. Nutrition is also an important consideration for parasitoids whose adults feed as well as oviposit on potential hosts. In Chapter 8, Jervis & Kidd discuss the ecological consequences of host-feeding and its impact on biological control strategies. They also review the approaches that have been taken to model host-feeding, and discuss discrepencies between theory and empirical data, a problem not unique to host-feeding.

This section closes with an important message that the dynamics of some phytophagous insects appear to be 'bistable', in the sense that their populations fluctuate between two conditions, a period of relatively stable low densities, interrupted by periods of outbreaks. 'Bistability' has long been appreciated by forest entomologists, but it has only recently come under the purview of formal theory. Understanding the dynamics of outbreak species represents a special challenge to both theoreticians and insect ecologists, but progress is being made. Complex dynamics (including chaos) can arise from very simple discrete-time models, but as models become more complex and realistic, the propensity for chaotic behavior increases in both discrete- and continuous time. As illustrated by Mills & Gutierrez and chapters in subsequent sections of this book, tritrophic models are now being applied to biological control systems. In Chapter 9, Cavalieri & Koçak explore the dynamics of a relatively simple tritrophic differential model. Although the terminology they use is mathematical, the message is clear. Incorporating an additional trophic level into a predator–prey model has two major effects on the dynamics: (i) populations can exhibit virtually all possible types of dynamics, depending on initial conditions, and (ii) transitions between periodic and chaotic dynamics occur in discrete 'jumps' that are difficult to reverse. These findings should be of more than theoretical interest to biological control workers.

4

The uniformity and density of pest exploitation as guides to success in biological control

MICHAEL E. HOCHBERG AND ROBERT D. HOLT

Introduction

Quantitative theory on biological control is usually so simple as to be justifiably criticized as 'unrealistic' and 'untestable'. The synthetic nature of parameters in simple models often leads to insurmountable difficulties in their accurate measurement. It is therefore hardly surprising that few field experiments (see Chapter 3), and only one relevant comparative analysis (Hawkins *et al.*, 1993) have been expressly designed to test biological control theory.

Difficulties in the precise interpretation and measurement of parameters are only part of the reason for a reticence in testing theory. A more chronic problem is that the large body of theoretical research on biological control lacks a conceptual synthesis. We believe that the synthesis developed here will be useful to biological control specialists, because population-specific parameters measurable in the field can be related to the two concepts we introduce below.

A survey

Insect parasitoids are without doubt the most commonly employed biological control agents both in practice and in theoretical developments. Table 4.1 presents a survey of modeling studies on insect parasitoids published since Hassell's seminal monograph on the subject (Hassell, 1978). This table provides a fairly complete catalog, reflecting how both topics and modeling approaches have evolved over the past 20 years. The criteria employed for selecting studies compiled in this list are the following:

1. The study must be published in a scientific journal.
2. The study must propose a new model structure, somehow extend previous models, or apply pre-existing models to a new biological problem.
3. One or more models must be explicitly presented.
4. The study must make explicit mention of insect parasitoids in relation to the model(s).
5. The model(s) must simulate or predict temporal population dynamics.

Two partially conflicting objectives of biological control

'Success' in biological control is ultimately assessed using economic criteria. An enemy highly effective at regulating a pest population to low densities may be a partial or complete failure if the economically permitted level of damage to the crop is low (Fig. 4.1). However, given no *a priori* reason for why there should be an association between realized depression of a pest population and economic objectives, it should be the case that greater depression of a pest population will be (on average) associated with attaining economic control objectives.

Unfortunately, cross-species comparisons of reduction in pest levels and economic gain have yet to be made, so we cannot say with certainty that population measures commonly gleaned from theoretical models are reliable enough indicators of the attainment of real economic targets. It is for these very reasons that one should be careful to distinguish population dynamic objectives from economic ones. In this chapter, we are interested in the former under the reasonable assumption that they are at least correlated with the attainment of the latter.

An incontestable indication of success in biological control programs is the local, or regional, eradication of a pest. Succinctly put, population dynamics theory says that to eradicate a population, a natural

Table 4.1. *Literature survey of parasitoid models*

Reference	Year	Generations	Spatial level	Subjects
Beddington *et al.*	1978	D	1	1,5
Hassell	1978	D	1,2	1,3,5,6
Hassell & Comins	1978	D	1	5
May	1978	D	1	1
Münster-Swendsen & Nachman	1978	D/C	1	1,2
Comins & Hassell	1979	D/C	2	1,5
Adams *et al.*	1980	C	1	5
Hassell	1980	D	1	1
Wang & Gutierrez	1980	D	1	2,3
May & Hassell	1981	D	1	1,5,6
May *et al.*	1981	D	1	1,2,5
Münster-Swendsen	1982	D	2,3	1
Reddingius *et al.*	1982	D	1	4
Hassell *et al.*	1983	D	1	1,3,6
Kidd & Mayer	1983	D	1,2	1
Gutierrez & Baumgaertner	1984	D/C	1	2,5,6
Hassell	1984a	D	2	1
Hassell	1984b	D	1	1,5
Hassell & Anderson	1984	D	1	1
Hogarth & Diamond	1984	D	1	1,2,6
Kakehashi *et al.*	1984	D	1	1,6
Murdoch *et al.*	1984	D	1	1,5
Barclay *et al.*	1985	D	1	1,3,5,6
Comins & Wellings	1985	D	1	3,5
Hassell	1985	D	1	1,4
Münster-Swendsen	1985	D	1	2,6
Murdoch *et al.*	1985	D	1	5,6
Shimada & Fujii	1985	D	1	2,6,7
Waage *et al.*	1985	D	1	5
Barclay	1986a	D	1	3,5
Barclay	1986b	D	1	1,3,5,6
Bernstein	1986	D	1	1,3,5
Chesson & Murdoch	1986	D	1,2	1
Dempster & Pollard	1986	D	1	1,4,5
Hassell & May	1986	D	1	1,5,6
Perry & Taylor	1986	D	1	1
Bernstein	1987	D	1	1
Comins & Hassell	1987	D	2	1,6
Godfray & Hassell	1987	C,D	1	1,2,5
Morrison & Barbosa	1987	D	1	1,4
Murdoch *et al.*	1987	C	1	2
Perry	1987	D	1	1
Ravlin & Haynes	1987	D/C	1	2,5,6
Bellows & Hassell	1988	C/D	1	2
Gutierrez *et al.*	1988	C	1	2,5,6

Table 4.1. (*cont.*)

Reference	Year	Generations	Spatial level	Subjects
Hassell & May	1988	D	2	1
May & Hassell	1988	D	1	1,5,6
Reeve	1988	D	2	1,4,5
Taylor	1988	D	1	1,3
Yamamura & Yano	1988	C	1	2,5,7
Godfray & Hassell	1989	D,C	1	1,2,5
Kidd & Jervis	1989	D,C	1	1,2,4,8
Murdoch & Stewart-Oaten	1989	C	1,2	1
Reeve *et al.*	1989	D	2	1
Yano	1989	C/D	1	2,5
Godfray & Chan	1990	C	1	1,5
Hassell & Pacala	1990	D,C	1,2	1
Hochberg *et al.*	1990	D/C	1	1,5,6
Hochberg & Lawton	1990	D	1	1,5
Pacala *et al.*	1990	D	1	1
Reeve	1990	D	2	1,4
Barclay	1991	D	1	2,3,5
Godfray & Waage	1991	C	1	1,2,5,6
Gordon *et al.*	1991	C	1	1,2
Hassell *et al.*	1991a	D	3	1
Hassell *et al.*	1991b	D	1,2	1
Baveco & Lingeman	1992	D/C	3	4,8
Comins *et al.*	1992	D	3	1
Godfray & Pacala	1992	C	2	1
Hochberg & Hawkins	1992	D	1	1,6
Ives	1992a	C	2	1,2,4
Ives	1992b	D	1	1,5
Mangel & Roitberg	1992	D/C	1	1,7,8
Murdoch *et al.*	1992a	C	1,2	1,4
Murdoch *et al.*	1992b	C	1	2,3
Solé *et al.*	1992	D	3	
Taylor	1992	D	1	1,4
Adler	1993	D	2	1,4
Barlow & Goldson	1993	D/C	1	2,3,5
Boerlijst *et al.*	1993	D	3	7
Briggs	1993	C	1	2,6
Briggs *et al.*	1993	C	1	1,2,6
Gutierrez *et al.*	1993	C	1	2,3,5,6
Heinz *et al.*	1993	C/D	1	2,5
Hochberg & Hawkins	1993	D	1	1,5,6
Holt & Hassell	1993	D	1,2	1,4
Holt & Lawton	1993	D	1	1,6
Hopper & Roush	1993	C	3	5
Jones *et al.*	1993	D	1	1,6
Taylor	1993b	D	1	1,3

Table 4.1. (*cont.*)

Reference	Year	Generations	Spatial level	Subjects
Axelsen	1994	D/C	1	1,2,5
DeGrandi-Hoffman *et al.*	1994	C/D	1	2–6
Godfray *et al.*	1994	D/C	1	1,2
Gutierrez *et al.*	1994	C	1	3,5,6
Hassell *et al.*	1994	D	3	6
Hearne *et al.*	1994	C	1	1,2,5
Lampo	1994	D	1	1,2
Meier *et al.*	1994	D	1	1
Reeve *et al.*	1994b	C	1	1–3
Rohani *et al.*	1994a	D/C	2	1,2
Rohani *et al.*	1994b	D	2	1,3
Briggs *et al.*	1995	C	1	2,3
Flinn & Hagstrum	1995	C/D	1	2,5
Hochberg & Holt	1995	D	1	1,5,7
Barlow *et al.*	1996	D	1	1,2,5,6
Briggs & Latto	1996	C	1	2,6
Getz & Mills	1996	D	1	1,5
Hochberg	1996a	D	1	5,6
Hochberg *et al.*	1996	D/C	2	1,4,6,8
Mills & Gutierrez	1996	C	1	2,4,5,6
Murdoch *et al.*	1996	C	1	1,2,3,6
Reed *et al.*	1996	C	1	5
Rohani *et al.*	1996	D	2	1
Ruxton & Rohani	1996	D	3	4,6
Weisser & Hassell	1996	C	2	1
Wilson *et al.*	1996	D	1	1,6
Hochberg	1997	D	1	2,7
van Roermund *et al.*	1997	C	3	2,4,5,8
Wilson & Rand	1997	D	3	4

Notes:
Generations: D/C, discrete generations with within generation dynamics; C/D, continuous generations with discrete time steps.
Spatial level: 1, none or non-discrete; 2, discrete patches; 3, spatially explicit (>2 patches).
Subjects: 1, non-behavioral parasitoid interference; 2, age or size structure; 3, explicit parasitoid intraspecific competition; 4, stochastics or temporal variability; 5, contribution of parasitoid to control; 6, multispecies systems; 7, evolution; 8, individual based.
The first author of this chapter would appreciate being notified of any omissions to this table.

enemy must either reduce the pest population to the point that stochastic processes mediate its local extinction, or lower the long-term population growth rate of the pest to a level such that it can no longer sustain itself. When eradication is not attain- able in either of its forms, a logical alternative is to reduce pest population densities to the lowest, and most constant, levels possible in both time and space. The goal of low, constant densities is incidentally concordant with one of the central aims of more

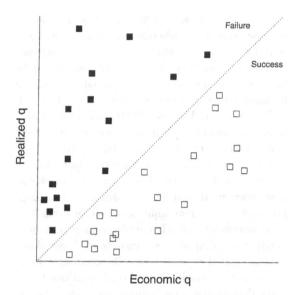

Figure 4.1. Hypothetical plot of outcomes of biological control programs in terms of realized level of pest suppression and the threshold level required for the program to be an economic success. q refers to the Beddington *et al.* (1975) index of equilibrium pest density after a biological control introduction, divided by equilibrium pest density prior to introduction.

fundamental modeling studies on host–parasitoid associations (Table 4.1): the search for conditions under which a natural enemy dominates the regulation of its host or prey population. In this chapter, we use the terms 'host' and 'pest' interchangeably.

As we shall see below, the theoretical properties of a parasitoid capable of eradicating its host do not quite border those for regulating parasitoids. Unless the objectives and constraints in biological control programs are carefully worked out, the differing attributes of what may be termed 'eradicators' and 'regulators' can lead biological control workers to introduce suboptimal enemies, or even worsen existing control (see Discussion, below).

Towards a unifying concept

Virtually all the studies presented in Table 4.1 either implicitly or explicitly model heterogeneous exploitation of hosts, and many consider in detail

how this phenomenon affects temporal population dynamics. This is not surprising since (following the seminal work of Nicholson & Bailey, 1935) many authors have shown that without recourse to other forms of density dependence, heterogeneity in host or prey exploitation by natural enemies generally promotes system persistence (Bailey *et al.*, 1962; Hassell & May, 1973; May, 1978; Chesson & Murdoch, 1986; Hassell *et al.*, 1991b; Ives, 1992a,b; Holt & Hassell, 1993; Rohani *et al.*, 1994a). Depending on the details of the model, heterogeneity may either stabilize (e.g., Hassell & May, 1973; Hassell *et al.*, 1991b) or destabilize dynamics (e.g., Murdoch & Stewart-Oaten, 1989; Murdoch, 1990; Taylor, 1993a,b), or bring other forms of density dependence into dynamic play (Hochberg & Lawton, 1990; Hochberg *et al.*, 1996).

The properties of the models surveyed in Table 4.1 converge to two major notions that encompass the many and varied population parameters often predicted to drive the outcomes of biological control. These notions are the *uniformity* and *density* of pest exploitation by the natural enemy.

Exploitation uniformity is the propensity for all individual hosts to be equally susceptible to parasitism. Exploitation uniformity is determined by spatial, temporal, and phenotypic components in (i) the capacity of parasitoid adults to locate and attack hosts, and in the ability of parasitoid larvae to develop within hosts, and (ii) in the capacity of host larvae to escape detection and attack, and to resist the development of parasitoid larvae. The complexity of the mechanisms determining exploitation distributions is largely what has fueled theoretical research on topics such as aggregation, spatial heterogeneity, and refuges in host–parasitoid systems (Table 4.1).

Exploitation density is the potential number of hosts an averagely fecund parasitoid female can parasitize and kill in the absence of competition with conspecifics. This quantity will notably be a function of the parasitoid's egg availability, the time and energy it takes to search for and to subdue hosts, and the density of the host population. Numerical (Hochberg & Lawton, 1990) and analytical (Getz & Mills, 1996) studies have shown that sufficient exploitation density is necessary for exploitation uniformity to affect the dynamics of host–parasitoid systems.

The important conceptual message we wish to convey below is that the parasitoid must enforce itself (*via* exploitation density) in order to obtain either pest eradication or regulation to low densities. In a nutshell: the parasitoid's imposition must be more uniform for eradication than for regulation, and if it is too heterogeneous, then other density dependent effects (such as other natural enemies) will contribute to, or even dominate, pest dynamics. Before illustrating these points with a simple mathematical model, we briefly discuss measures of these two important quantities.

Correlative measures of exploitation uniformity and density

Given the likely complexity of processes underlying exploitation uniformity and density, we suggest that their precise measurements will generally be unattainable. We argue for the use of two correlative quantities, both of which can be estimated in the field and easily incorporated into simple population dynamic models.

The simplest way to model exploitation uniformity is to assume that pest individuals are simply either vulnerable to, or protected from, parasitism. Protected hosts are said to have (or be in) *absolute refuges* to parasitism. (The term 'refuge' should not be taken in the strict sense of a physical hiding place. Our usage includes any form of protection whatsoever that reduces the likelihood that a subset of the host population will be parasitized as compared to the most susceptible individuals.) Absolute refuges are at best approximate measures of exploitation uniformity because they reflect both the distribution and density of parasitism.

When the absolute refuge is *proportional*, a constant fraction of the host population is protected from parasitism in each generation, whereas in a *constant number* refuge, a certain density of the host population is immune from parasitism, with the upper limit determined by the number of refuges in the environment. The major distinction between the two types is that in constant number refuges the proportion of hosts in the refuge is 100% at low host densities, decreasing to 0% as the host population becomes very large, whereas this proportion (obviously) remains constant for proportional refuges. Proportional refuges are generally thought to be more realistic approximations of variation in risk to parasitism than constant number refuges (Hassell, 1978; Holt and Hassell, 1993), and proportional refuges are more straightforward to estimate for most systems. In order of increasing precision, the proportional refuge is estimated from (i) the maximum observed level of parasitism, (ii) the estimated asymptotic maximum level of parasitism in a large group of independent samples, or (iii) the mean level of parasitism attainable (from a large number of independent samples) when parasitoids are super-abundant compared to their host. The crudest of these measures (i.e., maximum observed parasitism) suggests that proportional refuges are widespread in natural communities (Hochberg & Holt, 1995).

Exploitation density is most straightforwardly modeled as the maximum number of hosts a parasitoid can parasitize over its lifetime. This will be a function of a number of factors, such as egg load and/or rate of egg maturity, adult parasitoid lifespan, time taken in handling the host, and diverse adaptations to locate, subdue and develop within the host. Exploitation density can be estimated in three ways in order of increasing precision: (i) the maximum observed level of parasitoid population growth from one generation to the next, (ii) the estimated asymptotic maximum population growth based on a large number of independent samples, or (iii) the mean number of female parasitoid offspring produced (from a large number of independent samples) when a single, fecund parasitoid female is released into a previously unexposed host population. Data do exist on the first and second of these measures, but they are yet to be collated. Measures based on single parameters such as egg load, adult lifespan, etc., are bound to be poor descriptors on their own, due to the biological complexity of exploitation density.

An illustrative model

Consider a synchronous host–parasitoid system with discrete, non-overlapping generations, as would be the case for many temperate insect pests and their parasitoids. The host is at density N_t adult females at the beginning of generation t and they produce λN_t female offspring. A fraction, $1 - \alpha$ of these hosts are exposed to attack by a monophagous parasitoid species at adult female density P_t.

The antecedents of the mathematical model are numerous (e.g., Thompson, 1924; Nicholson & Bailey, 1935; Hassell, 1978). It takes the general form:

$$N_{t+1} = \lambda N_t g(\alpha + (1 - \alpha)f) \tag{4.1}$$

$$P_{t+1} = c\lambda N_t g(1 - \alpha)(1 - f) \tag{4.2}$$

where λ is the maximum population growth rate of the host (the number of female offspring surviving to maturity per female adult in the absence of density dependence), and c is the average number of female parasitoids produced per host attacked (assumed equal to 0.5). Finally, α is the proportion of hosts protected from parasitoid attack, that is, in the proportional refuge (see above). We will employ the quantity $1 - \alpha$ (the fraction of hosts exposed to parasitism) as a measure of the exploitation uniformity of the parasitoid population on the host population.

Equations (4.1) and (4.2) are characterized by two forms of density dependence, each modeled using a simple mathematical expression.

First, g is the proportion of hosts surviving non-explicit forms of density dependence (e.g., intraspecific competition, other natural enemies). Such 'self-limitation' can be modeled in a number of different ways, and its functional form (e.g., Hassell, 1978) and position in the host's life cycle with respect to other density-dependent mortalities (Wang & Gutierrez, 1980; May et al., 1981) can have important ramifications for temporal population dynamics. For the sake of simplicity, we assume that self-limitation acts before parasitism, and that it renders host populations constant over time when the parasitoid is absent from the system.

The function employed is:

$$g = \left[1 + \frac{(\lambda - 1)}{K} N_t \right]^{-1} \tag{4.3}$$

where K is the carrying capacity of the host population. This model reflects compensatory, stabilizing density dependence (Maynard-Smith & Slatkin, 1973; Hassell, 1978). To ensure host persistence in the absence of parasitism, we take $\lambda > 1$ and $K > 0$.

Second, the function for the proportion of exposed hosts escaping parasitoid attack, f, assumes the parasitoid searches at random over exposed hosts, and does not waste reproductive effort on protected hosts (either because they are simply never encountered, or because the parasitoid can discriminate them when encountered). f is given by

$$f = e^{(-XP_t/\zeta\{N_t\})} \tag{4.4}$$

where X is the functional response. We assume that parasitism comes at a cost to the parasitoid, either in terms of limited egg supply or in terms of limited time for host encounters. The Holling Type II function is the most widely applied to approximate costs to parasitism, particularly those associated with limited encounter rates. It is:

$$X = \frac{a\zeta\{N_t\}}{1 + a\zeta\{N_t\}/\eta} \tag{4.5}$$

a is a measure of parasitoid searching efficiency (assumed hereafter to equal 1). The function $\zeta\{N_t\}$ is the density of hosts over which the parasitoid expends reproductive effort (i.e., $\zeta\{N_t\} = \lambda N_t g(1 - \alpha)$), and η is the number of hosts a single adult female encounters and potentially parasitizes when hosts are not limiting (i.e., when $\zeta \gg \eta/a$). Because the function for ζ assumes that the parasitoid only expends reproductive effort on exposed hosts, this enables us to partially segregate the effects of exploitation uniformity and density in the discussion to follow. It should be stressed that the biological control models in which parasitoids waste reproductive effort on protected hosts are also interpretable in terms of these two conceptual measures.

In the context of this basic model, the exploitation density of the parasitoid is equivalent to its maximum population growth rate, R. R can be estimated from the partial derivative of eqn (4.2) with respect to adult parasitoid density, which gives:

$$R = \frac{ac\eta(1 - \alpha)K}{\eta + aK(1 - \alpha)} \qquad (4.6)$$

R increases with host carrying capacity (K), parasitoid searching efficiency (a), parasitoid reproductive capacity (η), and juvenile parasitoid survival (c). R also increases with increasing exploitation uniformity ($1 - \alpha$), but the association between the two is only pronounced as host carrying capacity becomes very small (i.e., $K \to \eta/\alpha(1 - \alpha)$).

In highly productive environments (i.e., when $K \gg \eta/\alpha(1 - \alpha)$), that is those which tend to define the host as a pest species, R simplifies to:

$$R \cong c\eta \qquad (4.7)$$

Note that the searching efficiency (a) of the parasitoid no longer influences its invasibility. This makes sense because searching area (a) scales how much of the K hosts can be encountered by the parasitoid.

Simulation methods

The patterns presented below are based on numerical simulations. We considered a system to have equilibrated if adult host density changed by no more than 10^{-7} between any two of a continuous string of 100 generations. Systems not meeting this criterion after 5000 generations were deemed non-equilibrated. We employ the q value of Beddington *et al.* (1975) to measure the ability of the introduced parasitoid species to depress or inflate the host population, relative to pre-introduction levels. q is defined as N^*/K, where N^* is the equilibrium level of hosts when the parasitoid is present in the system.

Model properties

The propensity for the parasitoid to reduce pest populations to extinction-menacing levels is clearly related to exploitation density and uniformity (Fig. 4.2). The effects of marginal changes in either or both of these quantities can have important consequences for q-levels, especially when exploitation density is low and uniformity high. For instance, when uniformity is near maximal (i.e., $\alpha \to 0$),

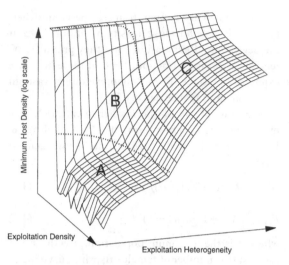

Figure 4.2. Minimum host population levels observed over the first 20 generations after the introduction of a single parasitoid into the host population at its carrying capacity. Region A, population dips below a density of 1 at least once. Region B, populations do not meet equilibrium criterion. Region C, populations meet equilibrium criterion. Other parameters: $\lambda = 2$, $c = 0.5$, $a = 1$, and $K = 100\,000$.

exploitation densities approximately greater than the pest's maximal population growth rate (λ) ensure the eventual extinction of the pest (Fig. 4.2: region A). If either or both exploitation types are not sufficiently pronounced, then long-term cycles may ensue (Fig. 4.2: region B). This is generally an undesirable result in biological control, because excursions of the pest population above the damage threshold may occur more or less predictably. Finally, insufficient uniformity will mean that although the parasitoid is able to depress the host population (perhaps below economically damaging levels), the parasitoid is not necessarily the dominating factor in host regulation (Fig. 4.2: region C).

The important and intriguing message conveyed in Fig. 4.2 is that an undesirable result (region B) divides two highly desirable alternatives (region A and parts of region C adjacent to region B). If our model were an accurate descriptor of real biological control, and if one could estimate the key parameters with sufficient precision, then the cyclic parameter space would be duly avoided. But this ideal is proba-

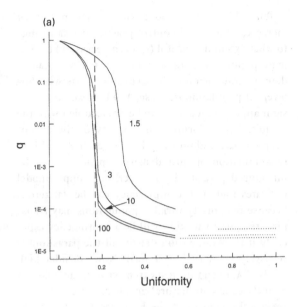

(a)

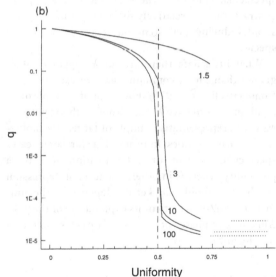

(b)

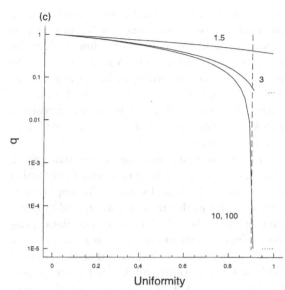

(c)

Figure 4.3. Effect of exploitation uniformity $(1 - \alpha)$ on the equilibrium level of depression of the host population by the parasitoid. q is calculated as N^*/K, where N^* is the parasitoid-enforced equilibrium of the host. Numbers next to curves show values of η, i.e., the approximate levels of the parasitoid's power of increase, R. (a) $\lambda = 1.2$. (b) $\lambda = 2$. (c) $\lambda = 10$. Dotted lines show exploitation uniformities resulting in the host population falling below a density of 1 in the 20 generations following the parasitoid introduction. Gaps between continuous lines represent systems that do not meet the equilibrium criterion. Only the continuous lines (i.e., stable systems) represent actual levels of host depression. For sufficiently large parasitoid powers of increase, the thick vertical line (at $\alpha\lambda = 1$) demarcates systems in which the parasitoid contributes to host density dependence, from those where it potentially dominates it. Other parameters as described in the legend to Fig. 4.2.

bly illusory for most systems. We suggest that ambitious aims of pest extinction or regulation to very low densities may require somewhat precise conditions (see below), and that model misspecification may lead one to chose an enemy that induces undesirable outbreaks in the pest (region B adjacent to region A). Such outbreaks are most likely when the exploitation density of the parasitoid approaches or exceeds the maximum productivity of the pest, or $R > \lambda$ and when exploitation uniformity is high.

Figure 4.3 considers these arguments in more detail. The target area for maximal depression under the host regulation scenario is delimited by a threshold at $1 - \alpha \approx 1 - 1/\lambda$. This means that if we know the productivity of the pest, then we can predict the level of uniformity necessary to obtain the lowest q-values.

The range of $1 - \alpha$ beyond the threshold yielding maximal depression increases with decreases in pest productivity, λ. This indicates that low productivity systems are the most likely to be amenable to the strong top-down regulation paradigm (compare Figs. 4.3(a)–(c)). It is also notable from Fig. 4.3 that the potential for maximal depression is only realised as $R > \lambda$, meaning again that pests with small productivities are easier to control.

In sum, the pest's maximum population growth rate, λ, emerges as a central parameter to predicting the success of biological control. The approximate conditions for exploitation uniformity and exploitation density to be sufficient for spectacular depression or induced extinction of the pest are $\lambda < 1/\alpha$ and $\lambda < R$, respectively. This means that the levels of exploitation uniformity and density leading to spectacular success will be contingent on the pest's maximum growth rate, λ.

Discussion

It is hardly surprising that the degree of success in biological control should be so dependent on the pest's power of increase, λ. But, without the aid of mathematical theory, it is not easy to see precisely why. Our conclusion is that population dynamic influences on the outcome of biological control employing a specialized natural enemy boil-down to levels in the mean and variance of the intensity of parasitism on the host population. This is the first time to our knowledge that theoretical models have been so simply interpreted in a conceptual framework for biological control.

The risk of introducing a natural enemy that induces pest outbreaks will depend importantly on the spatial scale of the interaction. This is because the arbitrary densities demarcating stochastic extinction from persistent cycles will be inversely related to the size of the system. In other words, systems become more and more invulnerable to stochastic extinctions as their spatial scale increases. At very small spatial scales, much of the parameter space denoted as 'cyclic' will in fact correspond to the extermination of the pest. At very large scales, the opposite will be the case: much of what was 'extinction' will become 'cyclic'.

With knowledge about the pest's power of increase, a biological control practitioner can gauge to what extent it is useful (or even necessary) to estimate population parameters associated with candidate natural enemies. First consider pests with low levels of population increase, $\lambda \to 1$. It would at first sight appear unnecessary to be preoccupied with the exploitation uniformity and density of the enemy, simply because almost any level of each should lead to domination of host density dependence by the introduced parasitoid. However, our simple model indicates that such complacency can be dangerous because temporally variable populations may transpire; temporal variation will be a contingency especially if the exploitation density of the parasitoid is relatively low (see Fig. 4.3(a)). Now consider high levels of λ. Even the most extensive program may not identify candidate natural enemies that can effect a spectacular control of the host (see Fig. 4.3(c)). In contrast, there is relatively little danger of the parasitoid inducing persistent outbreaks in such pest species.

What little data there are on λ suggests values greater than 2 to be commonplace. For example, data from Hassell *et al.* (1976) show 9 out of a sample of 11 lepidopterans to have $\lambda > 2$. Although this is not necessarily a representative sample of targets for biological control, it does indicate that for some cases spectacular control or extermination will be a possibility, whereas in others the level of depression by the parasitoid will keenly depend on the uniformity and/or density of its exploitation of the pest. Amassing estimates of λ for insect pests should be a worthwhile endeavor.

Theoretical antecedents

Our theoretical synthesis has a rich pedigree, some of the main points of which are recounted below (see also Hochberg, 1996b).

Natural enemies as regulators

This is the interface of regions B and C of Fig. 4.2. Beddington *et al.* (1978) presented six case studies of biological control releases resulting in pest population depression to extremely low levels. They examined the behavior of a series of mathematical models with the goal of discriminating the factors potentially responsible for intense population depression. They concluded that spatial heterogeneity in parasitism was the most likely explanation for host regulation dominated by a single natural enemy. Although spatial heterogeneity was already known to be a key parameter in top-down regulation (Bailey *et al.*, 1962; Hassell & May, 1973; Murdoch & Oaten, 1975), it was only Beddington *et al.*'s comparison and the concurrent works of Hassell (1978) and May (1978), that brought the applied importance of spatial heterogeneity in parasitism to the fore.

Much of the recent research on the impact of heterogeneity has concentrated on distinguishing components of this variation, notably spatial density dependence and density independent heterogeneity (Chesson & Murdoch, 1986; Pacala *et al.*, 1990; Hassell & Pacala, 1990; Hassell *et al.*, 1991b). Studies combining theory and data have since followed, examining the potential influence of heterogeneities in parasitism on population dynamics (Driessen & Hemerik, 1991; Pacala & Hassell, 1991; Jones *et al.*, 1993; Lampo, 1994; Reeve *et al.*, 1994a; Hochberg *et al.*, 1996).

Natural enemies as contributors

This is region C of Fig. 4.2. A separate body of theory has shown how parasitoids may contribute to host depression, but not be singly responsible for system dynamics. Two broad factors have been cited. First, density dependence acting on the host may lessen the regulatory impact of the parasitoid, and even be necessary for the parasitoid's persistence. Such density dependence may be produced by intraspecific competition (Beddington *et al.*, 1978; May *et al.*, 1981; Bernstein, 1986; Hochberg & Lawton, 1990; Ives, 1992b), host-feeding (Jervis & Kidd, 1986; Briggs *et al.*, 1995) or through the actions of other natural enemies (May & Hassell, 1981; Hassell & May, 1986; Hochberg *et al.*, 1990; Briggs, 1993; Hochberg, 1996a). Second, density dependence affecting the parasitoid itself can compromise its

influence on its host. Examples include mutual interference between parasitoid adults (e.g., Hassell, 1978), within-host competition between parasitoid larvae (Taylor, 1988), and density-dependent sex ratios (e.g., Comins & Wellings, 1985).

Numerous determinants of exploitation density and uniformity are already known to mediate the potential impact of the parasitoid on its host population. Both Hochberg & Lawton (1990) and Getz & Mills (1996) have shown how different forms of parasitoid exploitation affect the pertinence of spatial heterogeneity on system dynamics. In particular, Getz & Mills (1996) have elegantly shown how the potential population growth rate of the parasitoid must exceed that of the host for spatial heterogeneity to be of relevance to regulation. They identify egg limitation as an important constraint to the potential growth rate of the parasitoid. More generally, a diverse literature has identified how probabilistic host refuges (e.g., Hassell & May, 1973; Beddington *et al.*, 1978; Hassell, 1978; May, 1978; Perry & Taylor, 1986; Reeve *et al.*, 1989), and absolute host refuges (Hassell & May, 1973; Hassell, 1978; Murdoch *et al.*, 1987; Holt & Hassell, 1993; Hochberg & Holt, 1995) may lessen the impact of parasitoids on their hosts.

Natural enemies as exterminators

This is region A of Fig. 4.2. Nicholson (1933: conclusion 40) was the first to suggest that a host-parasitoid link could persist regionally, even if confronted with extinctions locally. Empirical arguments for why extinction/colonization dynamics could be a prevalent phenomenon were first made by Murdoch and colleagues (Murdoch *et al.*, 1984, 1985). They maintained that five of the six case studies employed by Beddington *et al.* (1978) are characterized by unstable population dynamics at local spatial scales. That locally doomed systems could persist regionally was explored in the context of parasitoids by Allen (1975) and then Münster-Swendsen (1982), and more recently by Hassell & May (1988), Reeve (1988), Murdoch *et al.* (1992a), Holt & Hassell (1993), Hassell *et al.* (1991a), Comins *et al.* (1992), Solé *et al.* (1992), and Hassell *et al.* (1994). So far, the implications of large-scale spatial structure have not been investigated in an explicit biological control context.

Evidence for our synthesis?

There is mounting evidence in support of the centrality of the two parasitoid exploitation potentials.

Comparative analyses in native and introduced systems

Hawkins *et al.* (1993) estimated host refuges from parasitism (which is the inverse of what we have called here 'exploitation uniformity') for a set of 74 biological control attempts. They employed the maximum observed level of parasitism by the introduced parasitoid as a measure of $1 - \alpha$. This measure, though potentially fraught with errors in its estimation (van Driesche *et al.* 1991), showed a highly statistically significant relationship with economic success in biological control programs. Hawkins and colleagues' analysis did not indicate that maximum parasitism could be employed as a precise estimator of population depression, but rather that as a proxy variable for the uniformity of exposure. Our synthesis lends theoretical support to their findings.

It could be argued that Hawkins *et al.*'s analysis is invalid, because maximum parasitism was estimated from the very control programs it was supposed to predict (Myers *et al.*, 1994; Hawkins *et al.*, 1994). To counter this criticism, Hawkins & Cornell (1994) compared maximum parasitism rates of natural enemies in their native habitats to economic success in biological control in target areas and found a highly significant relationship between the two. Together with the earlier analysis (Hawkins *et al.*, 1993) this indicates that maximum parasitism is an approximate predictor of success in biological control, based either on *a priori* information from native habitats or *a posteriori* data in the control arena. Intriguingly, in both the exotic data set (Hawkins & Cornell, 1994) and the control-site data set (Hawkins *et al.*, 1993), a cut-off exists in control outcomes at maximum parasitism rates (i.e., exploitation uniformity) of approximately 35%.

Below this level biological control very rarely achieves economic success. A challenge will be to explain this empirical phenomenon.

Host–parasitoid–pathogen interactions in experimental boxes

Recently, Begon *et al.* (1996) employed laboratory experiments to investigate the population dynamics of two and three species systems involving the Indian meal moth (*Plodia interpunctella*), a parasitoid (*Venturia canescens*) and a granulosis virus (PiGV). When the host is kept alone in experimental boxes with either the parasitoid or the virus, it exhibits cycles in adult moth numbers of approximately one generation interval. It would be misleading to call the populations 'unstable', since they are very regular in shape and amplitude from generation to generation, indicating that single natural enemy dynamics are relatively constant in time.

Consistent with our arguments for how such dynamics should arise, Begon and colleagues have shown limitations to what we call exploitation uniformity. For the virus, pathogen loads differ over the experimental arena (Sait *et al.*, 1994), meaning that larvae feeding in sparsely contaminated areas will be less prone to infection than those encountering pathogen-rich areas. As for the parasitoid, deeply feeding hosts are less likely to be found and parasitized by searching parasitoid females than are surface-feeders (Begon *et al.*, 1995). The major dynamic difference between the virus and parasitoid is that the former has very little impact on the amplitude of the host population cycles, whereas the latter's effect is intense. This is suggestive of high exploitation densities associated with the parasitoid.

Because of the spatial and temporal differences in exploitation tactics by the two natural enemies, total exploitation uniformity in the three-species system is greater than either of its two parts alone (see Hochberg, 1996c). Begon *et al.* (1996) found that these three-species systems ultimately went extinct. We suggest that the total exploitation uniformity and density were sufficient to generate either persistent cycles in the system as a transient phenomenon to extinction (region A in Fig. 4.2), or as a final state (region B), which terminated in stochastic extinction due to small population numbers in the troughs of the cycles. In other words, adding a second natural enemy takes the system from either 'contribution' by the pathogen, or 'regulation' by the parasitoid, to 'cycles' or 'extinction' by both natural enemies acting together.

Additional concerns

In debating whether contribution, regulation or eradication is the most attainable option, an additional, important concern is the repercussion of control for the community surrounding the pest. We do not advocate the use of the theory discussed above without the thorough study of possible adverse effects on the interacting community. It is important that the conservation of other species be respected, and that no new pest species should emerge as a result of any control measure.

References

Adams, V. D., DeAngelis, D. L. & Goldstein, R. A. (1980) Stability analysis of the time delay in a host–parasitoid model. *Journal of Theoretical Biology* 83: 43–62.

Adler, F. R. (1993) Migration alone can produce persistence of host–parasitoid models. *American Naturalist* 141: 642–50.

Allen, J. C. (1975) Mathematical models of species interactions in time and space. *American Naturalist* 109: 319–42.

Axelsen, J. A. (1994) Host–parasitoid interactions in an agricultural ecosystem: a computer simulation. *Ecological Modelling* 73: 189–203.

Bailey, V. A., Nicholson, A. J. & Williams, E. J. (1962) Interactions between hosts and parasites when some individuals are more difficult to find than others. *Journal of Theoretical Biology* 3: 1–18.

Barclay, H. J. (1986a) Host–parasitoid dynamics: effects of the position of density dependence. *Ecological Modelling* 32: 291–9.

Barclay, H. J. (1986b) Models of host–parasitoid interactions to determine the optimal instar of parasitization for pest control. *Natural Resource Modeling* 1: 81–103.

Barclay, H. J. (1991) Combining pheromone-baited and food-baited traps for insect pest control: effects of additional control by parasitoids. *Researches on Population Ecology* 33: 287–306.

Barclay, H. J., Otvos, I. S. & Thomson, A. J. (1985) Models of periodic inundation of parasitoids for pest control. *Canadian Entomologist* 117: 705–16.

Barlow, N. D. & Goldson, S. L. (1993) A modelling analysis of the successful biological control of *Sitona discoideus* (Coleoptera: Curculionidae) by *Microctonus aethiopoides* (Hymenoptera: Braconidae) in New Zealand. *Journal of Applied Ecology* 30: 165–78.

Barlow, N. D., Moller, H. & Beggs, J. R. (1996) A model for the effect of *Sphecophaga vesparum vesparum* as a biological control agent of the common wasp in New Zealand. *Journal of Applied Ecology* 33: 31–44.

Baveco, J. M. & Lingeman, R. (1992) An object-oriented tool for individual-oriented simulation: host–parasitoid system application. *Ecological Modelling* 61: 267–86.

Beddington, J. R., Free, C. A. & Lawton, J. H. (1975) Dynamic complexity in predator–prey models framed in difference equations. *Nature* 225: 58–60.

Beddington, J. R., Free, C. A. & Lawton, J. H. (1978). Modelling biological control: on the characteristics of successful natural enemies. *Nature* 273: 513–19.

Begon, M., Sait, S. M. & Thompson, D. J. (1995) Persistence of a parasitoid–host system: refuges and generation cycles. *Proceedings of the Royal Society of London, Series B* 260: 131–7.

Begon, M., Sait, S. M. & Thompson, D. J. (1996) predator–prey cycles with period shifts between two- and three-species systems. *Nature* 381: 311–15.

Bellows, T. S. & Hassell, M. P. (1988) The dynamics of age-structured host–parasitoid interactions. *Journal of Animal Ecology* 57: 259–68.

Bernstein, C. (1986) Density dependence and the stability of host–parasitoid systems. *Oikos* 47: 176–80

Bernstein, C. (1987) On assessing the role of spatially-heterogeneous density-independent host mortality on the stability of host–parasitoid systems. *Oikos* 49: 236–9.

Boerlijst, M. C., Lamers, M. E. & Hogeweg, P. (1993) Evolutionary consequences of spiral waves in a host-parasitoid system. *Proceedings of the Royal Society of London, Series B* 253: 15–18.

Briggs, C. J. (1993) Competition among parasitoid species on a stage-structured host and its effect on host suppression. *American Naturalist* 141: 372–97.

Briggs, C. J. & Latto, J. (1996) The window of vulnerability and its effect on relative parasitoid abundance. *Ecological Entomology* **21**: 128–40.

Briggs, C. J., Nisbet, R. M. & Murdoch, W. W. (1993) Coexistence of competing parasitoid species on a host with a variable life-cycle. *Theoretical Population Biology* **44**: 341–73.

Briggs, C. J., Nisbet, R. M., Murdoch, W. W., Collier, T. R. & Metz, J. A. J. (1995) Dynamical effects of host-feeding in parasitoids. *Journal of Animal Ecology* **64**: 403–16.

Chesson, P. L. & Murdoch, W. W. (1986) Aggregation of risk: Relationships among host–parasitoid models. *American Naturalist* **127**: 696–715.

Comins, H. N. & Hassell, M. P. (1979) The dynamics of optimally foraging predators and parasitoids. *Journal of Animal Ecology* **48**: 335–51.

Comins, H. N. & Hassell, M. P. (1987) The dynamics of predation and competition in patchy environments. *Theoretical Population Biology* **31**: 393–421.

Comins, H. N. & Wellings, P. W. (1985) Density-related parasitoid sex-ratio: influence on host–parasitoid dynamics. *Journal of Animal Ecology* **54**: 583–94.

Comins, H. N., Hassell, M. P. & May, R. M. (1992) The spatial dynamics of host-parasitoid systems. *Journal of Animal Ecology* **61**: 735–48.

DeGrandi-Hoffman, G., Diehl, J., Donghui, Li, Flexner, L., Jackson, G., Jones, W. and Debolt, J. (1994) BIOCONTROL-PARASITE: parasitoid–host and crop loss assessment simulation model. *Environmental Entomology* **23**: 1045–60.

Dempster, J. & Pollard, E. (1986) Spatial heterogeneity, stochasticity and the detection of density dependence in animal populations. *Oikos* **46**: 413–16.

Driessen, G. & Hemerik, L. (1991) Aggregative responses of parasitoids and parasitism in populations of *Drosophila* breeding on fungi. *Oikos* **61**: 96–107.

Flinn, P. W. & Hagstrum, D. W. (1995) Simulation model of *Cephalonomia waterstoni* (Hymenoptera: Bethylidae) parasitizing the rusty grain beetle (Coleoptera: Cucujidae). *Environmental Entomology* **24**: 1608–15.

Getz, W. M. & Mills, N. (1996) Host-parasitoid coexistence and egg-limited encounter rates. *American Naturalist* **148**: 333–47.

Godfray, H. C. J. & Chan, M. S. (1990) How insecticides trigger single-stage outbreaks in tropical pests. *Functional Ecology* **4**: 329–37.

Godfray, H. C. J. & Hassell, M. P. (1987) Natural enemies may be a cause of discrete generations in tropical insects. *Nature* **327**: 144–7.

Godfray, H. C. J. & Hassell, M. P. (1989) Discrete and continuous insect populations in tropical environments. *Journal of Animal Ecology* **58**: 153–74.

Godfray, H. C. J. & Pacala, S. W. (1992) Aggregation and the population dynamics of parasitoids and predators. *American Naturalist* **140**: 30–40.

Godfray, H. C. J. & Waage, J. K. (1991) Predictive modelling in biological control: the mango mealy bug (*Rastrococcus invadens*) and its parasitoids. *Journal of Applied Ecology* **28**: 434–53.

Godfray, H. C. J., Hassell, M. P. & Holt, R. D. (1994) The population dynamic consequences of phenological asynchrony between parasitoids and their hosts. *Journal of Animal Ecology* **63**: 1–10.

Gordon, D. M., Nisbet, R. M., deRoos, A., Gurney, W. S. C. & Stewart, R. K. (1991) Discrete generations in host–parasitoid models with contrasting life cycles. *Journal of Animal Ecology* **60**: 295–308.

Gutierrez, A. P. & Baumgaertner, J. U. (1984) II. A realistic model of plant–herbivore–parasitoid–predator interactions. *Canadian Entomologist* **116**: 933–49.

Gutierrez, A. P., Neuenschwander, P., Schulthess, F., Herren, H. R., Baumgaertner, J. U., Wermelinger, B., Lohr, B. & Ellis, C. K. (1988) Analysis of biological control of cassava pests in Africa. II. Cassava mealybug *Phenacoccus manihoti*. *Journal of Applied Ecology* **25**: 921–940.

Gutierrez, A. P., Neuenschwander, P. & van Alphen, J. J. M. (1993) Factors affecting biological control of cassava mealy bug by exotic parasitoids: a ratio-dependent supply-demand driven model. *Journal of Applied Ecology* **30**: 706–21.

Gutierrez, A. P., Mills, N. J., Schreiber, S. J. & Ellis, C. K. (1994) A physiological based tritrophic perspective on bottom-up–top-down regulation of populations. *Ecology* **75**: 2227–42.

Hassell, M. P. (1978) *The Dynamics of Arthropod Predator–Prey Systems*. Princeton: Princeton University Press.

Hassell, M. P. (1980) Foraging strategies, population models and biological control: a case study. *Journal of Animal Ecology* **49**: 603–28.

Hassell, M. P. (1984a) Parasitism in patchy environments: inverse density dependence can be stabilizing. *IMA Journal of Mathematics Applied in Medicine and Biology* **1**: 123–33.

Hassell, M. P. (1984b) Insecticides in host–parasitoid interactions. *Theoretical Population Biology* **26**: 378–86.

Hassell, M. P. (1985) Insect natural enemies as regulating factors. *Journal of Animal Ecology* **54**: 323–34.

Hassell, M. P. & Anderson, R. M. (1984) Host susceptibility as a component in host–parasitoid systems. *Journal of Animal Ecology* **53**: 611–21.

Hassell, M. P. & Comins, H. N. (1978) Sigmoid functional responses and population stability. *Theoretical Population Biology* **14**: 62–7.

Hassell, M. P. & May, R. M. (1973) Stability in insect host–parasite models. *Journal of Animal Ecology* **42**: 693–726.

Hassell, M. P. & May, R. M. (1986) Generalist and specialist natural enemies in insect predator–prey interactions. *Journal of Animal Ecology* **55**: 923–40.

Hassell, M. P. & May, R. M. (1988) Spatial heterogeneity and the dynamics of parasitoid–host systems. *Annals Zoologici Fennici* **25**: 55–61.

Hassell, M. P. & Pacala, S. W. (1990) Heterogeneity and the dynamics of host–parasitoid interactions. *Philosophical Transactions of the Royal Society of London, Series B* **330**: 203–20.

Hassell, M. P., Lawton, J. H. & May, R. M. (1976) Patterns of dynamical behaviour in single-species systems. *Journal of Animal Ecology* **45**: 471–86.

Hassell, M. P., Waage, J. K. & May, R. M. (1983) Variable parasitoid sex ratios and their effect on host–parasitoid dynamics. *Journal of Animal Ecology* **52**: 889–904.

Hassell, M. P, Comins, H. N. & May, R. M. (1991a) Spatial structure and chaos in insect population dynamics. *Nature* **353**: 255–8.

Hassell, M. P., May, R. M., Pacala, S. W. & Chesson, P. L. (1991b) The persistence of host–parasitoid association in patchy environments. I. A general criterion. *American Naturalist* **138**: 568–83.

Hassell, M. P, Comins, H. N. & May, R. M. (1994) Species coexistence via self-organizing spatial dynamics. *Nature* **370**: 290–2.

Hawkins, B. A. & Cornell, H. V. (1994) Maximum parasitism rates and successful biological control. *Science* **266**: 1886.

Hawkins, B. A., Thomas, M. B. & Hochberg, M. E. (1993) Refuge theory and biological control. *Science* **262**: 1429–32.

Hawkins, B. A., Hochberg, M. E. & Thomas, M. B. (1994) Biological control and refuge theory–reply. *Science* **265**: 812–13.

Hearne, J. W., van Coller, L. M. & Conlong, D. E. (1994) Determining strategies for the biological control of a sugarcane stalk borer. *Ecological Modelling* **73**: 117–33.

Heinz, K. M., Nunney, L. & Parrella, M. P. (1993) Toward predictable biological control of *Liriomyza trifolii* (Diptera: Agromyzidae) infesting greenhouse cut chrysanthemums. *Environmental Entomology* **22**: 1217–33.

Hochberg, M. E. (1996a) Consequences for host population levels of increasing natural enemy species richness in classical biological control. *American Naturalist* **147**: 307–18.

Hochberg, M. E. (1996b) An integrative paradigm for the dynamics of monophagous parasitoid–host interactions. *Oikos* **77**: 556–60.

Hochberg, M. E. (1996c) When three is a crowd. *Nature* **381**: 276–7.

Hochberg, M. E. (1997) Hide or fight? The competitive coevolution of concealment and encapsulation in host–parasitoid systems. *Oikos* **80**: 342–52.

Hochberg, M. E. & Hawkins, B. A. (1992) Refuges as a predictor of parasitoid diversity. *Science* **255**: 973–6.

Hochberg, M. E. & Hawkins, B. A. (1993) Predicting parasitoid species richness. *American Naturalist* **142**: 671–93.

Hochberg, M. E. & Holt, R. D. (1995) Refuge evolution and the population dynamics of coupled host–parasitoid associations. *Evolutionary Ecology* **9**: 633–61.

Hochberg, M. E. & Lawton, J. H. (1990) Spatial heterogeneities in parasitism and population dynamics. *Oikos* **59**: 9–14.

Hochberg, M. E., Hassell, M. P. & May, R. M. (1990) The dynamics of host–parasitoid–pathogen interactions. *American Naturalist* **135**: 74–94.

Hochberg, M. E., Elmes, G. W., Thomas, J. A. & Clarke, R. T. (1996) The population dynamics of monophagous parasitoids: a case model of *Ichneumon euremus* (Hymenoptera:ichneumonidae). *Philosophical Transactions of the Royal Society of London, Series B* **351**: 1713–24.

Hogarth, W. L. & Diamond, P. (1984) Interspecific competition in larvae between entomophagous parasitoids. *American Naturalist* **124**: 552–60.

Holt, R. D. & Hassell, M. P. (1993) Environmental heterogeneity and the stability of host–parasitoid interactions. *Journal of Animal Ecology* **62**: 89–100.

Holt, R. D. & Lawton, J. H. (1993) Apparent competition and enemy-free space in insect host–parasitoid communities. *American Naturalist* **142**: 623–45.

Hopper, K. R. & Roush, R. T. (1993) Mate finding, dispersal, number released, and the success of biological control introductions. *Ecological Entomology* 18: 321–31.

Ives, A. R. (1992a) Continuous-time models of host–parasitoid interactions. *American Naturalist* 140: 1–29.

Ives, A. R. (1992b). Density-dependent and density-independent parasitoid aggregation in model host–parasitoid systems. *American Naturalist* 140: 912–37.

Jervis, M. A. & Kidd, N. A. C. (1986) Host-feeding strategies in Hymenopteran parasitoids. *Biological Reviews* 61: 395–434.

Jones, T. H., Hassell, M. P. & Pacala, S. W. (1993) Spatial heterogeneity and the population dynamics of a host–parasitoid system. *Journal of Animal Ecology* 62: 251–62.

Kakehashi, N., Suzuki, Y. & Iwasa, Y. (1984) Niche overlap of parasitoids in host-parasitoid systems: its consequence to single versus multiple introduction controversy in biological control. *Journal of Applied Ecology* 21: 115–31.

Kidd, N. A. C. & Jervis, M. A. (1989) The effects of host-feeding behaviour on the dynamics of parasitoid–host interactions, and the implications for biological control. *Researches on Population Ecology* 31: 235–74.

Kidd, N. A. C. & Mayer, A. D. (1983) The effect of escape responses on the stability of insect host–parasite models. *Journal of Theoretical Biology* 104: 275–87.

Lampo, M. (1994) The importance of refuges in the interaction between *Contarinia sorghicola* and its parasitic wasp *Aprostocetus diplosidis*. *Journal of Animal Ecology* 63: 176–86.

Mangel, M. & Roitberg, B. D. (1992) Behavioral stabilization of host–parasite population dynamics. *Theoretical Population Biology* 42: 308–20.

May, R. M. (1978) Host–parasitoid systems in patchy environments: a phenomenological model. *Journal of Animal Ecology* 47: 833–43.

May, R. M. & Hassell, M. P. (1981) The dynamics of multiparasitoid–host interactions. *American Naturalist* 117: 234–61.

May, R. M. & Hassell, M. P. (1988) Population dynamics and biological control. *Philosophical Transactions of the Royal Society of London, Series B* 318: 129–69.

May, R. M., Hassell, M. P., Anderson, R. M. & Tonkyn, D. W. (1981) Density dependence in host-parasitoid models. *Journal of Animal Ecology* 50: 855–65.

Maynard-Smith J. & Slatkin, M. (1973) The stability of predator–prey systems. *Ecology* 54: 384–91.

Meier, C., Senn, W., Hauser, R. & Zimmermann, M. (1994) Strange limits of stability in host–parasitoid systems. *Journal of Mathematical Biology* 32: 563–72.

Mills, N. J. & Gutierrez, A. P. (1996) Prospective modelling in biological control: an analysis of the dynamics of heteronomous hyperparasitism in a cotton–whitefly–parasitoid system. *Journal of Applied Ecology* 33: 1379–94.

Morrison, G. & Barbosa, P. (1987) Spatial heterogeneity, population 'regulation' and local extinction in simulated host–parasitoid interactions. *Oecologia* 73: 609–14.

Münster-Swendsen, M. (1982) Interactions within a one-host-two-parasitoids system, studied by simulations of spatial patterning. *Journal of Animal Ecology* 51: 97–110.

Münster-Swendsen, M. (1985) A simulation study of primary-, clepto- and hyperparasitism in *Epinotia tedella* (Lepidoptera, Tortricidae). *Journal of Animal Ecology* 54: 683–695.

Münster-Swendsen, M. & Nachman, G. (1978) Asynchrony in insect host–parasite interaction and its effect on stability, studied by a simulation model. *Journal of Animal Ecology* 47: 159–71.

Murdoch, W. W. (1990) The relevance of pest–enemy models to biological control. In *Critical Issues in Biological Control*, ed. M. Mackauer, L. H. Ehler & J. Roland, pp. 1–24. Andover: Intercept.

Murdoch, W. W. & Oaten, A. (1975) Predation and population stability. *Advances in Ecological Research* 9: 2–131.

Murdoch, W. W. & Stewart-Oaten, A. (1989) Aggregation by parasitoids and predators: effects on equilibrium and stability. *American Naturalist* 134: 288–310.

Murdoch, W. W., Reeve, J. D., Huffaker, C. B. & Kennett, C. E. (1984) Biological control of olive scale and its relevance to ecological theory. *American Naturalist* 123: 371–92.

Murdoch, W. W., Chesson, J. & Chesson, P. L. (1985) Biological control in theory and practice. *American Naturalist* 125: 344–66.

Murdoch, W. W., Nisbet, R. M., Blythe, S. P., Gurney, W. S. C. & Reeve, J. D. (1987) An invulnerable age class and stability in delay-differential parasitoid–host models. *American Naturalist* 129: 263–82.

Murdoch, W. W., Briggs, C. J., Nisbet, R. M., Gurney, W. S. C. & Stewart-Oaten, A. (1992a) Aggregation and stability in metapopulation models. *American Naturalist* 140: 41–58.

Murdoch, W. W., Nisbet, R. M., Luck, R. F., Godfray, H. C. J. & Gurney W. S. C. (1992b) Size-selective sex-allocation and host feeding in a parasitoid–host model. *Journal of Animal Ecology* 61: 533–41.

Murdoch, W. W., Briggs, C. J. & Nisbet, R. M. (1996) Competitive displacement and biological control in parasitoids: a model. *American Naturalist* 148: 807–26.

Myers, J. H., Smith, J. N. M. & Elkinton, J. S. (1994) Biological control and refuge theory. *Science* 265: 811.

Nicholson, A. J. (1933) The balance of animal populations. *Journal of Animal Ecology* 2: 131–78.

Nicholson, A. J. & Bailey, V. A. (1935) The balance of animal populations. Part 1. *Proceedings of the Zoological Society of London* 1935: 551–98.

Pacala, S. W. & Hassell, M. P. (1991) The persistence of host–parasitoid associations in patchy environments. II. Evaluation of field data. *American Naturalist* 138: 584–605.

Pacala, S. W., Hassell, M. P. & May, R. M. (1990) Host–parasitoid interactions in patchy environments. *Nature* 344: 150–3.

Perry, J. N. (1987) Host–parasitoid models of intermediate complexity. *American Naturalist* 130: 955–7.

Perry, J. N. & Taylor, L. R. (1986) Stability of real interacting populations in space and time: implications, alternatives and the negative binomial k_c. *Journal of Animal Ecology* 55: 1053–68.

Ravlin, F. W. & Haynes, D. L. (1987) Simulation of interactions and management of parasitoids in a multiple host system. *Environmental Entomology* 16: 1255–65.

Reddingius, J., de Vries, S.A & Stam, A. J. (1982) On a stochastic version of a modified Nicholson–Bailey model. *Acta Biotheoretica* 31: 93–108.

Reed, D. J., Begon, M. & Thompson, D. J. (1996) Differential cannibalism and population dynamics in a host–parasitoid system. *Oecologia* 105: 189–93.

Reeve, J. D. (1988) Environmental variability, migration, and persistence in host-parasitoid systems. *American Naturalist* 132: 810–36.

Reeve, J. D. (1990) Stability, variability, and persistence in host–parasitoid systems. *Ecology* 71: 422–6.

Reeve, J. D., Kerans, B. L. & Chesson, P. L. (1989) Combining different forms of parasitoid aggregation: effects on stability and patterns of parasitism. *Oikos* 56: 233–9.

Reeve, J. D., Cronin, J. T. & Strong, D. R. (1994a) Parasitoid aggregation and the stabilization of a salt marsh host–parasitoid system. *Ecology* 75: 288–95.

Reeve, J. D., Cronin, J. T. & Strong, D. R. (1994b) Parasitism and generation cycles in a salt-marsh planthopper. *Journal of Animal Ecology* 63: 912–920.

Rohani, P., Godfray, H. C. J. & Hassell, M. P. (1994a) Aggregation and the dynamics of host–parasitoid systems: a discrete-generation model with within-generation redistribution. *American Naturalist* 144: 491–509.

Rohani, P., Miramontes, O. & Hassell, M. P. (1994b) Quasiperiodicity and chaos in population models. *Proceedings of the Royal Society of London, Series B* 258: 17–22.

Rohani, P., May, R. M.. & Hassell, M. P. (1996) Metapopulation and equilibrium stability: the effects of spatial structure. *Journal of Theoretical Biology* 181: 97–109.

Ruxton, G. D. & Rohani, P. (1996) The consequences of stochasticity for self-organized spatial dynamics, persistence and coexistence in spatially extended host–parasitoid communities. *Proceedings of the Royal Society of London, Series B* 263: 625–31.

Sait, S. M., Begon, M. & Thompson, D. J. (1994) Long-term population dynamics of the Indian meal moth *Plodia interpunctella* and its granulosis virus. *Journal of Animal Ecology* 63: 861–70.

Shimada, M. & Fujii, K. (1985) Niche modification and stability of competitive systems. III. Simulation model analysis. *Researches on Population Ecology* 27: 185–201.

Solé, R. V., Valls, J. & Bascompte, J. (1992) Spiral waves, chaos and multiple attractors in lattice models of interacting populations. *Physics Letters A* 166: 123–8.

Taylor, A. D. (1988) Parasitoid competition and the dynamics of host–parasitoid models. *American Naturalist* 132: 417–37.

Taylor, A. D. (1992) Deterministic stability analysis can predict the dynamics of some stochastic population models. *Journal of Animal Ecology* 61: 241–8.

Taylor, A. D. (1993a) Heterogeneity in host–parasitoid interactions: 'Aggregation of risk' and the '$CV^2 > 1$ rule'. *Trends in Ecology and Evolution* 8: 400–5.

Taylor, A. D. (1993b) Aggregation, competition and host–parasitoid dynamics: stability conditions don't tell it all. *American Naturalist* **141**: 501–6.

Thompson, W. R. (1924) La théorie mathématique de l'action des parasites entomophages et le facteur du hasard. *Annales Faculté des Sciences de Marseille* **2**: 69–89.

van Driesche, R. G., Bellows, T. S. Jr, Elkinton, J. S., Gould, J. R. & Ferro, D. N. (1991) The meaning of percentage parasitism revisited: solutions to the problem of accurately estimating total losses from parasitism. *Environmental Entomology* **20**: 1–7.

van Roermund, H. J. W., van Lenteren, J. C. & Rabbinge, R. (1997) Biological control of greenhouse whitefly with the parasitoid *Encarsia formosa* on tomato: an individual-based simulation approach. *Biological Control* **9**: 25–47.

Waage, J. K., Hassell, M. P. & Godfray, H. C. J. (1985) The dynamics of pest–parasitoid–insecticide interactions. *Journal of Applied Ecology* **22**: 825–38.

Wang, Y. H. & Gutierrez, A. P. (1980) An assessment of the use of stability analyses in population ecology. *Journal of Animal Ecology* **49**: 435–52.

Weisser, W. W. & Hassell, M. P. (1996) Animals 'on the move' stabilize host–parasitoid systems. *Proceedings of the Royal Society of London, Series B* **263**: 749–54.

Wilson, H. B. & Rand, D. A. (1997) Reconstructing the dynamics of unobserved variables in spatially extended systems. *Proceedings of the Royal Society of London, Series B* **264**: 625–30.

Wilson, H. B., Hassell, M. P., & Godfray, H. C. J. (1996) Host–parasitoid food webs: dynamics, persistence, and invasion. *American Naturalist* **148**: 787–806.

Yamamura, N. & Yano, E. (1988) A simple model of host–parasitoid interaction with host-feeding. *Researches on Population Ecology* **30**: 353–69.

Yano, E. (1989) A simulation study of population interaction between the greenhouse whitefly, *Trialeurodes vaporariorum* Westwood (Homoptera: Aleyrodidae), and the parasitoid *Encarsia formosa* Gahan (Hymenoptera: Aphelinidae). I. Description of the model. *Researches on Population Ecology* **31**: 73–88.

5

Biological control of insect pests: a tritrophic perspective

NICK J. MILLS AND ANDREW P. GUTIERREZ

Introduction

The success of classical biological control in providing spectacular examples of the impact that natural enemies can have on insect pest populations is rarely questioned. The introduction of the vedalia beetle from Australia for control of the cottony-cushion scale in California and the release of *Apoanagyrus lopezi* across central Africa for control of the cassava mealybug demonstrate that both predators and parasitoids can reduce damaging pests to innocuous levels of abundance (DeBach & Rosen, 1991). Successful biological control projects against insect pests have also proven, in a number of cases, to be repeatable in different geographic regions of the world and almost all truly successful projects have provided lasting control of the target pest (Clausen, 1978). It is obvious from the biological control record that some target pests are more suitable and more likely to result in further success than others. There have been more frequent successes in the control of homopteran pests than in any other taxon of insect pests (Clausen, 1978; Greathead, 1986; Greathead & Greathead, 1992) and so any general theory for the biological control of insect pests would do well to address the biological features of a crop–homopteran–natural enemy system. Despite the practical successes achieved in classical biological control, we have been rather less successful in developing a general theory for biological control and a mechanistic explanation for the effective top-down action of the natural enemies in successful biological control programs.

As discussed by Berryman (see Chapter 1), the theory of biological control has a long history that began with the simple mathematical models of Thompson, Lotka–Volterra and Nicholson–Bailey

(Mills & Getz, 1996). The permanence of successful biological control provided the impetus for a detailed examination of local stability in natural enemy–host models through the 1970s and 1980s (Beddington *et al.*, 1978; May & Hassell, 1988; Murdoch, 1990). There was a tremendous proliferation of ideas and variations on the early models during this period, none of which proved valuable to the biological control practitioner (Murdoch *et al.*, 1985; Waage, 1990), and which culminated in the paradox of biological control (Arditi & Berryman, 1991). The paradox states that although the addition of density dependence into simple natural enemy–host models induces the 'desirable' property of local stability, it does so at the expense of suppression of host density. In other words, the greater the stability the higher the host population density.

The paradox of biological control raised serious concerns about a general theory for biological control (Murdoch, 1990; Waage, 1990), and the value of *point stability* in explaining the observed dynamics of real systems was in question (Wang & Gutierrez, 1980; Gilbert, 1984). Murdoch and colleagues have further examined whether real natural enemy–host systems are characterized by stability at the local population level, and have suggested that metapopulation processes may hold the key to success in biological control (Murdoch *et al.*, 1985; Murdoch, 1990; Murdoch & Briggs, 1996). In other words, local extinctions may in fact occur, but metapopulation stability is achieved *via* the process of hide-and-seek that occurs among resource patches. In addition to the misleading emphasis on the stability of models used to describe biological control systems, Mills & Getz (1996), in a recent review, point out the contradictory properties generated by discrete-time and continuous-time models and suggest that we should no longer ignore the realities of age-structure and within-generation dynamics in developing our understanding of biological control systems. They also highlight the fact that although real biological control systems are inevitably tritrophic in nature, and the effects of the plant on the herbivore population and higher trophic levels have been amply demonstrated (Price, 1994), the only modeling approach to consistently include the plant population in biological control systems is that of Gutierrez and colleagues (Gutierrez *et al.*, 1984; 1988a, 1993, 1994). Briggs *et al.* (Chapter 2) provide a good discussion of the importance of age-structure, and the critical roles of the plant and within-generation dynamics are well illustrated in the recent tritrophic model of van Lenteren and van Roermund (Chapter 7).

So can we seriously expect to understand the dynamics of biological control systems if we ignore the plant trophic level? Apparently, for many theoreticians the answer is, yes, but for the rest of this chapter we will emphasize why a tritrophic perspective is so important in biological control. In our analysis of a general model for biological control we make use of phase plane analysis (i.e., point stability) in a qualitative sense to illustrate the well-accepted notion that consumer-resource interactions tend toward average or expected population levels that are characteristic of a particular environment. However, we recognize that in reality there are merely upper and lower bounds to the fluctuations generating the average levels of abundance (*sensu* Wang & Gutierrez, 1980) and that local extinctions can occur (*sensu* Murdoch *et al.*, 1985; see also Chapter 10).

Some characteristics of homopteran biological control systems

Let us go back, for a moment, to the homopteran systems that have been so successful as targets for biological control and examine some of the important biological characteristics of these systems. The most important features of these systems can be listed as:

- Simplified communities with at least the herbivore and natural enemy trophic levels occupied

by exotic species with reduced connectance to other organisms in the environment.
- Development in Homoptera, like all insects, is characterized by a series of distinct stages, a natural form of age-structure, that has a particularly important influence on the host selection and sex allocation behavior of parasitoids (e.g., Luck & Podoler, 1985; Mackauer, 1986; Gutierrez *et al.*, 1988a; van Dijken *et al.*, 1991; Godfray & Waage, 1991; and see Chapter 16).
- Homoptera are characterized by asynchronous development rates and overlapping generations, and so exhibit continuous population development throughout the year (Hughes, 1972; Gilbert *et al.*, 1976; McClure, 1990; Byrne & Bellows, 1991).
- The Homoptera often have short generation times and are able to respond rapidly to changes in the quality of the host plant (Carter *et al.*, 1980; Gutierrez *et al.*, 1988a; Hare *et al.*, 1990).

With these characteristics in mind, for a more general model to provide a better understanding of biological control it must incorporate a sufficient amount of biological detail to be able to represent the homopteran systems that provide the best examples of success. Thus it is sufficient for the model to represent a simple food chain, but it must include stage-structure, host selection, variable development rates, and changes in plant quality to capture the essential details of tritrophic interactions (as discussed by Gilbert *et al.*, 1976). In addition, although not stated explicitly, the general goal of biological control is to protect plants from damage, rather than to suppress pest population abundance *per se*. This latter aspect of biological control can only be appreciated through the use of a tritrophic model that can represent plant damage as well as the cascading effect of plant resources through the food chain.

A simple population dynamics model of a tritrophic system

The simplest biological control system is a vertical food chain that consists of a host plant, an insect herbivore and a natural enemy. The herbivore occupies an intermediate position in the trophic chain to incorporate the essential linkage between resources acquired from the plant and those lost to the natural enemy. In this section we will generally follow the methods and models outlined in Gutierrez *et al.* (1994), Mills & Gutierrez (1996) and Gutierrez (1996). The simplest form of this tritrophic system is represented by:

$$\frac{dC}{dt} = h_c\{f_c(L,C)C\} - f_H(C,H)H$$

$$\frac{dH}{dt} = h_H\{f_H(C,H)H\} - f_P(H,P)P \qquad (5.1)$$

$$\frac{dP}{dt} = h_P\{f_P(H,P)P\} - \mu P$$

where C, H, and P represent the population abundance of the crop or host plant, insect herbivore, and parasitoid populations, respectively; L is the population equivalent of the incident light available to the plants for photosynthesis; $f_i(\cdot)$ is the *per capita* functional response that determines the number of individuals from the $i-1$ trophic level that are acquired by an individual of the i trophic level, $h_i(\cdot)$ is the *per capita* numerical response that converts individuals from the $i-1$ trophic level to individuals of the i trophic level, and μ is the intrinsic death rate of the parasitoid population. The model is homogeneous at all trophic levels (i.e., the populations at each trophic level are governed by the same set of functional submodels) and incorporates the basic biology of resource acquisition and conversion between trophic levels without specifying an exact form for this biology.

The functional response

The functional response provides the linkage between trophic levels that determines the *per capita* rate of resource acquisition by the consumer population and hence the rate of loss from the resource population. The realized *per capita* rate of resource acquisition is also usefully defined as the supply rate $S_i = f_i(\cdot)$, and we will return to this later when we consider the general tritrophic model below. As pointed out by Royama (1971) the *per capita* functional

response $f_i(\cdot)$ is an instantaneous hunting function and, in the strictest sense, the form of this function should differ between predation and parasitism due to the fact that prey disappear as they are preyed upon, whereas parasitized hosts remain available for re-encounter and attack. In the context of the general model above, the insect herbivore is clearly a predator as the plant material disappears as it is consumed by the herbivore, in contrast to the parasitoid which leaves attacked herbivores to be re-encountered. However, the question of whether the plant predates or parasitizes its resources is less obvious. To resolve the correct form of a plant's functional response we need to consider the way in which plants use their resources. Sunlight, CO_2, and water are all passively extracted from the environment by plants and although it is hard to imagine that plants deplete the first two resources, the small quantities used, relative to their abundance in the environment, clearly disappear through the action of photosynthesis and are thus predated (see Gutierrez, 1996).

Although, for simplicity, a linear functional response is still frequently used in host–parasitoid models (e.g., Godfray & Waage, 1991; Godfray & Pacala, 1992; Briggs *et al.*, 1993, 1995; Murdoch *et al.*, 1996), it does not adequately describe the *per capita* limitation on resource acquisition by a consumer population. In fact, Holling (1959) described four different forms for the functional response, but the asymptotic or Type II functional response is the one most frequently encountered in laboratory and field observations over a full range of resource densities (Hassell *et al.*, 1976). Based on an assumption of a linear decrease in feeding rate as the rate of resource acquisition approaches a maximum, Ivlev (1955) developed one of the first expressions for a Type II functional response:

$$f_H(C,H) = D_H(1 - e^{-uC}) \qquad (5.2)$$

with u the attack rate when hungry and D_H the maximum *per capita* rate of resource acquisition or the demand rate of the consumer. This *per capita* functional response, like many others, ignores the influence of interference competition between consumers on the rate of acquisition of resources by individual consumers. As is implied in our general expression for the functional response $f_H(C,H)$, both resource and consumer densities are likely to be determinants of the *per capita* rate of resource acquisition in relation to resource density. The Gutierrez–Baumgärtner functional response equation (Gutierrez *et al.*, 1981), built indirectly on the framework of Ivlev's (1955) Type II response, incorporates the influence of competition between consumers:

$$f_H(C,H)=(1-e^{(\alpha_H C/D_H H)})D_H \qquad (5.3)$$

where D_H is the *per capita* demand rate (or maximum *per capita* rate of resource acquisition). If we define $\alpha_H=uD_H H$ as the demand related attack rate and substitute into eqn (5.3), we get Ivlev's model (eqn 5.2). Gutierrez (1996) provides a theoretical basis for this instantaneous hunting equation (5.3), shows that it is also closely related to Watt's (1959) model, and derives its more specific predator and parasitoid forms. Equation (5.3) is an example of a ratio-dependent functional response (Berryman *et al.*, 1995a; and see Chapter 1) and assumes that the acquisition of resources is driven by the demand of the consumer, such that in the presence of an infinite supply of resources the maximum *per capita* acquisition ($f_H(C, H)$) approaches the *per capita* demand for resources (D_H). Although Abrams (1994) and Murdoch & Briggs (1996) consider ratio-dependence to have no feasible biological foundation for continuous-time models (but see Berryman *et al.*, 1995a) we use eqn (5.3) to facilitate its extension to particular biological control systems where simulation requires the introduction of discrete time-steps. In this context eqn (5.3) incorporates two essential mechanistic features; an asymptote set by the *per capita* consumer demand rate (satiation), and a *per capita* resource acquisition rate that is proportional to the resource-to-consumer ratio.

So far we have assumed that the consumer has only a single resource, which is often true for the specialized parasitoids and their target hosts in biological control systems, but the model can easily be extended to food webs (Schreiber & Gutierrez, 1998). At the crop or plant trophic level, there is clearly a need to consider multiple resources as, in addition to sunlight, plants require carbon dioxide, water, nitrogen, etc. The *per capita* functional response can readily be adapted to deal with the acquisition of multiple resources and interested readers are referred to Gutierrez *et al.* (1994) and Gutierrez (1996) for further details.

The numerical response

Whenever an organism acquires resources the principles of ecological energetics govern the efficiency with which these resources are assimilated and converted to growth and reproduction. The process of conversion and allocation of resources within the consumer population provides us with the numerical response that is again common to all trophic levels. The essential components of the numerical response have been developed by Gutierrez and colleagues as the metabolic pool model (Gutierrez & Wang, 1977; Gutierrez *et al.*, 1984). If we again use the herbivore population, for purposes of illustration, the rate of production (growth and/or reproduction) of the herbivore population, P_H, is determined by:

$$P_H= \lambda_H H\{a_H f_H(C,H)-r_H H^{b_H}\} \qquad (5.4)$$

where a_H is the proportion of the resource assimilated, r_H is the *per capita* rate of respiration, $b_H>0$ is the rate at which the population respiration rate increases with herbivore density, and λ_H is the conversion efficiency of resources into consumers. It is worth noting here that although competition between herbivores reduces the *per capita* rate of resource acquisition through the functional response, self limitation is specifically included in the population respiration rate to take into account the increased cost of acquiring resources in the face of resource limitation. This expression can be simplified by multiplying through by $\lambda_H H$ to give:

$$P_H= \theta_H f(C,H)H - m_H H^{1+b_H} \qquad (5.5)$$

where the conversion coefficient $\theta_H= \lambda_H a_H$ and the *per capita* maintenance cost $m_H= \lambda_H r_H$. This numerical response differs from that used in most other host–parasitoid models in that it more explicitly includes the mechanism of resource conversion rather than subsuming this biology into a single proportionality constant. Further details of the metabolic pool model can be found in Gutierrez (1996).

A general tritrophic model for biological control

We obtain a general mechanistic model for the three trophic levels of a simple biological control system by combining the details of eqn (5.3) for the functional responses with those of eqn (5.5) for the numerical responses into the framework of eqn (5.1). The general tritrophic model is then:

$$
\frac{dC}{dt} = \phi_C \theta_C \left(1 - e^{-(\alpha_C R/D_C C)}\right) D_C C - m_C C^{1+b_C} \\
- \left(1 - e^{-(\alpha_H C/D_H H)}\right) D_H H
$$

$$
\frac{dH}{dt} = \phi_H \theta_H \left(1 - e^{-(\alpha_H C/D_H C)}\right) D_H H - m_H H^{1+b_H} \\
- \left(1 - e^{-(\alpha_P H/D_P P)}\right) D_P P \qquad (5.6)
$$

$$
\frac{dP}{dt} = \phi_P \phi_H \theta_P \left(1 - e^{-(\alpha_P H/D_P P)}\right) D_P P - m_P P^{1+b_P} - \mu P
$$

where $\phi_i = S_i/D_i$, and $i = C, H, P$ is the supply–demand ratio that scales the potential population growth rates for the ith trophic level to the realized rates. Note parasitized herbivores remain in the herbivore population during parasitoid development, and the same supply–demand ratio (ϕ_H) affects both the healthy and the parasitized herbivores. Although this model is framed as the rates of change of individuals in the three populations, it is often more convenient to use biomass as the currency of population change and trophic linkage as both the attack rate (α) and demand (D) vary with the size of an individual. A more detailed discussion of the inter-relationship of the individual and mass-based dynamics of the tritrophic model can be found in Gutierrez *et al.* (1994) and Gutierrez (1996).

Several of the parameters used in this general tritrophic model are obviously not constant, but in reality are likely to vary with time and/or environmental conditions. For example, the conversion coefficient (θ), the demand (D), and the attack rate (α) will vary with time (age), and the demand (D), the attack rate (α), and the maintenance costs (m) will vary with temperature. In applying this general tritrophic model to specific organisms, there is clearly a need to include the more specific details of age-

structure, time delays for maturation rates, and the temperature-dependence of resource acquisition and allocation. We will not elaborate on these aspects of the model here, and the reader is referred to Gutierrez (1996) for a more detailed discussion of these features of the tritrophic model.

Properties of the general tritrophic model for biological control

As pointed out by Royama (1992), a general model is a compromise between the intractability and unique nature of the real world and the more manageable general principles of the systems that we seek to understand. The general tritrophic model for biological control, presented above (eqn (5.6)), is relatively simple and captures the biological details of resource acquisition and conversion that are common to all trophic levels, using submodels that do not generate anomolous values at extremes of population densities. As such the model satisfies the two major considerations posed by Royama (1992) in developing a more general model, (i) that the model is parsimonious in structure (see Gutierrez *et al.*, 1994; Gutierrez, 1996) and (ii) that there are no unbounded biological properties (see Berryman *et al.*, 1995b for details of realistic bounds for tritrophic models).

It is of interest to explore the basic properties of this model to see whether there are any qualitative differences between the tritrophic model and a typical two-trophic level or bitrophic model. The properties of the model can be conveniently examined using phase–plane analysis, in which the dynamics of any two interacting populations are represented by their zero-growth isoclines. Setting the effects of other factors $\phi_i = 1$, we illustrate how this is done by setting the herbivore model to zero growth, $dH/dt = 0$, and solving the herbivore isocline in relation to plant abundance (C) we get:

$$
C = \frac{-D_H H}{\alpha_H} \log_e
$$

$$
\frac{\theta_H D_H H - m_H H^{1+b_H} - D_P (1 - e^{-\alpha_P H/D_P P}) P}{\theta_H D_H H} \qquad (5.7)
$$

Similarly, the herbivore isocline can be solved in relation to parasitoid abundance and the parasitoid isocline in relation to herbivore abundance. Note that this isocline has all three trophic levels represented. However, the plant isocline has no analytical solution and can only be evaluated numerically in relation to herbivore abundance (Gutierrez *et al.*, 1994).

Interestingly, the plant and the herbivore share an intermediate position in the trophic chain, and their isoclines can take two different forms (a condition previously documented by Arditi & Ginsberg, 1989 for a bitrophic model). The dichotomy of the two forms is based upon the efficiency of the consumer in acquiring resources (more specifically the *per capita* attack rate of the consumer population, α_{i+1}) in relation to the productivity of the resource (the potential *per capita* growth rate of the resource population, $\theta_i D_i$). If we consider the plant–herbivore phase plane, first of all, we find a humped plant isocline (Fig. 5.1(a)) with the plant axis intersected at the origin and again at the carrying capacity of the plant population in the absence of herbivory ($[\theta D/m]^{1/b}$) when the herbivore is efficient relative to the productivity of the plant ($\alpha_H > \theta_C D_C$). However, if the reverse is true ($\alpha_H < \theta_C D_C$), the plant isocline intersects the plant axis at the same upper limit, but rises to a vertical asymptote as plant abundance declines (Fig. 5.1(b)). In this case, the asymptote indicates that the plant population has a refuge below which it cannot be driven by herbivory (Crawley, 1992). The same two isocline forms are generated by the herbivore isocline viewed in the herbivore–parasitoid phase plane (Figs 5.1(e) and (f)).

A humped resource isocline results from nonlinear density dependence that changes from predominantly inverse at low resource population abundance to predominantly direct at higher resource population abundance (Crawley, 1992). This form of isocline is common in bitrophic models that combine a Type II functional response with nonlinear self limitation (Berryman, 1992), as is the case for our tritrophic model. The resource refuge effect, seen in the asymptotic form of the resource isocline, is unique, and clearly indicates that the resource population has an effective numerical refuge when the resource population growth rate exceeds the consumer attack rate. However, this form of isocline could also be generated from a bitrophic model with

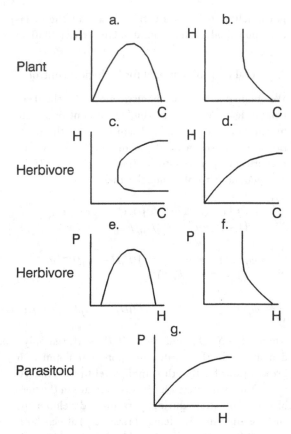

Figure 5.1. The shapes of the zero-growth isoclines for each of the interacting populations in the general tritrophic model for biological control (after Gutierrez 1996). See the text for details. H, herbivore; C, crop or host plant; P, parasitoid. (a) and (b) plant isocline; (c)–(f) herbivore isocline; and (g) parasitoid isocline.

constant input for the resource population and a ratio-dependent functional response (Arditi & Ginsberg, 1989).

The herbivore isocline, viewed in the plant–herbivore phase plane, changes from a C-shape with both upper and lower asymptotes (Fig. 5.1(c)) to an asymptotic curve with a single upper asymptote at the herbivore's carrying capacity (Fig. 5.1(d)) as the efficiency of the parasitoid is reduced relative to the growth rate of the herbivore population (i.e., as $\alpha_P > \theta_H D_H \rightarrow \alpha_P < \theta_H D_H$). However, the parasitoid isocline, representing the top of the trophic chain, has no dichotomy and takes the form of an asymptotic curve in this same phase plane (Fig. 5.1(g)). An asymptotic consumer isocline has been noted in bitrophic models and results from a ratio-dependent functional response (generates a linear isocline) together with non-linear self limitation on the consumer population (Berryman, 1992). The C-shaped consumer isocline, seen only at the intermediate (herbivore) trophic level in our model, is more unusual (but previously noted by Metzgar & Boyd, 1988). The lower asymptote of the C-shape is a threshold level of abundance below which the herbivore population cannot persist when under attack from an efficient parasitoid population. This lower threshold for persistence does not exist in simple bitrophic models and highlights the need to include the dynamics of the plant population as a component of a more general model of biological control.

Phase-plane analysis is an instructive ·way to examine the significance of two interacting isoclines, the points of intersection of the isoclines indicating the average densities that the two populations arc drawn toward, and the regions of phase space being characterized by specific vectors of population change. In this analysis we first of all confine our attention to the plant–herbivore phase plane, as these are the two interacting populations that are of greatest concern in biological control. We further restrict our attention to the humped plant isocline, as biological control systems are characterized by herbivore populations that have a devastating impact on the plant population (i.e., $\alpha_H > \theta_C D_C$). This leaves us with two possible scenarios for the interaction of the plant and herbivore isoclines in a biological control system (Figs 5.2(a) and (d)) in which there are four regions of phase space (labeled A, B, C, and D) classified by the directions of the population vectors (see Fig. 5.2(a)):

Region A – both plant and herbivore populations decreasing, leading to local extinction of the herbivore population in A*.
Region B – both plant and herbivore populations increasing.
Region C – plant population decreasing, but herbivore population increasing.

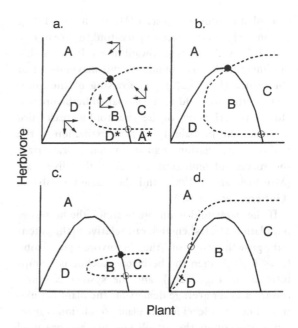

Figure 5.2. Combinations of interacting plant (—) and herbivore (- - - -) population isoclines, representing different scenarios for biological control (●, are stable equilibria; and ○, possibly unstable equilibria, in a point stability context). See the text for details.

Region D – plant population increasing, but herbivore population decreasing, leading to local extinction of the herbivore population in D*.

For an effective parasitoid, that has a high rate of attack on the herbivore population ($\alpha_P > \theta_H D_H$), the herbivore isocline is C-shaped as in Fig. 5.2(a). In this scenario the system is drawn toward an average plant population density that is close to the carrying capacity for the environment and, therefore, represents successful biological control. From a metapopulation perspective, the C-shaped herbivore isocline also creates a threshold for the herbivore population below which local extinction can occur, as can be seen from regions D* and A* of the phase space. The efficiency of the parasitoid (α_P) determines the average densities of the interacting populations. A moderately effective parasitoid ($\alpha_P \approx \theta_H D_H$) may support a relatively high herbivore population (Fig. 5.2(b)) with very low probability of local extinction, as appears to be the case for the biological control of California red scale (Murdoch *et al.*, 1985). In contrast, a very effective parasitoid ($\alpha_P \gg \theta_H D_H$) would draw the system toward mean densities that minimize herbivore abundance and maximize plant abundance (Fig. 5.2(c)). The raising of the lower asymptotic threshold of the C-shaped herbivore isocline as the efficiency of the parasitoid increases also increases the probability of local extinction of the herbivore population, a situation that characterizes the successful biological control of the olive scale (Murdoch *et al.* 1985) and the cassava mealybug (Gutierrez *et al.* 1993).

If the parasitoid is unable to attack the herbivore population at a fast enough rate relative to the potential growth rate of the herbivore population ($\alpha_P < \theta_H D_H$), then the herbivore isocline is asymptotic, as in Fig. 5.2(d), and the system is drawn toward a lower average density for the plant population. The exact level for the plant population is again dependent upon the attack rate of the parasitoid (α_P), but the reduction in plant abundance that is characterized by this scenario is unlikely to be judged a biological control success. It is also possible, in a metapopulation context, that local extinction of the plant population could occur when a patch has plant–herbivore densities that fall within region A of the phase space.

It is important to note that changes in the plant population can disrupt the success of biological control. If the plant is stressed by lack of nutrients or water (recall $\phi_C < 1$ in eqn 5.6), the carrying capacity of the plant population is reduced because ϕ_C factors $\theta_C D_C$ causing the plant isocline in Fig. 5.2(a) and (b) to shrink toward the origin in both height and width. Lack of soil nutrients has a cascading effect through the cassava system, reducing mealybug size and hence the production of females by the two parasitoids, and leading to disruption of the successful biological control observed in regions with sufficient soil nutrients (Gutierrez *et al.*, 1988a; Neuenschwander *et al.*, 1989). In contrast, enrichment of the environment may increase plant quality and the efficiency of conversion of the herbivore (θ_H), causing a widening of the C-shaped herbivore isocline or even a switch to an asymptotic isocline ($\alpha_P > \theta_H D_H \rightarrow \alpha_P < \theta_H D_H$), and a shifting of average

population densities to the left side of the plant isocline (as in Fig. 5.2(d)). A mite outbreak caused by an increase in foliar nitrogen was predicted from a tritrophic model of the cassava green mite and was subsequently supported by experimental field observations (Gutierrez *et al.*, 1988b; Wermelinger *et al.*, 1991). Thus for parasitoids, a decline in reproduction can be induced by a decrease in plant quality *via* host size effects, and for predatory mites an increase in plant quality affects reproduction of the herbivorous mites directly and hence their capacity to regulate their prey.

These insights into the influence of plant and parasitoid characteristics on the probable outcome of biological control are not intuitively obvious and can only be derived from a model that includes the plant as a variable resource and that captures the essential biology of resource acquisition and metabolic conversion.

Biological control of whiteflies – a prospective analysis

We have used the general tritrophic model to examine a particularly interesting and unusual twist to classical biological control that results from the unique biology of autoparasitism (Mills & Gutierrez, 1996). Several genera of aphelinid parasitoids have the ability to extend their host range to include competing parasitoid individuals through autoparasitism. Males of an obligate autoparasitoid are produced only by hyperparasitism of conspecific females within the primary host, whereas those of a facultative autoparasitoid result from hyperparasitism of any internal parasitoid within the primary host. Autoparasitism is common in four aphelinid genera, including *Encarsia*, that are parasitoids of whiteflies and scales, and this unusual form of host selection for male production needs to be taken into consideration in the development of classical biological control programs.

We used the general tritrophic model for biological control to ask the question: is autoparasitism compatible with the goals of biological control? This question was analyzed in the context of the biological control of whitefly by parasitoids in a cotton ecosystem. An age-structured simulation model, an extension of the general model (eqn 5.7), developed for the cassava mealybug system (Gutierrez *et al.*, 1993), provided the biological detail needed for this specific biological control system. We considered three types of parasitoid, a typical primary parasitoid (P_1), an obligate autoparasitoid (P_2) and a facultative autoparasitoid (P_3) acting alone on the whitefly population or in combination. The functional response of the parasitoid, that determines the rate of parasitism of the whitefly, needs to be modified to take into account the different range of hosts of the three types of parasitoid:

$$f_P(W,P) = (1 - s_1 s_2 s_3) \qquad (5.8)$$

where

$$s_i = e^{-\{D_i P_i^f [1 - e^{-(\alpha_i H_i / D_i P_i^f)}] / H_i\}} \qquad (5.9)$$

W is whitefly abundance, P_i^f is the abundance of females of the i-th parasitoid, and H_i is the abundance of all hosts (primary and secondary) of the i-th parasitoid. Note that in this more detailed model we use the parasitoid form of the functional response (eqn 5.3). The set of hosts that can be used by the different parasitoids is then:

$$H_1 = W$$

$$H_2 = (W + P_2^f) \qquad (5.10)$$

$$H_3 = (W + P_1^f + P_1^m + P_2^f + P_2^m + P_3^f)$$

which assumes that the parasitoids show no preference for particular host types and that they lay female eggs on encountering a primary host (healthy whitefly) and a male egg on encountering a secondary host (previously parasitized). There is little experimental evidence for this simplifying assumption, but a recent study of secondary host choice by the facultative autoparasitoid, *Encarsia pergandiella*, supports this view (Pedata & Hunter, 1996).

The cotton–whitefly–aphelinid parasitoids model has no analytical solution, but from a series of 1400 simulations of the outcome of the model, run over a single growing season using experimentally determined parameter values taken from the literature (see Mills & Gutierrez, 1996 for further details), we

obtained the general result that facultative autoparasitism can interfere with successful biological control. From a series of typical simulation runs of the model (Fig. 5.3) we can see that the primary parasitoid can reduce whitefly abundance substantially and that obligate autoparasitism is compatible with primary parasitism, but that facultative autoparasitism prevents a primary parasitoid from achieving its potential in biological control. These results offer a note of caution in using *Encarsia* species for the classical biological control of whiteflies, as the competitive superiority and the self-limitation of facultative autoparasitoids can have a disruptive influence on the biological control potential of primary parasitoids.

The most well-known example of the biological control of whiteflies is the control of the greenhouse whitefly, *Trialeurodes vaporariorum*, by augmentative releases of the parasitoid, *Encarsia formosa* (van Lenteren, 1995). In this case *E. formosa* is an obligate autoparasitoid that is unable to produce males (thelytokous) due to infection of females by sex ratio distorting bacteria (Zchori-Fein *et al.*, 1992). van Lenteren and van Roermund (see Chapter 7) discuss the development and validation of a detailed simulation model of this very successful biological control system. Heinz & Nelson (1996) also showed that weekly augmentative releases of both the facultative autoparasitoid, *E. pergandiella* and the thelytokous obligate autoparasitoid, *E. formosa*, improved control of the silverleaf whitefly, *Bemisia argentifolii*, in large greenhouse cages in comparison to releases of *E. formosa* alone. However, Williams (1996) noted that introduction of a facultative autoparasitoid, *E. tricolor*, into cages containing cabbage whitefly, *Aleyrodes proletella*, and the conventional parasitoid, *E. inaron*, resulted in a marked reduction in the abundance of the conventional parasitoid and a variable effect on whitefly abundance. In a set of reciprocal experiments *E. inaron* was unable to invade a system of the cabbage whitefly and *E. tricolor*. Williams (1996) further provides evidence of a wide variety of scale and whitefly systems in which a facultative autoparasitoid dominates the parasitoid complex, although there are few examples of the impact of this dominance on the degree of control of the whitefly hosts.

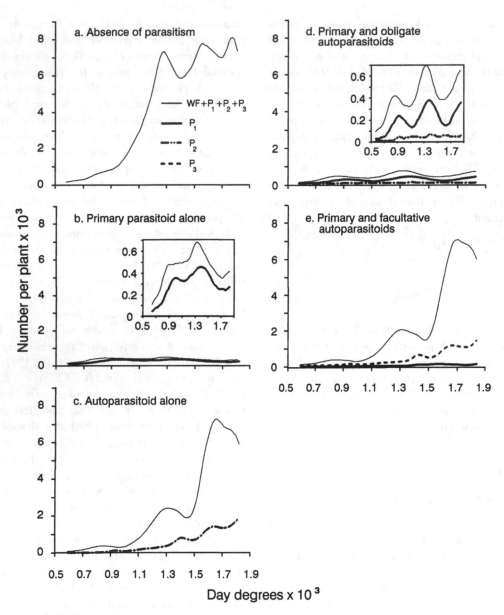

Figure 5.3. The number of whitefly and parasitoids per cotton plant, in the absence of parasitism (a) or in the presence of different combinations of parasitoids (b)–(e), in relation to a physiological time-scale representing the course of a growing season (after Mills & Gutierrez, 1996). For all figures: —, $WF+P_1+P_2+P_3$; —, P_1; –·–, P_2; – – –, P_3.

Conclusions

There has been enormous interest in the development of a strong theoretical basis for biological control, beginning with the simple models of Thompson and Nicholson–Bailey and extending to more detailed models incorporating age-structure and heterogeneity of natural enemy attack (Mills & Getz, 1996; and see Chapter 2). Almost all of these models have shared a similar approach, to explain the continued persistence of successful biological control by exploring features of a bitrophic interaction that confer local stability. Murdoch *et al.* (1985) shed doubt on the local stability of biological control systems and Karieva (1990) considered whether theoretical models have addressed the right questions in biological control. Here we have focused on two aspects of biological control systems that we consider to be particularly important and yet generally neglected in the context of biological control theory: the development of a more general model that explicitly includes the plant population, and the extension of this model to address crucial questions about the short-term, non-equilibrium dynamics of specific biological control systems.

It seems unlikely to us that we will be able to explain the successes and failures in the biological control record from the general properties of predator–prey or host–parasitoid models. All biological control systems are characterized by the interaction of at least three trophic levels and the plant population as a specific variable resource for the pest can no longer be ignored as an essential component of a more general tritrophic model for biological control. Although a general tritrophic model can provide some general principles that can be used to guide the reconstruction of natural enemy complexes in classical biological control projects, there are likely to be crucial details of the biology of specific biological control systems that are not easily encapsulated by a general model. One approach to this problem is to develop a simulation model that exactly describes the empirical observations made on the specific system. A very successful example of this approach is the tomato–whitefly–*Encarsia* model described by van Lenteren and van Roermund (see Chapter 7).

However, such models lack generality and it is not clear how many of the insights gained from this approach can be applied to other biological control systems. Models of intermediate complexity that retain a strong theoretical basis, but capture more specific details of an individual system seem to provide the best opportunity for further analysis of biological control. The general tritrophic model for biological control that we have presented here, has a sound theoretical framework, can readily be extended to incorporate the essential details of specific systems, and has successfully been used to examine the short-term dynamics of several tritrophic systems (Gutierrez *et al.*, 1994; Gutierrez, 1996). The recent extension of this paradigm to a full stochastic metapopulation setting (Gutierrez *et al.* unpublished results) reinforces the importance of between-patch movement but yields the same predictions as the single patch model with small rates of stochastic immigration (Gutierrez *et al.*, 1993).

References

Abrams, P. A. (1994) The fallacies of 'ratio-dependent' predation. *Ecology* 75: 1842–50.

Arditi, R. & Berryman, A. A. (1991) The biological control paradox. *Trends in Ecology and Evolution* 6: 32.

Arditi, R. & Ginsberg, L. R. (1989) Coupling in predator–prey dynamics: ratio dependence. *Journal of Theoretical Biology* 139: 311–26.

Beddington, J. R., Free, C. A. & Lawton, J. H. (1978) Characteristics of successful natural enemies in models of biological control of insect pests. *Nature* 273: 513–19.

Berryman, A. A. (1992) The origins and evolution of predator–prey theory. *Ecology* 73: 1530–5.

Berryman, A. A., Guitierrez, A. P. & Arditi, R. (1995a) Credible, parsimonious and useful predator–prey models – a reply to Abrams, Gleeson, and Sarnelle. *Ecology* 76: 1980–5.

Berryman, A. A., Michalski, J., Gutierrez, A. P. & Arditi, R. (1995b) Logistic theory of food web dynamics. *Ecology* 76: 336–43.

Briggs, C. J., Nisbet, R. M. & Murdoch, W. W. (1993) Coexistence of competing parasitoid species on a host with a variable life cycle. *Theoretical Population Biology* 44: 341–73.

Briggs, C. J., Nisbet, R. M., Murdoch, W. W., Collier, T. R. & Metz, J. A. J. (1995) Dynamical effects of host-feeding in parasitoids. *Journal of Animal Ecology* 64: 403–16.

Byrne, D. N. & Bellows, T. S. (1991) Whitefly biology. *Annual Review of Entomology* 36: 431–57.

Carter, N., McLean, I. F. G., Watt, A. D. & Dixon, A. F. G. (1980) Cereal aphids: a case study and review. In *Appied Biology, Volume 5*, ed. T.H. Coaker, pp. 271–348. London: Academic Press.

Clausen, C. P. (1978) *Introduced Parasites and Predators of Arthropod Pests and Weeds: A World Review*. USDA Handbook no. 480. Washington, DC: United States Department of Agriculture.

Crawley, M. J. (1992) Population dynamics of natural enemies and their prey. In *Natural Enemies, the Population Biology of Predators, Parasites and Diseases*, ed. M.J. Crawley, pp. 40–89. Oxford: Blackwell Scientific.

DeBach, P. & Rosen, D. (1991) *Biological Control by Natural Enemies*, 2nd edn. Cambridge: Cambridge University Press.

Gilbert, N. (1984) What they didn't tell you about limit cycles. *Oecologia* 65: 112–13.

Gilbert, N., Gutierrez, A. P., Frazer, B. D. & Jones, R. E. (1976) *Ecological Relationships*. New York: Freeman.

Godfray, H. C. J. & Pacala, S. W. (1992) Aggregation and the population dynamics of parasitoids and predators. *American Naturalist* 140: 30–40.

Godfray, H. C. J. & Waage, J. K. (1991) Predictive modelling in biological control: the mango mealybug (*Rastrococcus invadens*) and its parasitoids. *Journal of Applied Ecology* 28: 434–53.

Greathead, D. J. (1986). Parasitoids in classical biological control. In *Insect Parasitoids*, ed. J. K. Waage & D. J. Greathead, pp. 289–317. London: Academic Press.

Greathead, D. J. & Greathead, A. (1992). Biological control of insect pests by insect parasitoids and predators: the BIOCAT database. *Biocontrol News and Information* 13: 61N–68N.

Gutierrez, A. P. (1996) *Applied Population Ecology: A Supply–Demand Approach*. New York: Wiley.

Gutierrez, A. P. & Wang, Y. H. (1977) Applied population ecology: models for crop production and pest management. In *Pest Management*, ed. G.A. Norton & C.S. Holling, pp. 255–80. Oxford: Pergamon.

Gutierrez, A. P., Baumgärtner, J. U. & Hagen, K. S. (1981) A conceptual model for growth, development and reproduction in the ladybird beetle *Hippodamia convergens* G.-M. (Coccinellidae: Coleoptera). *Canadian Entomologist* 113: 21–33.

Gutierrez, A. P., Baumgärtner, J. U. & Summers, C. G. (1984) Multitrophic level models of predator–prey energetics. III. A case study of an alfalfa ecosystem. *Canadian Entomologist* 116: 950–63.

Gutierrez, A. P., Neuenschwander, P., Schultess, F., Herren, H. R., Baumgärtner, J. U., Wermelinger, B., Löhr, B. & Ellis, C. K. (1988a) Analysis of the biological control of cassava pests in Africa. II. Cassava mealybug *Phenacoccus manihoti*. *Journal of Applied Ecology* 25: 921–40.

Gutierrez, A. P., Yaninek, S. J., Wermelinger, B., Herren, H. R. & Ellis, C. K. (1988b) Analysis of the biological control of cassava pests in West Africa. III. The interaction of cassava and cassava green mite. *Journal of Applied Ecology* 25: 941–50.

Gutierrez, A. P., Neuenschwander, P. & van Alphen, J. J. M. (1993) Factors affecting biological control of cassava mealybug by exotic parasitoids: a ratio-dependent supply-demand driven model. *Journal of Applied Ecology* 30: 706–21.

Gutierrez, A. P., Mills, N. J., Schreiber, S. J. & Ellis, C. K. (1994) A physiological perspective on bottom up-top down regulation of populations. *Ecology* 75: 2227–42.

Hare, J. D., Yu, D. S. & Luck, R. F. (1990) Variation in life history parameters of California red scale on different citrus cultivars. *Ecology* 71: 1451–60.

Hassell, M. P., Lawton, J. H. & Beddington, J. R. (1976) The components of arthropod predation. I. The prey death rate. *Journal of Animal Ecology* 45: 135–64.

Heinz, K. M. & Nelson, J. M. (1996) Interspecific interactions among natural enemies of *Bemisia* in an inundative biological control program. *Biological Control* 6: 384–93.

Holling, C. S. (1959) The components of predation as revealed by a study of small mammal predation of the European pine sawfly. *Canadian Entomologist* 91: 293–320.

Hughes, R. D. (1972) Population dynamics. In *Aphid Technology*, ed. H.F. van Emden, pp. 275–93. London: Academic.

Ivlev, V. S. (1955) *Experimental Ecology of the Feeding of Fishes*. New Haven: Yale University Press.

Karieva, P. (1990) Establishing a foothold for theory in biocontrol practice: using models to guide experimental design and release protocols. In *New Directions in Biological Control: Alternatives for Suppressing Agricultural Pests and Diseases*, ed. R. Baker & P. Dunn, pp. 65–81. New York: A. Liss.

Luck, R. F. & Podoler, H. (1985) Competitive exclusion of *Aphytis lignanensis* by *A. melinus*: potential role of host size. *Ecology* 66: 904–13.

Mackauer, M. (1986) Growth and development interactions in some aphids and their hymenopteran parasites. *Journal of Insect Physiology* 32: 275–80.

May, R. M. & Hassell, M. P. (1988) Population dynamics and biological control. *Philosophical Transactions of the Royal Society of London, Series B* 318: 129–69.

McClure, M. S. (1990) Seasonal history. In *Armored Scale Insects: Their Biology, Natural Enemies and Control*, ed. D. Rosen, pp. 315–18. Amsterdam: Elsevier.

Metzgar, L. H. & Boyd, E. (1988) Stability properties in a model of forage-ungulate-predator interactions. *Natural Resource Modeling* 3: 3–43.

Mills, N. J. & Getz, W. M. (1996) Modelling the biological control of insect pests: a review of host–parasitoid models. *Ecological Modelling* 92: 121–43.

Mills, N. J. & Gutierrez, A. P. (1996) Prospective modelling in biological control: an analysis of the dynamics of heteronomous hyperparasitism in a cotton–whitefly–parasitoid system. *Journal of Applied Ecology* 33: 1379–94.

Murdoch, W. W. (1990) The relevance of pest–enemy models to biological control. In *Critical Issues in Biological Control*, ed. M. Mackauer, L. E. Ehler & J. Roland, pp. 1–24. Andover: Intercept.

Murdoch, W. W. & Briggs, C. J. (1996) Theory for biological control: recent developments. *Ecology* 77: 2001–13.

Murdoch, W. W., Chesson, J. & Chesson, P. L. (1985) Biological control in theory and practice. *American Naturalist* 125: 344–66.

Murdoch, W. W., Briggs, C. J. & Nisbet, R. M. (1996) Competitive displacement and biological control in parasitoids: a model. *American Naturalist* 148: 807–26.

Neuenschwander, P., Hammond, W. N. O., Gutierrez, A. P., Cudjoe, A. R., Baumgärtner, J. U., Regev, U. & Adjakloe, R. (1989) Impact assessment of the biological control of the cassava mealybug, *Phenacoccus manihoti* Matile Ferrero (Hemiptera: Pseudococcidae) by the introduced parasitoid, *Epidinocarsis lopezi* (DeSantis) (Hymenoptera: Encyrtidae). *Bulletin of Entomological Research* 79: 579–94.

Pedata, P. A. & Hunter, M. S. (1996) Secondary host choice by the autoparasitoid *Encarsia pergandiella*. *Entomologia Experimentalis et Applicata* 81: 207–14.

Price, P. W. (1994) Plant resources as the mechanistic basis for insect herbivore population dynamics. In *Effects of Resource Distribution on Animal–Plant Interactions*, ed. M.D. Hunter, T. Ohgushi & P.W. Price, pp. 139–73. San Diego: Academic Press.

Royama, T. (1971) A comparative study of models for predation and parasitism. *Researches on Population Ecology* Supplement 1: 1–91.

Royama, T. (1992) *Analytical Population Dynamics*. London: Chapman and Hall.

Schreiber, S. R. & Gutierrez, A. P. (1998) A supply–demand perspective of species invasions and coexistence: applications to biological control. *Ecological Modelling* 106: 27–45.

van Dijken, M. J., Neuenschwander, P., van Alphen, J. J. M. & Hammond, W. N. O. (1991) Sex ratios in field populations of *Epidinocarsis lopezi*, an exotic parasitoid of the cassava mealybug in Africa. *Ecological Entomology* 16: 233–40.

Van Lenteren, J. C. (1995) Integrated pest management in protected crops. In *Integrated Pest Management*, ed. D. Dent, pp. 311–43. London: Chapman and Hall.

Waage, J. K. (1990) Ecological theory and the selection of biological control agents. In *Critical Issues in Biological Control*, ed. M. Mackauer, L. E. Ehler & J. Roland, pp. 135–57. Andover: Intercept.

Wang, Y. H. & Gutierrez, A. P. (1980) An assessment of the use of stability analyses in population ecology. *Journal of Animal Ecology* 49: 435–52.

Watt, K. E. F. (1959) A mathematical model for the effects of densities of attacked and attacking species on the number attacked. *Canadian Entomologist* 91: 129–44.

Wermelinger, B., Oertli, J. J. & Baumgärtner, J. (1991) Environmental factors affecting the life table statistics of *Tetranychus urticae* (Acari: Tetranychidae). III. Host plant nutrition. *Experimental and Applied Acarology* 12: 259–74.

Williams, T. (1996) Invasion and displacement of experimental populations of a conventional parasitoid by a heteronomous hyperparasitoid. *Biocontrol Science and Technology* 6: 603–18.

Zchori-Fein, E., Roush, R. T. & Hunter, M. S. (1992) Male production induced by antibiotic treatment in *Encarsia formosa* (Hymenoptera: Aphelinidae), an asexual species. *Experientia* 48: 102–5.

6

The case for indigenous generalists in biological control

GARY C. CHANG AND PETER KAREIVA

A history of neglecting 'generalists' as biological control agents?

Although opportunity usually overrides strategic planning when it comes to pest control, practitioners of biological control have often speculated about the features possessed by a 'perfect' control agent. Many of the hypothetical life-history traits for this idealized agent are self-evident – characteristics such as adaptability to new climates, good dispersal ability, and high reproductive rate. One attribute that is not so clear-cut is feeding habit, or specificity. A common belief has been that specialists should be favored over generalists when searching for, selecting, or promoting biocontrol agents. Indeed, in one of the earliest texts on insect control, Wardle & Buckle (1923) identify specificity as one of the four keys to effective control. This promotion of specificity is still strongly represented in modern writing – for instance Hoy (1994) lists specificity as one of three desirable traits to be sought in biocontrol agents. Of course, not everyone has advocated specialists as ideal biocontrol agents, and there has been a longstanding exchange of ideas about specialists *versus* generalists in theory and practice (Sweetman, 1936; Watt, 1965; Huffaker, 1971; Ehler, 1977; Hassell, 1978; Murdoch *et al.*, 1985; Wilson *et al.*, 1996). In this chapter we revisit this debate. We argue that specialists and generalists each have advantages, and that the question should not be 'which is better?' The more useful question concerns what trade-offs correlate with specialization *versus* generalization, and how do these trade-offs influence biocontrol effectiveness. We use a simple model to highlight the trade-off we think is central to the specialist/generalist choice – increased efficiency of specialists at the expense of not being able to endure periods of pest scarcity, *versus* early

arrival of generalists at the onset of a pest problem, at the expense of lower efficiency as pest 'killers'.

How much attention is given to generalists in the current literature?

For animals, there exists a trophic continuum from feeding on only one or a small number of species, to exploiting many different species. Specialists and generalists are found at the ends of the continuum. Depending upon who is defining them, intermediates may be grouped as either generalists or specialists, or as their own 'oligophagous' category. While the generalist/specialist distinction may be useful in explaining and anticipating outcomes of biocontrol operations, it is not easy to assign a particular natural enemy species to one group or the other (Wardle & Buckle, 1923; Sweetman, 1936). Because even published host lists may contain serious mistakes (Askew & Shaw, 1986; Ramadan *et al.*, 1994), any review that compares specialist *versus* generalists is fraught with errors (Shaw, 1994). In addition, even if there were no errors in the literature regarding host lists, the assignment of a species to a feeding character can be ambiguous. For example, the potential and realized host range of a parasitoid often differs substantially (Shaw, 1994). This type of complication is illustrated by Brower's (1991) study of the parasitoid *Pteromalus cerealellae*, a potential agent for biocontrol of the lepidopteran stored product pest *Sitotroga cerealella*. For decades, *P. cerealellae* had been listed as specific to *S. cerealella*. However, when a laboratory culture of *P. cerealellae* that had been collected from and cultured on *S. cerealella* was tested for its ability to develop on 12 species of beetles found in stored products, adults were obtained from all of the hosts tested. Now it remains to be seen how commonly these beetles are parasitized by *P. cerealellae* outside the laboratory. Nonetheless, a specialist facing sub-optimal conditions may take unusual prey (Hattingh & Samways, 1992) and thereby become much more 'generalized.'

Another source of confusion is that morphologically indistinguishable species may actually contain different strains with different host ranges. Holler (1991) found this in the parasitoid *Aphidius rhopalosiphi*. Five sympatric strains of *A. rhopalosiphi* originated from field collections in Germany (but had been cultured in the laboratory for several generations), while a sixth strain was obtained from France. Three strains were limited to aphids of the genus *Sitobion* as hosts, whereas the other three strains of *A. rhopalosiphi* could develop on three additional aphid genera. Another example of geographic variation was reported by Smith & Hubbes (1986), who tested the ability of the egg parasitoid *Trichogramma minutum* from Maine and five locations in Ontario to parasitize spruce budworm, *Choristoneura fumiferana*. The strain from Maine had from one-seventh to one-third the fecundity of the Ontario strains. Thus, the geographic scale at which biocontrol agents are 'lumped' can determine whether a species is called a specialist or a generalist (Fox & Morrow, 1981). Clearly, with so many considerations, a consensus classification scheme of generalists and specialists is difficult to generate. Yet, keeping these difficulties in mind, ecologists do draw lines between generalists and specialists in the hope of understanding communities, and there is value in making the distinction.

One of our goals for this chapter was to quantify the use of generalist and specialist natural enemies as biocontrol agents. Unfortunately, many (probably most) published biocontrol reports do not define the feeding habits of the species under study. Without such information, we could not satisfactorily determine whether the agent under study should be classified as a generalist or a specialist. However, our impression from scanning the literature (described below) is that specialists garner more attention from biocontrol workers. It is certainly clear that very few experiments compare the contribution of generalists and specialists to biocontrol.

We tallied the number of recent articles involving arthropod biocontrol agents employed against arthropod pests in the following journals: *Environmental Entomology*, *Annals of the Entomological Society of America*, *Entomophaga*, *Bulletin of Entomological Research*, and *Biological Control*. Our survey included any article between the years 1990 and 1994 that alluded to biocontrol, 'economic significance', or 'pests'. Subject areas ranged from toxicology and developmental biology to community and population ecology. Although this 5 year period and portfolio of journals is certainly not all-inclusive, it generated a sampling of 616 articles, which is large enough to support our conclusion. Only six papers (less than 1% of the sample) reported experiments that could compare the contribution of generalists *versus* specialists on pest populations (Table 6.1). Since the few experiments that are reported do not provide compelling evidence for the superiority of specialists, it is clear that relegating generalists to the status of poor control agents cannot be supported by data.

A second dimension to the efficiency of generalists *versus* specialists is the use of native *versus* foreign natural enemies. Classical biological control emphasizes the importance of finding and releasing natural enemies from abroad to control exotic pests that have also come from abroad. If one is committed to classical biological control, generalist biocontrol agents may be undesirable because of their ecological side-effects on non-target species (Simberloff & Stiling 1996). However, there is no reason that biocontrol has to rely on imported species. Before the inception of classical biocontrol, encouraging or augmenting native natural enemies was commonly advocated as a means of accomplishing biological control (Sweetman, 1936). Riley (1931) considered Erasmus Darwin to have been a forerunner of the economic entomologist. Darwin (1800) sketched several proposed aphid control strategies, including the following plan using syrphid fly larvae:

> The most ingenious manner of destroying the aphis would be effected by the propagation of its greatest enemy, the larva of the aphidivorous fly . . . If the eggs could be collected and carefully

Table 6.1. *Experimental studies that have measured the contributions of generalist and specialist natural enemies to biological control. They have ruled in favor of, against, and neutrally on the effectiveness of generalists relative to specialists*

Citation	Generalist sp.	Specialist sp.	Pest sp.	Summary
Cloutier & Johnson, 1993	*Orius tristicolor*	*Phytosieulus persimilis*	*Frankliniella occidentalis*	Non-target effects of the generalist detracted from control
Heinz & Parrella, 1990	*Chrysoperla carnea*	*Diglyphus begini, Encarsia formosa*	*Liriomyza trifolii* and four others	Generalist promoted for role in controlling secondary pests
Heinz & Parrella, 1994a	*Delphastus pusillus*	*Encarsia luteola*	*Bemisia argentifolii*	Both natural enemy species contributed to control
Heinz & Parrella, 1994b	*D. pusillus*	*E. luteola, E. formosa,* and two others	*B. argentifolii*	*D. pusillus* and *E. luteola* deemed the most promising of the potential control agents
Kring *et al.*, 1993	Unspecified (five or fewer)	Unspecified (one)	*Helicoverpa zea*	All natural enemies combined caused <2% mortality
Stam & Elmosa, 1990	At least 21 species	At least 5 species	*Earias insulana, Heliothis armigera,* and *Bemisia tabaci*	High predator populations early in the season played a greater role in controlling *E. insulana* and *H. armigera* than parasitoids

preserved during the winter, and properly disposed on nectarine and peach trees in the early spring, or protected from injury in hot-houses; it is probable, that this plague of the aphis might be counteracted by the natural means of devouring one insect by another; as the serpent of Moses devoured those of the magicians' (Darwin, 1800: p. 356).

Using modern terms, Erasmus Darwin's favored scheme combines elements of conservation and augmentation of a generalist predator for biocontrol. Several other scientists have studied and refined this type of biocontrol (for examples, see Ridgway & Vinson, 1977). However, in the twentieth century, insecticides and classical biocontrol have achieved dramatic successes to become leading pest management options (Hoy, 1994; Metcalf, 1994). The greater amount of effort put into finding specialist classical biocontrol agents may contribute to native generalists being under-utilized for pest management. The neglect of native enemies in biocontrol programs has been noted before (e.g., Tisdell, 1990), but this situation may soon change because of a worldwide heightened awareness about the ecological risks of exotic species (Howarth, 1991; Lockwood, 1993a,b; Nafus, 1993; Vitousek *et al.*, 1996; Simberloff & Stiling, 1996).

What is the verdict from studies assessing generalists in biocontrol?

Before discussing the benefits of generalist enemies, it is worth recognizing some of their unique risks. One risk is the possibility that generalists may become invasive and displace native biodiversity. For example, the seven-spotted ladybeetle, *Coccinella septempunctata*, is a generalist that was introduced to North America by the United States Department of Agriculture (USDA) for control of aphid pests. This imported predator has been so successful in terms of its population expansion (not necessarily aphid control) that some entomologists worry that it is replacing native coccinellids (Elliot *et al.*, 1996). A second risk of generalists is that they might attack other natural enemies and thereby interfere with biocontrol efforts. For instance, hemipteran predators feed on larvae of lacewings as well as on aphids, and can actually increase aphid populations (Rosenheim *et al.*, 1993). Clearly the non-target effects of generalist predators must be considered before they are successfully employed in biocontrol.

Despite their drawbacks, generalists often have been successful biocontrol agents. In some cases generalists are the only biocontrol option, because no specialists attacking a particular pest are known. Such was the case with the psyllid, *Heterapsylla cubana*, in New Caledonia. The generalist coccinellid *Olla v-nigrum* was the only available natural enemy in New Caledonia, where the psyllid was a pest on *Leucaena* (Chazeau *et al.*, 1991). Similarly, over the entire world, only the generalist mirid predator *Stethoconus japonicus*, has been found to prey upon the azalea lace bug, *Stephanitis pyrioides* (Neal *et al.*, 1991).

In some settings where both generalist and specialist natural enemies are attacking a pest, empirical evidence has found that the generalists contribute more to control. One way to obtain this type of evidence is to apply an insecticide that differentially affect generalists and specialists. For example, DeBach (1946) serendipitously produced data on the contributions of generalists and specialists to biocontrol *via* different insecticide treatments. Citrus trees infested with the long-tailed mealybug, *Pseudococcus longispinus*, were sprayed with either DDT (dichlorodiphenyltrichloroethane) or talc, while unsprayed trees served as controls. After treatments, mealybugs, predators, and percent parasitism were monitored from April to July. DDT effectively eliminated predators for 2 months but percent parasitism remained high. In contrast, talc allowed predators to remain on trees but severely reduced the parasitism rate. The populations of *P. longispinus* were higher on DDT-treated trees (generalist enemies removed) in July than on talc-treated (specialist enemies removed) or control trees (no enemies removed). Furthermore, mealybug numbers had increased over the 4 months on the DDT-treated trees but declined during this time on the other trees.

DeBach concluded that generalists were the major factor limiting *P. longispinus* populations.

Later experiments have increased in scale and sophistication. Some have found generalists to cause more pest mortality than specialists, others studies have found the opposite. Six years of observations, compiling and analyzing life tables, and performing insecticide checks led Ehler (1977) to conclude that generalists are more important than specialists in controlling the cabbage looper, *Trichoplusia ni*, on cotton in California. In particular, a complex of generalist predators, especially hemipterans, killed more *T. ni* than parasitoids or pathogens. Cotton is grown as an annual in California; thus it is a regularly disturbed agroecosystem. Ehler suggested that in such systems, generalists are essential because of their ability to colonize the habitat. An assortment of prey items allows generalist predators to keep high populations during times when *T. ni* populations are low. When *T. ni* populations begin increasing, the generalists are immediately available to act as population checks.

Murdoch *et al.* (1985) also emphasize that natural enemies that can persist in an area in the absence of the pest could be effective in biocontrol. They noted that the desirable attributes of biocontrol agents will vary based on the nature of the pest problem. Thus, when a low, stable pest equilibrium is the management goal, many mathematical models predict that specialists will provide better control (Hassell, 1978). In contrast, Murdoch *et al.* (1985) presented case studies showing that, in practice, successful control can be achieved without a stable equilibrium. They offered two alternative biocontrol strategies, dubbed 'lying-in-wait' and 'search-and-destroy'. The search-and-destroy strategy entails local extinction but global persistence of natural enemies and pests. The features of specialists are desirable in such a system. The lying-in-wait strategy promotes preventative pest control *via* the perpetual local presence of a natural enemy.

The persistence of generalists when pests are not present depends largely on alternative food sources. Plant products, detritus, and sugar solutions provided by humans can serve as alternate food for certain natural enemies (Butler & Ritchie, 1971; Ben

Saad & Bishop, 1976; Hagen, 1976). Furthermore, many generalist arthropods can survive periods when no other species are present by engaging in cannibalism. For instance, the lifetime reproductive output of lacewings is great enough that in the absence of any prey five of a female's progeny can survive to adulthood by eating only their siblings (Canard *et al.*, 1984). Of course, to be an effective biocontrol agent, a species needs to be able to do more than simply persist.

Various physiological and behavioral adaptations allow spiders to persist for long periods without prey (Riechert, 1992). For example, generally low metabolic rates can be decreased even further during food shortages, while very large meals are exploited by external digestion to minimize waste and maximal storage is achieved in a highly branching midgut and distended abdomen (Riechert, 1992). Yet the value of spiders in biocontrol is debatable. In part, they have been neglected due to poor knowledge of their ecology (Nyffeler *et al.*, 1994). Observations of pest outbreaks despite the presence of spiders have also caused some to disregard them as biocontrol agents (Foelix, 1996). On the other hand, spiders often possess the traits of 'lying-in-wait' biocontrol agents (Riechert & Lockley, 1984; Nyffeler *et al.*, 1994). Indeed, 30 laboratory or field studies have been compiled by Riechert & Lockley (1984) that recorded significant spider predation on pests in various agroecosystems.

Tabulating mortality is not the most compelling evidence, and experimental manipulation of spider populations would provide greater insight. A good example of such an experiment was performed by Riechert & Bishop (1990), who manipulated spider populations in potato, cabbage, and brussels sprout gardens in Tennessee. They augmented spider populations by an order of magnitude primarily by laying mulch around crop plants. This treatment raised the humidity around the plants, making the habitat more hospitable for the spiders. Pest insect populations fell by as much as 80% and damage to plants plummeted as much as 90% in plots with more spiders. Beyond showing that spiders can be useful in biocontrol, another lesson from Riechert & Bishop's study is that even when generalist natural enemies

are already found lying-in-wait, simple actions can be taken to enhance their populations and their efficacy.

Conversely, disrupting biocontrol agents that were lying-in-wait has contributed to secondary pest outbreaks. One example involves the rice brown planthopper, *Nilaparvata lugens*, which emerged as a pest in rice in the 1970s. In Indonesian rice, Settle *et al.* (1996) have demonstrated that a complex of indigenous generalist predators is responsible for naturally controlling *N. lugens*. Pesticide applications lower early-season predator abundance; subsequently, pests undergo a sudden population increase later in the growing season. Without pesticide use, the predator population feeds on detritivores and increases prior to the rise in herbivores. The maintenance of a larger predator population in rice paddies prevents the pests from achieving outbreak levels. Settle and his coworkers have also identified other land use practices that can boost predacious arthropod populations in rice, such as asynchronous, small-scale planting. Thus, anthropogenic disturbances are reduced to allow natural enemies to control pests.

A model that focuses on specialist versus generalist trade-offs

In order to evaluate the relative merits of generalists versus specialists, we need models that formalize the advantages hypothesized to be associated with each strategy. The major advantage of specialization typically is an assumed enhanced efficiency at locating and killing prey or hosts compared to generalist enemies (Rogers & Hubbard, 1974). Wilson *et al.* (1996) have shown that the degree of difference in searching efficiency can be a critical parameter in community models of coexistence and invasibility. Specialization could also produce more efficient conversion of consumed prey into growth or reproduction. The major advantage of a generalist strategy is the possibility of alternative food sources, so that generalist populations can hang on through periods of pest scarcity or even absence (e.g., Stam & Elmosa, 1990; Settle *et al.*, 1996). Thus, at the onset of a pest outbreak, generalists may already be lurking in the area, making their living from alternative food, whereas specialists probably have to migrate into the area from afar. We sought to incorporate these ideas into the simplest possible model. We begin by describing pest population dynamics (with V denoting pest density) as simple exponential growth (at rate r) minus mortality due to predators or parasites (with P denoting predator or parasite density) at a rate m:

$$\frac{\mathrm{d}V}{\mathrm{d}t} = rV - mPV \tag{6.1}$$

The equation for predator or parasite dynamics is slightly more complicated because it is intended to describe the recruitment of new enemies to a field as well as emigration out of a field:

$$\frac{\mathrm{d}P}{\mathrm{d}t} = \frac{iV}{(s+V)} - EP \tag{6.2}$$

where the positive term is a saturating recruitment curve, such that more enemies join the population per unit time as pest density (V) increases. The parameter i of this recruitment function defines the asymptotic rate of natural enemy arrival as V approaches infinity, while s establishes the number of prey at which the enemy arrival rate is half of its maximum. There is assumed to be a steady emigration rate, E, describing the departure of enemies or density-independent mortality. This simple model then describes an exponentially growing prey population attacked by enemies that migrate into the field at a rate that increases with pest density and that depart the field at a fixed rate. Increased specialization is represented in this model by larger values of m, the net killing rate coefficient. Increased generalization is reflected in lower values of m and higher initial natural enemy populations.

To implement the model we numerically solved the differential equations and tracked populations over single growing seasons, a departure from the traditional approach to predator–prey modeling (which emphasizes dynamics over many years and the stability of interactions, see Hassell, 1978). We chose a single-season timespan because in biocontrol situations, insect populations in one year are often poorly correlated with populations the previous year, and the concerns of a farmer are typically expressed on the basis of a single growing season. In addition,

Table 6.2. *Parameter combinations for single season simulations of biocontrol*

Parameter	High fecundity pest value	Low fecundity pest value
r (day^{-1})	0.4	0.2
initial V (individuals)	50	50
i (day^{-1})	10	10
s (individuals)	30	15
E (day^{-1})	4	8
range of m (individuals^{-1} day^{-1})	0.001–0.601	0.001–0.601
range of initial P (individuals)	0–24	0–24

in temperate agriculture, between-season insect population dynamics are often unlikely to be controlled by biotic factors such as predator density. Indeed, the factors that cause pest mortality during the non-growing season in North America are often weather related. For example, for the Russian wheat aphid, *Diuraphis noxia*, winter survival is primarily influenced by soil surface temperatures, accumulated subzero temperatures, and duration of snow cover (Armstrong & Peairs, 1996). In contrast, predators and parasitoids play a major role in *D. noxia* population dynamics during the growing season (see e.g., Hopper *et al.*, 1995).

In Table 6.2 we present the combination of parameters used for our numerical simulations. Of these parameters, only pest population growth rate (r), was obtained from the literature (see Wennergren, 1992). Two sets of parameter values representing different but plausible biocontrol scenarios were used. The first set of parameters represents a system with a high fecundity pest and relatively low-dispersive natural enemies, while the second set depicts a lower fecundity pest with more mobile natural enemies. In the second parameter set, the natural enemies are more likely to leave a field with fewer prey as well as to respond faster to an abundance of prey.

To summarize the results for each run, we used total accumulated 'pest days' at the end of the growing season as the response variable. This

(a)

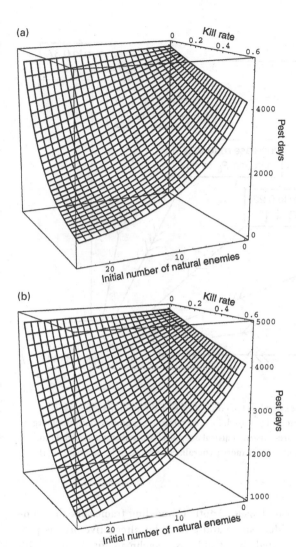

(b)

Table 6.3. *Consequences of doubling either* m *or* P_0 *on pest days*

	$m=0.121$	$m=0.241$
$P_0=6$	4491 (100%)	3763 (84%)
$P_0=12$	3939 (88%)	

Note:

Increasing m gives a greater decline in pests, but P_0 also has an effect.

number was obtained by summing the pest population for each day over the course of the simulated growing season. The number represents the total feeding damage a crop would have experienced, and makes good sense from an economic perspective. To represent changes in pest damage as a function of 'generalist' *versus* 'specialist' life histories, we plotted pest days *versus* m and P_0. Recall that high m represents specialization and high P_0 represents generalization. Importantly, the resulting three-dimensional graphs (Fig. 6.1) were qualitatively similar for both parameter sets corresponding to the two different biocontrol scenarios. The greatest amount of pest damage occurs where m and P_0 are lowest. Increasing either m or P_0 decreases pest days, but changing m leads to a more rapid drop in pest days. This can be illustrated by selecting a point from the output and moving up along the m axis. This simulates the outcome of replacing a pre-existing natural enemy with a new, more efficient biocontrol agent. This is then compared with tracing a path from the initially selected point to a higher value of P_0, which represents either replacement with a more generalized agent, or augmenting the numbers of the natural enemy that is already present. For example, from the simulation with the first parameter set, 4491 pest days are accumulated when $m=0.121$ and $P_0=6$ (Table 6.3). When P_0 is held steady, but m is doubled, the number of pest days falls to 3763, or 84% of the reference value. If m is held steady and P_0 is doubled, the number of pest days falls to 88% of what it was. However, not surprisingly, increasing P_0 can compensate for a low m. That is, enough generalists can always equal a specialist. To illustrate this point,

Figure 6.1. Results of simulations examining the trade-off between a lying-in-wait strategy *versus* a search-and-destroy strategy. The x-axis is the value of m, the mortality each natural enemy causes in the pest population. The z-axis is the number of natural enemies in the field at the start of a season (P_0). Increasing m represents increasingly specialized natural enemies, while increasing P_0 represents more generalized natural enemies. The response variable, 'pest days,' is on the y-axis and reflects the total pest load over the entire growing season. (a) Plots the results from the simulation of a high fecundity pest and less dispersive natural enemies, while (b) is from the simulation of a low fecundity pest with more mobile natural enemies. The graphs are qualitatively similar.

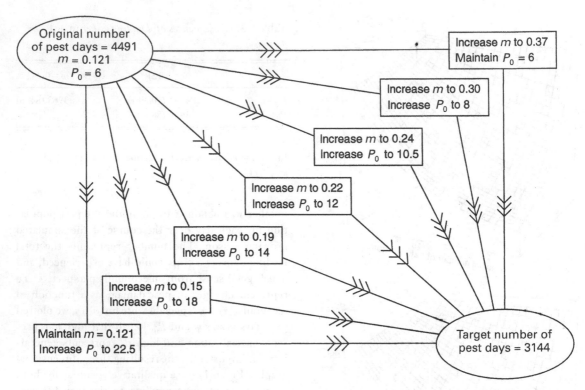

Figure 6.2. Simulation results illustrating different routes to the same level of control. When a reduction in the pest is desired, the model requires m and/or P_0 of the biocontrol agent to be increased. Several parameter combinations for a target reduction of 30% are shown. Increasing m reflects replacing the pre-existing natural enemies with a more specialized biocontrol agent. Increasing P_0 can be accomplished either by replacement with a more generalized species, or by augmenting the natural enemies already present.

consider once again the 4491 pest day reference point from $m = 0.121$ and $P_0 = 6$. A desired 30% decrease (to 3144) in pest days can be achieved by finding a different natural enemy with $m = 0.37$, or by increasing P_0 to 22.5 (Fig. 6.2). Obviously, particular numerical results are not the point of this model; rather, the model illustrates the key trade-off (m *versus* P_0) that determines the relative advantages of specialist *versus* generalist biocontrol agents. More importantly, this model directs our attention to attributes of biocontrol agents that should be measured under field conditions: killing efficiency and immigration rate. Only when these parameters are measured, can we really deduce anything about generalists *versus* specialists.

This model is a simple exploration of a compli-

cated issue. Another modeling framework that considers space and a term for alternative prey is presented by Walde & Nachman in Chapter 10. Alternative prey could be added to our model in the form of an additional equation(s). It would be important to ascertain what happens to these populations, whether they are considered additional pests or non-target species. Alternative prey could also affect the level of control of the original target pest by a generalist. If the target pest 'wins' in apparent competition against the alternative prey, biocontrol would diminish; if the alternative prey 'wins,' biocontrol by a generalist would improve (Holt & Lawton, 1993, 1994).

Another general consideration in biocontrol is the type and magnitude of refuge from natural enemies for the pest species (Hawkins *et al.*, 1993). Including a refuge parameter into our model could result in additional benefits from using a generalist biocontrol agent. This is easiest to envision when an absolute number of prey are always invulnerable to natural enemies. If pest refuges were equal for a generalist and a specialist natural enemy, then the generalist might provide superior biocontrol. This is because while the pest would always be present in a refuge, its natural enemies could go extinct. Specialist natural enemies would be more likely than generalists to go extinct from not being able to secure enough prey, and the surviving pest population could surge forth from their refuge. In contrast, generalists would remain present and able to stunt such refuge-centered pest population increases. However, it may also be the case that prey species would be afforded smaller refuges from specialist enemies. Then it becomes another trade-off, this time between the persistence of the generalist and the smaller refuge from the specialist.

Implications

Generalist natural enemies may not be as effective *per capita* as specialists, but they can compensate for this deficiency in pest control schemes by being present earlier in the season (Nyffeler *et al.*, 1994; Settle *et al.*, 1996). We have presented one mathematical explanation for why this might be the case. The crucial biological information required when comparing a species of generalist to a specialist for biocontrol is the trade-off between P_0 and m. For example, to attain a 20% increase in m, what reduction in P_0 is likely? Unfortunately, this key information is totally lacking from the literature.

Of course, not all constraints on biocontrol are ecological. Although the ease of culturing an agent should not be a primary selection criterion (Waage, 1990), it is a factor to be considered. A model like ours might suggest whether it is worth devoting additional effort in promoting a specialist if a more generalized species is easier to rear. An example with which we are familiar involves *Coccinella septempunctata* and the lacewing *Chrysoperla plorabunda*. In standardized laboratory trials, individual *C. septem-*

punctata larvae consume more bean aphids, *Aphis fabae*, than individual *Ch. plorabunda* (Chang, 1996). However, *C. septempunctata* is not necessarily the better biocontrol agent, because it is much more difficult to culture. *Ch. plorabunda* adults are not predatory and reproduce readily on a simple diet of honey, whey, and yeast. In contrast, *C. septempunctata* are predaceous throughout their lives and require aphids to lay eggs. Here again, the biocontrol agent with a lower m may prove to be preferable.

A different perspective on the issue of generalists and specialists in biocontrol emphasizes that instead of choosing one over the other, the best solution might be to have both working together. Our model has little to say on this issue, but it is a promising direction for further research. For example, the larvae of the syrphid *Metasyrphus corollae* display a preference for prey with a thin cuticle (Ruzicka, 1976), which results in their avoiding parasitized aphids near mummification and mummies (Kindlmann & Ruzicka, 1992). Kindlmann & Ruzicka (1992) explored the syrphid–parasitoid–aphid system using an experiment with which they parameterized a model. First, they set out two cages in a field of brussels sprouts that excluded syrphids but allowed passage of aphids and parasitoids. After aphid populations reached 20 000 in the experimental cage, syrphids were introduced. The other cage served as a control (no syrphid) treatment. They monitored aphid populations and percent parasitism.

Kindlmann & Ruzicka (1992) then fitted a simulation model to the data from the cages with percent parasitism as the output variable. In their model, aphid populations began on day 0 without natural enemies. On day 7, parasitoids entered the system. Syrphids fed for only 17 days after being introduced into the system; after that time, aphids and parasitoids remained in the system. After the model was parameterized, it was further explored by varying the timing of syrphid release. This greatly affected the model's results. When syrphids were released on day 26, aphid populations rapidly reached 100% parasitism (extinction presumably follows). The aphid population at 26 days is small enough so that when the syrphids are introduced, all non-parasitized aphids are consumed by the predators. If syrphids are released on day 27, the aphid population is too large to be completely consumed by them. But the parasitoids in the system can parasitize all of the aphids that remain after the predator feeding period, so again, aphids are eliminated from the system. A syrphid release on day 28, however, resulted in only about 10% parasitism, allowing the persistence of aphids in the system. The aphid population that escapes predation is too large for the parasitoids to suppress.

Kindlmann & Ruzicka (1992) assumed that syrphids entirely avoided parasitized aphids, but this may not be so. Interference between natural enemies, and generalists and specialists in particular, has been quantified in a few systems. When such interference occurs, it obviously lowers the level of control that would otherwise be expected when using natural enemies in combination. For example, Cloutier & Johnson (1993) found that the insect predator *Orius tristicolor* significantly interfered with a predatory mite, *Phytoseiulus persimilis*, which controls spider mites in greenhouses. Field observations of parasitoid *Aphytis* spp. in almonds found that they faced a considerable risk of predation from spiders, ants, and the assassin bug, *Zelus renardii*, all generalist predators (Heimpel *et al.*, 1997). Thus, the debate over whether multiple species of biocontrol agents provide better control is unresolved (Hassell, 1978; Ehler & Hall, 1982; Kakehashi *et al.*, 1984; Hassell & May, 1986; Myers *et al.*, 1989; Briggs, 1993; Hochberg, 1996), and the effectiveness of generalists with specialists is an open area of research.

Conclusions

Generalists may too often be discounted because too much attention is given to their liabilities, without allowing for their 'lie-in-wait' advantages. It is clear that we require empirical data on the trade-off between early season abundances and kill rates of natural enemies. This information will be useful for determining the relative merits of generalist and specialist biocontrol agents in specific cases. Indeed, the imbalance in research devoted to generalists and specialists may arise from forcing generalists into search-and-destroy situations, when they are more

appropriately used as preventative 'lie-in-wait insurance' against pest outbreaks.

Considering the 'lie-in-wait' benefits of generalists also enhances the comparison of augmentative and classical biocontrol approaches. Indigenous predators have received less attention than exotic parasitoids in biocontrol, although they have proven capable of providing excellent pest control in some agricultural systems. We need to consider whether it is worth the effort to search for exotic specialists for biocontrol when, with a few adjustments in management practices, native generalists may suffice.

Acknowledgements

We thank Bradford Hawkins and two anonymous reviewers for comments that improved the manuscript.

References

Askew, R. R. & Shaw, M. R. (1986) Parasitoid communities: their size, structure and development. In *Insect Parasitoids*, ed. J. K. Waage & D. J. Greathead, pp. 225–64. San Diego: Academic Press.

Armstrong, J. S. & Peairs, F. B. (1996) Environmental parameters related to winter mortality of the Russian wheat aphid (Homoptera: Aphididae): basis for predicting mortality. *Journal of Economic Entomology* 89: 1281–7.

Ben Saad, A. A. & Bishop, G. W. (1976) Effect of artificial honeydews on insect communities in potato fields. *Environmental Entomology* 5: 453–7.

Briggs, C. J. (1993) Competition among parasitoid species on a stage-structured host and its effect on host suppression. *American Naturalist* 141: 372–97.

Brower, J. H. (1991) Potential host range and performance of a reportedly monophagous parasitoid, *Pteromalus cerealellae* (Hymenoptera: Pteromalidae). *Entomological News* 102: 231–5.

Butler G. D., Jr & Ritchie, P. L., Jr (1971) Feed Wheast and the abundance and fecundity of *Chrysopa carnea*. *Journal of Economic Entomology* 64: 933–4.

Canard, M., Semeria, Y. & New, T. R. (eds) (1984) *Biology of Chrysopidae*. The Hague: Junk.

Chang, G. C. (1996) Comparison of single versus multiple species of generalist predators for biological control. *Environmental Entomology* **25**: 207–12.

Chazeau, J., Bouyé, E. & Bonnet de Larbogne, L. (1991) Cycle de développement et table de vie d'*Olla v-nigrum* [*Col.: Coccinellidae*] ennemi naturel d'*Heteropsylla cubana* [*Hom.: Psyllidae*] introduit en Nouvell-Calédonie. *Entomophaga* **36**: 275–85.

Cloutier, C. & Johnson, S. G. (1993) Predation by *Orius tristicolor* (Hemiptera: Anthocoridae) on *Phytoseiulus persimilis* (Acarina: Phytoseiidae): testing for compatibility between biocontrol agents. *Environmental Entomology* **22**: 477–82.

Darwin, E. (1800) *Phytologia; or the Philosophy of Agriculture and Gardening.* London: J. Johnson.

DeBach, P. (1946) An insecticidal check method for measuring the efficacy of entomophagous insects. *Journal of Economic Entomology* **39**: 695–7.

Ehler, L. E. (1977) Natural enemies of cabbage looper on cotton in the San Joaquin Valley. *Hilgardia* **45**: 73–106.

Ehler, L. E. & Hall, R. W. (1982) Evidence for competitive exclusion of introduced natural enemies in biological control. *Environmental Entomology* **11**: 1–4.

Elliot, N., Kieckhefer, R. & Kauffman, W. (1996) Effects of an invading coccinellid on native coccinellids in an agricultural landscape. *Oecologia* **105**: 537–44.

Foelix, R. F. (1996) *Biology of Spiders*, 2nd edn. New York: Oxford University Press.

Fox, L. R. & Morrow, P. A. (1981) Specialization: species property or local phenomenon? *Science* **211**: 887–93.

Hagen, K. S. (1976) Role of nutrition in insect management. *Proceedings Tall Timbers Conference on Ecological Animal Control and Habitat Management* **6**: 221–61.

Hassell, M. P. (1978) *The Dynamics of Arthropod Predator–Prey Systems.* Princeton: Princeton University Press.

Hassell, M. P. & May, R. M. (1986) Generalist and specialist natural enemies in insect predator–prey interactions. *Journal of Animal Ecology* **55**: 923–40.

Hattingh, V. & Samways, M. J. (1992) Prey choice and substitution in *Chilocorus* spp. (Coleoptera: Coccinellidae). *Bulletin of Entomological Research* **82**: 327–34.

Hawkins, B. A., Thomas, M. B. & Hochberg, M. E. (1993) Refuge theory and biological control. *Science* **262**: 1429–32.

Heinz, K. M. & Parrella, M. P. (1990) Biological control of insect pests on greenhouse marigolds. *Environmental Entomology* **19**: 825–35.

Heinz, K. M. & Parrella, M. P. (1994a) Biological control of *Bemisia argentifolii* (Homoptera: Aleyrodidae) infesting *Euphorbia pulcherrima*: evaluations of releases of *Encarsia luteola* (Hymenoptera: Aphelinidae) and *Delphastus pusillus* (Coleoptera: Coccinellidae). *Environmental Entomology* **23**: 1346–53.

Heinz, K. M. & Parrella, M. P. (1994b) Poinsettia (*Euphorbia pulcherrima* Willd. ex Koltz.) cultivar-mediated differences in performance of five natural enemies of *Bemisia argentifolii* Bellows and Perring, n. sp. (Homoptera: Aleyrodidae). *Biological Control* **4**: 305–18.

Heimpel, G. E., Rosenheim, J. A. & Mangel, M. (1997) Predation on adult *Aphytis* parasitoids in the field. *Oecologia* **110**: 346–52.

Hochberg, M. E. (1996) Consequences for host population levels of increasing natural enemy species richness in classical biological control. *American Naturalist* **147**: 307–18.

Holler, C. (1991) Evidence for the existence of a species closely related to the cereal aphid parasitoid *Aphidius rhopalosiphi* De Stefani-Perez based on host ranges, morphological characters, isoelectric focusing banding patterns, cross-breeding experiments and sex pheromone specificities (Hymenoptera, Braconidae, Aphidiinae). *Systematic Entomology* **16**: 15–28.

Holt, R. D. & Lawton, J. H. (1993) Apparent competition and enemy-free space in insect host–parasitoid communities. *American Naturalist* **142**: 623–45.

Holt, R. D. & Lawton, J. H. (1994) The ecological consequences of shared natural enemies. *Annual Review of Ecology and Systematics* **25**: 495–520.

Hopper, K. R., Aidara, S., Agret, S., Cabal, J., Coutinot, D., Dabire, R., Lesieux, C., Kirk, G., Reichert, S., Tronchetti, F. & Vidal, J. (1995) Natural enemy impact on the abundance of *Diuraphis noxia* (Homoptera: Aphididae) in wheat in southern France. *Environmental Entomology* **24**: 402–8.

Howarth, F. G. (1991) Environmental impacts of classical biological control. *Annual Review of Entomology* **36**: 485–509.

Hoy, M. A. (1994) Parasitoids and predators in management of arthropod pests. In *Introduction to Insect Pest Management*, ed. R. L. Metcalf & W. H. Luckmann, pp. 129–98. New York: John Wiley & Sons.

Huffaker, C. B., Messenger, P. S. & DeBach, P. (1971) The natural enemy component in natural control and the theory of biological control. In *Biological Control*, ed. C. B. Huffaker, pp. 16–67. New York: Plenum Press.

Kakehashi, N., Suzuki, Y. & Iwasa, Y. (1984) Niche overlap of parasitoids in host–parasitoid systems: its consequence to single versus multiple introduction controversy in biological control. *Journal of Applied Ecology* **21**: 115–31.

Kindlmann, P. & Ruzicka, Z. (1992) Possible consequences of a specific interaction between predators and parasites of aphids. *Ecological Modelling* **61**: 253–65.

Kring, T. J., Ruberson, J. R., Steinkraus, D. C. & Jacobson, D. A. (1993) Mortality of *Helicoverpa zea* (Lepidoptera: Noctuidae) pupae in ear-stage field corn. *Environmental Entomology* **22**: 1338–43.

Lockwood, J. A. (1993a) Environmental issues involved in biological control of rangeland grasshoppers (Orthoptera: Acrididae) with exotic agents. *Environmental Entomology* **22**: 503–18.

Lockwood, J. A. (1993b) Benefits and costs of controlling rangeland grasshoppers (Orthoptera: Acrididae) with exotic organisms: search for a null hypothesis and regulatory compromise. *Environmental Entomology* **22**: 904–14.

Metcalf, R. L. (1994) Parasitoids and predators in management of arthropod pests. In *Introduction to Insect Pest Management*, ed. R. L. Metcalf & W. H. Luckmann, pp. 245–314. New York: John Wiley & Sons.

Murdoch, W. M., Chesson, J. & Chesson, P. L. (1985) Biological control in theory and practice. *American Naturalist* **125**: 344–66.

Myers, J. H., Higgins, C. & Kovacs, E. (1989) How many insect species are necessary for the biological control of insects? *Environmental Entomology* **18**: 541–7.

Nafus, D. M. (1993) Movement of introduced biological control agents onto nontarget butterflies, *Hypolimnas* spp. (Lepidoptera: Nymphalidae). *Environmental Entomology* **22**: 265–72.

Neal, J. W. Jr, Haldemann, R. H. & Henry, T. J. (1991) Biological control potential of a Japanese plant bug *Stethoconus japonicus* (Heteroptera: Miridae), an adventitive predator of the azalea lace bug (Heteroptera: Tingidae). *Annals of the Entomological Society of America* **84**: 287–93.

Nyffeler, M., Sterling, W. L. & Dean, D. A. (1994) How spiders make a living. *Environmental Entomology* **23**: 1357–67.

Ramadan, M. M., Wong, T. T. Y. & Herr, J. C. (1994) Is the oriental fruit fly (Diptera: Tephritidae) a natural host for the opiine parasitoid *Diachasmimorpha tryoni* (Hymenoptera: Braconidae)? *Environmental Entomology* **23**: 761–9.

Ridgway, R. L. & Vinson, S. B. (eds) (1977) *Biological Control by Augmentation of Natural Enemies*. New York: Plenum Press.

Riechert, S. E. (1992) Spiders as representative 'sit-and-wait' predators. In *Natural Enemies: The Population Biology of Predators, Parasites and Diseases*, ed. M. J. Crawley, pp. 313–28. Oxford: Blackwell Scientific.

Riechert, S. E. & Bishop, L. (1990) Prey control by an assemblage of generalist predators: spiders in garden test systems. *Ecology* **71**: 1441–50.

Riechert, S. E. & Lockley, T. (1984) Spiders as biological control agents. *Annual Review of Entomology* **29**: 299–320.

Riley, W. A. (1931) Erasmus Darwin and the biologic control of insects. *Science* **73**: 475–6.

Rogers, D. & Hubbard, S. (1974) How the behaviour of parasites and predators promotes population stability. In *Ecological Stability*, ed. M. B. Usher & M. H. Williamson, pp. 99–119. London: Chapman and Hall.

Rosenheim, J. A., Wilhoit, L. R. & Armer, C. A. (1993) Influence of intraguild predation among generalist insect predators on the suppression of an herbivore population. *Oecologia* **96**: 439–49.

Ruzicka, Z. (1976) Prey selection by larvae of *Metasyrphus corollae* (Diptera, Syrphidae). *Acta Entomologica Bohemoslovaca* **73**: 305–11.

Settle, W. H., Ariawan, H., Astuti, E. T., Cahyana, W., Hakim, A. L., Hindayana, D., Lestari, A. S., Pajarningsih & Sartanto (1996) Managing tropical rice pests through conservation of generalist natural enemies and alternative prey. *Ecology* **77**: 1975–88.

Shaw, M. R. (1994) Parasitoid host ranges. In *Parasitoid Community Ecology*, ed. B. A. Hawkins & W. Sheehan, pp. 111–44. New York: Oxford University Press.

Simberloff, D. & Stiling, P. (1996) How risky is biological control? *Ecology* **77**: 1965–74.

Smith, S. M. & Hubbes, M. (1986) Strains of the egg parasitoid *Trichogramma minutum* Riley. *Journal of Applied Entomology* **101**: 223–39.

Stam, P. A. & Elmosa, H. (1990) The role of predators and parasites in controlling populations of *Earias insulana*, *Heliothis armigera* and *Bemisia tabaci* on cotton in the Syrian Arab Republic. *Entomophaga* **35**: 315–27.

Sweetman, H. L. (1936) *The Biological Control of Insects*. Ithaca, NY: Comstock.

Tisdell, C. A. (1990) Economic impact of biological control of weeds and insects. In *Critical Issues in Biological Control*, ed. M. Mackauer, L. E. Ehler & J. Roland, pp. 301–16. Andover: Intercept.

Vitousek, P. M., D'Antonio, C. M., Loope, L. L. & Westbrooks, R. (1996) Biological invasions as global environmental change. *American Scientist* **84**: 468–78.

Waage, J. K. (1990) Ecological theory and the selection of biological control agents. In *Critical Issues in Biological Control*, ed. M. Mackauer, L. E. Ehler & J. Roland, pp. 135–57. Andover: Intercept.

Wardle, R. A. & Buckle, P. (1923) *The Principles of Insect Control*. New York: University of Manchester Press.

Watt, K. E. F. (1965) Community stability and the strategy of biological control. *Canadian Entomologist* **97**: 887–95.

Wennergren, U. (1992) *Population Growth and Structure in a Variable Environment*. Linkoping Studies in Science and Technology Dissertations no. 293. Vimmerby, Sweden: VTT-Grafiska.

Wilson, H. B., Hassell, M. P. & Godfray, H. C. J. (1996) Host–parasitoid food webs: dynamics, persistence, and invasion. *American Naturalist* **148**: 787–806.

7

Why is the parasitoid *Encarsia formosa* so successful in controlling whiteflies?

JOOP C. VAN LENTEREN AND HERMAN J. W. VAN ROERMUND

Introduction

Of the 1200 described whitefly species, only some 20 may be considered as potential pests. Until recently most research on control of whiteflies was directed towards the greenhouse whitefly (*Trialeurodes vaporariorum* (Westwood)) (van Lenteren & Woets, 1988). However, since the mid-1980s, whiteflies of the genus *Bemisia* have created problems of such a large scale both in the field and in large, commercial greenhouses that research has shifted to this pest (Gerling, 1990; Gerling & Mayer, 1996; van Lenteren *et al.*, 1997). Greenhouse whitefly (*Trialeurodes vaporariorum*) and silverleaf whitefly (*Bemisia argentifolii*) are very common, highly polyphagous pest insects worldwide. Biological control of greenhouse whitefly with the parasitoid *Encarsia formosa* was first applied in 1926 and has since been used with great commercial success.

Usually, in natural ecosystems and agroecosystems where pesticides are not (or selectively) used, an array of natural enemies keeps the number of whiteflies at very low levels: predators, parasitoids, and pathogens all take their toll. Examples from cotton and tomato show that both whitefly species can be kept at densities well below the economic threshold (van Lenteren *et al.*, 1996). If natural control is insufficient, inoculative or inundative releases with natural enemies can be made. Commercial biological control of greenhouse whitefly through releases of the parasitoid *Encarsia formosa* Gahan is used at present on about 5000 ha of greenhouse crops and in most countries with an important greenhouse industry (van Lenteren, 1995). Although other types of natural enemies of greenhouse whitefly are known, such as predators and entomopathogenic fungi (Gerling, 1990), it is

mainly introductions of parasitoids that have led to economically feasible control.

Biological control of *Bemisia* species is still problematic and a worldwide search for effective natural enemies is currently being undertaken (Gerling & Mayer, 1996). Much of this research is opportunistic, and evaluation of the control potential of natural enemies is often a purely empirical process. Neither the farmer nor the scientist is much helped by such an approach, because many projects are terminated prematurely if success is not obtained quickly, and thus, scientific insight is prevented. Therefore, our biological control approach is directed towards obtaining insight into the functioning of natural enemies in agroecosystems and in developing criteria for scientifically sound selection of natural enemies (Vet & Dicke, 1992; van Lenteren, 1993). It is often claimed by biological control practitioners that such an approach slows down the implementation of practical biological control programs. That this is not the case, is illustrated by the successes we have obtained in greenhouses where, during the past 25 years, some 40 natural enemy species have been evaluated that are now used commercially for pest control (van Lenteren, 1995). Modeling has always formed an essential part of our approach (e.g., Hulspas-Jordaan & van Lenteren 1989, Yano *et al.*, 1989; van Roermund, 1995; van Steenis, 1995).

One of the problems we faced with whitefly control by *E. formosa* was that in some greenhouse crops (e.g., tomato) results were very reliable, while on other important crops (e.g., cucumber) control often failed. Another, more recent problem, is that *E. formosa* does not seem to be a suitable parasitoid for the control of *Bemisia* and new natural enemies need to be evaluated (Drost *et al.*, 1996). In the 1980s a long-term project was started in which modeling and experimentation were integrated. The aim was to obtain a quantitative understanding of this tritrophic system in order to explain failure or success of biological control. Because of the multitude of relationships between trophic levels, the most important factors for successful control can only be traced after integration of all relevant processes. Systems analysis and simulation are powerful tools for this purpose. This systems analysis and simulation approach bridges the gap between knowledge at the individual level and understanding at the population level.

The commercial success obtained with biological control of whitefly can, for a large part, be attributed to extensive fundamental behavioral and ecological research, which was done concurrently with the development of practical application methods. Modeling has played a role in the process of selecting and improving the efficacy of releases of natural enemies, but often biologically unrealistic simplifications were part of these models, which strongly limited their predictive value.

In this chapter we will describe a model that is unique in that it is individual-based and simulates the local searching and oviposition behavior of individual parasitoids in a whitefly-infested crop. This model includes stochasticity and spatial structure, and it comprises several submodels for: (i) the parasitoid's foraging behavior, (ii) the whitefly and parasitoid population development, (ii) the spatial distribution of whitefly and parasitoid within and between plants in the crop, and (iv) leaf production. The population dynamics of whitefly and parasitoid are linked through simulation of the foraging behavior of individual parasitoids in the crop. Barlow (Chapter 3, Table 3.1) classified the model as an individual-based, temperature-dependent simulation model with age-structure; the model uses a mechanistic behavioral base for the ecological processes and incorporates considerably more biological detail (i.e., behavior) than most other models of insect–parasitoid interaction.

With the model we can simulate temporal and spatial dynamics of pest and natural enemy. We will illustrate how the model helps us in understanding how biological control in greenhouses works and how the model can be used to test the effect of various parasitoid release schemes. We will also explain how the model assists us in developing criteria for the selection of natural enemies.

Methodology

Life history of host and parasitoid

When we started modeling, many data on life-history parameters of the two insect species, such as immature development rate, immature mortality, adult longevity, sex ratio, fecundity, and oviposition frequency, were already available. For whitefly these data were reviewed in Hulspas-Jordaan & van Lenteren (1989), for *E. formosa* data were reviewed in van Roermund & van Lenteren (1992b). Furthermore, extensive studies had been done on the whitefly's host-plant preference and performance, on selection of feeding and oviposition sites and on spatial distribution patterns (reviewed in van Lenteren & Noldus, 1990). Also, comprehensive work had been done on biology and foraging behavior of the parasitoid when it was confined to small experimental arenas (reviewed in Noldus & van Lenteren, 1990). A critical analysis of all this earlier work resulted in a selection of data on life-history parameters of greenhouse whitefly and *E. formosa*. With these data the relationships between life-history parameters and temperature were estimated by non-linear regression (van Roermund & van Lenteren, 1992a,b).

For each relationship, five mathematical equations were fitted, the best being selected on the basis of the coefficient of determination (r^2) and on visual comparison of the curves. For all immature stages of whitefly and parasitoid the relationship between development rate (1/development duration) and temperature was best described by the Logan curve. Just above the lower threshold temperature the development rate increases exponentially to an optimum, thereafter it declines sharply until the upper lethal temperature has been reached. The relationships of longevity, fecundity, and oviposition frequency with temperature are best described by the Weibull curve: they increase exponentially from the lower lethal temperature to an optimum, thereafter they decrease exponentially. No relationship with temperature was found for the immature mortality and the sex ratio (van Roermund & Van Lenteren, 1992a,b).

Time allocation of the parasitoid on tomato leaflets

When developing a realistic model based on behavior of individual parasitoids, information is needed on the time allocated to searching for and oviposition in hosts by the parasitoids on tomato leaflets. Leaflets form the basic unit for host searching and oviposition, and the time spent on these leaflets as well as the type of activities undertaken, largely determines the outcome of the model. Information was collected, gaps in knowledge identified, and a number of experiments were run to obtain missing information. The experiments consisted of direct observations of parasitoid behavior in a number of different search situations, such as clean leaves, leaves with low/high host densities, leaves with unparasitized/parasitized hosts, etc. This work provided information on the parasitoid's walking speed, walking activity and host handling behavior. The tendency to depart from leaves, which was related to the quality and number of hosts on the leaves, was also determined (van Roermund & van Lenteren, 1995a,b; van Roermund et al., 1994).

The following conclusions could be drawn. *Encarsia formosa* searches at random on leaves. Parasitoids are arrested on the leaf by encounters with, and especially by ovipositions in, unparasitized hosts, by encounters with parasitized hosts and by contact with honeydew. The giving up time (GUT) on clean tomato leaflets or after the last host encounter is about 20 min. This increases to 40 min after the first oviposition in an unparasitized host. This patch-leaving behavior can be described by a stochastic threshold mechanism, which is characterized by a certain tendency (probability per time) to leave.

Simulation of foraging behaviour of the parasitoid

The foraging data were used as input in a stochastic Monte Carlo simulation model of the parasitoid's behavior. Earlier research made clear that *E. formosa* does not use either volatile or visual cues to locate infested plants. Once on a leaf, they search at random. After finding a host, host exuviae or honeydew, the walking speed and tortuousness of the search path does not change (van Lenteren et al., 1976). However, after contact with hosts or host-related materials the parasitoid does stay much longer on such a leaflet (Noldus & van Lenteren, 1990; van Roermund et al., 1994).

The complexity of the model was increased step by step. The searching behavior was first simulated for a small experimental arena, then for a tomato leaflet and finally for a tomato plant. For each level of complexity, the model was first validated with independent experimental data before the next step was made. The model of foraging behavior at the plant level, simulates the parasitization process of *E. formosa* from the moment of landing on the plant until departure from that plant (van Roermund et al., 1996a,b, 1997a,b).

The time-step of this model is only 0.00005 day (4.3 s), which is smaller than the smallest time coefficient in the model (handling time for antennal rejection of a parasitized host), needed for accurate numerical integration. The physiological state of the parasitoid (i.e., egg load) is simulated as well. After oviposition, new eggs mature at a rate depending on actual egg load and temperature. Maximum egg maturation rate was found to be 0.7 eggs/h for *E. formosa* at 25 °C.

According to the model, the number of host encounters and ovipositions on the plant increased with host density according to a Type II functional response. Under host conditions usually found on plants (low density, clustered distribution over leaves), the number of ovipositions was mostly affected by leaf area, the parasitoids' walking speed and activity, and the probability of oviposition after encountering a host.

Population dynamics of pest and natural enemy in a crop

After simulation of the foraging behavior a model was designed that simulates the population dynam-

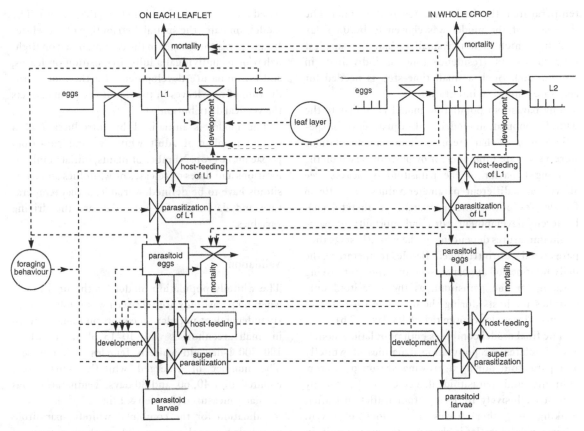

Figure 7.1. Relational diagram for the population growth of greenhouse whitefly and *Encarsia formosa* with details on the interaction between host and parasitoid: the state variables or integrals (insect numbers) are presented within rectangles; the rates of change within valve symbols, auxiliary variables within circles; the flow of material is indicated by solid lines plus arrows, the flow of information by dotted lines plus arrows; small bars in a rectangle indicate series of integrals (boxcar train); L1, L2 . . . are the first, second, . . . etc larval stages of whitefly.

ics of the whitefly–parasitoid interaction in a tomato crop at realistic and varying temperature conditions. This model comprises several sub-models. One of these is the (stochastic) model of the parasitoid's foraging behavior on tomato leaflets. Other submodels simulate the pest and parasitoid population development, the spatial distribution of whitefly and parasitoid in the canopy based on the dispersion of adult whiteflies and parasitoids from leaf to leaf, and finally the production of new leaves in the canopy. For a comprehensive description of the model, the reader is referred to van Roermund (1995). A simplified scheme is given in Fig. 7.1.

Information on the most important submodels follows below.

Development of whitefly and parasitoid is modeled by using a boxcar-train procedure (Goudriaan & van Roermund, 1993), which simulates development of successive stages including variation within each stage. The boxcar-train procedure is a powerful method to realistically mimic insect development.

The whitefly population model is an explanatory state-variable model. Whitefly stages are represented by boxcars. Since development, adult reproduction and longevity are affected by temperature,

temperature is the driving variable in the model. The time-step of the model was chosen to be 0.05 day (1.2 h), which is much smaller than the smallest time-coefficient (residence time of individuals in one boxcar). Such a small time-step is needed for accurate numerical integration.

The parasitoid population model is similar to the whitefly population model, with adjustments for specific *E. formosa* characteristics. The main adaptations were a separate submodel for oviposition based on the foraging behavior of the parasitoid on leaves (see above), and different parameter values for a 100% female sex ratio (the parasitoid is thelytokous). Different development rates and mortality of parasitoid immatures depending on the whitefly stage they parasitized were introduced. Besides temperature, the daily light period appeared to be an important driving variable for the parasitoid, as the parasitoid only searches for hosts during the day. The time-step of this model was also chosen to be 0.05 day (1.2 h).

The final model combines both population models and simulates the population dynamics of whitefly and parasitoid during a growing season of a crop. Whitefly and parasitoid disperse in the canopy almost exclusively by flying from leaflet to leaflet, walking to other leaves is seldom observed. Therefore, the leaflet is chosen as the spatial unit. In the model, the canopy is divided into plants with a certain number of leaves, and production of new leaves is simulated. The coordinates of each plant, leaf, and leaflet represent their position in the canopy. The interaction between both populations in the greenhouse is achieved by the local foraging behavior of the parasitoid adults on leaves (Fig. 7.1). The submodel of the foraging behavior of the parasitoid on leaves ultimately determines the outcome of events in the whitefly and parasitoid populations. New parasitoids are generated after ovipositions, while the whitefly population decreases in size as result of ovipositions. Host-feeding by parasitoids kills additional whiteflies.

The foraging efficiency of the parasitoid is strongly affected by whitefly density and distribution, and both are simulated in the model. Vertical distribution within the plant and horizontal distribution between plants of whiteflies have been modeled based on research by Noldus *et al.* (1985, 1986). The model simulates the spatial distribution of the whitefly and parasitoid stages in the canopy, based on flight behavior of individual adults, oviposition on leaves, and on immature development. Furthermore, production of new leaves in the top of the plants affects the vertical distribution.

The model is initialized by introduction of a certain number of adult whiteflies and parasitoid pupae in the crop. Number of plants, initial numbers, locations and dates of releases of whiteflies and parasitoids have to be defined. Variable, actual temperature and daylength for each day are the driving variables.

Validation

The whitefly population model (or the final model when no parasitoids are released) was validated with six independent population counts on tomato grown in small greenhouses with 10–15 plants, in which 100–300 1-day old whitefly females were released. The number of emerged whitefly pupae were counted after 40, 60, and 80 days. Temperature was regularly measured (Joosten & Elings, 1985).

Validation for the combined whitefly–parasitoid population model was possible with data from an experiment in a large commercial greenhouse with 18 000 tomato plants. Absolute counts of white (unparasitized) and black (parasitized) pupae were made during a 112 day period (Eggenkamp-Rotteveel Mansveld *et al.*, 1982a, b).

Results

Validation

Population increase of greenhouse whitefly when simulated in the absence of parasitoids agrees well with the observations. One of the six comparisons between model simulation and greenhouse counts is shown in Fig. 7.2. The excellent fit can be explained to a large extent by the accurate estimates of the life-history parameters, which are based on many experiments at a wide range of temperatures (van Roermund & van Lenteren, 1992a).

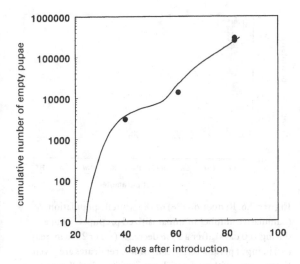

Figure 7.2. Simulated (continuous line) and observed (filled circles) number of empty whitefly pupae in a small greenhouse with tomato plants (on day 1, 100 whitefly females were released on 10 tomato plants, whitefly pupal counts made ca 40, 60, and 80 days after release).

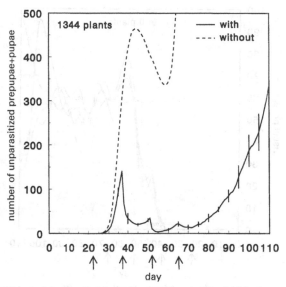

Figure 7.3. Simulated numbers of unparasitized whitefly prepupae + pupae in a large patch (1344 tomato plants) in a commercial greenhouse, with (—)and without (- - -)parasitoid releases; bars represent standard deviations (*n* = 5), arrows indicate releases of parasitoids.

For validation of the model when parasitoids were introduced, one of the large whitefly patches in the greenhouse comprising 1344 tomato plants was used. The number of white and black pupae was simulated for a 112 day period, the same period during which absolute counts were made in the greenhouse. Simulations were run with and without introductions of parasitoids. At the end of the simulation, the whitefly population without parasitoid introduction was more than 300 times larger than when parasitoids were released (Fig. 7.3). Subsequently validation was performed for an area representing 1/20 of the total greenhouse (900 plants). The total number of prepupae plus pupae for five identical simulation runs with the stochastic model is given in Fig. 7.4. The simulated numbers of prepupae plus pupae agree well with actually counted numbers in the greenhouse. The variation between the simulations is largely due to the low initial whitefly density. Also percent visual parasitism (Fig. 7.5) shows good agreement with observed percentages. Apparently, the foraging behavior of the parasitoids, as well as the life history of both insect species, were correctly modeled.

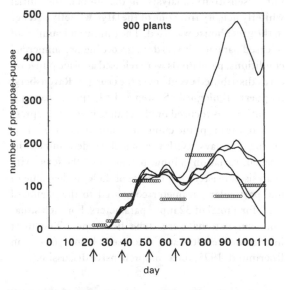

Figure 7.4. Simulated (continuous lines; 5 replicates) and observed (open circles) total numbers of white and black prepupae + pupae on 900 tomato plants (1/20 of total greenhouse), with parasitoid releases, indicated by arrows.

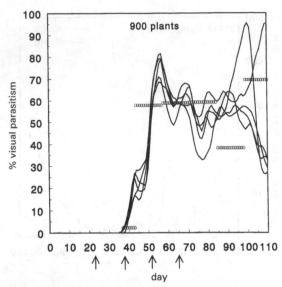

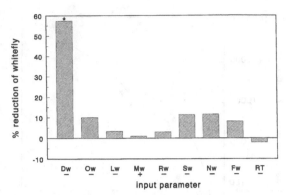

Figure 7.5. Simulated (continuous lines; 5 replicates) and observed (open circles) percent of visual parasitism (% black pupae) on 900 tomato plants, with parasitoid releases, indicated by arrows.

Figure 7.6. Reduction (%) of the whitefly population (cumulative number of unparasitized prepupae + pupae during 100 days) after a 25% decrease ($-$) or 25% increase ($+$) in input parameter. Means of five replicates are given. Bars marked with an asterisk are significantly different from 0 (Student t-test on population mean; $p = 0.05$). Whitefly parameters are: Dw, immature development rate; Ow, daily oviposition (or fecundity); Lw, longevity; Mw, immature mortality; Rw, relative dispersion (coefficient of variation of stage duration); Sw, sex ratio; Nw, initial number; Fw, average flight distance from one plant to the next; RT, average residence time on leaf of dead immatures and empty pupal cases.

Sensitivity analysis

For a sensitivity analysis of the model, the initial whitefly density from one of the large whitefly patches in the greenhouse was used. The timing and number of released parasitoids was done according to commercial conditions: parasitoids were released as black pupae on cards distributed evenly over the crop (W. Ravensberg, Koppert Biological Systems Ltd, pers. comm.). Attention was focused on the change in whitefly 'pressure' in the crop: the cumulative number of whiteflies during 100 days. With the model we determined the effect of variation in input parameters. The sensitivity analysis consisted of a change of 25% in the value of one particular parameter compared to the 'standard run', for a total of 32 input parameters. For each situation, simulations were done five times. The most important results are discussed below (see van Roermund, 1995, for a comprehensive discussion).

Effect of variation in whitefly parameters

The influence of changes of the following parameters on whitefly numbers was determined: immature

development rate (Dw), daily oviposition (Ow), longevity (Lw), immature mortality (Mw), degree of variation in stage duration among individuals (Rw), sex ratio (Sw), initial population size (Nw), average flight distance from plant to plant (Fw), and average residence time on leaf of dead hosts and empty pupal cases (RT). For whiteflies, only a change in the development rate significantly influenced the whitefly population: a 25% decrease in development rate resulted in a 57% decrease of the whitefly population (Fig. 7.6).

Effect of variation in parasitoid parameters

The influence of changes in the following parameters of the parasitoid on whitefly numbers was tested: immature development rate (Dp), longevity (Lp), immature mortality (Mp), degree of variation in stage duration (Rp), number of released parasitoids (Np), and average flight distance from plant to plant (Fp). For the parasitoid, variation in two parameters

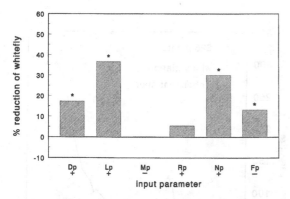

Figure 7.7. Reduction (%) of the whitefly population after a 25% decrease (−) or 25% increase (+) in input parameter. For more details see the legend to Fig. 7.6. Parasitoid parameters are: Dp, immature development rate; Lp, longevity; Mp, immature mortality; Rp, relative dispersion (coefficient of variation of stage duration); Np, number of released adults; Fp, average flight distance from one plant to the next. Bars marked with an asterisk are significantly different from 0 (student's *t*-test on population mean; $p = 0.05$).

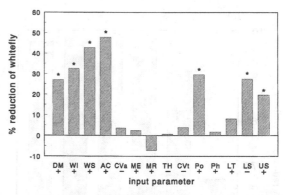

Figure 7.8. Reduction (%) of the whitefly population after a 25% decrease (−) or 25% increase (+) in input parameter. For more details see the legend to Fig. 7.6. Input parameters are: DM, diameter of host stages; WI, width of parasitoids' searching path; WS, parasitoids' walking speed; AC, walking activity; CVa, coefficient of variation of walking activity; ME, maximum egg load; MR, egg maturation rate; TH, host handling time; CVt, coefficient of variation of handling time; Po, probability of oviposition after encountering a host; Ph, probability of host feeding after encountering a host; LT, parasitoids' leaving tendency; LS, tendency of changing from the lower leaf side to the upper; US, tendency of changing from the upper leaf side to the lower. Bars marked with an asterisk are significantly different from 0 (student's *t*-test on population mean; $p = 0.05$).

resulted in the most important changes to the whitefly population: longevity and number of released parasitoids. The strongest influence was from a 25% increase in longevity, which reduced the whitefly population by 36% (Fig. 7.7). A 25% increase in number of parasitoids reduced whiteflies by 30%.

Effect of variation in parasitoid foraging parameters

Fourteen parameters related to the parasitoid's foraging behavior were tested in the sensitivity analysis. Variation (a 25% increase) in five of these resulted in the most important decreases in the whitefly population: walking activity (AC, 48%), walking speed (WS, 43%), width of searching path (WI, 33%), probability of oviposition in host (Po, 30%), search time on lower (infested) leaf side compared to search time on upper side (LS, 23%). In addition, variation in host size (DM, a 25% increase) resulted in a 27% decrease in the whitefly population (Fig. 7.8). Except for the probability of oviposition, these parameter changes all resulted in an increased probability of encountering hosts.

The observed variation of the parasitoid's giving up time (GUT = time until departure from a leaf after landing, or after the latest host encounter), strongly influences whitefly population reduction. First, the minimum value of the GUT is important. When it is too short, parasitoids leave 'too soon' from infested leaves and whitefly populations are not sufficiently reduced, even though time would be saved on the majority of leaves that are uninfested (Fig. 7.9(a)). Then, the existence of an arrestment effect appeared very important, as a longer period spent searching after contact with a host leads to increased parasitism of other hosts on that leaf. Finally, the observed variation in GUT appeared essential for a long-lasting relationship between pest and parasitoid. For *E. formosa*, the GUT after landing on the tomato leaflet, or if it occurred, from the latest host encounter until leaving varied to a great extent (van Roermund *et al.*,

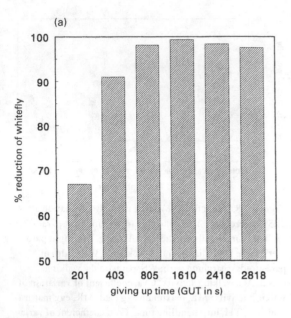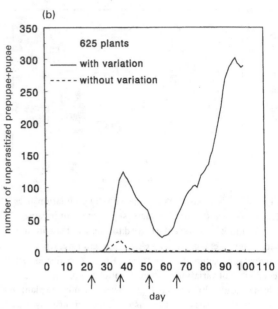

Figure 7.9. (a) Reduction (%) of the whitefly population at different (constant) giving up times (GUT) of the parasitoids, (b) simulated number of unparasitized prepupae + pupae with (−) and without (- - -) variation in GUT, on 625 tomato plants.

1994). Median GUT was about 20 min, but increased to 40 min after the first oviposition in an unparasitized host. In the model, each GUT of a parasitoid was drawn from an exponential distribution to mimic its variation. When variation in GUT was excluded in the model, the whitefly population became less stable and nearly went extinct (Fig. 7.9(b)).

Another important point to stress here is that at low pest densities, which is what the grower aims at with biological control, a high rate of reproduction of the parasitoid is of limited value. The model showed that changes in maximum egg load and egg maturation rate of the parasitoid do not affect the whitefly population. In the literature on evaluation of control capacity of natural enemies, often much attention is paid to the maximum reproduction capacity of natural enemies, instead of searching efficiency.

Effect of plant growth and leaf size

Leaf or leaflet size (or total leaf area) has a large effect on the probability that whiteflies are found by the parasitoid: when leaflet size is decreased by 25%

the whitefly population decreased by 57% (Fig. 7.10).

Recently emerged whitefly adults migrate to young leaves at the top of the plant to feed and oviposit. Slower leaf production results in a longer stay and more ovipositions of whitefly adults on a particular leaflet, which leads to a more aggregated host distribution. In such situations, whiteflies are suppressed to much lower numbers than on faster growing plants. A higher whitefly aggregation increases the probability of being found by the parasitoid, due to the stronger parasitoid arrestment. A 25% reduction in leaf production rate results in a 38% reduction of the whitefly population (Fig. 7.10).

Effect of variation in the driving variables temperature and daylength

When the temperature was reduced by 1 °C, the whitefly population was reduced by 30% when no parasitoids were released. In the presence of parasitoids, the whitefly population size was hardly affected by a 1 °C decrease in temperature.

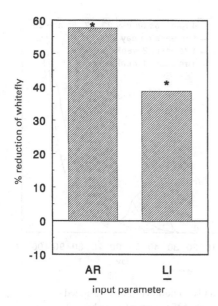

Figure 7.10. Reduction (%) of the whitefly population after a 25% decrease (−) or 25% increase (+) in input parameter. For more details see the legend to Fig. 7.6. Tomato parameters are: AR, leaflet area; LI, leaf production rate. Asterisks indicate a significant difference from 0 (student's t-test on population mean; $p < 0.05$).

Apparently, the lower whitefly population growth (30%) was compensated for by less efficient biological control at the lower temperature, due to a lower walking speed of the parasitoids. Thus, the whitefly population under biological control is less sensitive to temperature than when not controlled.

A reduction in daylength of 1 h when parasitoids were present, resulted in a whitefly population increase of 19%, simply because parasitoids do not search in darkness and, therefore, parasitize fewer whiteflies.

Effects of environmental factors such as temperature and daylength on biological control are now being studied in combination with effects on tomato yield to study the optimal climate regime. For this, different climate regimes will be used as input for the current model and for the tomato crop growth model of Heuvelink (1996).

Effect of variation in parasitoid releases

The location of release sites and the number of released parasitoids were both important factors in the reduction of the whitefly population. When all parasitoids were released in the center of the plot where the first whitefly adults were observed, the whitefly population was reduced by 63 % compared to releases at regular spatial intervals as is normally done in commercial releases. When the number of released parasitoids was increased by 25% at regular spatial intervals, the whitefly population decreased by 30% (Fig. 7.11(a)).

If all parasitoids are released at once instead of 4 times, timing of release is very important to obtain a sufficient reduction of the whitefly population, and in general single introductions result in unstable whitefly populations when compared with multiple releases. Timing of multiple releases (4 times with a 2 week interval, which is the procedure followed by the commercial biological control company, Koppert Biological Systems Ltd) is less important, although early releases resulted in less stable populations than later releases (Figs 7.11(b) and (c)).

Discussion

Successful biological control of whitefly by *E. formosa* is, contrary to what we expected, not the result of a reaction of the parasitoid to long-distance cues emitted by the whiteflies or the infested tomato plants. The high percent parasitism observed in commercial greenhouses is the end result of a random search process by the parasitoid, whereby parasitoids will search longer on leaves where they have found hosts, host exuviae or honeydew. Extended searching in such situations leads to an increased encounter probability of hosts that show a clustered distribution.

The principal theoretical explanation for successful biological control has long been that effective natural enemies operate by creating a stable pest–enemy equilibrium at low densities and that aggregation of natural enemies at patches with high host or prey densities is the critical feature that results in stability (Waage & Hassell, 1982). The

(a)
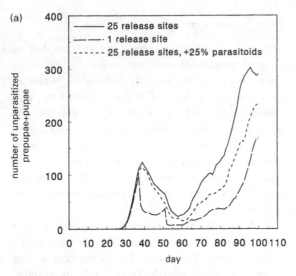

(b)
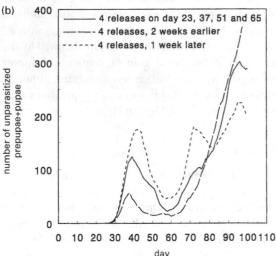

(c)
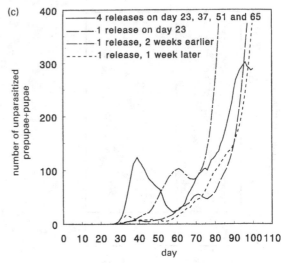

Figure 7.11. Simulated number of unparasitized prepupae + pupae on 625 tomato plants when (a) all parasitoids were released in the centre of the plot where the first adult whiteflies were observed or when 25% more parasitoids were released, (b) when the schedule of four parasitoid releases started 2 weeks earlier or 1 week later, (c) when the schedule of four parasitoid releases was changed to one release and when that release started 2 weeks earlier or 1 week later.

central role of this low, stable pest equilibrium has recently been challenged and different explanations for the functioning of biological control have been put forward (see e.g., Murdoch, 1990; and see Chapter 2). Our present analysis indicates that the parasitoid *E. formosa* does not really create a stable pest–enemy equilibrium during the growing season of a crop, but that it suppresses the whitefly population sufficiently to prevent it from causing yield losses. With the practice of releasing parasitoids in 4 waves with a 2 week interval at the start of the

growing season, the simulated whitefly population shows one peak during the 250 day simulation and then decreases to very low densities. Whitefly numbers do not increase above the economic threshold density during the growing season (van Roermund, 1995; van Roermund *et al.*, 1997a, b).

The model described in this chapter links the population dynamics of whitefly and the parasitoid through simulation of the foraging behavior of individual parasitoids in a crop where whiteflies are present in common distribution patterns and age classes. Most of the models on the population dynamics of host–parasitoid or predator–prey relationships developed so far that include a component of searching, use experimentally determined functional response curves as input (see e.g., Hassell, 1986; and see Chapter 2). Functional response curves describe the relationship between the host density and number of hosts attacked per natural enemy per unit of time. Such curves are generally based on experiments on leaves or in small arenas. They are then extrapolated to the crop level by deriving rates of parasitism from the average host density in the crop. This implicitly assumes that the observed relationship for small units such as leaves is also valid at larger spatial scales, which is rather unrealistic, particularly when hosts show a strongly clustered distribution in the crop (which is the case for whitefly) and when functional response curves are non-linear (which is the case for many natural enemies, including *E. formosa*).

The present model does not manifest the inherent problems of functional-response-based models. Judson (1994) wrote that individual-based models such as this one are a necessity when local interactions and stochasticity are important and, as far as we know, this is the first individual-based host–parasitoid model. Another problem with many models of population dynamics is that they treat insects usually as identical, synchronized individuals. In our model the so-called boxcar train technique for simulation of development was used, which is able to handle all possible development stages simultaneously as well as the variation in developmental time for each stage. As a result, development of whitefly and parasitoid could be simulated very realistically.

With the model, characteristics of the crops were identified that strongly influence the feasibility of biological control, such as the effect of leaf surface, leaf structure, and leaf size on the foraging efficiency of the parasitoid. For example, the walking speed of *E. formosa* is 0.31 mm/s on tomato leaves at 20 °C (van Roermund & van Lenteren, 1995a), whereas on hairy cucumber leaves it is only 0.21 mm/s (van Lenteren *et al.*, 1995). This difference resulted in a 55% reduction of the whitefly population during 100 days according to the model. The importance of the parasitoids' walking speed for whitefly control is also shown in Fig. 7.8. Other differences between tomato and cucumber are not included in this example. It is particularly the combined effect of these factors that can be evaluated with this model for different crops or plant varieties.

Also the effect of differences in crops or cultivars on the rate of pest development and the effect on bio-

logical control could be evaluated. Simulation results indicate, for example, that breeding cultivars for resistance against whitefly is much more profitable when it concentrates on an increase of the immature developmental time of whitefly instead of aiming at an increase of immature mortality or reduced fecundity (see Fig. 7.6).

At present, the model is being used to evaluate the effects of temperature and daylength on biological control and yield of a tomato crop. As a result, greenhouse climate control may not only be based on crop characteristics (yield, quality, timing) and production costs, but also on prevention of production risks such as the environmentally friendly control of pests (Challa & van Straten, 1993). Recent problems with the use of *E. formosa* in greenhouses during autumn/winter will be addressed using the model to evaluate the effects of increased temperatures during short periods of the day. The model will also be used to improve biological control on the ornamental *Gerbera jamesonii*. In another project the combination of two natural enemies, an entomophagous fungus (*Aschersonia* spp, *Paecilomyces* spp) and the parasitoid *E. formosa*, is being evaluated with the model. Further, it is used to evaluate several parasitoids of the silverleaf whitefly (Drost *et al.*, 1996).

In addition, the model pointed to weak points in previous evaluation programs of natural enemies. In biological control research, parasitoids are usually tested in small-scale experiments at high host densities before introduction in the field (Waage, 1990). As a result, maximum daily oviposition of parasitoids is measured, whereas the model showed that for field performance at low pest densities, effective host searching is the most important factor in determining the number of encounters and ovipositions. With the model we were able to identify a number of parasitoid characteristics that can easily be measured in the laboratory and provide important information about its searching efficiency, such as walking speed, walking activity, probability of host acceptance and the parasitoid's arrestment effect when searching on clean and infested leaves. Another important characteristic is the parasitoids' longevity. The model thus assists in identification of the characteristics that compose an efficient natural enemy for various types of biological control. As a result, selection of natural enemies can be improved.

For a first selection, there are good evaluation criteria available to allow for a choice between useless and potentially promising natural enemies (van Lenteren, 1986; Pak, 1988; van Lenteren & Woets, 1988; Minkenberg, 1990; van Roermund, 1995; van Steenis, 1995; Drost *et al.*, 1996). Such a choice prevents research on and introduction of inefficient natural enemies. With a gradual improvement and integration of evaluation criteria, ranking promising natural enemies will be possible. Systems analysis and simulation are important tools for such an integration, but go beyond the mere comparison of parasitoids, as they also take characteristics of the pest, the crop, and the environment into account. In such a complete framework and with models including sufficient biological detail, the role of different natural enemies can be analysed.

Acknowledgements

H. J. W. v. R. was financially supported by the Netherlands Technology Foundation (NWO – STW) and a post-doctoral grant provided by the C.T. de Wit Graduate School Production Ecology. R.F. Luck is gratefully acknowledged for reviewing an earlier version of this chapter.

References

Challa, H. & van Straten, G. (1993) Optimal diurnal climate control in greenhouses as related to greenhouse management and crop requirements. In *The Computerized Greenhouse. Automatic Control Application in Plant Production*, ed. H.Hashimoto, G. P. A. Bot, W. Day, H. J. Tantau & H. Nonami, pp. 119–37. San Diego, CA: Academic Press.

Drost, Y. C., Fadl Elmula, A., Posthuma-Doodeman, C. J. A. M. & van Lenteren, J. C. (1996) Development of selection criteria for natural enemies in biological control: parasitoids of *Bemisia argentifolii*. *Proceedings of the section of Experimental and Applied Entomology of the Netherlands Entomological Society, Amsterdam* 7: 165–70.

Eggenkamp-Rotteveel Mansveld, M. H., Ellenbroek, F. J. M. & van Lenteren, J. C. (1982a) The parasite–host relationship between *Encarsia formosa* (Hymenoptera: Aphelinidae) and *Trialeurodes vaporariorum* (Homoptera: Aleyrodidae). XII. Population dynamics of parasite and host in a large, commercial glasshouse and test of the parasite-introduction method used in the Netherlands, part 1. *Journal of Applied Entomololgy* 93: 113–30.

Eggenkamp-Rotteveel Mansveld, M. H., Ellenbroek, F. J. M. & van Lenteren, J. C. (1982b) The parasite-host relationship between *Encarsia formosa* (Hymenoptera: Aphelinidae) and *Trialeurodes vaporariorum* (Homoptera: Aleyrodidae). XII. Population dynamics of parasite and host in a large, commercial glasshouse and test of the parasite-introduction method used in the Netherlands, part 2. *Journal of Applied Entomology* 93: 258–79.

Gerling, D. (ed.) (1990) *Whiteflies: Their Bionomics, Pest Status and Management.* Andover: Intercept

Gerling, D. & Mayer, R. T. (eds) (1996) *Bemisia 1995: Taxonomy, Biology, Damage, Control and Management.* Andover: Intercept.

Goudriaan, J. & van Roermund, H. J. W. (1993) Modelling of ageing, development, delays and dispersion. In *On Systems Analysis and Simulation of Ecological Processes*, ed. P. A. Leffelaar, pp. 89–126. Dordrecht: Kluwer Academic Publishers.

Hassell, M. P. (1986) Parasitoids and population regulation. In *Insect Parasitoids*, ed. J. K. Waage & D. J. Greathead, pp. 201–24. London: Academic Press.

Heuvelink, E. (1996) Tomato Growth and Yield: Quantitative Analysis and Synthesis. PhD thesis: Wageningen Agricultural University, The Netherlands.

Hulspas-Jordaan, P. M. & van Lenteren, J. C. (1989) The parasite–host relationship between *Encarsia formosa* (Hymenoptera: Aphelinidae) and *Trialeurodes vaporariorum* (Homoptera: Aleyrodidae). XXX. Modelling population growth of greenhouse whitefly on tomato. *Wageningen Agricultural University Papers* 89.2: 1–54.

Joosten, J. & Elings, A. (1985) Onderzoek naar de Toetsmethoden voor Resistentie tegen Kaswittevlieg (*Trialeurodes vaporariorum* (Westwood)) in Tomaat (*Lycopersicon esculentum* Mill.). MSc thesis: Wageningen Agricultural University, The Netherlands.

Judson, O. P. (1994) The rise of the individual-based model in ecology. *Trends in Ecology and Evolution* 9: 9–14.

Minkenberg, O. P. J. M. (1990) On Seasonal Inoculative Biological Control. PhD thesis: Wageningen Agricultural University, The Netherlands.

Murdoch, W. W. (1990) The relevance of pest–enemy models to biological control. In *Critical Issues in Biological Control*, ed. M. Mackauer, L. E. Ehler & J. Roland, pp. 1–24. Andover: Intercept.

Noldus, L. P. J. J. & van Lenteren, J. C. (1990) Host aggregation and parasitoid behaviour: biological control in a closed system. In *Critical Issues in Biological Control*, ed. M. Mackauer, L. E. Ehler & J. Roland, pp. 229–62. Andover: Intercept.

Noldus, L. P. J. J., Xu Rumei & van Lenteren, J. C. (1985) The parasite–host relationship between *Encarsia formosa* (Hymenoptera: Aphelinidae) and *Trialeurodes vaporariorum* (Homoptera: Aleyrodidae). XVII. Within-plant movement of adult greenhouse whiteflies. *Journal of Applied Entomology* 100: 494–503.

Noldus, L. P. J. J., Xu Rumei & van Lenteren, J. C. (1986) The parasite–host relationship between *Encarsia formosa* (Hymenoptera: Aphelinidae) and *Trialeurodes vaporariorum* (Homoptera: Aleyrodidae). XVIII. Between-plant movement of adult greenhouse whiteflies. *Journal of Applied Entomology* 101: 159–76.

Pak, G. A. (1988) Selection of *Trichogramma* for Inundative Biological Control. PhD thesis: Wageningen Agricultural University, The Netherlands.

van Lenteren, J. C. (1986) Parasitoids in the greenhouse: successes with seasonal inoculative release systems. In *Insect Parasitoids*, ed. J. K Waage & D. J. Greathead, pp. 342–74. London: Academic Press.

van Lenteren, J. C. (1993) Parasites and predators play a paramount role in pest management. In *Pest Management: Biologically Based* Technologies, ed. R. D. Lumsden & J. L. Vaughn, pp. 68–81. Washington, DC: American Chemical Society.

van Lenteren, J. C. (1995) Integrated pest management in protected crops. In *Integrated Pest Management*, ed. D. Dent, pp. 311–43. London: Chapman and Hall.

van Lenteren, J. C. & Noldus L. P. P. J. (1990) Whitefly–plant relationships: behavioural and ecological aspects. In *Whiteflies: Their Bionomics, Pest Status and Management*, ed. D. Gerling, pp. 47–89. Andover: Intercept.

van Lenteren, J. C. & Woets, J. (1988) Biological and integrated control in greenhouses. *Annual Review of Entomology* 33: 239–69.

van Lenteren, J. C., Nell, H. W., Sevenster-van der Lelie, L. A. & Woets, J. (1976) The parasite–host relationship between *Encarsia formosa* (Hymenoptera: Aphelinidae) and *Trialeurodes vaporariorum* (Homoptera: Aleyrodidae). I. Host finding by the parasite. *Entomologia Experimentalis et Applicata* 20: 123–30.

van Lenteren, J. C., Li Zhao Hua, Kamerman, J. W. & Xu Rumei (1995) The parasite–host relationship between *Encarsia formosa* (Hym., Aphelinidae) and *Trialeurodes vaporariorum* (Hom., Aleyrodidae) XXVI. Leaf hairs reduce the capacity of *Encarsia* to control greenhouse whitefly on cucumber. *Journal of Applied Entomology* 119: 553–9.

van Lenteren, J. C., van Roermund, H. J. W. & Sütterlin, S. (1996) Biological control of greenhouse whitefly (*Trialeurodes vaporariorum*): how does it work? *Biological Control* 6: 1–10.

van Lenteren, J. C., Drost, Y. C., van Roermund, H. J. W. & Posthuma-Doodeman, C. J. A. M. (1997) Aphelinid parasitoids as sustainable biological control agents in greenhouses. *Journal of Applied Entomology* 121: 473–85.

van Roermund, H. J. W. (1995) Understanding Biological Control of Greenhouse Whitefly with the Parasitoid *Encarsia formosa*: From Individual Behaviour to Population Dynamics. PhD thesis: Wageningen Agricultural University, The Netherlands.

van Roermund, H. J. W. & van Lenteren, J. C. (1992a) The parasite–host relationship between *Encarsia formosa* (Hymenoptera: Aphelinidae) and *Trialeurodes vaporariorum* (Homoptera: Aleyrodidae). XXXIV. Life-history parameters of the greenhouse whitefly, *Trialeurodes vaporariorum* as a function of host plant and temperature. *Wageningen Agricultural University Papers* 92.3: 1–102.

van Roermund, H. J. W. & van Lenteren, J. C. (1992b) The parasite–host relationship between *Encarsia formosa* (Hymenoptera: Aphelinidae) and *Trialeurodes vaporariorum* (Homoptera: Aleyrodidae). XXXV. Life-history parameters of the greenhouse whitefly parasitoid *Encarsia formosa* as a function of host stage and temperature. *Wageningen Agricultural University Papers* 92.3: 103–47.

van Roermund, H. J. W. & van Lenteren, J. C. (1995a) Foraging behaviour of the whitefly parasitoid *Encarsia formosa* on tomato leaflets. *Entomologia Experimentalis et Applicata* **76**: 313–24.

van Roermund, H. J. W. & van Lenteren, J. C. (1995b) Residence times of the whitefly parasitoid *Encarsia formosa* Gahan (Hym. Aphelinidae) on tomato leaflets. *Journal of Applied Entomology* **119**: 465–71.

van Roermund, H. J. W., Hemerik, L. & van Lenteren, J. C. (1994) The influence of intra-patch experiences and temperature on the time allocation of the whitefly parasitoid *Encarsia formosa*. *Journal of Insect Behaviour* **7**: 483–501.

van Roermund, H. J. W., van Lenteren, J. C. & Rabbinge, R. (1996a) Analysis of the foraging behaviour of the whitefly parasitoid *Encarsia formosa* in an experimental arena: a simulation study. *Journal of Insect Behaviour* **9**: 771–97.

van Roermund, H. J. W., van Lenteren, J. C. & Rabbinge, R. (1996b) Analysis of the foraging behaviour of the whitefly parasitoid *Encarsia formosa* on a leaf: a simulation study. *Biological Control* **8**: 22–6.

van Roermund, H. J. W., van Lenteren, J. C. & Rabbinge, R. (1997a) Analysis of the foraging behaviour of the whitefly parasitoid *Encarsia formosa* on a plant: a simulation study. *Biocontrol Science and Technology* **7**: 131–51.

van Roermund, H. J. W., van Lenteren, J. C. & Rabbinge, R. (1997b) Biological control of greenhouse whitefly with the parasitoid *Encarsia formosa* on tomato: an individual-based simulation approach. *Biological Control* **9**: 25–47.

van Steenis, M. (1995) Evaluation and Application of Parasitoids for Biological Control of *Aphis gossypii* in Glasshouse Cucumber Crops. PhD thesis: Wageningen Agricultural University, The Netherlands.

Vet, L. E. M. & Dicke, M. (1992) Ecology of infochemical use by natural enemies in a tritrophic context. *Annual Review of Entomology* **37**: 141–72.

Waage, J. K. (1990) Biological theory and the selection of biological control agents. In *Critical Issues in Biological Control*, ed. M. Mackauer, L. E. Ehler & J. Roland, pp. 135–157. Andover: Intercept.

Waage, J. K. & Hassell, M. P. (1982) Parasitoids as biological control agents – a fundamental approach. *Parasitology* **84**: 241–68.

Yano, E., van Lenteren, J. C., Rabbinge, R. & Hulspas-Jordaan, P. M. (1989) The parasite–host relationship between *Encarsia formosa* (Hymenoptera: Aphelinidae) and *Trialeurodes vaporariorum* (Homoptera: Aleyrodidae). XXXI. Simulation studies of population growth of greenhouse whitefly on tomato. *Wageningen Agricultural University Papers* **89.2**: 55–73.

8

Parasitoid adult nutritional ecology: implications for biological control

MARK A. JERVIS AND NEIL A. C. KIDD

Introduction

Biological control practitioners have for long appreciated that food consumption by adults is likely to affect the impact of parasitoids on pest populations (see Jervis *et al.*, 1996a,b). Over the past decade, parasitoid feeding behavior has aroused the interest of ecologists, and new insights have been gained into parasitoid feeding ecology that have important implications for biological control. In the light of theory, we discuss parasitoid feeding behavior and ecology with respect to biological control strategies.

Feeding behavior

'Ovigeny' and dietary requirements

The females of pro-ovigenic species emerge with their near-to-full or full lifetime complement of mature eggs and therefore require little or no materials for egg production. However, those species that deposit their eggs over several days following eclosion are likely to require materials to fuel their locomotory and somatic maintenance requirements, and so their lifetime reproductive success will depend on feeding. The females of synovigenic species eclose with either no mature eggs or just a fraction of the number that can potentially be matured during a lifetime, so they require materials for egg production as well as maintenance (Flanders, 1950; Jervis & Kidd, 1986; van Lenteren *et al.*, 1987). Whether a species is pro-ovigenic or synovigenic will depend both on the extent of 'carry-over' of resources from the last larval instar to the pupa, and on the duration and metabolic costs of pupal development.

Various plant materials and host blood are exploited by parasitoids to supply their maintenance and reproductive needs; these foods differ in their fecundity and survival value (see below).

Honeydew, nectar and pollen feeding

Occurrence among parasitoids

Both male and female parasitoids, belonging to a wide diversity of Hymenoptera and Diptera, consume 'non-host foods'. Honeydew and nectar (either floral or extrafloral) appear to be the most commonly exploited materials under field conditions; pollen has often been purported to be a major food, but gut dissections and direct behavioral observations on feeding parasitoids carried out so far provide little evidence for this in parasitoids other than Bombyliidae (Jervis *et al.*, 1993; Gilbert & Jervis, 1998; Jervis 1998). Even host-feeding parasitoids (those that consume host blood and tissues) consume non-host foods (Jervis & Kidd, 1986).

Effects on fitness

In laboratory experiments carried out on a wide variety of species, females given sugar-rich non-host foods were usually found to be more fecund and longer-lived than adults that were given water and no sugar-rich food; this also applied to some host-feeding parasitoids both when allowed to host-feed and when prevented from doing so (literature reviewed by Jervis & Kidd, 1986; van Lenteren *et al.*, 1987; Gilbert & Jervis, 1998, Jervis, 1998). In *Aphytis melinus* DeBach the consumption of sugar-rich non-host food interacts with host-feeding in influencing lifespan and fecundity (Heimpel *et al.* 1997a). A sugar source needs to be available for the parasitoid to experience the fitness benefits of host-feeding (see

Heimpel & Collier, 1996, for a discussion of the physiological basis of this phenomenon).

Some synovigenic parasitoids, including probably the majority of host-feeding species, are able to resorb eggs when hosts are absent or scarce. Consumption of non-host foods can delay the onset of, and decrease the rate of, egg resorption (Edwards, 1954; Heimpel *et al.*, 1997a).

Both fecundity and longevity tend to vary according to the type of non-host food consumed (see review by Jervis *et al.*, 1996b)

Influence on key foraging decisions

The following predictions can be intuitively derived concerning the role of non-host foods in parasitoid foraging behavior (Jervis *et al.*, 1996b): (i) a pro-ovigenic parasitoid faced with a choice of visiting either a host patch or a food patch should visit the host patch when her energy reserves are high and visit the food patch when her energy reserves are low; (ii) as she approaches the end of her life, oviposition becomes the more likely option; (iii) with synovigenic parasitoids, similar predictions apply, except that the optimal choice will also be influenced by the number of mature eggs (her 'egg load') and developing eggs the female has. Clear empirical support for predictions (i) and (iii) comes from the results of wind tunnel (Lewis & Takasu, 1990) and Y-tube olfactometer (Wackers, 1994) odour choice experiments, and examination of the egg loads of field-collected females (van Emden, 1963; and see discussion in Jervis *et al.*, 1996b).

Host-feeding

Occurrence among parasitoids

Host-feeding behavior has been recorded in the females of 17 families of parasitoid Hymenoptera (Jervis & Kidd, 1986) and two families of Diptera (Clausen, 1940; Nettles, 1987; Disney, 1994). As far as is known, all species are synovigenic. Some species may consume host tissues in addition to host blood, but most consume only blood.

Fitness benefits of host-feeding

The widespread occurrence of host-feeding can be attributed to the benefits it confers over the consumption of other materials (Jervis & Kidd, 1986; Heimpel & Collier, 1996):

1. Compared with plant-derived foods such as nectar and honeydew, host blood is superior as a source of proteinaceous materials and essential vitamins and salts for egg development (see 'Ovigeny' and dietary requirements, above) (Jervis & Kidd, 1986; Heimpel & Collier, 1996). In general, host-feeding parasitoids are more fecund when given hosts (with or without water) than when provided with liquid non-host foods only (see references in Jervis & Kidd, 1986; and in van Lenteren *et al.*, 1987). However, as we have mentioned above (see Honeydew, nectar and pollen feeding section), in *Aphytis melinus* sugar-rich food is required for host-feeding to increase fecundity. The blood of some host species has a more beneficial effect on parasitoid fecundity than that of others (Leius, 1962; Smith & Pimentel, 1968).

2. Hosts are, for some parasitoids, a valuable source of nutrients for maintenance metabolism. It has been shown for several species that host-feeding can, by itself, promote longevity. In others such as *A. melinus* it is insufficient to maintain insects either at all or for more than a few days (see references in Heimpel & Collier, 1996; and Heimpel *et al.*, 1997a).

3. Hosts are a more convenient source of nutrients. For many parasitoids non-host foods are spatially separated from hosts in the field, so searching for them is likely to involve the expenditure of significant amounts of energy and time – two potentially important resources for foraging females (Kidd & Jervis, 1989). Parasitoids may also incur a higher predation risk when foraging for non-host foods, particularly floral materials (Jervis, 1990)

Fitness costs of host-feeding

Host-feeding has costs as well as benefits (Jervis & Kidd, 1986; Kidd & Jervis, 1991a; Heimpel & Collier,

1996). It may bring about a reduction in the quality of the host as an oviposition resource. It may cause the host to die; hosts are killed as a result either of feeding itself or of associated behavior (stinging, mutilation). Destructive host-feeders usually use different host individuals for feeding and oviposition ('non-concurrent destructive host-feeding'). Alternatively, the host may survive the feeding attack ('non-destructive host-feeding') but the quantity of resource available to parasitoid offspring laid in or on it may be reduced, thus adversely affecting progeny fitness (e.g., reduced larval growth, adult size).

If there is such a cost, it may be minimized by the use of lower quality hosts for host-feeding and higher quality hosts for oviposition. In species using the same host stage for host-feeding or oviposition, smaller host individuals are used for feeding (e.g., see Rosenheim & Rosen, 1992). Among destructively host-feeding species the trend is for females to host-feed preferentially or exclusively on earlier host developmental stages (which tend to be smaller in size) and to oviposit preferentially or exclusively in later host stages (which tend to be larger in size) (Kidd & Jervis, 1991a). Such size-selective feeding and oviposition appears to be most common among destructively host-feeding parasitoids of Homoptera. If feeding is restricted to smaller hosts and oviposition to larger ones, mortality of progeny (either from accidental host-feeding by the parent or other mortality factors) is minimized and the size of the oviposition resource used is maximized (Kidd & Jervis, 1991a) (bear in mind the positive relationship between host size and the size and fecundity of progeny in many parasitoids (King, 1989; Liu, 1985; see also references in Godfray, 1994, and in Jervis & Copland,1996). As Godfray (1994) has pointed out, large hosts potentially provide larger meals, but it is likely that large hosts are relatively more valuable as oviposition resources than as adult food resources. Female parasitoids are probably restricted in the amount of blood they can remove from larger hosts, owing to satiation or gut-limitation.

Another potential fitness cost of host-feeding is handling time, which is usually longer for host-feeding than for oviposition (Kidd & Jervis, 1991a, Heimpel & Collier, 1996). Increased handling time

may be an important fitness constraint for parasitoids that are time-limited in terms of lifetime oviposition opportunities, and it may also be costly in terms of predation risk (see Heimpel & Collier, 1996, and references therein; also Heimpel *et al.*, 1997c).

Modeling host-feeding behavior

For non-concurrent destructive host-feeders, the decision whether to oviposit in a host or feed upon it represents a trade-off between current reproduction (oviposition) and investment in future reproduction (feeding) (Collier, 1995; Heimpel & Rosenheim, 1995), and modelers consider host-feeding behavior from this standpoint.

Host-feeding behavior has been modeled in different ways (Jervis & Kidd, 1986; Heimpel & Collier, 1996). Recently, valuable insights have been gained from the dynamic state variable approach (Mangel & Clark, 1988). Dynamic state variable models include several state variables such as the quantity of the parasitoid's nutrient reserves, and/or its egg load, that change when feeding or oviposition occur. In this modeling technique, the individual parasitoid's lifetime is divided into many time-steps within each of which there is a given probability of an encounter with a host. The optimal decision by the parasitoid female maximizes the sum of current fitness gains (from oviposition) and future fitness gains (from feeding). Future fitness gains depend both on changes in the state variable(s) and on the probability of surviving the time-step. With each time-step, the female ages and approaches a 'time horizon' representing either its maximum lifespan or the end of the season (i.e., parasitoid age is implicitly included as a state variable).

Some early dynamic state variable models assume the female parasitoid to be pro-ovigenic, yet no pro-ovigenic parasitoids are known to host-feed. The more realistic models assume nutrients to be used for both egg production and maintenance metabolism, i.e., synovigeny. The empirical finding that in several parasitoid species the fraction of hosts used for host-feeding is a decreasing function of host availability (Fig. 8.1) indicates such a dual function of host-feeding (Godfray, 1994). When hosts are scarce, a

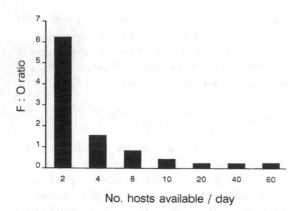

Figure 8.1. The ratio of the number of feeding (F) to oviposition (O) attacks by *Dicondylus indianus* Olmi on *Nilaparvata lugens* (Stål) at different levels of host availability over the lifetime of the wasps. Reproduced from Sahragard *et al.* (1991) with the permission of Blackwell Wissenschafts-Verlag GmbH.

larger fraction of them has to be used for feeding so that the parasitoid can meet its locomotory and other maintenance requirements (Jervis & Kidd, 1986). Models assuming a dual function ('resource pool models': Chan & Godfray, 1993) predict that host-feeding can occur at egg loads greater than zero, and that feeding is more likely at low egg loads. Critical egg loads at and below which host-feeding occurs depend on the female's age or closeness of the time horizon. The propensity to host-feed declines as the female approaches the end of her reproductive life and is thus unable to achieve much in terms of future survival or egg production. The incorporation into models of a realistically long egg maturation time delay (i.e., the period of time taken for a blood meal to be converted to mature eggs) generally raises the critical egg load at which host-feeding is predicted to occur (Collier, 1995). With a delay, the female avoids the risk of being without eggs to lay for an extended period of time.

In one of his models, Collier (1995) examined the effect of egg resorption on the decision to host-feed. Like Jervis & Kidd (1986, 1992) he regarded egg resorption as a 'last resort survival tactic' on the part of the female, with the nutrients from eggs being used to maintain the female until she is able to resume oviposition. Because resorption incurs a metabolic cost, i.e., is less than 100% efficient, host-feeding is predicted by Collier's model to occur at higher critical egg loads.

The prediction that females should host-feed when egg loads are low is supported empirically, although caution has to be exercised in inferring an effect of egg load, because a female's experience can be a confounding factor (Rosenheim & Rosen, 1992; Heimpel & Rosenheim, 1995). The effects of egg load on host-feeding behavior have therefore been examined independently of the parasitoid's history of host encounter. To achieve this, Rosenheim & Rosen (1992) utilized adult size-related differences in egg load, and also manipulated egg load by holding newly eclosed females at different temperatures, thereby varying the rate of ovigenesis. Heimpel & Rosenheim (1995) likewise took advantage of size-related differences in egg load and also used diet and age treatments to manipulate the rate of egg resorption by females. Rosenheim & Rosen (1992) did not detect an effect of egg load in *Aphytis lingnanensis* Compere, whereas Heimpel & Rosenheim (1995) did detect such an effect in *A. melinus*. In a field study, Heimpel *et al.* (1997b) made observations on the behavior of individual free-ranging female *A. aonidiae* (Mercet). After they were observed to handle hosts, wasps were captured and later dissected, to allow their egg load to be measured. Heimpel *et al.* found that, in accordance with the predictions of dynamic state variable models, the likelihood of a host being used for oviposition as opposed to feeding increased with female egg load. They also found that in the field the propensity to oviposit increased with host size (see above).

By using the technique of 'forward iteration' (Mangel & Clark, 1988), predictions can be made with respect to the fraction of available hosts that are used for feeding (see also the models of Jervis & Kidd, 1986). The prediction of most models is that the relationship between the fraction of hosts fed upon and host availability should be dome-shaped (Fig. 8.2). The monotonic decline in the relationship over the upper range (i.e., middle to highest levels) of host availability is, as we have mentioned, supported empirically (Fig. 8.1; for further discussion of empirical work, see Heimpel & Collier, 1996). The func-

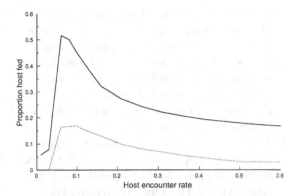

Figure 8.2. The graphical relationship predicted by Chan & Godfray's (1993) 'pro-ovigenic' (dotted line) and 'resource pool' (continuous line) models. Reproduced with the permission of Kluwer Academic and Lippincott Raven Publishers.

tional explanation for the decline in the relationship over the *lowest* levels of host availability is that at such levels the female adopts a 'cutting of losses' tactic; the encounter rate is too low to meet *(via* host-feeding) the insect's energy requirements, and the female oviposits in every host encountered (Jervis & Kidd, 1986; Heimpel *et al.*, 1994).

Models predict that host-feeding is more likely when nutrient reserves/gut contents are at or below a critical level: low levels of nutrient reserves and low gut contents presumably warn of the impending risk of starvation and/or egg limitation (Heimpel & Collier, 1996, see also Jervis & Kidd, 1986). In general, the critical level of nutrient reserves/gut contents depends on the current egg load and *vice versa* (Heimpel & Collier, 1996). Figure 8.3 shows that increasing egg load lowers the nutrient reserves/gut contents threshold at or below which host-feeding occurs.

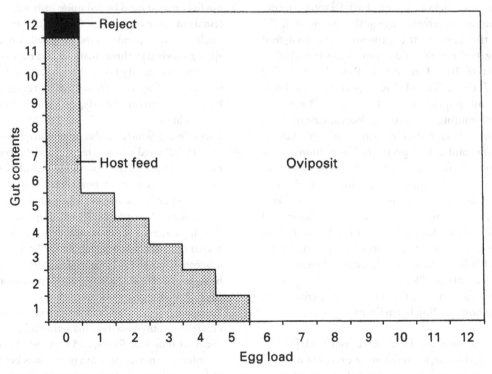

Figure 8.3. The predicted propensity of females to host-feed, as a function of gut contents and egg load (from Heimpel & Collier, 1996, based on Collier's (1995) 'basic model'). Reproduced with the permission of Cambridge University Press.

In experimental studies, caution has to be exercised in inferring any effect of nutritional status on the propensity to host-feed because of the potentially confounding effects of egg load and other factors (Heimpel & Collier, 1996). Heimpel & Rosenheim (1995) separated the effects of diet, egg load and other factors statistically, using a multiple regression model applied to data obtained from laboratory experiments in which females were deprived of hosts prior to host contact and were given a diet of either sucrose and water or yeast and water. Whilst diet treatment was found to covary with egg load in its effect upon behavior, an effect of diet was recorded in the youngest females in which an influence of diet on egg load could not yet be detected. What little empirical evidence exists (see references in Heimpel & Collier, 1996) not surprisingly supports the prediction that the propensity to host-feed is highest when gut contents are low.

As we have seen, models predict that critical egg loads at and below which host-feeding occurs depend upon the female's age or the closeness of the end of the season, the propensity to host-feed declining as the parasitoid approaches the end of its reproductive life. Heimpel & Rosenheim (1995) examined the effect of life expectancy on host-feeding and oviposition in *A. melinus*. The potentially confounding effects of egg load and nutritional status were separated from parasitoid age statistically, using multiple regression. They found from their experiments that, contrary to theoretical predictions, younger parasitoids are no more likely to host-feed than older parasitoids. The most likely explanation for this mismatch between theory and empirical data is that life expectancy in the field is low compared to what is recorded in the laboratory (Heimpel & Rosenheim, 1995; see also Hardy *et al.*, 1992, and Godfray, 1994).

Other factors that need to be considered when modeling host-feeding behavior are:

1. *Storage of nutrients.* In resource pool models developed so far, all nutrients in excess of metabolic demand are directed towards egg production. However, in reality some of the nutrients are likely to be allocated to storage in the fat body

(Collier, 1995a; see also Heimpel *et al.*, 1997a).

2. *Mortality risk independent of aging.* Starvation is one such risk. Not surprisingly, host-feeding is favored as the risk of starvation increases! By contrast, increasing the risk of mortality from stochastic factors that are independent of starvation, e.g., predation, favors oviposition (Chan & Godfray, 1993; Collier, 1995a).

3. *Variation in host quality.* Most models assume the host population to be homogeneous with respect to potential parasitoid fitness gain. Chan (1992, cited in Heimpel & Collier, 1996) relaxed this assumption and modeled the situation where the female encounters hosts that vary in quality for both feeding and oviposition: hosts were assumed to be either of low quality (low number of eggs maturable from host-feeding, low probability of offspring survival) or of high quality. Her models predicted that host-feeding should generally be confined to lower quality hosts, and that higher quality hosts should be fed upon only when the female has no mature eggs. Oviposition in low quality hosts depends on the probability of offspring survival in those hosts and on the availability of higher quality hosts. The divergence between feeding and oviposition is greatest when hosts are abundant and offspring survival probability is high.

4. *Variation in spatial distribution of hosts.* Houston *et al.* (1992) used a pro-ovigenic model incorporating a gamma distribution to encapsulate the higher variance in inter-host intervals that would pertain when hosts are patchily, rather than randomly (or regularly) distributed. High variance (i.e., high degree of host patchiness) results in a higher critical energy threshold for host-feeding: it becomes worthwhile for a parasitoid to feed on hosts in order to stay alive in the expectation of another 'run' of host encounters.

5. *Variation in quality and spatial distribution of non-host foods.* Many host-feeding parasitoids consume non-host foods, and some need to do so in order to experience the major fitness benefits of host-feeding (see Heimpel *et al.*'s (1997a) study of *A. melinus*). So, a valuable next step in modeling host-feeding behavior would be to take

account of the quality and spatial distribution pattern of sugar-rich food sources (Jervis & Kidd, 1995). Models would more closely approximate the natural situation if they also incorporated factor (4) above. The models would need to be of the 'resource pool' type.

Population dynamics

Empirical evidence

It is quite reasonable to assume that, given the effects of diet on parasitoid fecundity, lifespan, and foraging decisions, and the direct effect destructive host-feeding has on host survival, adult feeding behavior plays an important role in influencing host–parasitoid population interactions (compare Chapter 15).

The long-held view of biological control practitioners is that consumption of non-host foods should lead to increases in levels of parasitism and concomitant decreases in host densities. Empirical support for such a relationship comes from several studies, conducted either in the laboratory or in the field, that have demonstrated a higher percentage parasitism of hosts when non-host foods were available, compared with when they were either absent or less available (Allen & Smith, 1958; Telenga, 1958; Chumakova, 1960; Kopvillem, 1960; Guly, 1963; Leius, 1967; National Academy of Sciences, 1969; Foster & Ruesink, 1974; Topham & Beardsley, 1975; Lingren & Lukefahr, 1977; Gallego *et al.*, 1983; Treacy *et al.*, 1987). However, the field evidence is equivocal and there is a need for rigorous field testing of the hypothesis that consumption of non-host foods ultimately results in improved host population suppression (Jervis *et al.*, 1996b).

Except for that relating to parasitoid recruitment rate in cultures, empirical evidence regarding the population dynamic importance of host-feeding behavior (i.e., effects on parasitoid and host equilibrium levels and on stability/persistence of the host–parasitoid interaction) is equivocal (for further discussion see below and see the Biological control, Classical biological control, section below).

What theory tells us

The role of non-host foods in host–parasitoid population dynamics has so far received little attention from theoreticians, and so far only one model explicitly incorporating the consumption of non-host foods has been constructed (Briggs *et al.*, 1995, for a destructive host-feeding parasitoid, see below). Theoreticians have been mainly concerned with the population dynamic consequences of host-feeding behavior.

With concurrent non-destructive host-feeding, oviposition may or may not be accompanied by feeding on the same host individuals. If feeding has no effect on the survival of either the host or the parasitoid's progeny, host–parasitoid population dynamics should be identical to those exhibited by a parasitoid species that does not host-feed, all else being equal (Kidd & Jervis, 1989).

Because of its direct effects on host survival rate, non-concurrent destructive host-feeding behavior has so far attracted most attention from theoreticians. Jervis & Kidd (1986) developed a simple analytical model that relates the number of eggs deposited during the lifetime of an individual, destructively host-feeding, parasitoid to the number of hosts attacked. Kidd & Jervis (1989) incorporated this model into the Nicholson–Bailey population model (see Chapter 3) and a modified (phenomenological) version of the same. So far as population *equilibria* are concerned (H^*, host; P^*, parasitoid), in stabilized versions of the aforementioned population models, H^* is raised and P^* lowered compared with a corresponding stabilized Nicholson–Bailey model not incorporating destructive host-feeding (Fig. 8.4). Note that the models were stabilized using a density-dependent function for the host population; destructive host-feeding *per se* does not affect stability (see below). Also, the models were given the same value for searching efficiency. Therefore, compared with other parasitoids taken as a whole (i.e., including concurrent non-destructive host-feeders and parasitoids that do not host-feed), destructive host-feeders ought to have a less depressive effect on H^*. This prediction is conservative, given that the fecundities of destructive

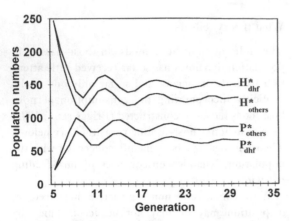

Figure 8.4. Schematic representation of the relative positions of host and parasitoid equilibria predicted by theoretical modeling, for destructive host-feeding parasitoids and 'other' parasitoids. H^*_{dhf} and H^*_{others} are the host population equilibria for interactions involving destructive host-feeders and 'other' parasitoids, respectively, P^*_{dhf} and P^*_{others} are the parasitoid population equilibria for destructive host-feeders and 'other' parasitoids, respectively. Reproduced from Jervis *et al.* (1996b) with the permission of CAB International.

host-feeders are, on average, lower than those of other parasitoids taken as a whole (see Biological control, Classical biological control, below). The precise degree of depression depends on the values of two other parameters in the host-feeding submodel, the greatest depression occurring when the parasitoid's maintenance costs or egg production costs are low.

Using other forms of analytical modeling, Yamamura & Yano (1988) (Lotka–Volterra framework), Huffaker & Gutierrez (1990), Murdoch *et al.* (1992) and Briggs *et al.* (1995) (delay-differential framework) similarly showed that a larger host population is required to maintain the destructively host-feeding parasitoid population, all else (e.g., parasitoid searching efficiency) being equal. Put another way, destructive host-feeders are predicted to have a lower numerical response, i.e., lower rate of recruitment per female, compared with other parasitoids. Note that host-feeding does not provide nutrients for future egg production in the host-feeding model of Murdoch *et al.* (1992), whereas it does in the models of

Kidd & Jervis (1989) and Briggs *et al.* (1995), although the conversion of hosts to eggs is imperfect.

A lower rate of progeny recruitment in destructive host-feeding parasitoids is a phenomenon well-known to workers who maintain cultures of such parasitoids (House, 1980; Waage *et al.*, 1985) (see Biological control, Culturing of parasitoids, below).

Murdoch *et al.* (1992), used a delay-differential stage-structured analytical model (the 'basic model' of Murdoch *et al.* (1987), which collapses to a Lotka–Volterra analog) to explore the population dynamic effects of size-selective host-feeding, coupled with another common behavior among destructively host-feeding parasitoids: size-selective sex allocation (male eggs laid in smaller, i.e., younger hosts, and female eggs in larger, i.e., older, hosts). Size-selective feeding and sex allocation were included as a single component in their model; this is a reasonable first step in modeling, but the two behaviors may in reality be uncoupled to a significant degree. Differential sex allocation was found to accentuate the aforementioned effects of destructive host-feeding on host equilibria, i.e., H^* was higher and P^* was lower.

Population models must contain some degree of physiological realism to determine whether and how destructive host-feeding affects the *stability* and *persistence* of host–parasitoid interactions (Kidd & Jervis, 1989). As pointed out by Kidd & Jervis (1989) and Jervis & Kidd (1992) destructive host-feeding needs to be considered together with its physiological correlates. Accordingly, Kidd & Jervis (1989) incorporated synovigeny (egg limitation, egg resorption, variable longevity) in a stage-structured simulation version of the Nicholson–Bailey model, which had energy as the key modeling currency. The model predicted that if the host population reaches a very low level, destructive host-feeding, acting in conjunction with synovigeny, reduces the likelihood that the parasitoid population will persist for very long. Female parasitoids respond to high host scarcity by resorbing eggs (i.e., parasitoid recruitment rate drops to zero). Egg resorption is insufficient to supply all maintenance and egg production needs (females keep the egg production machinery going simultaneously while practising resorption), so females continue feeding on hosts and force the host population to even lower levels. Eventually, when the encounter rate with hosts is so low that host-feeding can no longer sustain females, the parasitoid population becomes extinct. This mechanism resulting in parasitoid extinction was shown by Kidd & Jervis (1989) to override supposedly powerful stabilizing and persistence-promoting processes such as survival of hosts in refugia.

Briggs *et al.* (1995) achieved a degree of physiological realism in their delay-differential analytical population model by assuming that the population comprises a mix of individuals with different histories of encountering hosts. This variability was expressed in terms of egg load, which controlled the decision whether to host-feed or oviposit (as egg load increases, the propensity to host-feed decreases). Protein was the chosen modeling currency, but it was employed in one model in a similar way to the energy in Kidd & Jervis's (1989) models, inasmuch as materials obtained through feeding were used for maintenance as well as reproduction. The following simplifying assumptions were made so as to allow egg load to control all behavior (and physiology): (i) ingested nutrients move instantaneously from gut to eggs; (ii) they then either drain at a constant rate to supply maintenance requirements or are available for oviposition.

In the 'parasitoid cohort' models of Briggs *et al.* (1995), egg load was a continuous variable (as opposed to taking integer values; this allowed for the continuous use of protein for maintenance purposes), and parasitoid death rate was a continuously decreasing function of egg load. Where there is no drain on materials towards maintenance (as we have seen above, in the Modeling host-feeding behavior section, this is an unrealistic scenario), H^* and P^* are predicted to be stabilized (Fig. 8.5(a)). Where there is a continuous drain of protein for use in maintenance, and this drain is coupled with a dependence on egg load of either parasitoid death rate or birth rate, H^* and P^* are destabilized (Fig. 8.5(b)). Thus, as Kidd & Jervis (1989) showed, it is not destructive host-feeding *per se*, but such feeding together with its physiological correlates that affects the stability/persistence of host–parasitoid population interactions

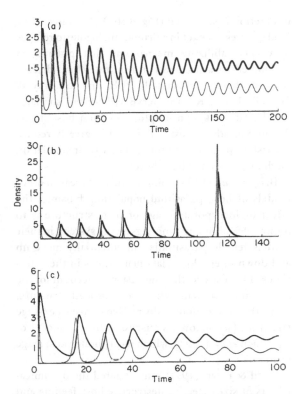

Figure 8.5. Briggs *et al.*'s (1995), simulations from their 'parasitoid cohort' model in which egg load was a continuous variable and parasitoid death rate was a continuously decreasing function of egg load. (a) If there is no drain on materials towards maintenance, the equilibrium is stable; (b) if there is a continuous drain, the equilibrium is unstable; (c) if the parasitoid also feeds on a non-host food source, and materials are allocated to egg production as well as maintenance, the equilibrium is stable. Thin continuous line, hosts; thick continuous line, parasitoids (Note that host equilibria are lower than parasitoid equilibria, because host numbers shown are those that follow parasitoid-induced mortality). Reproduced with the permission of Blackwell Scientific Ltd.

(Briggs *et al.*'s integer egg load model also showed this).

Briggs *et al.* (1995) also showed with their 'parasitoid cohort' model that an input of materials from a non-host source satisfying all maintenance requirements has no effect upon stability, but that if this input satisfies or partly satisfies egg production needs in addition to maintenance needs, the population interaction is stabilized (Fig. 8.5(c)). The alternative food sources thus act in a manner analogous to refuges or other stabilizing mechanisms that provide an uncoupled parasitoid recruitment rate (for a discussion of these, see Hassell, 1978; and Godfray & Hassell, 1994). Kidd & Jervis (1989) had similarly concluded that consumption of non-host foods during periods of host scarcity could greatly reduce the risk of parasitoid extinction that their simulation modeling predicted (see above).

Briggs *et al.* (1995) point out that whereas in other models of host–parasitoid population dynamics the effect of incorporating age or stage-structure is to impose destabilizing time-lags on the system, in their models female parasitoids can potentially move up and down in egg load class many times in their lifetimes. This feature, they suggest, may account for the sometimes stabilizing effects of egg load structure. Note that their models do not have host age/stage structure (see below for discussion of the model of Murdoch *et al.* (1997), which incorporates host stage structure).

Kidd & Jervis (1991b) investigated the dynamical effects of size-selective destructive host-feeding and oviposition using a series of stage-structured simulation models based on the Kidd & Jervis (1989) computer model. With discrete generations, stage discrimination had no effect on stability/persistence, irrespective of either the degree of density dependence acting on the host population or the initial age structure of the host population. With continuous generations, stage discrimination affected stability/persistence at certain parasitoid development times (relative to those of the host) but not at others.

We have already mentioned Murdoch *et al.*'s (1992) modeling of the dynamic effects of size-selective host-feeding and sex allocation. As far as stability is concerned, size-selective behavior results in delayed 'pseudo-density-dependence' in the parasitoid recruitment rate. An increase in the density of searching female parasitoids causes more small immature hosts to be killed rather than older ones. Consequently, the future number of larger immature hosts available for parasitism decreases, and the future per head rate of parasitoid recruitment is further reduced. This mechanism, which operates with a time-lag, causes either instability or stability, the relative extents of the locally stable and unstable parameter space depending on the relative durations of the immature hosts, the adult hosts and the parasitoid immatures (Murdoch *et al.*, 1992; Murdoch, 1994).

Murdoch *et al.* (1997) also explored the consequences of combining realism in parasitoid physiology with host stage-structure and size-selective sex allocation and host-feeding. The model contains the basic behavioral and physiological features of Briggs *et al.*'s (1995) models: the parasitoid population comprises a mix of individuals with different egg loads, and the propensity to host-feed declines with increasing egg load; however, the probability of host-feeding on larger hosts falls off more rapidly than on small hosts with increasing egg load (i.e., smaller hosts are preferred for feeding). The latter feature, as in the Murdoch *et al.* (1992) model, induces delayed density-dependence in parasitoid recruitment, but the magnitude of the effect depends on the value of the parameter that determines the probability with which early host immatures are fed upon (or used for oviposition) relative to the value of that which determines the probability with which late host immatures are fed upon (or used for oviposition). This model was developed further by taking account of egg limitation. Their model allows the rate of host-feeding to increase linearly with the rate of encounter with hosts. They note, however, that in reality the rate at which new eggs can be produced from blood meals must have an upper limit, for example due to gut limitation. They therefore made feeding rate a decelerating function of host density (i.e., a saturating, Type 2 'functional response'). The main effect of this is to limit the rate of feeding on small host stages, thereby limiting the disparity in the recruitment of female offspring from large *versus* small hosts, but no essentially new dynamics arise.

Murdoch *et al.* (1997) also showed that size-selective host-feeding reinforces the delayed density-dependence in the parasitoid recruitment rate that is induced by size-selective clutch size allocation.

Some destructive host-feeding parasitoids feed on non-host (i.e., prey *sensu stricto*) insect species as well as host species (Bartlett, 1964; Cole, 1967; Sandlan,

1979; Chua & Dyck, 1982; Schaupp & Kulman, 1992; A. Sahragard, unpublished results). Briggs *et al.* (1995) point out that feeding on the non-host species will be equivalent in its population dynamic effects to consumption of non-host food, the proviso being that the dynamics of the non-host population are independent of those of the host–parasitoid interaction.

A valuable, but undoubtedly difficult, next step in modeling the effects of host-feeding behavior on host–parasitoid population dynamics would be to incorporate variation in the spatial distribution of both hosts and non-host food sources, with the latter varying also in quality.

Biological control

Classical biological control

Is destructive host-feeding a positive attribute of parasitoids?

Many biocontrol practitioners would answer yes to this question, both because that type of host-feeding is an additional source of host mortality to parasitism and because some parasitoid species may at times kill more hosts by feeding than by parasitism (an observation first made by Johnston, 1915). Presumably because it has strong intuitive appeal, this assumption has rarely been challenged (Flanders, 1953; Kidd & Jervis, 1989; Jervis *et al.*, 1996a); this does *not* mean it is correct, however.

Two conventional measures of parasitoid efficacy in classical biocontrol, establishment rate and success rate, can be used to compare destructive host-feeders with 'other' parasitoids and so provide an answer to the above question. We deal with these in turn below.

Establishment rate

At least two verbal models of parasitoid establishment probability come to mind (Jervis *et al.*, 1996a; for a critique of the assumptions of the following models, see that paper).

Model 1. Biological control practitioners who consider destructive host-feeding to be a desirable attribute (e.g., Chandra, 1980; Greathead, 1986; Olmi, 1994) evidently assume that, compared with 'other' parasitoids (those that either feed non-destructively or do not host-feed at all), destructive host-feeders oviposit in an equivalent proportion of the hosts available, and that those used for feeding contribute an additional fraction of the hosts killed. However, no empirical evidence for equivalence in fecundity/ searching efficiency between the two types of parasitoid has ever been given in support of this hypothesis. Assuming this model to reflect reality, the probability of progeny being present in the next host generation (i.e., the probability of establishment) ought, all else (e.g., host density) being equal, to be the same for the two types of parasitoid.

Model 2. Takes into account the diversity in reproductive biology, in particular fecundity, that exists among parasitoids (see Jervis *et al.*, 1996a for details). All else being equal, destructive host-feeders can be expected to have, on average, a much lower probability of establishment than 'other' parasitoids, because of their lower lifetime fecundities: whereas destructive host-feeders comprise mainly anhydropic–synovigenic species (which mostly have low lifetime fecundities), 'other' parasitoids also include many hydropic–synovigenic species (which mostly have much higher lifetime fecundities) (Jervis *et al.* 1996a). (In parasitoid wasps, yolk-deficient eggs are termed hydropic, whereas yolk-rich eggs are termed anhydropic).

These predictions were tested by analyses of historical data (the BIOCAT database: Greathead & Greathead, 1992) on classical biological control introductions (Jervis *et al.*, 1996a). The 'total introductions' and 'average fraction' methods were applied both to the data set on introductions of hymenopteran parasitoids against Homoptera and to a smaller data set on introductions against mealybugs and scale insects. With the 'total introductions' method, data on the total number of introductions are pooled without regard to the distribution of

Table 8.1. *Rates of parasitoid establishment and success, calculated (as percentages) from data on parasitoid introductions against homopteran pests contained in the BIOCAT database*

	Establishment rate		Success rate	
	DHF	Others	DHF	Others
1. *Total introductions method*				
All Homoptera	64	55	45	30
		*		***
			$71^\dagger$	$55^\dagger$
				***
Scale insects and mealybugs only	67	45	39	21
		***		***
			$59^\dagger$	$47^\dagger$
				*
2. *Average fraction method*				
All Homoptera	55	43	30	22
		*		NS
Scale insects and mealybugs only	49	36	30	17
		NS		**

Notes:

DHF, destructive host-feeders; NS, not significant; $*p<0.05$; $**p<0.01$; $***p<0.001$.
†success rate calculated as a proportion of all establishments as opposed to being calculated as a proportion of introductions. See the text for further explanation.

attempts against particular species; whereas with the 'average fraction' method, the data are weighted by accounting for repeated introductions against particular hosts (Stiling, 1990). The aforementioned analyses showed destructive host-feeders to be either just as likely or *more* likely, to become established as 'other' parasitoids considered as a whole (Jervis *et al.*, 1996a: Table 8.1).

Why, as far as establishment rate is concerned, might destructive host-feeders be *better than*, rather than as good as (Model 1) or worse than (Model 2) 'other' parasitoids? One obvious reason is that Models 1 and 2 are poor descriptions of real population processes. There are, indeed, grounds for regarding the models as lacking in realism (Jervis *et al.*, 1996a):

1. 'Other parasitoids' may not have as high a probability of establishment as predicted by Model 2, because their higher fecundity is likely to be offset to a large extent by a higher mortality of progeny (a well-known trade-off, see Price, 1975); that is, the assumption in this model of 'all else being equal' almost certainly does not apply.

2. Model 2 may not apply when only parasitoids of scale insects and mealybugs are considered. Destructively host-feeding parasitoids cannot necessarily be expected to be poorer at establishing, given that the 'other parasitoids' in this case (Aphelinidae and Encyrtidae) are mostly anhydropic–synovigenic, and their fecundity should be the same (Model 1) or in a few cases even lower (Jervis *et al.*, 1996a). Note, however, that when scale insect/mealybug data only are analyzed, use of the total introductions method indicates superiority whereas use of the average fraction method indicates equality.

3. In classical biological control, parasitoids are usually introduced into a high density pest population. As we have mentioned above

(Feeding behavior, Host-feeding, section), destructive host-feeding parasitoids oviposit in a larger fraction of hosts at higher levels of host availability (DeBach, 1943; Collins *et al.*, 1981; Lohr *et al.*, 1988; Bai & Mackauer, 1990; Sahragard *et al.*, 1991). Therefore, it is likely that when they are introduced, their fecundity will be at its highest and their feeding will be at its lowest; non-destructive host-feeders likewise vary their feeding-to-oviposition ratio in relation to changing host availability (Jervis & Kidd, 1986), but they constitute a small proportion of 'other' parasitoids, so the fecundity and therefore the establishment probability of 'other' parasitoids taken as a whole will not be raised significantly as a result of such behavior.

It is difficult to see how the above factors could account for the overall trend in superiority of destructive host-feeders with respect to the establishment rate recorded in Jervis *et al.*'s (1996) analyses (see Table 8.1).

Success rate

Biological control practitioners measure the success of introduced biocontrol agents in controlling pests in economic terms: introduction must, at the very least, result in a reduction in the frequency of pest outbreaks and/or the frequency of pesticide applications (DeBach, 1971). From an ecological perspective, the introduced parasitoid has both to depress the pest population below a certain level and to prevent it from again reaching that level by promoting stability (Waage & Hassell, 1982).

As we have seen (see Population Dynamics, above), population models predict that destructive host-feeders depress H^* less strongly than do 'other' parasitoids. The predictions of these models with respect to stability/parasitoid persistence are, unfortunately, equivocal, because they depend on the nature of the components incorporated. However, successful pest control by natural enemies can result from mere suppression. For example, a density-independent mortality by natural enemies may suppress pest numbers sufficiently for them to be regulated by other, density-dependent, factors (e.g., see Roland, 1994, and Kidd & Jervis, 1997).

We can reasonably conclude from the predicted equilibrium effects alone that destructive host-feeders ought to be less successful biocontrol agents compared to 'other' parasitoids. However, analyses (the total introductions, and the average fraction methods, see above) of the BIOCAT database carried out by Jervis *et al.* (1996a) show that when introductions of Hymenoptera against either Homoptera in general or just mealybugs and scale insects are considered, destructive host-feeders show a significantly *higher*, not lower, success rate than 'other' parasitoids taken as a whole (see Table 8.1).

This disparity between the predictions of population dynamics theory and historical data on introductions suggests that with destructive host-feeders: (i) H^* is depressed to a greater extent than predicted, and/or (ii) the host population is not destabilized. In *Aphytis melinus*, under laboratory conditions, host-feeding can contribute as much as 75% to the lifetime realized fecundity (Heimpel *et al.*, 1997a) suggesting that models seriously underestimate the contribution of host-feeding to *per capita* recruitment rate and therefore underestimate the suppressive power of destructive host-feeders. However, despite the contribution of destructive host-feeding, the average lifetime fecundity of destructive host-feeding parasitoids is significantly lower than that of 'other' parasitoids (see above), so models do not seriously misrepresent the field situation. Perhaps under field conditions females of destructive host-feeding species rely less upon hosts as a food source when non-host foods are available (Kidd & Jervis, 1989; Jervis *et al.*, 1996a); laboratory studies suggest so (House, 1980; Heimpel & Rosenheim, 1995). The parasitoid population, by consuming nectar and honeydew, would require a smaller host population to maintain itself i.e., H^* would be lower than predicted. Because, however, these foods contain very small quantities of proteinaceous materials, which are presumably important for egg production in host-feeding parasitoids, it is unlikely that the host equilibrium would be reduced to the level achieved in interactions involving 'other' parasitoids. By the same token, consumption of non-host foods could stabilize an otherwise unstable host–parasitoid interaction, either in the manner predicted by the Kidd & Jervis (1989) model or in the manner predicted by the Briggs *et al.* (1995) model.

Regarding persistence of the parasitoid population, the lower destructive host-feeding parasitoid equilibrium (P^*_{dhf}, Fig. 8.5) ought to make the risk of extinction more likely and thus jeopardize long-term pest control. This prediction is, however, unlikely to be realized in nature because the aforementioned population models are concerned with *local populations*, not the parasitoid and pest *metapopulations* that exist under field conditions. Local extinction of parasitoid populations is unlikely to jeopardize long-term biological control provided the parasitoid survives in other local populations of the pest metapopulation and can re-invade localities where previously it has become extinct (Murdoch, *et al.*, 1985; and see Chapter 2) (there is the proviso, however, that a time-delay in re-invasion could increase pest population fluctuations that may prove detrimental to control). Waage (1990) pointed out that in classical biological control no failure appears to have occurred through the lack of persistence of an established agent, thus lending support to the view that field metapopulation dynamic processes render the predictions of local dynamics-based population models invalid.

Destructive host-feeders, despite the predictions of population dynamics theory, appear to be superior biological control agents compared with 'other' parasitoids; they are certainly no worse. However, because the BIOCAT database may (like any other such database) be flawed (see Waage & Mills, 1992), Jervis *et al.* (1996a) concluded that it would be unwise to employ destructive host-feeding as the sole, or even primary selection criterion when seeking agents for introduction.

Finally, it is perhaps a mistake to seek an explanation for the disparity between theory and database analysis in terms only of parasitoid reproductive biology, feeding ecology, or database reliability. Destructive host-feeders may be intrinsically superior with respect to some other, as yet unidentified, life-history attributes.

Other claims regarding the role of non-host foods in classical biological control

Biological control practitioners have for long held the view that in classical biological control programs parasitoids should have a ready supply of non-host foods in areas where they are released, in order to increase the prospects of the agents becoming established (d'Emmerez de Charmoy, 1917; Box, 1927; Wolcott, 1942; Leius, 1960), although it appears from the biological control literature that few workers have put it into practice.

Lack of non-host food ranks 10th among 13 cited reasons for failure (i.e., failure to control the pest adequately: this includes failure in establishing) in classical biological control programs involving parasitoids (analysis performed on the database in Stiling, 1993, using only his information on parasitoids). This probably undervalues to some extent the real importance of food availability in determining the outcomes of programs; the very low ranking probably reflects in part the subjective biases of those reporting on outcomes, and it ignores the possible role of food availability as a cofactor.

Focusing on a particular agroecosystem, success rates in Mediterranean olives are much lower than expected for an orchard system, while in forest systems worldwide both establishment and success rate are lower than in orchard and amenity systems. It has been hypothesized in both cases that a deficiency of non-host food sources has contributed to the poor performance of agents (Jervis *et al.*, 1992; Kidd & Jervis, 1997).

Conservation and augmentation

As we have shown, the body of evidence gathered so far suggests that the availability of non-host foods is likely to play an important role in host–parasitoid population dynamics, through its effects on fecundity, longevity, key foraging decisions and, ultimately, searching efficiency (see Chapter 15). This applies not only to introduced but also to indigenous parasitoids.

Field provision of non-host food sources normally falls under the heading of 'conservation', according to DeBach & Rosen (1991). However, since it can potentially increase parasitoid recruitment rate by increasing fecundity and, in most cases, also increase parasitoid immigration, it should also be seen as an augmentation technique (Jervis *et al.* 1996b). Jervis *et al.* (1996b) provide advice on protocol and also identify potential pitfalls of the technique. The essential point arising from their discussion is that provision of non-host food, even if the food is highly suitable nutritionally, may not improve pest control.

Inundative release

With inundative release, in contrast to classical biological control, the actual or potential effects of parasitoids on host equilibrium levels are not a consideration. Given both their (on average) lower fecundity compared with other parasitoids and their tendency to host-feed less at high levels of host availability (i.e., host outbreak conditions), destructive host-feeding parasitoids should, in theory, tend to be less effective pest control agents when employed in inundative release.

Parasitoid release strategy

The realization that female nutritional state influences key foraging decisions stimulated Takasu & Lewis (1995) to explore the implications for field release protocol. Their findings suggest that in both classical biocontrol and inundative release programs parasitoids should be released when experiencing a degree of hunger (for an exception, see below), provided that non-host foods are readily available in the release areas. Female *Microplitis croceipes* (Cresson) that were either fed on honey for 2 days post-eclosion ('well fed') or were not fed at all for the same period of time ('unfed'), were introduced into corn and soybean plots containing either: (i) some plants with a host caterpillar attached to them and others with an artificial food source (a droplet of honey) attached to them, or (ii) some plants that bore a host and others

that were 'clean'. Prior to release, all females, whether 'well fed' or 'unfed' were provided with a honey droplet for 10 s and then allowed to sting two host larvae on a host-damaged plant; this enabled females to learn to associate particular olfactory stimuli with food and hosts.

In plots where food was absent 'well fed' females parasitized more hosts than 'unfed' females, and spent more of the time available actively searching for hosts. In plots where food was provided, 'unfed' females quickly found food and rapidly commenced searching for hosts, parasitizing as many hosts as (soybean plots) or more hosts than (corn plots) 'well fed' females. Takasu & Lewis suggested that the on average higher parasitism by 'unfed' females in the latter case can be attributed to the wasps learning to associate plant-related olfactory or physical stimuli with the artificial food; plants bearing the latter occurred in close proximity to host-bearing plants. 'Well fed' females, they point out, had learned to associate particular olfactory stimuli with feeding only in the laboratory prior to release. So, Takasu & Lewis argue, by manipulating the hunger state of females prior to release (the insects must be given water) into a crop where food is present, both better retention of parasitoids in the crop following release and higher levels of parasitism can be achieved.

Takasu & Lewis (1995) identified a further need for foods to be readily available to released parasitoids: 'unfed' females introduced into corn plots lacking food showed a greater tendency to superparasitize hosts than 'well fed' females. Fletcher *et al.* (1994) working on *Venturia canescens* (Gravenhorst), had similarly shown that starved females superparasitize more frequently than females given a non-host food. These findings support the following prediction of foraging models (Iwasa *et al.*, 1984; Charnov & Skinner, 1985; Mangel, 1987; 1989a,b; Charnov & Stephens, 1988; Visser *et al.*, 1992a,b; Weisser & Houston, 1993): if host handling times and mortality risks whilst handling do not differ for unparasitized and parasitized hosts, females ought to be less choosy if their life expectancy is low.

Destructive host-feeding parasitoids, by contrast, should be released when well fed, not hungry, in classical biological control programs unless, that is, one can ensure that females have attained a high egg load (see Fig. 8.3). If females are released in a physiological condition predisposing them to host-feed rather than oviposit (e.g., low gut contents, low egg load), there will be a lower likelihood of the species becoming established (see Biological control, Classical biological control, above).

Seasonal inoculative release

Destructive host-feeding by parasitoids presents a particular problem in seasonal inoculative release. If *Encarsia formosa* Gahan is released into the glasshouse crop during the very early stages of an infestation by the whitefly *Trialeurodes vaporariorum* (Westwood), i.e., when early immature stages predominate, the mortality from destructive host-feeding may prevent the whitefly population from increasing in subsequent generations (Hussey, 1985). Consequently, the risk of parasitoid extinction will increase (just as likely, the lack of later host stages for oviposition will contribute to parasitoid extinction by seriously constraining the rate of parasitoid recruitment into the second generation (Kidd & Jervis, 1991b)). In order to circumvent this problem it is therefore important to precisely time parasitoid releases to coincide with an appropriate host population age structure and to do so using an appropriate host:parasitoid population ratio (Hussey, 1985).

Culturing of parasitoids

Parasitoid productivity problems with cultures of destructive host-feeding parasitoids are well known to parasitoid workers (e.g., see Waage *et al.*, 1985). As predicted by population models (see Population Dynamics, above), compared with other parasitoids, a higher input of hosts is required for destructive host-feeders in order for there to be an equivalent level of culture productivity (i.e., parasitoid recruitment rate). Productivity will be least (and the risk of culture extinction greatest) if host densities reach low levels, due to a higher ratio of feeding-to-oviposition attacks by females. Productivity problems may be alleviated or overcome by the provision of nonhost foods. For example, Heimpel & Rosenheim

(1995) showed that female *Aphytis melinus* provided with a 50% sucrose solution containing 5% (by weight) of yeast hydrolysate had both high egg loads and a significantly lower tendency to host-feed. This suggests that in cultures of *A. melinus* host mortality resulting from destructive host-feeding can be significantly reduced. In *Itoplectis conquisitor* (Say) a fourfold increase in parasitoid population was obtained after parasitoids in culture had been provided for more than a year with a simple artificial adult diet. Prior to the artificial diet being used, substantial numbers of hosts were required to satisfy the parasitoids' host-feeding requirement (around a third of all the host pupae supplied for culturing: House, 1980). Such a marked increase in parasitoid numbers suggests that the artificial diet supplied to *I. conquisitor* was nutritionally far superior to host blood.

There is also a need, in cultures of destructively host-feeding parasitoids of Homoptera, for an appropriate starting host population age structure as well as host:parasitoid population ratio (see Seasonal inoculative release, above).

Concluding remarks

Parasitoid adult feeding ecology is a discipline currently undergoing a rapid transition from being based largely on intuition to being based on a significant body of theory and supporting empirical data; mathematical modeling has provided much of the impetus for this change. However, we are still far from having a feeding ecology-based *practical rationale* for the manipulation and management of parasitoids in biological control. Biological control practitioners have, for whatever reasons: (i) often not implemented (many appear not to have even considered implementing) the recommendations of theoreticians (it is noteworthy that in a recent symposium concerned with opportunities for improving biological control (Waage, 1996), feeding ecology is mentioned only very cursorily); and (ii) contributed little to the development of theoretical models of parasitoid adult feeding behavior and ecology. It is imperative that they and theoreticians combine and coordinate their efforts in order that the goal of a practical rationale can be more rapidly achieved.

Acknowledgements

We are very grateful to George Heimpel for allowing us to quote from several of his manuscripts prior to their publication, and for sharing his thoughts on parasitoid feeding ecology generally. We are also very grateful to Bill Murdoch for allowing us to quote from one of his manuscripts prior to publication, and to Buck Cornell and two anonymous reviewers for their comments on our manuscript.

References

Allen, W. W. & Smith, R. F. (1958) Some factors influencing the efficiency of *Apanteles medicaginis* Muesebeck (Hymenoptera: Braconidae) as a parasite of the alfalfa caterpillar, *Colias philodice eurytheme* Boisduval. *Hilgardia* 28: 1–42.

Bai, B. & Mackauer, M. (1990) Oviposition and host-feeding patterns in *Aphelinus asychis* (Hymenoptera: Aphelinidae) at different host densities. *Ecological Entomology* 15: 9–16.

Bartlett, B. L. (1964) Patterns in the host-feeding habit of adult parasitic Hymenoptera. *Annals of the Entomological Society of America* 57: 344–50.

Box, H. E. (1927) The introduction of braconid parasites of *Diatraea saccharalis* Fabr., into certain of the West Indian islands. *Bulletin of Entomological Research* 28: 365–70.

Briggs, C. J., Nisbet, R. M., Murdoch, W. W., Collier, T. R. & Metz, J. A. J. (1995) Dynamical effects of host feeding in parasitoids. *Journal of Animal Ecology* 64: 403–16.

Chan, M. S. & Godfray, H. C. J. (1993) Host-feeding strategies of parasitoid wasps. *Evolutionary Ecology* 7: 593–604.

Chandra, K. (1980) Taxonomy and bionomics of the insect parasites of rice leafhoppers and planthoppers in the Philippines and their importance in natural biological control. *Philippine Entomologist* 4: 119–39.

Charnov, E. L. & Skinner, S. W. (1985) Complementary approaches to the understanding of parasitoid oviposition decisions. *Environmental Entomology* 14: 383–91.

Charnov, E. L. & Stephens, D. W. (1988) On the evolution of host selection in solitary parasitoids. *American Naturalist* 132: 707–22.

Chua, T. H. & Dyck, V. A. (1982) Assessment of *Pseudogonatopus flavifemur* E. & H. (Hymenoptera: Dryinidae) as a biocontrol agent of the rice brown planthopper. *Proceedings of the International Conference on Plant Protection in the Tropics, Kuala Lumpur, March 1982*, pp. 253–365.

Chumakova, B. M. (1960) Supplementary feeding as a factor increasing the activity of parasites of harmful insects. *Trudy Vsesoyuznogo Nauchnoissledovatelscogo Instituta Zaschchity Rastenii* 15: 57–70.

Clausen, C. P. (1940) *Entomophagous Insects*. New York: McGraw-Hill.

Cole, L. R. (1967) A study of the life-cycles and hosts of some Ichneumonidae attacking pupae of the green oak-leaf roller moth, *Tortrix viridana* (L.) (Lepidoptera: Tortricidae) in England. *Transactions of the Royal Entomological Society of London* 119: 267–81.

Collier, T. R. (1995) Adding physiological realism to dynamic state variable models of parasitoid host feeding. *Evolutionary Ecology* 9: 217–35.

Collins, M. D., Ward, S. A. & Dixon, A. F. G. (1981) Handling time and the functional response of *Aphelinus thomsoni*, a predator and parasite of the aphid *Drepanosiphum platanoidis*. *Journal of Animal Ecology* 50: 479–87.

DeBach, P. (1943) The importance of host feeding by adult parasites in the reduction of host populations. *Journal of Economic Entomology* 36: 647–58.

DeBach, P. (1971) The use of natural enemies in insect pest management ecology. *Proceedings of the Tall Timbers Conference on Ecological Animal Control by Habitat Management* 3: 211–33.

DeBach, P. & Rosen, D. (1991) *Biological Control by Natural Enemies*. Cambridge: Cambridge University Press.

Disney, R. H. L. (1994) *Scuttle Flies: The Phoridae*. London: Chapman and Hall.

Edwards, R. L. (1954) The effect of diet on egg maturation and resorption in *Mormoniella vitripennis* (Hymenoptera, Pteromalidae). *Quarterly Journal of Microscopical Science* 95: 459–68.

d'Emmerez de Charmoy, D. (1917) Notes relative to the importation of *Tiphia parallela* Smith, from Barbados to Mauritius for the control of *Phytalus smithi* Arrow. *Bulletin of Entomological Research* 8: 93–102.

Flanders, S. E. (1950) Regulation of ovulation and egg disposal in the parasitic Hymenoptera. *Canadian Entomologist* 82: 134–40.

Flanders, S. E. (1953) Predatism by the adult hymenopterous parasite and its role in biological control. *Journal of Economic Entomology* **46**: 541–4.

Fletcher, J. P., Hughes, J. P. & Harvey, I. F. (1994) Life expectancy and egg load affect oviposition decisions of a solitary parasitoid. *Proceedings of the Royal Society of London, Series B* **258**: 163–7.

Foster, M. A. & Ruesink, W. G. (1974) Influence of flowering weeds associated with reduced tillage in corn on a black cutworm (Lepidoptera: Noctuidae) parasitoid, *Meteorus rubens* (Ness von Esenbeck). *Environmental Entomology* **13**: 664–8.

Gallego, V. C., Baltazar, C. R., Cadapan, E. P. & Abad, R. G. (1983) Some ecological studies on the coconut leafminer, *Promecotheca cumingii* Baly (Coleoptera: Hispidae) and its hymenopterous parasitoids in the Philippines. *Philippine Entomologist* **5**: 417–94.

Gilbert, F. S. & Jervis, M. A. (1998) Functional, evolutionary and ecological aspects of feeding-related mouthpart specializations in parasitoid flies. *Biological Journal of the Linnean Society* **63**: 495–535.

Godfray, H. C. J. (1994) *Parasitoids: Behavioural and Evolutionary Ecology*. Princeton: Princeton University Press.

Godfray, H. C. J. & Hassell, M. P. (1994) How can parasitoids regulate the population densities of their hosts? *Norwegian Journal of Agricultural Sciences, Supplement* **16**: 41–57.

Greathead, D. J. (1986) Parasitoids in biological control. In *Insect Parasitoids*, ed. J. K. Waage & D. J. Greathead, pp. 289–318. London: Academic Press.

Greathead, D. J. & Greathead, A. H. (1992) Biological control of insect pests by insect parasitoids and predators: the BIOCAT database. *Biocontrol News and Information* **13**: 61N-8N.

Guly, V. V. (1963) On the increase in the efficiency of parasites of hibernating shoots of *Evetria buoliana* Schiff. in pine stands on northern slopes of the central Caucasus. *Zoologicheskii Zhurnal* **42**: 1414–15.

Hardy, I. C. W., Griffiths, N. T. & Godfray, H. C. J. (1992) Clutch size in a parasitoid wasp: a manipulation experiment. *Journal of Animal Ecology* **61**: 121–9.

Hassell, M. P. (1978) *The Dynamics of Arthropod Predator–Prey Systems*. Princeton: Princeton University Press.

Heimpel, G. E. & Collier, T. R. (1996) The evolution of host-feeding behaviour in insect parasitoids. *Biological Reviews* **71**: 373–400.

Heimpel, G. E. & Rosenheim, J. A. (1995) Dynamic host feeding in the parasitoid *Aphytis melinus*: the balance between current and future reproduction. *Journal of Animal Ecology* **64**: 153–67.

Heimpel, G. E., Rosenheim, J. A. & Adams, J. M. (1994) Behavioural ecology of host feeding in *Aphytis* parasitoids. *Norwegian Journal of Agricultural Sciences, Supplement* **16**: 101–15.

Heimpel, G. E., Rosenheim, J. A. & Kattari, D. (1997a) Adult feeding and lifetime reproductive success in the parasitoid *Aphytis melinus*. *Entomologia Experimentalis et Applicata* **83**: 305–15.

Heimpel, G. E., Rosenheim, J. A. & Mangel, M. (1997b) Egg limitation, host quality and dynamic behaviour by a parasitoid in the field. *Ecology* **77**: 2410–20.

Heimpel, G. E., Rosenheim, J. A. & Mangel, M. (1997c) Predation on adult *Aphytis* parasitoids in the field. *Oecologia* **110**: 346–52.

House, H. L. (1980) Artificial diets for the adult parasitoid *Itoplectis conquisitor* (Hymenoptera: Ichneumonidae). *Canadian Entomologist* **112**: 315–20.

Houston, A. I., McNamara, J. M. & Godfray, H. C. J. (1992) The effect of variability on host feeding and reproductive success in parasitoids. *Bulletin of Mathematical Biology* **54**: 465–76.

Huffaker, C. B. & Gutierrez, A. D. (1990) Evaluation of the efficiency of natural enemies in biological control. In *Armored Scale Insects: Their Biology, Natural Enemies and Control*, ed. D. Rosen, pp. 473–95. Amsterdam: Elsevier.

Hussey, N. W. (1985) Whitefly control by parasites. In *Biological Control: The Glasshouse Experience*, ed. N. W. Hussey & N. Scopes, pp. 104–15. Poole: Blandford Press.

Iwasa, Y., Suzuki, Y. & Matsuda, H. (1984) Theory of oviposition strategy of parasitoids. I. Effects of mortality and limited egg number. *Theoretical Population Biology* **26**: 205–27.

Jervis, M. A. (1990) Predation of *Lissonota coracinus* (Gmelin) (Hymenoptera: Ichneumonidae) by *Dolichonabis limbatus* (Dahlbom) (Hemiptera: Nabidae). *Entomologist's Gazette* **41**: 231–3.

Jervis, M. A. (1998) Functional and evolutionary aspects of mouthpart structure in parasitoid wasps. *Biological Journal of the Linnean Society* **63**: 461–93.

Jervis, M. A. & Copland, M. J. W. (1996) The life cycle. In *Insect Natural Enemies: Practical Approaches to their Study and Evaluation*, ed. M. A. Jervis & N. A. C. Kidd, pp. 63–161. London: Chapman and Hall.

Jervis, M. A. & Kidd, N. A. C. (1986) Host-feeding strategies in hymenopteran parasitoids. *Biological Reviews* **61**: 395–434.

Jervis, M. A. & Kidd, N. A. C. (1992) The dynamic significance of host-feeding by parasitoids – what modellers ought to consider. *Oikos* **62**: 97–9.

Jervis, M. A. & Kidd, N. A. C. (1995) Incorporating physiological realism into models of parasitoid feeding behaviour. *Trends in Ecology and Evolution* **10**: 434–6.

Jervis, M. A., Kidd, N. A. C., McEwen, P., Campos, M. & Lozano, C. (1992) Biological control strategies in olive pest management. In *Research Collaboration in European IPM Systems*, ed. P. T. Haskell, pp. 31–9. BCPC Monographs no. 52. Farnham: British Crop Protection Council.

Jervis, M. A., Kidd, N. A. C., Fitton, M. G., Huddleston, T. & Dawah, H. A. (1993) Flower-visiting by hymenopteran parasitoids. *Journal of Natural History* **27**: 67–105.

Jervis, M. A., Hawkins, B. A. & Kidd, N. A. C. (1996a) The usefulness of destructive host feeding parasitoids in classical biological control: theory and observation conflict. *Ecological Entomology* **21**: 41–6.

Jervis, M. A., Kidd, N. A. C. & Heimpel, G. E. (1996b) Parasitoid adult feeding behaviour and biocontrol – a review. *Biocontrol News and Information* **17**: 11N–26N.

Johnston, F. A. (1915) Asparagus-beetle egg parasite. *Journal of Agricultural Research* **4**: 303–14.

Kidd, N. A. C. & Jervis, M. A. (1989) The effects of host-feeding behaviour on the dynamics of parasitoid-host interactions, and the implications for biological control. *Researches on Population Ecology* **31**: 235–74.

Kidd, N. A. C. & Jervis, M. A. (1991a) Host-feeding and oviposition strategies of parasitoids in relation to host stage. *Researches on Population Ecology* **33**: 13–28.

Kidd, N. A. C. & Jervis, M. A. (1991b) Host-feeding and oviposition by parasitoids in relation to host stage: consequences for parasitoid–host population dynamics. *Researches on Population Ecology* **33**: 87–99.

Kidd, N. A. C. & Jervis, M. A. (1997) The impact of natural enemies on forest insect populations. In *Forests and Insects*, ed. A. D. Watt, N. Stork & M. D. Hunter, pp. 49–68. London: Academic Press.

King, B. H. (1989) Host-size dependent sex ratios among parasitoid wasps: does host growth matter? *Oecologia* **78**: 420–6.

Kopvillem, H. G. (1960) Nectar plants for the attraction of entomophagous insects. *Zaschchita Rastenii ot Vreditelei i Boleznii* **5**: 33–4.

Leius, K. (1960) Attractiveness of different foods and flowers to the adults of some hymenopterous parasites. *Canadian Entomologist* **92**: 369–76.

Leius, K. (1962) Effects of body fluids of various host larvae on fecundity of females of *Scambus buolianae* (Htg.) (Hymenoptera: Ichneumonidae). *Canadian Entomologist* **94**: 1078–82.

Leius, K. (1967) Influence of wild flowers on parasitism of tent caterpillar and codling moth. *Canadian Entomologist* **99**: 444–6.

Lewis, W. J. & Takasu, K. (1990) Use of learned odours by a parasitic wasp in accordance with host and food needs. *Nature* **348**: 635–6.

Lingren, P. D. & Lukefahr, M. J. (1977) Effects of nectariless cotton on caged populations of *Campoletis sonorensis*. *Environmental Entomology* **6**: 586–8.

Liu, S.-S. (1985) Development, adult size and fecundity of *Aphidius sonchi* reared in two instars of its aphid host, *Hyperomyzus lactucae*. *Entomologia Experimentalis et Applicata* **37**: 41–8.

Lohr, B., Neuenschwander, P., Varela, A. M. & Santos, B. (1988) Interactions between the female parasitoid *Epidinocarsis lopezi* De Santis (Hym.: Encyrtidae) and its host the cassava mealybug *Phenacoccus manihoti* Matile-Ferrero (Hom.: Pseudococcidae). *Journal of Applied Entomology* **105**: 403–12.

Mangel, M. (1987) Oviposition site selection and clutch size in insects. *Journal of Mathematical Biology* **25**: 1–22.

Mangel, M. (1989a) An evolutionary interpretation of the 'motivation to oviposit'. *Journal of Evolutionary Biology* **2**: 157–72.

Mangel, M. (1989b) Evolution of host selection in parasitoids: does the state of the parasitoid matter? *American Naturalist* **133**: 688–705.

Mangel, M. & Clark, C. W. (1988) *Dynamic Modeling in Behavioural Ecology*. Princeton: Princeton University Press.

Murdoch, W. W. (1994) Population regulation in theory and practice. *Ecology* **75**: 271–87.

Murdoch, W. W., Chesson, J. & Chesson, P. L. (1985) Biological control theory and practice. *American Naturalist* **125**: 344–66.

Murdoch, W. W., Nisbet, R. M., Blythe, S. P. & Gurney, W. S. C. 1987) An invulnerable age class and stability in delay-differential parasitoid-host models. *American Naturalist* **129**: 263–82.

Murdoch, W. W., Nisbet, R. M., Luck, R. F. & Godfray, H. C. J. (1992) Size-selective sex allocation and host-feeding in a parasitoid–host model. *Journal of Animal Ecology* **61**: 533–41.

Murdoch, W. W., Briggs, C. J. & Nisbet, R. M. (1997) Dynamical effects of parasitoid attacks that depend on host size and parasitoid state. *Journal of Animal Ecology*, **66**: 542–56.

National Academy of Sciences (1969) *Principles of Plant and Animal Control. Volume 3. Insect Pest Management and Control*, pp. 100–64. Washington DC: National Academy of Sciences.

Nettles, W. C. (1987) *Eucelatoria bryani* (Diptera: Tachinidae): effect on fecundity of feeding on hosts. *Environmental Entomology* **16**: 437–40.

Olmi, M. (1994) The Dryinidae and Embolemidae (Hymenoptera: Chrysidoidea) of Fennoscandia and Denmark. *Fauna Entomologica Scandinavica* **30**:1–100.

Price, P. W. (1975) Reproductive strategies of parasitoids. In *Evolutionary Strategies of Parasitic Insects and Mites*, ed. P. W. Price, pp. 87–111. New York: Plenum.

Roland, J. (1994) After the decline: what maintains low winter moth density after successful biological control? *Journal of Animal Ecology* **63**: 392–8.

Rosenheim, J. A. & Rosen, D. (1992) Influence of egg load and host size on host feeding behaviour by the parasitoid *Aphytis lingnanensis. Ecological Entomology* **17**: 263–72.

Sahragard, A., Jervis, M. A. & Kidd, N. A. C. (1991) Influence of host availability on rates of oviposition and host-feeding, and on longevity in *Dicondylus indianus* Olmi (Hym., Dryinidae), a parasitoid of the rice brown planthopper, *Nilaparvata lugens* Stål (Hem., Delphacidae). *Journal of Applied Entomology* **112**: 153–62.

Sandlan, K. P. (1979) Host-feeding and its effects on the physiology and behaviour of the ichneumonid parasitoid *Coccygomimus turionellae. Physiological Entomology* **4**: 383–92.

Schaupp, W. C. & Kulman, H. M. (1992) Behaviour and host utilization of *Coccygomimus disparis* (Hymenoptera: Ichneumonidae) in the laboratory. *Environmental Entomology* **21**: 401–8.

Smith, G. C. J. & Pimentel, D. (1968) The effect of two host species on the longevity and fecundity of *Nasonia vitripennis. Annals of the Entomological Society of America* **62**: 305–8.

Stiling, P. D. (1990) Calculating the establishment rates of parasitoids in biological control. *American Entomologist* **36**: 225–30.

Stiling, P. D. (1993) Why do natural enemies fail in classical biological control programs? *American Entomologist* **39**: 31–7.

Takasu, K. & Lewis, W. J. (1995) Importance of adult food sources to host-searching of the larval parasitoid *Microplitis croceipes. Biological Control* **5**: 25–30.

Telenga, N. A. (1958) Biological methods of pest control in crops and forest plantations in the USSR. *Report of the Soviet Delegation, Ninth International Conference on Quarantine and Plant Protection, 1958*, pp. 1–15.

Topham, M. & Beardsley, J. W. (1975) Influence of nectar source plants on the New Guinea sugarcane weevil parasite, *Lixophaga sphenophori* (Villeneuve). *Proceedings of the Hawaiian Entomological Society* **22**: 145–54.

Treacy, M. F., Benedict, J. H., Walmsley, M. H., Lopez, J. D. & Morrison, R. K. (1987) Parasitism of bollworm (Lepidoptera: Noctuidae) eggs on nectaried and nectariless cotton. *Environmental Entomology* **16**: 420–3.

van Emden, H. F. (1963) Observations on the effect of flowers on the activity of parasitic Hymenoptera. *Entomologist's Monthly Magazine* **98**: 265–70.

van Lenteren, J. C., van Vianen, A., Gast, H. F. & Kortenhoff, A. (1987) The parasite–host relationship between *Encarsia formosa* Gahan (Hymenoptera: Aphelinidae) and *Trialeurodes vaporariorum* (Westwood) (Homoptera: Aleyrodidae). XVI. Food effects on oogenesis, life span and fecundity of *Encarsia formosa* and other hymenopterous parasites. *Zeitschrift für Angewandte Entomologie* **103**: 69–84.

Visser, M. E., Luykx, B., Nell, H. W. & Boskamp, G. J. F. (1992a) Adaptive superparasitism in solitary parasitoids: marking of parasitized hosts in relation to the pay-off from superparasitism. *Ecological Entomology* **17**: 76–82.

Visser, M. E., van Alphen, J. J. M. & Hemerik, L. (1992b) Adaptive superparasitism and patch time allocation on solitary parasitoids: an ESS model. *Journal of Animal Ecology* 61: 93–101.

Waage, J. K. (1990) Ecological theory and the selection of biological control agents. In *Critical Issues in Biological Control*, ed. M. Mackauer, L. E. Ehler & J. Roland, pp. 135–57. Andover: Intercept.

Waage, J. K. (ed.) (1996) *Biological Control Introductions: Opportunities for Improved Crop Production*. BCPC Monographs no. 67. Farnham: British Crop Protection Council.

Waage, J. K. & Hassell, M. P. (1982) Parasitoids as biological control agents: a fundamental approach. *Parasitology* 84: 241–68.

Waage, J. K. & Mills, N. J. (1992) Biological control. In *Natural Enemies*, ed. M. J. Crawley, pp. 412–30. Oxford: Blackwell Scientific.

Waage, J. K., Carl, K. P., Mills, N. J. & Greathead, D. J. (1985) Rearing entomophagous insects. In *Handbook of Insect Rearing. Volume I*, ed. P. Singh & R. F. Moore, pp. 45–66. Amsterdam: Elsevier.

Wackers, F. L. (1994) The effect of food deprivation on the innate visual and olfactory preferences in the parasitoid *Cotesia rubecula*. *Journal of Insect Physiology* 40: 641–9.

Weisser, W. W. & Houston, A. I. (1993) Host discrimination in parasitic wasps: when is it advantageous? *Functional Ecology* 7: 27–39.

Wolcott, G. N. (1942) The requirements of parasites for more than hosts. *Science* 96: 317–18.

Yamamura, N. & Yano, E. (1988) A simple model of host-parasitoid interaction with host–feeding. *Researches on Population Ecology* 30: 353–69.

9

Coexistence of multiple attractors and its consequences for a three-species food chain

LIEBE F. CAVALIERI AND HUSEYIN KOÇAK

Introduction

In this chapter we will consider the dynamics of a three-species food chain and the nature of the attractor relevant to this system. An attractor is a geometric object to which trajectories from a given initial state are attracted. For concretness, we may imagine an initial population size as representing an initial state. In the present case, the attractor serves as the final destination for the three trajectories that represent the three players in this drama. Knowledge of the nature of the attractor is crucial if biological control is to be successful: the attractor may be a fixed point as in the rest point of a pendulum, a regularly fluctuating cycle as in a heart beat, or some sort of aperiodic behavior. An important feature with which we will be concerned is the basins of attraction. The basins of attraction represent those collective points whose trajectories, after a period of elapsed time, end up on an attractor. Let us anticipate our discussion and mention at the outset that the attractors for this system are complicated. There may be coexisting chaotic and periodic attractors, or coexisting periodic attractors, depending on the initial conditions. Which attractor is the final destination of a trajectory depends intricately on the initial conditions, e.g., on initial population sizes. They will determine whether the trajectory will land on a chaotic attractor – in which case the resulting population sizes as they evolve may reach, in an erratic manner, extraordinarily high values – or whether the populations will oscillate in a periodically predictable manner. Even if the oscillations are periodic, the amplitudes, e.g., the sizes of the populations, may not be predictable. These important questions emphasize the need for understanding the structure of the attractor, if biological control is to be achieved. We will discuss this aspect presently.

Another important feature of this system is hysteresis. Briefly, hysteresis represents an irreversibility phenomenon. It has been observed in physical, mechanical, and ecological settings. Thus, in field or laboratory work dealing with animal populations, for example, if a population has increased, say because of an increase in growth rate, hysteresis is evident when the original size of the population is not restored by lowering the growth rate to its original value. The initial population size can only be restored when the growth rate is reduced far below its original value, at which point the system does not undergo a smooth transition but rather jumps to the original population size. Hysteresis, then, is characterized by these so-called jumps. In the parlance of dynamics, these jumps represent sudden transitions from one attractor to another. As an example, hysteresis has been shown to play an important role in certain insect populations. An early and now classical study dealt with the spruce budworm that defoliated balsam fir trees in Canada (Ludwig *et al.*, 1978). The mathematical analysis of Ludwig *et al.* showed the existence of a cusp catastrophe surface that involved a sudden transition from one surface to another, i.e., a jump from a stable budworm population to an outbreak that occurred when the budworm population reached a high, critical level which was no longer controllable by the predator. In the case at hand of a three-species food web, we observe a jump from a periodic to a chaotic attractor. This is of interest from the viewpoint of biological control: a jump is undesirable, because, in general, a chaotic state may produce erratic insect-pest population sizes that may reach very high values. Hastings (Chapter 12) deals with insect outbreaks.

Models of food webs

Although recognition of the complexity of food webs dates back to the early part of this century, it has been only in the recent past that detailed analyses of specific food chains have been undertaken (see, for example, Pimm, *et al.*, 1991; Cohen *et al.*, 1993; Abrams & Roth, 1994; Klebanoff & Hastings, 1994; Kuzenetsov & Rinaldi, 1996). Early research by Rosensweig & MacArthur (1963) dealt with simple predator–prey models. Later, Rosensweig (1973) and Wollkind (1976) extended the research to trophic levels. Gilpin (1979) showed for the first time that the parameter values of a two prey–one predator model could be chosen so that deterministic chaos resulted.

Recent advances in mathematical ecology have ushered in a much more sophisticated and relevant endeavor wherein field and laboratory results are not only technically tractable from a mathematical point of view but are more realistic ecologically, taking into account real world problems. In the present climate of environmental awareness such studies are increasingly assuming a sense of urgency. For example, there is heightened concern about the increase of toxic environmental pollutants at higher trophic levels of the food chain; or the ecological effects of decreasing a species in a particular community, say by deforestation, leading to unknown consequences.

In a noteworthy paper, Hastings & Powell (1991) recently reported the occurrence of chaotic dynamics in a continuous-time model of a three-species food chain. Their model consisted of the three differential equations:

$$\frac{dx}{dt} = x(1-x) - \frac{a_1 x y}{(1+b_1 x)}$$

$$\frac{dy}{dt} = \frac{a_1 x y}{(1+b_1 x)} - \frac{a_2 x z}{(1+b_2 y)} - d_1 y \qquad (9.1)$$

$$\frac{dz}{dt} = \frac{a_2 y z}{(1+b_2 y)} - d_2 z$$

where x is the bottom species, representing prey; y is the middle species that is both predator of x and prey for the top species z. a_1, a_2, b_1, b_2, d_1 and d_2 are non-dimensionalized parameters. The first four parametrize the functional response; the last two refer to the

constant death rates of y and z, respectively. In a related study McCann & Yodzis (1994, 1995) examined the Hastings & Powell model and also found chaotic dynamics using a set of parameter values that were based on body sizes and metabolic characteristics of the animals.

It is implicit in Hastings & Powell's simulations that for certain ranges of the parameter values the system above can have coexisting attractors: on the one hand, the system can have two periodic attractors; on the other, a chaotic and a periodic one. The coexistence of two distinct attractors has important theoretical and practical consequences in ecology, specifically in biological control. In this chapter we investigate the geometry of the basins of coexisting attractors and report on the presence of hysteresis in the three-species food chain model of Hastings & Powell. We conclude that knowledge of basins of attraction and hysteresis have relevance to questions of biological control.

Analysis and results

In a dissipative deterministic model such as the one described above, solutions of the equations define the approach to the global attractor; moreover, the evolution of the system is ultimately determined by the dynamics on the attractor. As indicated above, an attractor can be in an equilibrium state, a periodic orbit or in a more complicated state. A most remarkable feature of the above model is that for biologically reasonable parameter values the attractor consists of two distinct pieces with different dynamics. Therefore, it is crucial to decide which solutions approach a given piece of the attractor for this will determine the outcome.

Heretofore, we will refer to each (connected) piece of the global attractor as a separate attractor. The subset of the initial conditions in the three-dimensional state space from which the solutions approach a particular attractor is called the basin of attraction of that attractor. In the ordinary parlance of dynamics, initial conditions that lead to an attractor are said to reside in the basin of attraction of that attractor. If an initial condition is sufficiently close to an attractor, the solution starting from this initial condition

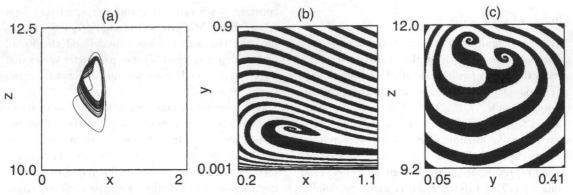

Figure 9.1. (a) Phase portrait showing coexistence of chaotic and periodic attractors. (b) and (c) are basins of attraction of the attractors in (a). The attractors in (a) were computed using the value $b_1 = 2.46$. These are the only attractors and almost all initial conditions approach either one of these attractors. The figure is the projection of these attractors on to the (x,z)-plane. In (b) the three-space is sliced at $z = 11.36$, and 160 000 (400×400) points on this plane (i.e., initial conditions of the form $(x,y, 11.36)$) are used as the initial conditions. The orbit from each initial condition is computed and observed as to whether the orbit approaches the chaotic or periodic attractor. The initial data whose orbits approach the chaotic attractor are white, and the initial data whose orbits approach the periodic attractor are black. The fraction of the black portion is 39% of the total. (c) is the basin of attraction of the attractors in (a). In this case, the three-space is sliced at $x = 0.56$. The form is $(0.56, y, z)$ and computation was as described for (b). The fraction of black is 42.1%.

certainly approaches the attractor. However, away from the attractor, it is a difficult mathematical problem to decide which of the two attractors a trajectory might approach. Our task was to determine the configuration and the boundaries of the two basins of attraction. For certain representative parameter values, we have determined the geometry of the basins of attraction of attractors through rather extensive numerical calculations.

Since this is a three-dimensional system, it is computationally impractical to map the basins in a three-dimensional region. The matter is simplified by reducing the system to two dimensions, which can be achieved by slicing the three-space that contains the attractor with a plane and considering the orbits starting on that plane. On each plane we took 160 000 points (400×400) and by numerical integration determined the attractor of the solution starting from each point. The basic computations for each plane required about 55 h of dedicated computing on a DEC Alpha workstation. We have used the parameter values of Hastings & Powell (1991): $a_1 = 5.0$, $a_2 = 0.1$, $b_2 = 2.0$, $d_1 = 0.4$, $d_2 = 0.01$, varying only the value for b_1, which relates to the carrying capacity of the prey population.

For $b_1 = 2.46$ (Fig. 9.1(a)), where the attractors are projected onto the (x,z)-plane, we see the coexistence of a chaotic and a periodic attractor. The basins of attraction of these attractors are visualized in Figs 9.1(b) and (c). In Fig. 9.1(b) the z-plane is fixed at 11.36, so that the projection is onto the (x,y)-plane. The initial conditions are, therefore, of the form $(x,y, 11.36)$. A total of 160 000 points (400×400) in the (x,y)-plane are used as initial conditions. The orbit from each initial condition is computed and assigned a final destination. Orbits whose initial data approach the chaotic attractor appear as white; those that approach the periodic attractor appear black. The black portion is about 39% of the total. In Fig. 9.1(c) the x-plane is fixed at 0.56 (rather than the z-plane, as above), so that the initial conditions are of the form $(0.56, y, z)$. The data are computed as in Fig. 9.1(b). In this case the black (periodic) portion is about 42%. We calculated the basins on other planar sections and obtained similar pictures.

It is important to notice the configuration of the basins of attraction: they interweave. Near the boundaries of these domains of attraction the system becomes particularly sensitive to the initial condi-

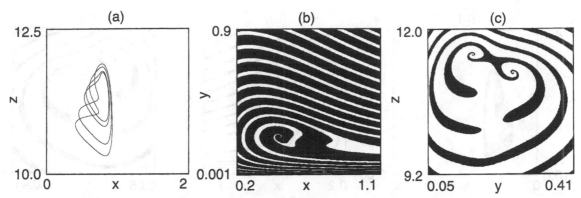

Figure 9.2. (a) Phase portrait showing two periodic attractors. (b) and (c) are the basins of attraction of attractors in (a). Computation for (a) was at $b_1 = 2.51$. These are two period-2 attractors, the two smaller inner loops are one periodic attractor and the two outer loops are the other attractor. As in Fig. 9.1, in (a) the periodic orbits are projected onto the (x,z)-plane. In (b) the plane at $z = 11.36$ is considered. Protocol was as described in the legend to Fig. 9.1. The fraction of black is 69.8%. (c) this time the plane is at $x = 0.56$. The fraction of black is 26.8%.

tions. For example, while the initial conditions $x = 1.04611$, $y = 1.0551$, $z = 11.36$ lead to a predictable periodic state, the nearby initial conditions $x = 1.0551$, $y = 0.20$, $z = 11.36$ lead to a chaotic state. Herein lies the practical significance of these boundaries. In field work it is important to try to control the initial condition(s) (e.g., population size), for the outcome may be chaotic or periodic, depending on these conditions.

Figure 9.2(a) shows the coexistence of two periodic attractors that result when the b_1 value is increased from 2.46 (as in Fig. 9.1) to 2.51. The periodic orbits are projected onto the (x,z)-plane, as in Fig. 9.1(a). The amplitudes of these periodic attractors are quite different. Notice that the small increase in this parameter value has resulted in the transformation of a chaotic attractor into a periodic one. The basins of these periodic attractors in Figs 9.2(b) and (c) are computed in a manner similar to that described in Figs 9.1(b) and (c). In this case, although both attractors are periodic so that the populations will ultimately fluctuate periodically, the amplitudes of the oscillations do depend on the initial sizes of the populations in a rather complicated way.

Figures 9.3(a),(b) and (c) were computed similarly, with the value of b_1 in this case being increased from 2.51 to 2.53. In this case, as for $b_1 = 2.46$, there are

coexisting periodic and chaotic attractors. Notice, however, that the periodic orbit in Fig. 9.1(a) is larger than that in Fig. 9.3(a), pointing to the sensitivity of these attractors to small changes in parameter values.

In some dynamical systems such as the forced damped pendulum (see, for example, Moon & Guanng-Xuan Li, 1985, and Nusse & Yorke, 1994) the basin boundaries show a pattern of ever increasing detail: close examination of the boundary reveals fine details which, upon still closer examination, provide yet more detail, *ad infinitum*; that is, there seems to be no end-point, and the boundary remains jagged, for example. This defines a fractal set. However, such does not appear to be the case in the present system. Successive enlargements of selected portions of the basin boundaries revealed only smooth boundaries and no fractal structure.

So far, we have examined the attractors and their basins for three representative values for the parameter b_1. Since the dynamics of the system vary considerably, it is important to investigate the dependence of the dynamics on the parameter b_1 more carefully. For this purpose, we have computed a bifurcation diagram for the system, which is depicted in Fig. 9.4. In this diagram local maxima (peaks) of the top predator species, z, *versus* the carrying capacity, b_1, of the prey species were plotted. As can be seen from

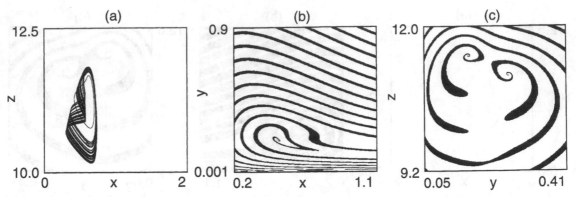

Figure 9.3. (a) Phase portrait showing the coexistence of chaotic and periodic attractors. (b) and (c) show the basins of attraction for the two attractors, one chaotic, one periodic in (a). (a) shows the projection of these attractors onto the (x,z)-plane. Computation for (a) was at $b_1 = 2.53$. It is interesting to compare this figure with Fig. 9.1(a). In this case the periodic orbit is smaller than the chaotic attractor. (b) basin of attraction of the chaotic and periodic attractors in (a). The plane at $z = 11.36$ is considered. The fraction of black is 19.8%. (c) basins of attraction of the chaotic and periodic attractors in (a). This time the plane at $x = 0.56$ is considered. The fraction of black is 17.6%.

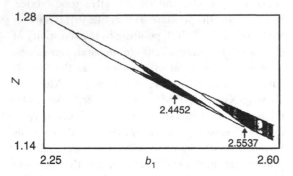

Figure 9.4. Bifurcation diagram of the three-species food chain model. The ordinate is the local maxima (peaks) of the top predator species, z; on the abscissa the parameter b_1, referring to the carrying capacity of the prey population, is varied while the other parameters in the model are held fixed at $a_1 = 5.0$, $a_2 = 0.1$, $b_2 = 2.0$, $d_1 = 0.4$ and $d_2 = 0.01$. The range of b_1 from 2.4452 to 2.5537, where two attractors coexist (the region of hysteresis), is marked with arrows.

this diagram, starting with $b_1 = 2.25$ the system has a globally attracting periodic orbit. As the parameter is increased, this periodic orbit appears to undergo several period-doubling bifurcations and eventually becomes a global chaotic attractor around $b_1 = 2.41$. As the parameter passes $b_1 = 2.4452$, while the origi-

nal chaotic attractor continues to exist, a second attractor (upper right in the figure), which is periodic, comes into existence. In the parameter range 2.4452–2.5537 both attractors exist: the (lower left) chaotic attractor through period-halving bifurcations becomes a periodic one and ceases to exist at $b_1 = 2.5537$; meanwhile, the (upper) periodic attractor undergoes period doubling and eventually becomes a chaotic one. Finally, for parameter values from 2.5537 until 2.60 the upper chaotic attractor remains the sole global attractor.

The portion of the bifurcation diagram for the values of the parameter b_1 in the range 2.4452 to 2.5537 where two attractors coexist is particularly noteworthy. When two attractors coexist there can be present an important dynamical phenomenon known as hysteresis (see, for example, Murray, 1989, and Hale & Koçak, 1991): a dynamical change that the system undergoes at a certain value of the parameter while the parameter is increased but that does not reverse itself at the same parameter value when the parameter is gradually decreased.

Let us re-examine the bifurcation diagram from the viewpoint of hysteresis. Imagine a point where $b_1 = 2.42$. Whatever the initial conditions are the system eventually will be at the chaotic attractor which is globally attracting. Under these circumstances an insect pest population, for example, may rise to unusually high values, as noted earlier (Cavalieri & Koçak, 1994, 1995). Let us suppose that sufficient time has elapsed so that the system is indeed at the chaotic attractor. If the parameter b_1 is increased slowly, the system will follow this attractor, which eventually becomes periodic, until it ceases to exist at $b_1 = 2.5537$. Notice that although this attractor is changing its dynamics the maximum value of the top predator species, z, varies continuously with respect to the parameter b_1. When the parameter is increased further the system has no choice but to jump to the upper right attractor at $b_1 = 2.5537$. Notice that at this point not only is the new attractor chaotic, but its maximum value is considerably larger than the deceased (lower) periodic attractor. Consequently, this sudden increase (and the fluctuations) in the maximum value(s) of the species z will be observable. This means that the jump, a discontinuity, has resulted in a significant increase in the size of the population of species z, which has to be considered in the light of biological control. Attempts at reversing this jump by decreasing the parameter gradually will not succeed until the value of $b_1 = 2.4452$, at which point the system will jump back to the lower left attractor. At this point there will be a noticeable drop in the maximum value of the species z. This is hysteresis: whereas the system had jumped from the lower left attractor to the upper right attractor at a b_1 value of 2.5537, it now jumps from the upper right attractor back to the lower left attractor at the quite different value of 2.4452, which represents a chaotic state with lower values of the top predictor species z.

How was the diagram in Fig. 9.4. constructed? First, the parameter b_1 was increased from 2.25 to 2.60 and then decreased from 2.60 to 2.25. For the first parameter value $b_1 = 2.25$, an initial condition was set and the solution was followed (for some long fixed time) until it reached the attractor. Thereafter, the parameter was increased in steps of 0.0005 and whatever the solution was for the previous value of the parameter it is now used as the initial condition at the new parameter value. Since the parameter is changed by a very small amount, the attractor moves

a small amount also. Therefore, the initial condition obtained during the previous parameter value is very near the attractor at the present parameter value. Consequently, if the parameter is varied gradually one follows a particular attractor (as long as the attractor persists). The setting for this computer experiment is a model for one in which one tries to control the population of the top species in the field or laboratory by actively inducing gradual changes in the bottom species in a food chain.

Currently, there is a great deal of interest, and concern, in the possibility of chaotic behavior in ecological systems. Although it is implicit in the foregoing discussion of hysteresis, it is important to emphasize a practical point arising from the coexistence of periodic and a chaotic attractor in the three-species model. Imagine that $b_1 = 2.46$ and the initial population sizes are such that the system is at the (upper) right periodic attractor, as seen in Fig. 9.4. If one gradually decreases the b_1 parameter the top attractor jumps to the lower one at $b_1 = 2.4452$, as just noted, at which the upper periodic attractor has ceased to exist and the system is now chaotic. If this chaotic state is undesirable (and it probably is) and one tries to reverse the change by gradually increasing the parameter, the top predator population will not become (singly) periodic until about $b_1 = 2.52$. Therefore, to undo this potentially undesirable chaotic effect in the predator species decreasing the prey species will require a considerable increment in the prey population (in terms of b_1, nearly 25 times as much as inducing it in the first place) in order to return to a predictable periodic state. This may not be cost effective or even feasible.

Discussion

The coexistence of chaotic and periodic attractors in the three-species food chain model is of interest both from a theoretical and a practical point of view. From a theoretical point of view, we have shown that the basins of attraction of these attractors possess a rather intricate geometry in the phase space. While the existence of chaotic attractors signals the sensitive dependence of the system upon the initial data, it is evident from Fig. 9.2 that even when the system has no chaotic state, but only two coexisting periodic attractors, great care must be exercised in setting the initial conditions (population sizes) in order to achieve a predictable outcome: which attractor is approached does, indeed, depend on the initial population sizes. Also, detailed computations reveal only smooth basin boundaries rather than fractal ones as is common in many mechanical systems with multiple attractors. From a practical point of view, this is of some comfort because a fractal basin boundary would preclude the possibility of predicting the dynamics of populations with initial sizes near the boundary. Such a boundary is, after all, in a sense ephemeral.

A second point of interest is the phenomenon known as hysteresis, which this system exhibits. The fact that an increase in a parameter (for example, in the present case b_1) causes the system to jump from one attractor to the other adds a complication to biological control which, although difficult to rectify, can be managed if the underlying theory is understood.

It is interesting that May (1977) recorded a number of ecosystems that showed sharp 'breakpoints', which are, in fact another way of defining hysteresis. In insect populations, he observed, for example, that the spruce budworm population showed sharp 'breakpoints' in their life cycle (Ludwig *et al.*, 1978, implied this). That is, at a critical value of the total leaf area per tree, there would be a dramatic jump from a low number of insect pests to a new, higher equilibrium position. Most importantly from a practical point of view, May observed, presciently, that the use of pesticides to control insect pests had to be approached very cautiously: the outcome, he warned, could be a kind of treadmill, leading to the endless use of insecticides. In the three-species model hysteresis has similar important practical implications for field work when one is trying to control the growth of an organism in a food chain, say by controlling the growth of the bottom prey species.

Acknowledgements

We would like to thank Chris Cosner and Donald Olson for their critical readings of the manuscript. H. Koçak was supported in part by the National Science Foundation.

References

Abrams, P. A. & Roth, J. D. (1994) The effects of enrichment of three-species food chains with non-linear functional responses. *Ecology* 75: 1118–30.

Cavalieri, L. F. & Koçak, H. (1994) Chaos in biological control systems. *Journal of Theoretical Biology* 169: 179–87.

Cavalieri, L. F. & Koçak, H. (1995) Chaos: a potential problem in the biological control of insect pests. *Mathematical Biosciences* 127: 1–17.

Cohen, J. E., Beaver, R. A., Cousins, S. H., DeAngelis, D. L., Goldwasser, L., Heong, K. L., Holt, R. D., Kohn, A. J., Lawton, J. H., Martinez, N., O'Malley, R., Page, L. M., Patten, B. C., Pimm, S. L., Polis, G. A., Rejmánek, M., Schoener, T. W., Schoenly, K., Sprules, W. G., Teal, J. M., Ulanowicz, R. E., Warren, P. H., Wilbur, H. M. & Yodzis, P. (1993) Improving food webs. *Ecology* 74: 252–8.

Gilpin, M. (1979) Spiral chaos in a predator–prey model. *American Naturalist* 113: 306–8.

Hastings, A. & Powell, T. (1991) Chaos in a three-species food chain. *Ecology* 72: 896–903.

Hale, J. & Koçak, H. (1991) *Dynamics and Bifurcations*. New York: Springer-Verlag.

Klebanoff, A. & Hastings, A. (1994) Chaos in three species food chains. *Journal of Mathematical Biology* 32: 427–71.

Kuzenetsov, Y. A. & Rinaldi, S. (1996) Remarks on food chain dynamics. *Mathematical Biosciences* 134: 1–33.

Ludwig, D., Jones, D. D. & Hollings, C. S. (1978) Qualitative analysis of insect outbreak systems: the spruce budworm and forest. *Journal of Animal Ecology* 47: 315–32.

May, R. M. (1977) Thresholds and breakpoints in ecosystems with a multiplicity of stable states. *Nature* 269: 471–7.

McCann, K. & Yodzis, P. (1994) Nonlinear dynamics and population disappearances. *American Naturalist* 144: 873–9.

McCann, K. & Yodzis, P. (1995) Bifurcation structure of a three-species food chain model. *Theoretical Population Biology* 48: 92.

Moon, F. C. & Guanng-Xuan, Li (1985) The fractal dimension of the two-well potential strange attractor. *Physica* 17D: 99–100.

Murray, J. D. (1989) *Mathematical Biology*. New York: Springer-Verlag.

Nusse, H. E. & Yorke, J. A. (1994) *Dynamics: Numerical Explorations*. New York: Springer-Verlag.

Pimm, S. L., Lawton, J. H. & Cohen, J. E. (1991) Food web patterns and their consequences. *Nature* **350**: 669–89.

Rosensweig, M. L. (1973) Exploitation in three trophic levels. *American Naturalist* **107**: 275–94.

Rosensweig, M. L. & MacArthur, R. H. (1963) Graphical representation and stability conditions of predator–prey interactions. *American Naturalist* **97**: 209–53.

Wollkind, D. J. (1976) Exploitation in three trophic levels: an extension allowing intraspecies carnivore interaction. *American Naturalist* **110**: 431–47.

PART III

Spatial considerations

The ability of spatial structure to stabilize predator–prey interactions was demonstrated in Huffaker's classic mite–mite–orange experiment 40 years ago. Yet, atypically theory lagged behind the empirical data, and it was not until the 1970s that spatially explicit models began to appear, and only more recently have they been applied to biological control problems. Although the issue of spatial structure arises in other sections of this book, this section comprises three chapters whose primary focus is on the role of space on population dynamics, enemy–victim interactions, and biological control.

It is not surprising that an understanding of spatial phenomena is most developed for mite biological control; mites were Huffaker's original subjects, they are small enough to maintain numerous populations on a desk-top, and they mostly disperse by walking. In Chapter 10, Walde & Nachman review the use of spatially explicit models to explain and improve the control of spider mites. They emphasize the integration of theoretical predictions and experimental data on mite populations in greenhouses and apple orchards. These two systems represent fundamentally different environmental situations, yet they share enough attributes that the same conceptual tools apply to both. The key to success in spatially structured pest systems is dispersal, and biological control workers have long appreciated its importance. However, the developing theory allows this complex variable to be dissected into its components to identify which are most important to pest control. Walde & Nachman also draw on the differences between specialist and generalist predators introduced by Chang & Kareiva in the previous section to account for the differences that are found in greenhouse and orchard systems.

In Chapter 11, Tscharntke & Kreuss discuss the ramifications of habitat fragmentation to biological control. As they point out, population and community ecologists and conservation biologists have devoted much attention to the impact of habitat fragmentation on population dynamics and community size and structure. In contrast, the importance of fragmentation to pest management in general, and biological control in particular, is less well documented. Tscharntke & Kreuss review the evidence that habitat insularization typically generated by agriculture affects biological control agents as predicted by island biogeography and metapopulation theories.

Using outbreak species as exemplars, Hastings (Chapter 12) argues that theory's traditional emphasis on stability may be inappropriate for describing the population dynamics of many insect pests. He reviews the use of spatially explicit models for understanding insect outbreaks, but further suggests that the key to understanding dynamics lies in the non-equilibrium, transient behavior of populations. Because this idea has not been emphasized by either theoreticians or biological control workers in the past, Hasting's approach has not had time to be applied to any specific biological control systems. Nevertheless, he outlines a protocol for future research that may, in time, improve the chances of controlling a very difficult class of insect pests.

10

Dynamics of spatially structured spider mite populations

SANDRA J. WALDE AND GÖSTA NACHMAN

Introduction

Laboratory studies of phytophagous and predacious mites were among the first to suggest that spatial structure could dramatically affect predator–prey dynamics (Huffaker, 1958; Huffaker *et al.*, 1963). Space is now a hot topic in much of ecology, and it comes in many guises: dispersal behavior, predator aggregation, environmental heterogeneity, meta-populations, source–sink dynamics, invasions. In the first part of this chapter we present an overview of predator–prey models with spatial structure. The aim is not to provide an all-inclusive or in-depth review of these models. Rather we give a sketch of current approaches, summarizing predicted effects of dispersal on aspects of predator–prey dynamics relevant to the biological control of spider mites. Thus we are primarily concerned with the predictions these models make with respect to prey and predator abundances, prey and predator distributions in space, and persistence of the predator–prey interaction. We then present some predictions based on a more specific simulation model of acarine predator–prey dynamics in a greenhouse environment. Finally we discuss how well the models reflect biological control of phytophagous mites by predacious mites in greenhouses and orchards, and examine the relationship between model predictions and observed dynamics of these systems.

Spatially structured predator–prey models

The incorporation of spatial structure into predator–prey models in the 1970s and 1980s used one of two approaches, reaction–diffusion or patch models (Hastings, 1990). Patch models were of two types (Taylor, 1988): (i) Levins (1969) type models where within-cell dynamics consisted of species presence/absence, or (ii) models where patches had more explicit dynamics. These models have been reviewed several times (Taylor, 1988, 1990; Hastings, 1990; Kareiva, 1990), and there is a consensus about several points. Reaction–diffusion modeling has been less concerned with stability than with spatial patterning, but increasing movement rates in these models is usually destabilizing (Hastings, 1990). Dispersal can stabilize patch models under certain conditions, namely asynchronous dynamics among cells, appropriate migration rates and a sufficient number of cells. Levins-type models are usually stabilized by high migration rates, while patch models that have within-cell dynamics are generally most stable with low migration rates (Taylor, 1988). Asynchrony in these models can be generated by stochasticity in parameters such as birth, death or dispersal rates, or by fixed differences among patches.

Since the late 1980s there has been a veritable explosion of predator–prey models incorporating spatial structure, and here we concentrate on what these new models have to offer. The models can be classified in a variety of ways (Taylor, 1988; Hastings, 1990 and see Chapter 12; Caswell & Etter, 1993). Here we consider only population-based models. We first divide the models into those that treat space as continuous *versus* discrete (Hastings, 1990). We then look at four categories of discrete space or patch models: (i) two-patch models, (ii) multi-patch models with global coupling, (iii) multi-patch models with local coupling, and (iv) multi-patch models with three trophic levels.

Continuous space

Models of dispersal in continuous space have been more concerned with equilibrium distributions of predator and prey than with persistence of the predator–prey interaction *per se*. The most interesting general result is that movement can generate heterogeneous distributions of predator and prey in homogeneous space (Segel & Jackson, 1972; Levin & Segel, 1976), including traveling waves of prey and predators when predator mobility exceeds that of the prey (Dunbar, 1983). The ecological conditions for these instabilities to occur are quite restrictive (e.g., an Allee effect for the prey and density-dependent mortality for the predator), but less so if the predator–prey interaction is modeled as a discrete time process (Kot, 1992; Neubert *et al.*, 1995). Additional interesting pattern formation is seen if the environment is assumed to be heterogeneous. If diffusion is biased toward favorable locations, high rates of diffusion lead to correlated distributions of predator and prey, with both found in the favorable locations. With low diffusion the prey become concentrated in less favorable locations (Comins & Blatt, 1974). If dispersal is not biased, the distribution of prey always reflects the underlying environmental heterogeneity (McLaughlin & Roughgarden, 1991). Increasing predator diffusion relative to prey exaggerates this reflection, while greater prey diffusion reduces it.

Diffusion tends to destabilize continuous-time models in the absence of environmental heterogeneity (Hastings, 1990). However, moderate prey dispersal can allow predator persistence in an otherwise unstable discrete-time model (Neubert *et al.*, 1995). If space is heterogeneous, predator dispersal is stabilizing but increasing prey dispersal is destabilizing (McLaughlin & Roughgarden, 1991), and biased dispersal can also be stabilizing (Comins & Blatt, 1974).

Average densities can also be affected by diffusion rate, and in particular, by the relative movement rates of predator and prey (Levin & Segel, 1976; McLaughlin & Roughgarden, 1991, 1992). Details are in Table 10.1.

Discrete space

Two-patch models

Two-patch models can be considered a first approximation to multi-patch models, with the advantage that analytical results can often be obtained for models of moderate complexity. Logistic growth of the prey population, Type II predator responses, density-dependent prey movement, predator aggregation and age structure are some of the factors that have been incorporated into these models. Two-patch models with Lotka–Volterra assumptions (exponential prey growth, Type I functional response), identical patches and symmetrical dispersal are neutrally stable. However, these models can easily be stabilized by environmental heterogeneity (different parameter values for the two patches) or by non-symmetrical dispersal (Murdoch & Oaten, 1975; Godfray & Pacala, 1992; Ives, 1992; Nisbet *et al.*, 1993). Density-dependent dispersal by the prey is destabilizing (Murdoch *et al.*, 1992) and aggregation by the predator to high prey densities may be stabilizing, destabilizing or have little effect on dynamics (Ives, 1992; Murdoch *et al.*, 1992). Dispersal of predators to a sink (patch with only predators) can also be stabilizing (Holt, 1985; Schaeffer & de Boer, 1995) although this effect may depend on selection of functional response (Holt, 1985).

Two-patch discrete-time models have received much less attention than continuous-time versions. It appears that it is more difficult to get stability by simply having different parameter values in the two patches in these models, but variation in attack rate or in predator conversion rate can be stabilizing (Holt & Hassell, 1993). Different initial conditions can also be stabilizing, and this is enhanced by low host reproduction, overall low migration rates, and parasitoid migration rates that are lower than that of the host (Adler, 1993).

Multi-patch models with global coupling

Most of the early multi-patch models had identical patches linked by dispersal, where all patches were equally accessible from all other patches (global dispersal). The key to stability in these models is maintaining asynchrony of local patch dynamics, *via* environmental stochasticity or fixed differences among patches. In both continuous and discrete-time versions, where within-patch dynamics are included,

Table 10.1. *Summary of predictions relevant to biological control extracted from spatially structured predator-prey models*

Regional persistence/stability

Increases persistence:

Dispersal

 Localized dispersal[1,2]

 Intermediate dispersal rates[3,4,5,6]

 Prey dispersal higher than predator[1,7]

 Predator aggregation (sometimes)[8]

 Prey dispersal from predator patches[9]

More stable local dynamics [10]

High plant K per patch [4,19]

Density

Lower prey density

 Low predator aggregation [13]

 High predator dispersal [14]

 Low prey dispersal [14]

 Variable prey reproductive rate [10]

 Low variation in predator attack rate [10]

 Prey dispersal lower than predator [10]

Decreases persistence:

Dispersal

 Density-dependent prey dispersal [11]

 Predator dispersal at end of interaction [9]

 Presence of a second predator [12]

 Predator aggregation (sometimes) [11]

 'Killer' strategy for predator [3]

 Within-generation predator movement [13]

Higher predator density

 High predator dispersal [15]

 High prey dispersal [14]

Distribution

Patterns in homogenous environments (continuous or discrete)

 Allee effect, density-dependent predator mortality [15,16]

 Localized dispersal in discrete space [5]

 Spatial segregation of predators in two predator systems [17]

Patterns in heterogeneous environments

 Prey dispersal biased toward favorable locations [18]

  High dispersal rates: predator and prey both in favorable locations

  Low dispersal rates: prey in less favorable locations

 Dispersal unbiased: prey always in favorable locations [14]

  Higher predator dispersal: distribution pattern is exaggerated

Notes:

Note that these are predictions (simplified in some cases) of particular models and may be specific to those models.

Sources: [1]Nachman (1987b); [2]Comins *et al.* (1992); [3]Takafuji *et al.* (1983); [4]Fujita (1983); [5]Hassell *et al.* (1991); [6] Hassell *et al.* (1993); [7]Adler (1993); [8]Ives (1992); [9]Diekmann (1993); [10]Reeve (1988); [11]Murdoch *et al.* (1992); [12]Hassell *et al.* (1994); [13]Rohani & Miramontes (1995); [14]McLaughlin & Roughgarden (1991); [15]·Levin & Segel (1976); [16]Segel & Jackson (1972); [17]Ruxton & Rohani (1996); [18]Comins & Blatt (1974); [19]Jansen & Sabelis (1992).

dispersal must be low to ensure persistence since movement among patches tends to synchronize dynamics (Taylor, 1988). In these models (e.g., Reeve, 1988), dispersal does not favor good biological control. Higher dispersal rates tend to be destabilizing, and in addition, dispersal introduces a 'floor' to host densities.

Multi-patch models with local coupling

Local dispersal affects dynamics quite differently than does global dispersal (stepping stone *versus* island models). First, it is easier to get persistence when dispersal is localized (Maynard Smith, 1974; Nachman, 1987a,b; Woolhouse & Harmsen, 1987; Hassell *et al.*, 1991). Second, spatial patterning emerges, where predator and prey show heterogeneous distributions in homogeneous arenas (Hassell, *et al.*, 1991; Comins *et al.*, 1992). Different spatial patterns imply different temporal dynamics; spatial chaos is associated with bounded temporal chaos, spiral waves with limit cycles and a lattice structure with a stable equilibrium (Hassell *et al.*, 1993). The most ordered spatial patterns have the lowest amount of temporal fluctuation, but not the lowest host densities (Rohani & Miramontes, 1995). Coexistence is also possible in three–species versions, particularly if the competitors differ in mobility, and there is some trade-off associated with high mobility. Competitors spatially segregate in these systems (Comins & Hassell, 1996). Adding stochastic noise to such systems increases the probability of the more mobile competitor going extinct (Ruxton & Rohani, 1996).

Tritrophic multi-patch models

When multi-patch models have been used to look at spider mite dynamics, they have often been expanded to include three trophic levels. Most assume global dispersal (Fujita, 1983; Takafuji *et al.*, 1983; Jansen & Sabelis, 1992), but one looked at local dispersal (Nachman, 1987a,b). Localized dispersal tends to be more stabilizing since it enhances asynchrony among patches, but the increase in plant damage associated with low dispersal rates makes it detrimental to biological control. An important general result is that

the outcome of the tritrophic interaction depends strongly on the nature of the prey–plant interaction (Fujita, 1983; Takafuji *et al.*, 1983; Jansen & Sabelis, 1992).

The simplest version of these models (Lotka–Volterra dynamics for all three trophic levels, no carrying capacities, global and instantaneous dispersal) is neutrally stable (Sabelis *et al.*, 1991). When the plant has a carrying capacity (K), the presence of a stable three-species equilibrium depends on a sufficiently high K (i.e., large individual patches) (Jansen & Sabelis, 1992). Higher predator efficiency both increases stability, and results in higher plant biomass with fewer prey.

Using a transition-state model, Takafuji *et al.* (1983) showed that the ability of the host plant to recover from over-exploitation affects the period and the amplitude of the predator–prey oscillations. They also found that the important properties of the predator are those that are effective in suppressing prey within patches (predacious and reproductive capacities), thereby affecting the migration pattern and extinction rate of the prey. Dispersal rate of the predator mattered little. Similarly, van Baalen & Sabelis (1995) compared predators that quickly eliminate their prey from patches after invasion ('killers') with predators that only limit the increase of the prey ('milkers'), and found that the former tend to be destabilizing and the latter stabilizing. Nachman (1987a) found that higher predator efficiency and higher dispersal rates, especially long-distance dispersal, increased the amplitude of the predator–prey oscillations, which reduced persistence.

Summary of predictions

Our aim in this chapter is to compare the dynamics of spider mite populations with the predictions of some mathematical models. Before doing so, we need to consider what type of match we should be aiming for. A general objective of population dynamics models is prediction, either prediction of the type of dynamics the system will exhibit at equilibrium or prediction of how population densities will change over time. Model predictions of particular population trajectories have rarely been very accurate, and despite recalibration of models using new data or information, good prediction has remained elusive for most systems. A number of phenomena contribute to the unpredictability of ecological systems. First, environmental stochasticity, i.e., fluctuations due to factors external to the system, may make prediction of particular trajectories impossible. However, outcomes can sometimes be framed in terms of probabilities. Second, systems may be chaotic, that is, they may exhibit fluctuations that never settle into repeated patterns, the trajectories of which are sensitive to initial conditions. To the extent that these systems are represented in nature or in biological control systems (e.g., Hanski *et al.*, 1993; Ellner & Turchin, 1995) they will, in practice, be unpredictable.

When one moves from non-spatial to spatial models, additional factors hinder predictability. First, demographic stochasticity (internally generated noise associated with chance events such as birth, death, and migration), which is usually irrelevant in large populations, can become important in subdivided populations where the individual units are small (e.g., Chesson, 1978; Nachman, 1987a,b). Second, transient times in spatial models may be inordinately long, and systems may shift unexpectedly from one type of dynamics to another after thousands of generations (Hastings, 1993; Hastings & Higgins, 1994; White *et al.*, 1996; and see Chapter 12). Third, the outcome of these models may be very sensitive to the magnitude of dispersal, a parameter for which it is usually very difficult to obtain accurate data (Wennergren *et al.*, 1995).

So at what level should we be aiming for predictability? Here we suggest that we can expect quite a good qualitative match (e.g., increases *versus* decreases in density), and that when we do not get this level of agreement, we need to look closer at the biology of the system. We generate qualitative predictions from the above spatial models (Table 10.1) and from a more detailed model (see below). We then compare these predictions to greenhouse and orchard spider mite dynamics. Finally we discuss what the model and experimental results tell us about the differences between spider mite dynamics in greenhouse and orchard systems.

Spider mites in greenhouses

Biology of greenhouse mites

The two-spotted spider mite (*Tetranychus urticae* Koch) is a serious pest of greenhouse crops as well as many fields crops and fruit orchards (e.g., Helle & Sabelis, 1985). The mites damage the plants by inserting needle-shaped mouthparts into the epidermal leaf cells and sucking out the cell contents (Liesering, 1960). High levels of attack lead to wilting, which reduces the productivity of the plant, and may ultimately kill it (Tomczyk & Kropczynska, 1985). Greenhouses offer *T. urticae* nearly perfect conditions for becoming a pest: temperature and humidity are usually near optimal, food is almost unlimited, natural enemies are often absent, and the mites can easily spread from plant to plant. Population doubling times can be on the order of 2–3 days (Ohnesorge, 1981), forcing the grower to intervene very soon after spider mites are discovered, either by the application of acaricides or, preferably, by the release of natural enemies.

The natural enemy used most frequently for the control of *T. urticae* is the specialist predatory mite *Phytoseiulus persimilis* Athias-Henriot (see e.g., van Lenteren & Woets, 1988). Its success can be attributed to a combination of high reproductive rate, short developmental time and high search efficiency for prey (e.g., Amano & Chant, 1977; Sabelis, 1981; Zhang & Sanderson, 1995). Invading predatory mites and their progeny usually remain in a colony of *T. urticae* until all prey have been killed, leading to a highly unstable predator–prey interaction (e.g., Bernstein, 1984; Zhang *et al.*, 1992). Like most other phytoseiids (see Sabelis, 1985b, for a review), *P. persimilis* has a Holling's (1959) Type II functional response (Takafuji & Chant, 1976; Fernando & Hassell, 1980; Eveleigh & Chant, 1981), which is also destabilizing (Hassell, 1978). All prey stages are attacked, but there is a weak preference for eggs (Fernando & Hassell, 1980).

In spite of frequent local extinctions there is some empirical evidence that the two species are able to coexist at larger spatial scales (Nachman, 1981; Gough, 1991; van de Klashorst *et al.*, 1992), due to shifting mosaic dynamics and spatial asynchrony (Nachman, 1988, 1991; Nihoul, 1993). In fields and orchards, mites may move among plants by crawling, phoresis, or they may be carried by air currents (Kennedy & Smitley, 1985). Spider mites produce silk threads, which are used for web construction and also for aerial transport ('ballooning'). In greenhouses, however, a combination of low wind speeds and plant proximity makes crawling from leaf to leaf the dominant mode of dispersal among plants. Most dispersal is thus short-distance, and as a result, neighboring plants tend to have more similar densities of mites than distant plants (Nachman, 1991). *Tetranychus urticae* emigrates mainly as a response to poor food plant quality (Hussey & Parr, 1963a), while *P. persimilis* moves away when the prey are depleted (Bernstein, 1984).

Tetranychus urticae and *P. persimilis* have very heterogeneous spatial distributions (Nachman, 1981). Aggregation of *T. urticae* is likely due to low dispersal rates, coupled with high growth rates (Nachman, 1988). Aggregation of *P. persimilis* is caused by its tendency to concentrate searching effort in areas of high prey density (Eveleigh & Chant, 1982a,b), and its distribution therefore usually reflects that of the prey, that is, the correlation between numbers of prey and predators per patch is positive (Nachman, 1981). However, overexploitation of the prey may sometimes lead to a negative correlation, because predators are most abundant in patches where the prey has just been eliminated (Huffaker *et al.*, 1969; Nachman, 1985). Prey and/or plant emitted substances may be involved in the attraction and retention of predators in prey patches (e.g., Sabelis & van de Baan, 1983; Sabelis *et al.*, 1984; and see Chapter 15).

A spatial model of T. urticae–P. persimilis *dynamics*

The *T. urticae–P. persimilis* system is one where dispersal appears to be integral to understanding both the dynamics and the success/failure of biological control (Fujita, 1983; Takafuji *et al.*, 1983). To better understand how dispersal affects the interactions, Nachman (1987a) developed a spatially explicit model that included both within-plant dynamics and

inter-plant dispersal. Demographic stochasticity (May, 1974) was included to allow for random effects associated with births, deaths, immigration, and emigration, whereas the physical environment was considered constant. The populations were assumed to be invariant in time with respect to age distribution and sex ratio. The most important findings were that (i) persistence increases with the number of plants, (ii) persistence is highest for intermediate dispersal rates, (iii) short-distance dispersal favors persistence (Nachman, 1987b).

In the original version of the model it was assumed that prey and predators could perform either shortor long-distance dispersal. The fact that probability of successful dispersal usually declines with distance moved was ignored. Here the model is extended to look at the effect that varying inter-plant distance has on predator–prey dynamics, and ultimately the success of *P. persimilis* in controlling *T. urticae*.

Modeling distance-dependent dispersal

The objective of the simulations was to determine how increasing the distance between adjacent plants (reducing the rates of successful dispersal) affect mite dynamics. All parameters other than distance were kept constant across all runs (Table 10.2). It should be emphasized that many of the model's parameters are still qualified guesstimates. Figure 10.1 shows the agreement between the model and real greenhouse data. Note that the progressive discrepancy between observed and predicted dynamics is probably because the model does not adjust for the above-average temperatures during the summer period. Temperature influences the observed densities for most plant-feeding arthropods (e.g., see Chapter 7), and increases in temperature have been found to increase the amplitude and decrease the period of predator–prey oscillations (Burnett, 1970).

The effect of distance on dispersal was modeled as follows. Assume that an individual mite leaves a plant (the donor) in search of a new plant (the recipient). Mortality risk during dispersal is assumed to be constant per unit distance moved. The likelihood (p) of arriving at the ith plant at distance Ω_i from the donor plant is found as:

Table 10.2. *Values of parameters used to simulate greenhouse spider mite dynamics in Fig. 10.1*

Symbol	Description	Value
A_m	Maximum leaf area of a plant	18510 cm^2
r	Injury recovery rate	0.0071 cm^2 h^{-1}
D_m	Maximum tolerable leaf injury	0.6
s	Injury rate per prey	0.000137 cm^2 h^{-1}
b_X	Prey birth rate	0.0107 h^{-1}
d_X	Minimum death rate per prey	0.0021 h^{-1}
δ_X	Prey death rate coefficient	0.0162 h^{-1}
e_X	Minimum emigration rate per prey	0.0 h^{-1}
ϵ_X	Prey emigration rate coefficient	0.000021 h^{-1}
λ_X	Prey settling rate during dispersal	2.0 m^{-1}
μ_X	Mortality rate of prey during dispersal	1.0 m^{-1}
ψ	Prey conversion factor	0.178
C	Coefficient of spatial overlap	0.7
a'	Predator attack coefficient	3.274 cm^2 h^{-1}
t_h	Predator handling time per prey	0.775 h
b'	Predator encountering coefficient	1.195 cm^2
b_Y	Predator birth rate coefficient	0.0151
d_Y	Death rate per predator	0.0038 h^{-1}
δ_Y	Predator death rate coefficient	16.92 h
P_m	Predation rate threshold	0.0958 h^{-1}
λ_Y	Predator settling rate during dispersal	1.0 m^{-1}
μ_Y	Predator mortality rate during dispersal	0.5 m^{-1}

Note:
See Nachman (1987a) for a detailed description of the model and its parameters.

$$p_i = \frac{e^{-(\lambda+\mu)\Omega_i}}{\sum_{i=1}^{N-1} e^{-\lambda\Omega_i}} \tag{10.1}$$

where N is the number of plants in the system, μ is the mortality rate per distance moved, and λ is a positive constant (see Caley, 1991). The lower the value of λ, the further an individual moves, on average, before settling. Thus, $\lambda = 0$ produces global (island) dispersal, where all plants have the same chance of being found (provided $\mu = 0$). Likewise, $\lambda \to \infty$ leads to local (stepping stone) dispersal, where the nearest plants are most likely to be found. The effect of distance on transit time is assumed to be negligible; arrivals are assumed to occur in the next time-step of the model (1 hour) regardless of distance moved. Plants are assumed to be arranged in an array (as in a greenhouse) with distance ω_b between rows and distance ω_w between plants within a row. Hence, the distance Ω between two plants is found by means of the Pythagorean theorem.

For simplicity, it was assumed that the distance between plants and rows was the same, i.e., $\omega_b = \omega_w = \omega$, and the following values for ω were used: 0.5, 1, 2, 4, and 8 m. Each level was replicated five times, and simulation runs were terminated if either of the species went extinct or after 2000 days. A system consisted of 120 plants arranged in 8 rows. Systems were 'sampled' every week to obtain estimates of mean density of each mite species, variance of mites among plants, mean plant injury, number of plants killed, and number of plants occupied by prey, predator or both. At the end of a simulation the following systems attributes were calculated:

1. *Persistence*: time until the first species went extinct.
2. *Mean density of mites*: mean of densities from day 21 to day 196, recorded at intervals of 1 week.
3. *Temporal variation*: coefficient of variation (CV) of mean weekly densities (see McArdle & Gaston, 1992).
4. *Predator–prey ratio*: mean predator density divided by mean prey density.
5. *Plant injury*: mean value of the weekly plant injury index D (proportion of leaf chlorophyll destroyed (Nachman, 1987a)).
6. *Spatial overlap between prey and predators*: number of plants (or other spatial units such as leaves) hosting both mite species relative to the number of plants with either prey or predators. A plant can be in one of four states (Maynard Smith, 1974): (a) without mites; (b) with prey

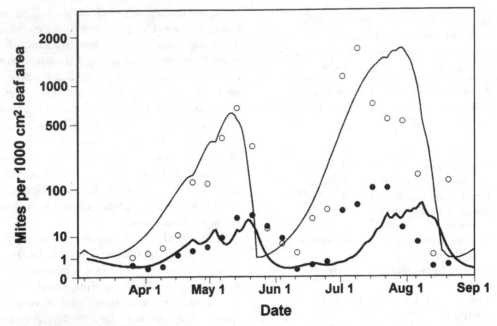

Figure 10.1. A simulation of a greenhouse experiment (Nachman, 1981). Six adult female *Tetranychus urticae* were introduced onto every second plant in early March followed 2 days later by three adult female *Phytoseiulus persimilis* on the same plants. Table 10.2 gives the parameter values of the model. Open circles and thin line, observed and predicted density of *T. urticae*, respectively; filled circles and heavy line, observed and predicted density of *P. persimilis*, respectively. The observed values are running averages of two subsequent samplings each consisting of three leaves picked from 36 randomly chosen plants.

only; (c) with both prey and predators; (d) with predators only. Spatial overlap is expressed by the Dice (1945) index $DI = 2N_c/(N_b + 2N_c + N_d)$, where N_b, N_c, and N_d are the numbers of plants in states b, c and d, respectively.

7. *Index of biological control (IBC)*: the probability that a plant (or other spatial unit) occupied by spider mites is also occupied by predatory mites. It is calculated as $IBC = N_c/(N_b + N_c)$.

8. *Spatial aggregation*: proportion of plants occupied (p) when the mean density per plant is $\bar{x}$. A general relationship between p and $\bar{x}$ is the Weibull (1951) distribution, given as:

$$p = 1 - e^{-\alpha\bar{x}^\beta} \tag{10.2}$$

where α and β are constants (see also Gerrard & Chiang, 1970; Torii, 1971; Nachman, 1981, 1984, for sampling applications of the Weibull

distribution). The parameters are estimated by linear regression of $\ln(-\ln(1-p))$ against $\ln \bar{x}$. Small values of α, β or both indicate, all else being equal, a high degree of spatial aggregation.

9. *Spatial asynchrony*: relationship between the spatial variance (s^2) (the variance of individuals per plant) and the mean density per plant $\bar{x}$. If all local populations are completely in phase, $s^2 = 0$; the more out of phase the populations are, the greater the value of s^2. Spatial variance was estimated by fitting:

$$\ln s^2 = \ln a + b \ln \bar{x} \tag{10.3}$$

to the data, where a and b are constants (Taylor, 1961). Large values of a, b or both indicate, all else being equal, a high degree of spatial asynchrony.

Attributes (2)–(8) were only calculated for runs in which both prey and predators persisted for at least 28 weeks. Results from the first 2 weeks were not used.

Results of simulations

The model produces fluctuations that are relatively regular, with a rather constant period and a more variable amplitude (sample runs in Figure 10.2). Since there are no external factors to trigger the cycles, the trajectories gradually diverge with time due to endogenously generated noise (phase-forgetting quasi-cycles (Nisbet & Gurney, 1982)). The principal effect of decreasing inter-plant distance is to increase the amplitude of the fluctuations, and at the same time make the cycles more regular. The strong coupling between prey and predator fluctuations disappears when plants are very distant. The period of the cycles is approximately 85 days, which is in good accordance with experimental data (Huffaker, 1958; Burnett, 1970, 1979; van de Klashorst *et al.*, 1992; but see also Gough, 1991, who found rather erratic fluctuations). Effects of decreasing inter-plant distance on predator–prey dynamics were as follows:

1. *Persistence:* persistence was maximal (all replicates persisted > 2000 days) when inter-plant distance was 1 m. At the shorter distance ($\omega = 0.5$ m), the prey went extinct in one run after 174 days and the predators in another after 733 days. When the distance between plants was 2 m or 4 m predators always went extinct before 2000 days, but never before 196 days. Average persistence time was 946 days for 2 m and 407 days for 4 m. At 8 m, average persistence time decreased to 154 days and no predators survived past 181 days.
2. *Mean density:* as inter-plant distance decreased (more dispersal), density of prey decreased and density of predators increased.
3. *Temporal variation:* as inter-plant distance decreased, temporal variability increased for both prey and predators, implying more pronounced predator–prey oscillations.
4. *Predator–prey ratio:* the ratio increased as the

inter-plant distance decreased, indicating better biological control at higher dispersal rates.
5. *Plant injury:* plant injury also decreased with decreasing inter-plant distance.
6. *Spatial overlap:* spatial overlap between prey and predators increased with decreasing inter-plant distance.
7. *Index of biological control:* IBC increased with decreasing inter-plant distance.
8. *Spatial aggregation:* β increased as the inter-plant distance decreased, but since α decreased, there was no overall effect on spatial aggregation. A plot of the relationships showed, however, that over most of the range of mite densities, both species tended to be less aggregated as inter-plant distance decreased.
9. *Spatial asynchrony:* as above, the two parameters were negatively correlated, since *b* decreased and *a* increased as the inter-plant distance decreased (see also Hanski, 1987). A plot of the relationships revealed that dispersal had little effect on spatial asynchrony for the prey. Local predator densities, however, tended to be more synchronized when distances were short.

Experimental dynamics of greenhouse spider mite

Experimental protocol

A series of experiments were conducted where between-plant dispersal rates of *T. urticae* and *P. persimilis* were varied (see Nachman, 1997, for details). Lima bean plants were arranged in a lattice design on a water-saturated blanket, which served as an almost complete barrier against dispersal. The distance between plants and rows was 25 cm and the plants did not touch each other. Dispersal among plants was primarily *via* bridges connecting neighboring plants. To connect two neighboring plants, a 50 cm metal stake was pressed into the soil of each pot, and bent at a 90° angle 6 cm from the top. The stakes were then connected with plastic tubing. Experimental treatments were: (A) no/low dispersal (no bridges); (B) moderate dispersal (bridges to 4 neighboring plants, cardinal directions); (C) high dispersal (bridges to 8 neighboring plants). The use of bridges

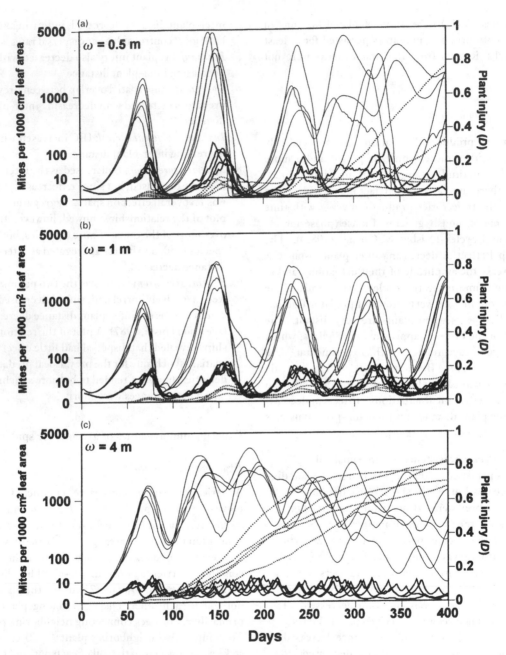

Figure 10.2. Examples of simulated population dynamics in systems differing with respect to distance between plants. Initial conditions and parameter values as described in the legend to Fig. 10.1. Each distance was replicated 5 times. Thin lines, mean density of *T. urticae*; bold lines, mean density of *P. persimilis*, dotted lines, mean plant injury.

Table 10.3. *Summary of experimental protocol for greenhouse spider mite experiments*

Year	Number of plants/system	Treatments
1991	30	A,B,C
1992	45	A,B,C
1993	45	B,B
1994	45	B,C

Note:
A, no bridges; B, few bridges; C, many bridges.

allowed dispersal rate to be manipulated without altering the growing conditions of the plants. Due to space limitations, a maximum of three systems could be run per year. The number of plants and initial densities varied from year to year, but were consistent across systems within years (Table 10.3).

Mites were introduced onto the plants in early April. Mite densities were assessed every 2 weeks by sampling one third of the plants (sample contained same number of plants per age group and plants were chosen randomly within age groups). Leaf injury was assessed visually using the leaf damage index (LDI) defined by Hussey & Parr (1963b). LDI values were then converted to chlorophyll per cm^2 by means of a calibration curve, and plant injury was the ratio of amount of chlorophyll removed by spider mites to the total chlorophyll available (Tomkiewicz *et al.*, 1993). The sampled plants were replaced by young plants, but the mites were not returned to the system.

Sampling continued until mid-December or until one of the species was not recorded for five consecutive sampling dates. The only exception to the protocol was in 1991, when predators were reintroduced into all systems after 10 weeks, since none had been observed in system C (many bridges) for the three previous sampling dates. At the end of an experiment, all plants were sampled. A species was deemed extinct if no individuals were found at this time. If the species had not been found in previous samplings as well, extinction was assumed to have occurred on the first date that no individuals were seen (see McArdle & Gaston, 1993, for the problem of distinguishing sampling zeros from structural zeros).

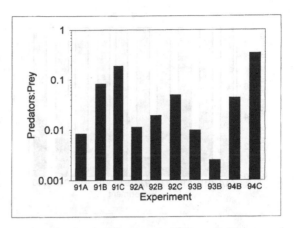

Figure 10.3. Overall predator–prey ratios in 10 experiments. Labels on the abscissa refer to year and type of system : A, no bridges; B, few bridges; C, many bridges.

Experimental results

1. *Persistence*: *Tetranychus urticae* never went extinct in the 7 systems with no and few bridges, but did go extinct in 2 out of the 3 systems with many bridges. Persistence times of the prey in these systems were 20 weeks (1992C) and 8 weeks (1994C). *Phytoseiulus persimilis* went extinct before its prey in 3 systems. Persistence times were 28 weeks (1991C), 26 weeks (1992A), and 10 weeks (1994B). Note that *P. persimilis* was reintroduced in 1991, which likely increased persistence time. Overall, persistence was greatest for systems with no bridges and lowest for systems with many bridges. The prey in particular seemed to benefit from the absence of bridges.
2. *Mean density*: the presence of bridges had no clear effect on mean density of either species.
3. *Temporal variability*: bridges also had no consistent effect on temporal variability.
4. *Predator–prey ratio*: predator–prey ratio increased with the number of bridges when systems were compared within years (Fig. 10.3). Since several aspects of the experimental protocol varied with year (number of plants, pattern of mite introduction, climatic conditions), the data were analysed as a two-factor design with year and treatment as factors. A non-parametric

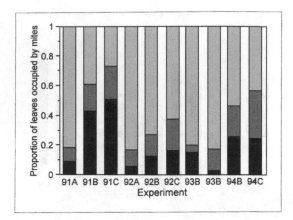

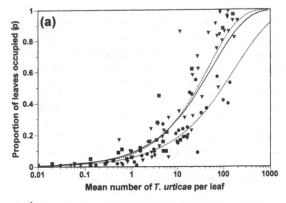

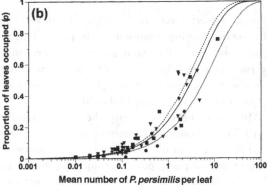

Figure 10.4. Overall distribution of leaves with only *T. urticae* (light grey), with only *P. persimilis* (black), and with both species (dark grey). Labels on the abscissa refer to year and type of system: A, no bridges; B, few bridges; C, many bridges.

Figure 10.5. Relationship between the mean density and the proportion of leaves occupied by (a) *T. urticae* and (b) *P. persimilis*. The lines are a fit of eqn (10.2) to data. (The circumflex indicates that the parameter value has been estimated.) The parameters for *T. urticae* are as follows for no bridges: $\hat{\alpha}=0.0596$, $\hat{\beta}=0.539$, $R^2=0.842$, d.f. = 30; few bridges: $\hat{\alpha}=0.0823$, $\hat{\beta}=0.649$, $R^2=0.910$, d.f. = 61; many bridges: $\hat{\alpha}=0.0912$, $\hat{\beta}=0.591$, $R^2=0.884$, d.f. = 24. For *P. persimilis* the parameters for no bridges are: $\hat{\alpha}=0.186$, $\hat{\beta}=0.728$, $R^2=0.917$, d.f. = 18; for few bridges: $\hat{\alpha}=0.333$, $\hat{\beta}=0.788$, $R^2=0.956$, d.f. = 51; and for many bridges: $\hat{\alpha}=0.279$, $\hat{\beta}=0.772$, $R^2=0.955$, d.f. = 23. Circles and dotted line, no bridges; triangles and broken line, few bridges; squares and continuous line, many bridges.

approach was used, where results were ordered within years and replaced by ranks. Considering the systems run in 1991, 1992, and 1994, there are 72 possible combinations of ranks, each equally likely if bridges had no effect. All results were in the expected direction, which is the most extreme possible outcome. Hence, the test is significant at the 5% level ($p_{\text{one-tailed}} = 1/72 = 0.0139$).

5. *Plant injury*: plant injury in the individual systems was closely related to the density of spider mites, but showed no relation to the number of bridges.

6. *Spatial overlap*: the proportion of leaves occupied by both prey and predators increased with the number of bridges (Fig. 10.4). The result was significant ($p_{\text{one-tailed}} = 0.0139$, same test as for predator–prey ratio). The proportion of leaves with *T. urticae* alone decreased and the proportion with *P. persimilis* alone increased as the number of bridges increased.

7. *Index of biological control*: IBC increased consistently with an increase in the number of bridges ($p_{\text{one-tailed}} = 0.0139$, same test as above).

8. *Spatial aggregation*: the relationship between proportion of occupied leaves and density was influenced by the number of bridges (Fig. 10.5). Data

were pooled across years, as there were no inter-year differences and analysed by means of an ANCOVA (PROC GLM in SAS, SAS Institute) using density as the covariate. Spatial aggregation was significantly higher ($p = 0.0038$ for *T. urticae* and $p = 0.0229$ for *P. persimilis*) in systems without bridges than in the two other types of

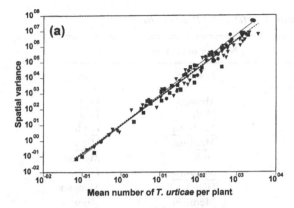

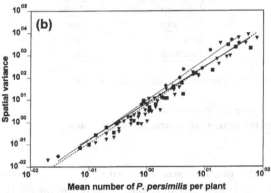

Figure 10.6. Relationship between the mean density and the spatial variance of (a) *T. urticae* and (b) *P. persimilis*. The lines are a fit of eqn (10.3) to data. (The circumflex denotes that the parameter value has been estimated.) The parameters for *T. urticae* are as follows for no bridges: $\hat{a} = 2.321$, $\hat{b} = 1.923$, $R^2 = 0.984$, d.f. = 30; for few bridges: $\hat{a} = 2.433$, $\hat{b} = 1.800$, $R^2 = 0.985$, d.f. = 62; for many bridges: $\hat{a} = 2.271$, $\hat{b} = 1.806$, $R^2 = 0.976$, d.f. = 24. For *P. persimilis* the parameters for no bridges are: $\hat{a} = 2.303$, $\hat{b} = 1.793$, $R^2 = 0.978$, d.f. = 19; for few bridges: $\hat{a} = 1.782$, $\hat{b} = 1.711$, $R^2 = 0.964$, d.f. = 51; and for many bridges: $\hat{a} = 2.004$, $\hat{b} = 1.634$, $R^2 = 0.977$, d.f. = 23. Circles and dotted line, no bridges; triangles and broken line, few bridges; squares and continuous line, many bridges.

system. The systems differed significantly only with respect to intercepts ($p < 0.0001$ for both species).

9. *Spatial asynchrony*: the relationship between spatial variance and density was influenced by the number of bridges (Fig. 10.6). Data were analysed

as above. Spatial asynchrony was significantly higher ($p = 0.0073$ for *T. urticae* and $p = 0.0295$ for *P. persimilis*) in systems without bridges than in the two other types of system. The systems differed significantly only with respect to intercepts ($p < 0.0001$ for *T. urticae* and $p = 0.0003$ for *P. persimilis*).

Comparison of predictions and experimental results for greenhouse spider mite

The results of the experiments were summarized by averaging the outcome of all experiments of the same type, regardless of year. Values were replaced by rank (low, medium, high), and bold underlined text indicates qualitative agreement between empirical and theoretical results (Table 10.4). Complete agreement was obtained for 7 out of 13 attributes. The model correctly predicted that increasing the number of bridges (i.e., increasing dispersal) should increase temporal variability, decrease spatial aggregation and increase spatial synchrony of the prey, increase predator density, predator–prey ratio, and spatial overlap of prey and predators. Low prey densities and persistence in systems with highest dispersal were also seen in both the model and experimental results. Finally, the model also correctly predicted that the system with no bridges should have the lowest predator coefficient of variation (CV), and highest spatial asynchrony and aggregation of the predator. Since agreement between observed and predicted results may occur by chance, the data in Table 10.3 were analysed using a one-sample G-test (see e.g., Fowler & Cohen, 1990), where the observed and expected distributions of complete, partial and no agreement were compared (7,5,1 *versus* 2.2,6.5,4.3). This yielded $G_{adj} = 10.33$; $p_{one-tailed} < 0.005$; df = 2, indicating that bridges had an overall effect on the measured attributes. Note, however, that the test requires independence among the various attributes, which may not be the case.

The major differences between model and experimental results were the effects of no bridges on prey density, level of plant injury and persistence of the systems. The number of bridges affected predator densities more than prey densities, perhaps because

Table 10.4. *Summary of greenhouse experiments with T. urticae and P. persimilis*

Attribute	Number of bridges		
	None	Few	Many
Persistence	High	Medium	**Low**
Density: prey	Medium	High	**Low**
: predator	**Low**	**Medium**	**High**
Temporal variation: prey	**Low**	**Medium**	**High**
: predator	**Low**	High	Medium
Predator–prey ratio	**Low**	**Medium**	**High**
Plant injury	**Low**	High	Medium
Spatial overlap	**Low**	**Medium**	**High**
Index of biological control	**Low**	**Medium**	**High**
Spatial aggregation: prey[1]	**High**	**Medium**	**Low**
: predator[1]	**High**	Low	Medium
Spatial asynchrony: prey[1]	**High**	**Medium**	**Low**
: predator[1]	**High**	Low	Medium

Notes:

Bold underlined results indicate agreement between prediction and observations.

[1] The result applies to the majority of observed densities, but pattern may reverse at extreme values.

P. persimilis tends to disperse much more readily than does *T. urticae*. Increased dispersal also led to higher predator–prey ratios and greater spatial overlap between predator and prey, which together should have led to better biological control. This was not observed in the experiments. A likely explanation is that plants were substituted during an experiment. A plant remained on average 6 weeks in a system (minimum 2 and maximum 10 weeks). The time until a new plant was invaded by spider mites was likely to be considerably longer in systems without bridges than otherwise. Consequently, the absence of bridges reduced the average time available to the spider mites to build up high population densities on a plant before it was removed.

Bridges decreased persistence time by making it easier for the predators to overexploit the prey. This was also reflected in a lower level of spatial asynchrony in systems with many bridges. Absence of bridges, on the other hand, reduced the risk that the predators could overexploit their prey, but made it more difficult for the predators to survive even when the prey was abundant. It was reflected by a decrease in the predator–prey ratio with time. This would probably have lead to regional extinction of the predators, if the experiments had lasted longer. Hence, even though the observed persistence times were not qualitatively in accordance with the predictions, the mechanisms leading to regional extinction were. The dispersal rates that maximize persistence are probably lower than those in systems with few bridges but higher than those in systems without bridges.

In general, the agreement between observed and predicted results was best for systems without bridges, indicating that the decisive factor was the presence of bridges, whereas the number of bridges was of minor importance. This accords with separate dispersal experiments showing that the rate of successful dispersal between two plants increases about 100 times (*T. urticae*) and 40 times (*P. persimilis*) if the plants are connected by a bridge, but only by a factor of 2 if the number of bridges is doubled (pers. obs.).

In conclusion, at least part of the qualitative outcome of an experiment aimed at looking at the dynamics of a spatially structured system can be correctly captured with a simulation model. The discrepancies between model and experimental results point to aspects that perhaps merit further investigation, for example, the rate of renewal of the host plants and the range of dispersal rates that promote persistence of predators and prey.

Practical implications for biological control.

Theory as well as experiments show that any measure that can promote between-plant dispersal (e.g., contacts between plants, connecting plants along and across rows by means of wires or actively transferring predators from one plant to another) will contribute to the success of biological control of greenhouse spider mites by means of predatory mites. Most growers (in Denmark at least) know this from personal experience, so in this respect theory just confirms common practice. The drawback of high dispersal rates is that the predators are more likely to go extinct due to lack of food. However, this risk declines sharply with the number of plants in a greenhouse (Nachman, 1987a, 1991).

Spider mites in apple orchards

Biology of orchard mites

Panonychus ulmi (Koch) is a spider mite pest of economic importance in most apple growing regions of the world. It feeds on the cell contents of apple leaves, and at high densities can lower the photosynthetic rate and ultimately the productivity of apple trees (Hardman *et al.*, 1985; Mobley & Marini, 1990). Compared with *T. urticae*, *P. ulmi* produces much less webbing and has a less aggregated distribution. It also has lower fecundity and slower development (Sabelis, 1985a). In the study area (Nova Scotia, Canada), overwintering eggs hatch in May, and three to five generations are completed by fall. Female mites begin laying winter eggs in August and leaves fall off the trees in mid to late October.

The pest status of *P. ulmi* is largely an artificial one; chemicals used to control other apple pests often kill the spider mite's natural enemies. Under appropriate management regimes, *P. ulmi* can be held

to economically insignificant levels by the predatory mite, *Typhlodromus pyri* Scheuten. *Typhlodromus pyri* consumes larval, protonymph, deutonymph, and adult stages of *P. ulmi*, but does not feed on eggs (Walde *et al.*, 1992). *Typhlodromus pyri* is a generalist predator. *Panonychus ulmi* is its preferred prey (Dicke, 1988; Dicke & deJong, 1988; Nyrop, 1988), but it will feed on other mites, including *Aculus schlechtendali* (Eriophyiidae) and *T. urticae*, as well as pollen and plant juices (Chant, 1959).

Typhlodromus pyri is quite mobile within trees, remaining only hours on individual leaves, and commonly leaving a leaf before all prey are consumed (Lawson & Walde, 1993). The extent to which *P. ulmi* disperses within trees is not known. Both predator and prey disperse among trees using air currents, but the dispersal rate of *T. pyri* appears to be considerably less than that of *P. ulmi* (Walde, 1994), and less than that of some other phytoseiids (Dunley & Croft, 1990, 1994). *Typhlodromus pyri* is also less mobile than *P. persimilis*. *Panonychus* species spin silk threads and descend in response to a deteriorating food supply (Johnson, 1983; Wanibuchi & Saito, 1983), but some *P. ulmi* disperse even when densities are low (Lawson *et al.*, 1996).

Other species of mites are also commonly present in apple orchards. In Nova Scotia, the most important two are apple rust mite (*A. schlechtendali*), also a plant-feeding mite, and *Zetzellia mali* (Stigmaeidae), another generalist predator.

Fit of experimental dynamics with model predictions for orchard spider mite

Local dynamics

The interaction between *P. ulmi* and *T. pyri* appears to be unstable at the local population level. Isolated populations of *P. ulmi* on young trees go extinct in the presence of the predator (Walde *et al.*, 1992; Walde, 1994). The probable sources of the instability include a Type II predator functional response (Walde *et al.*, 1992), a delayed density-dependent numerical response (Walde, 1995a) and the generalist habit of the predator. The latter allows the predator to persist even in the absence of its preferred prey.

Additional evidence for the instability of local populations of *P. ulmi* comes from experiments conducted over a 4 year period in Oregon, USA (Croft & MacRae, 1992, 1993; Croft, 1994). In these experiments, trees were far enough apart that dispersal among trees was probably insignificant. *Panonychus ulmi* was apparently absent from many trees at the end of the 4 year period, replaced by a competing plant-feeding mite, *A. schlechtendali* (Croft, 1994). *Typhlodromus pyri* also declined in numbers, possibly due to the decline of its preferred food, but more likely due to a competing predator, *Z. mali*. Thus interactions with other members of the acarine complex on apple may add to the inherent instability of the *P. ulmi–T. pyri* interaction.

Regional dynamics

In experimental apple orchards, immigration rate was manipulated by varying the number of interacting populations (number of trees within a patch) (Walde, 1991, 1994, 1995b). Results coincided with some, but not all of the expectations derived from our review of spatially structured models (see Table 10.1). Increased immigration of prey did lead to greater persistence of local prey populations. Greater local persistence should lead to greater regional persistence. Most models predict that intermediate dispersal rates maximize the probability of regional persistence (Table 10.1 and the *T. urticae–P. persimilis* model, above). Immigration rate did not affect the temporal variability of local populations, but there was a trend (however, statistically non-significant) for metapopulations to have lower temporal variability than isolated populations (Walde, 1995b). Higher prey dispersal rates are expected to increase prey densities (see Table 10.1), and this was also seen. However, contrary to the predictions of most spatial models, predator persistence and densities were unaffected by dispersal, probably due to the generalist feeding habit of *T. pyri* (Walde *et al.*, 1992).

Mechanisms

How does dispersal typically enhance persistence in spatially structured models, and how does this relate

to the above observations of orchard mite dynamics? First it is important to consider the effects that dispersal has on local *versus* regional dynamics, although the two are, of course, linked. Density-independent migration, as well as density-dependent immigration and emigration can contribute to the stability of individual populations (Nisbet *et al.*, 1993). Density-independent migration can stabilize by making the *per capita* immigration rate temporally density-dependent (Murdoch *et al.*, 1992). This is easy to visualize in systems with global dispersal. With density-independent migration (e.g., a constant proportion), low density populations will, on average, receive more immigrants relative to their density than high density populations. Density-dependent immigration (higher *per capita* immigration with low densities) and emigration (higher *per capita* emigration with high densities) tend to stabilize local populations; they act as density-dependent influences on local population growth rate. Therefore to the extent that increasing local stability increases regional stability, this type of dispersal is also stabilizing at the metapopulation level. However, density-dependent migration, as well as high rates of density-independent migration, also have a destabilizing effect at the regional scale, since they tend to synchronize fluctuations in the different local populations, and thus the net effect on the metapopulation can be stabilizing or destabilizing, depending on migration rate and the strength of the density-dependence.

The above set of mechanisms does not appear to account for increased local persistence of the prey in the orchard mite system, since neither *per capita* immigration rate nor *per capita* emigration rate of the prey was related to density (Walde, 1994; and unpublished results). The hypothesis cannot be completely rejected based on these data, however, since colonization is made up of two parts, immigration and establishment. If the likelihood of establishment for newly arrived *P. ulmi* is negatively related to the density of *T. pyri*, and *T. pyri* density increases with density of *P. ulmi*, then *colonization* rate could be density-dependent, i.e., higher in patches or trees where the density of *P. ulmi* is low. In addition, dispersal could have enhanced local stability *via* the predator, although this is highly speculative. *Per capita* immigration of *T. pyri* was density-dependent, and thus would have contributed both to the persistence of the predator and to the stability of the predator–prey interaction. However, *per capita* emigration of the predator was inversely density-dependent, and thus the net effect of predator dispersal on persistence is not clear. In sum, increased persistence of *P. ulmi* with greater dispersal could have been due to 'shifting mosaic' dynamics even if the predators do not go extinct. On the other hand, it might be argued that one should probably not expect colonization rate to be density-dependent when dispersal is localized. Local dispersal leads to patchiness, but patchiness at a scale greater than that of a single plant (e.g., Nachman, 1991). Low density populations thus tend to be closer to other low density populations; immigration is therefore going to be lower, and to offset this, establishment rates would have to be much higher in low density populations.

There is another plausible explanation for the enhanced persistence of *P. ulmi* due to dispersal in the orchard mite system. Increased persistence may be simply the result of increased average prey densities, which come about due to increased growth rate of the population on a tree because reproductive rate is supplemented by immigration (e.g., Conner *et al.*, 1983; Fahrig & Merriam, 1985). This explanation has two implications. First, further increases in successful dispersal should always lead to increased persistence, not to lower persistence as predicted by most models. Second, while this mechanism led to greater persistence over the experimental period, it is not stabilizing in the mathematical sense (does not lead to density-dependence in any vital rates), and thus these systems should not be stable over the long term. There was some evidence that all experimental systems, even the most persistent, were on a trajectory toward densities of zero (Walde, 1994), and so some other mechanism may be ultimately responsible for the persistence of *P. ulmi*. One possibility is that human disturbance is responsible – perhaps persistence in orchards requires the occasional creation of outbreaks by elimination of predators. This does not, of course, explain persistence on a global scale in the absence of human interference. At these larger scales, however, other mechanisms could well be operating, for example, *P. ulmi* could have a refuge in other host plants where *T. pyri* is absent or less effective.

Significance for biological control of orchard spider mite

The implications of this work for the biological control of *P. ulmi* in apple orchards are very simple, and we suggest that they may apply equally well to other systems with generalist predators as the key biological control agent. Prey persistence ceases to be a relevant issue for successful control with a generalist predator, and thus efforts should be directed at maximizing predator impact on the prey. From a metapopulation perspective, the critical factor is minimizing prey dispersal among populations since dispersal tends to prop up prey densities. Minimizing the potential for prey dispersal into the orchard from other sources might also be a consideration. In addition, measures that enhance predator persistence or densities are likely to be helpful, for example the provision of supplemental food.

Dynamics of spider mites in greenhouses *versus* orchards

A glance at Table 10.5 shows how different the effects of increasing dispersal were for the greenhouse *versus* orchard mite systems. Higher dispersal rates decreased persistence in the greenhouse system, at least in part by synchronizing prey dynamics. In contrast, prey persistence in the orchard was enhanced by higher dispersal, and synchrony of dynamics played no role in this outcome. Higher dispersal rates also led to better biological control in the greenhouse (lower prey densities, increased predator density and increased predator–prey ratios), but to poorer control in the orchard (higher prey densities, lower predator–prey ratios). There are a number of potentially important differences between the greenhouse and orchard systems. Below we discuss one difference in some detail, the diet breadth of the predator,

Table 10.5. *Comparison of outcomes for greenhouse, orchard and model spider mite systems as dispersal rate increases from low to high*

Attribute	Greenhouse	Orchard	Prediction
Persistence	Decrease	Increase	Increase then decrease
Density: prey	Decrease	Increase	Decrease
: predator	Increase	No change	Increase
Temporal variation: prey	Increase	Decrease (NS)	Increase
Predator–prey ratio	Increase	Decrease	Increase
Spatial asynchrony: prey	Decrease	No change	Decrease

Notes:
Note that not all the variables from Table 10.3 were assessed for the orchard mite system.
NS, not significant.

and then briefly consider several other potential causes of the differences in dynamics.

Generalist and specialist predators in metapopulation dynamics

A key difference between the greenhouse and orchard systems considered here lies in the diet breadth of the predators. *Phytoseiulus persimilis* is a specialist on *T. urticae*, while *T. pyri* uses a variety of foods in addition to *P. ulmi*. This has ramifications for the dynamics (especially persistence) of the two systems, as well as implications for biological control (i.e., prey density).

Stability or persistence is a key objective in virtually all modeling of biological control, and at times also in the practice of biological control. Local prey extinction is possible with either a generalist or specialist predator; the likelihood of prey persistence will depend on predator functional and numerical responses for both predator types, and on prey preferences for the generalist. Diet breadth will affect predator persistence, however, since persistence of a specialist predator is dependent on persistence of the pest, while persistence of a generalist may not be. Thus at the local scale, persistence of the prey may be either more or less likely with a generalist predator, but probability of predator persistence will almost certainly be greater for a generalist over a specialist predator.

Model of a generalist predator

This difference in local dynamics should have repercussions at the regional scale. Here we use the greenhouse spider mite model (above) to begin to explore what these might be. We wanted to know how critical the specialist food habit of the predator is to the model dynamics, that is, whether most of the differences in dynamics between orchard and greenhouse systems can be accounted for by the fact that *T. pyri* is a generalist predator with alternative food available.

We created a generalist by allowing the predator to feed, and reproduce, on alternative food. Alternative food was assumed to be limiting, so that, for example, a doubling of the number of predators would halve the amount of alternative food available per individual. With Y predators present on a plant, the amount of alternative food per predator thus becomes Z/Y, where Z is the total amount of alternative food available per time unit. Z is assumed to be independent of Y, which may be reasonable if the alternative food is pollen or fungi, but probably not if it is another prey species. It was assumed that alternative food would be eaten as a supplement to prey, and with a rate proportional to the fraction of time available. Hence, if P_1 and P_2 denote the rate of consumption of prey and alternative food, respectively, the total rate of consumption per predator (expressed in prey equivalents) becomes:

$$P = P_1 + vP_2 = P_1 + \xi(1 - t_h P_1) Z/Y \qquad (10.4)$$

Table 10.6. *Predictions for predator–prey metapopulation dynamics for a specialist* versus *generalist predator as dispersal rate is increased (distance between plants is decreased)*

Prediction attribute	Specialist	Generalist
Persistence	Increase then decrease	Increase
Density: prey	Decrease	Increase then decrease
: predator	Increase	Increase then no change
Temporal variation : prey	Increase	Increase
: predator	Increase	No change
Predator–prey ratio	Increase	Decrease then increase
Plant injury	Decrease	Increase then decrease
Spatial overlap	Increase	Increase
Index of biological control	Increase	No change
Spatial aggregation : prey[1]	Decrease	Decrease
: predator[1]	Decrease	Decrease
Spatial asynchrony : prey[1]	Decrease	Decrease
: predator[1]	Decrease	Decrease

Notes:
The attack coefficient (a') of the generalist is 20% of that for the specialist.
[1] The result applies to the majority of observed densities, but pattern may reverse at extreme values.

where t_h is the handling time per prey, v is a constant converting alternative food into prey equivalents, while ξ is another constant related to the predator's efficiency to exploit alternative food (ξ also includes v). Computation of P_1 follows eqn (19) in Nachman (1987a).

Incorporating alternative food into the model, while keeping all other things equal, increased the regional persistence of the predator but decreased persistence of the prey. As expected, increasing the amount of alternative food available (increasing ξ or Z), increased the mean density of predators. Alternative food *per se* was not able to stabilize the predator–prey interaction over the long term. Low amounts of alternative food did have a positive effect on persistence, as it prolonged the time until regional extinction of the predator. High amounts of alternative food allowed local (and thus also regional) persistence of the predator, but the local persistence of the predator prohibited the prey from colonizing patches, and thus the prey went extinct.

Long-term persistence of both predator and prey was only achieved if the predators were less efficient with respect to feeding on the prey. The lower the

attack coefficient (a'), the easier it was for the prey to establish in patches already occupied by predators. For example, reducing a' to one-fifth of its original value ($a' = 0.6548$ cm^2/h) resulted in long-term persistence of both predator and prey (> 2000 days) and low plant injury as long as plant distance was no greater than 4 m. Beyond that distance, the prey could no longer persist because the colonization rate was insufficient to balance the extinction rate. However further reducing a', to one-tenth of its original value, resulted in higher prey densities, greater plant injury, and ultimately to a decline in the number of surviving plants and regional extinction.

We further looked at the impact of the generalist food habit by determining how the various aspects of dynamics measured earlier respond to increases in dispersal rate (Table 10.6). As we anticipated, increasing dispersal in a system with a generalist predator consistently increases persistence. The generalist food habit of *T. pyri*, combined with its lower efficiency as compared with *P. persimilis*, can thus account for one of the major differences in the greenhouse *versus* orchard experimental results, the fact that increasing dispersal decreased persistence in the greenhouse and increased persistence in the orchard. The model also predicts that increasing dispersal should cause prey densities (and plant injury) to increase (over a range of dispersal rates) if the predator is a generalist but to decrease if the predator is a specialist. This also accords with what we observed experimentally in the two systems (Table 10.5). However, the model also suggests that a generalist food habit should not affect how several other attributes of dynamics respond to increasing dispersal rate, including spatial overlap of predator and prey, temporal variability of the prey, spatial aggregation and spatial asynchrony, and thus any differences seen in these attributes must be due to other differences between the systems.

Implications for biological control

Although it has been argued in the past that specialist natural enemies should be superior biological control agents (see Chapter 6), there is no doubt that generalist mite (phytoseiid) predators provide effective control of spider mites in a number of different crops (McMurtry, 1992). Especially significant is the fact that the importance of prey persistence matters much less to the success of a biological control system if the predator is generalist. Murdoch (1994) and Briggs *et al.* (Chapter 2) point out that persistence in host–parasitoid or predator models is often bought at the cost of higher prey densities. For example, predator aggregation can stabilize Nicholson–Bailey host–parasitoid models, but increases the equilibrium host density. Similar trade-offs are evident in some spatially structured models. Higher prey dispersal and low variability in prey reproduction lead to more constant but higher prey densities (Reeve, 1988). Models with more than one enemy tend to be less stable than the equivalent single predator model (Hassell *et al.*, 1994), but prey density may well be depressed. This trade-off may be necessary for systems with specialist predators, but the need for prey persistence disappears in systems with generalist predators. We suggest that, as a result, optimal distances between plants will usually be greater for systems with generalist predators than for systems with specialist predators. This comes about because, while low rates of prey dispersal are beneficial to bio-

logical control in both cases, low predator dispersal rates are detrimental to biological control by a specialist, but have little effect on a generalist predator. Thus while the best control of *T. urticae* is attained with closely spaced plants, management practices such as the recent trend toward high density plantings in apple orchards could potentially have negative impacts on control of *P. ulmi* by *T. pyri*.

Other differences between greenhouse and orchard systems

Several other predator attributes probably contribute to the different dynamics seen in greenhouse *versus* orchard systems, including tendency to eliminate prey patches, dispersal rate, and ability to aggregate in high density prey patches. Systems with 'killer' predators, those that rapidly eliminate patches of their prey (e.g., *P. persimilis*) tend to be less stable than systems with 'milkers', those that just limit prey growth rate (Takafuji *et al.*, 1983; van Baalen & Sabelis, 1995; and see Chapter 15). *Typhlodromus pyri* does seem to eventually eliminate its prey, but this is over a much longer time-scale than seen for *P. persimilis*, and thus *T. pyri* is closer to the milker end of the continuum. Predator dispersal is expected to have different effects for killer *versus* milker systems; low dispersal of killers enhances persistence, while high dispersal of milkers increases persistence (Takafuji *et al.*, 1983). *Phytoseiulus persimilis* disperses readily among plants, while the rate of dispersal among trees is very low for *T. pyri*. Thus in both cases predator dispersal might be expected to contribute to instability of the predator–prey interaction. This is probably the case. Reducing dispersal of *P. persimilis* among plants does enhance persistence, as it increases the time that prey patches escape detection. The low dispersal rate for *T. pyri* also arguably contributes to prey extinction, as it means that the predator does not leave the tree when its preferred prey is gone, and thus may still be present when and if the tree is reinvaded by the prey.

Biased dispersal of predators toward patches with high densities of prey can be either stabilizing or destabilizing in metapopulation models. Strong aggregation tends to destabilize some continuous-time models with spatial structure (Murdoch *et al.*, 1992), and in others destabilizes under certain reasonable conditions such as when there is among-patch variation in prey dispersal (Ives, 1992). For *P. persimilis*, aggregation increases its ability to suppress prey densities, as it can use cues such as odors and webbing to detect high density prey colonies (Zhang & Sanderson, 1993). Once a colony is invaded, the prey are driven to extinction. Aggregation thus contributes to the instability of this system. *Typhlodromus pyri* has a much more modest response, although it, too, responds behaviorally to host odors (Nyrop, 1988). In addition any aggregative behavior that it does exhibit is limited to within-tree movement, and thus any effect it has on dynamics will be *via* effects on local dynamics.

Another important difference between greenhouse and orchard mite dynamics is the probable importance of host plant dynamics. Models of *P. persimilis* and *T. urticae* have usually included the plant as an interactive part of the interaction. This reflects how critical the prey–plant interaction is to the outcome of this predator–prey interaction. The *T. urticae*–plant interaction is inherently very unstable and stability of the entire system requires balancing the tendency for *T. urticae* to overexploit the plant with the tendency for *P. persimilis* to overexploit *T. urticae*. This is less true for the orchard mite interaction. Over-exploitation of the plant with repercussions for prey reproductive rate can occur, but only at relatively high prey densities. Such high prey densities are not as easily reached in the orchard environment, perhaps due to the lower innate fecundity of *P. ulmi* compared with *T. urticae*, and possibly also due to the lower temperatures typically seen in orchards compared with greenhouses.

A final point of difference between the two systems is the spatial scale of the among-plant interactions. Plants that are cultivated in a greenhouse are usually much smaller than apple trees, even young apple trees. Thus local populations in an orchard have a much larger carrying capacity, or maximum size, which is in itself stabilizing (Fujita, 1983; Jansen & Sabelis, 1992). Second, due to its size there is much more potential for within-population structure, and thus within-tree movement and distribu-

tions are likely to play a greater role in local population dynamics in orchards.

Spider mites as spatially structured populations

Consideration of spatial structure is essential for understanding and predicting the dynamics of spider mite populations. Dispersal is an important feature of the taxon's life history, although behavioral details vary from species to species. Persistence, particularly in the presence of predators cannot be understood without considering dispersal, and both average densities and temporal variability in densities are affected by dispersal. Therefore any mathematical model relevant to spider mite dynamics must include spatial structure.

Explicit comparison of experimental dynamics with those predicted by spatially structured models has aided our understanding of spider mite dynamics, and thus biological control in at least two ways. The models have suggested that certain factors are particularly likely to be dynamically important, for example, emigration rates, distance moved, mortality during dispersal, and efficacy of the natural enemy. Second, differences between experimental results and model predictions have indicated areas where our interpretation of the system may not be correct or where more detailed data and analyses are required. This applies, for instance, to the role played by alternative prey in the persistence of a generalist predator.

It may also be relevant to include evolutionary considerations (see Chapter 15). Does increased plant distance select for increased or reduced dispersal (e.g., Margolies, 1993, 1995)? Do specialist and generalist predators do better in different habitat types (e.g., Didham *et al.*, 1996)? Does spatial structure affect the evolution of different foraging strategies, which then affects dynamics (e.g., killer *versus* milker types (van Baalen & Sabelis, 1993, 1995))? And how does dispersal behavior and the spatial structure of the population affect the rate at which pesticide resistance spreads in a population (Dunley & Croft, 1992; Caprio & Hoy, 1994)?

In conclusion, when selecting predatory mites for

biological control, it is necessary not only to consider prey preferences, functional response, developmental time, resistance to pesticides, etc, but also dispersal behavior. The way in which dispersal will affect predator–prey dynamics, and thus ultimately biological control, will depend in part on the biology (including distribution and dispersal characteristics) of the prey, and in part on the spatial structure of the environment. Dispersal behavior is, of course, subject to natural selection, and thus we should also be aware that the dynamics of a system could change over time, if there are selective advantages to different dispersal strategies.

References

Adler, F. R. (1993) Migration alone can produce persistence of host–parasitoid models. *American Naturalist* 141: 642–50.

Amano, H. & Chant, D. A. (1977) Life history and reproduction of two species of predacious mites, *Phytoseiulus persimilis* Athias-Henriot and *Amblyseius andersoni* (Chant) (Acarina: Phytoseiidae). *Canadian Journal of Zoology* 55: 1978–83.

Bernstein, C. (1984) Prey and predator emigration responses in the acarine system *Tetranychus urticae–Phytoseiulus persimilis*. *Oecologia* 61: 134–42.

Burnett, T. (1970) Effect of temperature on a greenhouse acarine predator–prey population. *Canadian Journal of Zoology* 48: 555–62.

Burnett, T. (1979). An acarine predator–prey population infesting roses. *Researches on Population Ecology* 20: 227–34.

Caley, M. J. (1991) A null model for testing distributions of dispersal distances. *American Naturalist* 138: 524–32.

Caprio, M. A. & Hoy, M. A. (1994) Metapopulation dynamics affect resistance development in the predatory mite, *Metaseiulus occidentalis* (Acari: Phytoseiidae). *Journal of Economic Entomology* 87: 525–34.

Caswell, H. I. & Etter, R. J. (1993) Ecological interactions in patchy environments: from patch-occupancy models to cellular automata. In *Patch Dynamics*, ed. S. A. Levin, T. M Powell & J. H. Steele, pp. 93–109. Berlin: Springer-Verlag.

Chant, D. A. (1959) Phytoseiid mites (Acarina: Phytoseiidae). Part I. Bionomics of seven species in southeastern England. *Canadian Entomologist* 12 (**Supplement**): 1–166.

Chesson, P. (1978) Predator–prey theory and variability. *Annual Review of Ecology and Systematics* 9: 323–47.

Comins, H. N. & Blatt, D. W. E. (1974) Prey–predator models in spatially heterogeneous environments. *Journal of Theoretical Biology* 48: 75–83.

Comins, H. N. & Hassell, M. P. (1996) Persistence of multispecies host–parasitoid interaction in spatially distributed models with local dispersal. *Journal of Theoretical Biology* 183: 19–28.

Comins, H. N., Hassell, M. P. & May, R. M. (1992) The spatial dynamics of host–parasitoid systems. *Journal of Animal Ecology* 61: 735–48.

Connor, E. F., Faeth, S. H. & Simberloff, D. (1983) Leafminers on oak: the role of immigration and in situ reproductive recruitment. *Ecology* 64: 191–204.

Croft, B. A. (1994) Biological control of apple mites by a phytoseiid mite complex and *Zetzellia mali* (Acari: Stigmaeidae): long-term effects and impact of azinphosmethyl on colonization by *Amblyseius andersoni* (Acari: Phytoseiidae). *Environmental Entomology* 23: 1317–25.

Croft, B. A. & MacRae, I. V. (1992) Persistence of *Typhlodromus pyri* and *Metaseiulus occidentalis* (Acari: Phytoseiidae) on apple after inoculative release and competition with *Zetzellia mali* (Acari: Stigmaeidae). *Environmental Entomology* 21: 1168–77.

Croft, B. A. & MacRae, I. V. (1993) Biological control of apple mites: impact of *Zetzellia mali* (Acari: Stigmaeidae) on *Typhlodromus pyri* and *Metaseiulus occidentalis* (Acari: Phytoseiidae). *Environmental Entomology* 22: 865–73.

Dice, L. R. (1945) Measures of the amount of ecological association between species. *Ecology* 26: 297–302.

Dicke, M. (1988) Prey preference of the phytoseiid mite *Typhlodromus pyri*. 1. Response to volatile kairomones. *Experimental and Applied Acarology* 4: 1–13.

Dicke, M. & deJong, M. (1988) Prey preference of the phytoseiid mite *Typhlodromus pyri*. 2. Electrophoretic diet analysis. *Experimental and Applied Acarology* 4: 15–25.

Didham, R. K., Ghazoul, J., Stork, N. E. & Davis, A. J. (1996) Insects in fragmented forests: a functional approach. *Trends in Evolution and Ecology* **11**: 255–60.

Diekmann, O. (1993) An invitation to structured (meta)population models. In *Patch Dynamics*, ed. S. A. Levin, T. M. Powell & J. H. Steele, pp. 162–75. Berlin: Springer-Verlag.

Dunbar, S. R. (1983) Travelling wave solutions of diffusive Lotka–Volterra equations. *Journal of Mathematical Biology* **17**: 11–32.

Dunley, J. E. & Croft, B. A. (1990) Dispersal between and colonization of apple *by Metaseiulus occidentalis* and *Typhlodromus pyri* (Acarina: Phytoseiidae). *Experimental and Applied Acarology* **10**: 137–49.

Dunley, J. E. & Croft, B. A. (1992) Dispersal and gene flow of pesticide resistance traits in phytoseiid and tetranychid mites. *Experimental and Applied Acarology* **14**: 313–25.

Dunley, J. E. & Croft, B. A. (1994) Gene flow measured by allozymic analysis in pesticide resistant *Typhlodromus pyri* occurring within and near apple orchards. *Experimental and Applied Acarology* **18**: 201–11.

Ellner, S. & Turchin, P. (1995) Chaos in a noisy world: new methods and evidence from time-series analysis. *American Naturalist* **145**: 343–75.

Eveleigh, E. S. & Chant, D. A. (1981) Experimental studies on acarine predator–prey interactions: effects of predator age and feeding history on prey consumption and functional response (Acarina: Phytoseiidae). *Canadian Journal of Zoology* **59**: 1387–1406.

Eveleigh, E. S. & Chant, D. A. (1982a) Experimental studies on acarine predator–prey interactions: the distribution of search effort and the functional and numerical responses of predators in a patchy environment (Acarina: Phytoseiidae). *Canadian Journal of Zoology* **60**: 2979–91.

Eveleigh, E. S. & Chant, D. A. (1982b) Experimental studies on acarine predator–prey interactions: effects of temporal changes in the environment on searching behaviour, predation rates, and fecundity (Acarina: Phytoseiidae). *Canadian Journal of Zoology* **60**: 2992–3000.

Fahrig, L. & Merriam, G. (1985) Habitat patch connectivity and population survival. *Ecology* **66**: 1762–8.

Fernando, M. H. J. P. & Hassell, M. P. (1980) Predator–prey responses in an acarine system. *Researches on Population Ecology* **22**: 301–22.

Fowler, J. & Cohen, L. (1990) *Practical Statistics for Field Biology*. Chichester: John Wiley and Sons.

Fujita, K. (1983) Systems analysis of an acarine predator–prey system. II: Interactions in discontinuous environment. *Researches on Population Ecology* **25**: 387–99.

Gerrard, D. J. & Chiang, H. C. (1970) Density estimation of corn rootworm egg populations based on frequency of occurrence. *Ecology* **51**: 237–45.

Godfray, H. C. J. & Pacala, S. W. (1992) Aggregation and the population dynamics of parasitoids and predators. *American Naturalist* **140**: 30–40.

Gough, N. (1991) Long-term stability in the interaction between *Tetranychus urticae* and *Phytoseiulus persimilis* producing successful integrated control on roses in southeast Queensland. *Experimental and Applied Acarology* **12**: 83–101.

Hanski, I. (1987) Cross-correlation in population dynamics and the slope of spatial variance – mean regressions. *Oikos* **50**: 148–51.

Hanski, I., Turchin, P., Korpimaki, E. & Henttonen, H. (1993) Population oscillations of boreal rodents: regulation by mustelid predators leads to chaos. *Nature* **364**: 232–5.

Hardman, J. M., Herbert, H. J., Sanford, K. H. & Hamilton, D. (1985) Effect of populations of the European red mite *Panonychus ulmi*, on the apple variety Red Delicious in Nova Scotia. *Canadian Entomologist* **117**: 1257–65.

Hassell, M. P. (1978) *The Dynamics of Arthropod Predator–prey Systems*. Princeton: Princeton University Press.

Hassell, M. P., Comins, H. N. & May, R. M. (1991) Spatial structure and chaos in insect population dynamics. *Nature* **353**: 255–8.

Hassell, M. P., Godfray, H. C. J. & Comins, H. N. (1993) Effects of global change on the dynamics of insect host–parasitoid interactions. In *Biotic Interactions and Global Change*, ed. P. M. Kareiva, J. G. Kingsolver & R. B. Huey, pp. 402–23. Sunderland: Sinauer.

Hassell, M. P., Comins, H. N. & May, R. M. (1994) Species coexistence and self-organizing spatial dynamics. *Nature* **370**: 190–2.

Hastings, A. (1990) Spatial heterogeneity and ecological models. *Ecology* **71**: 426–8.

Hastings, A. (1993) Complex interactions between dispersal and dynamics: lessons from coupled logistic equations. *Ecology* **74**: 1362–72.

Hastings, A. & Higgins, K. (1994) Persistence of transients in spatially structured ecological models. *Science* **263**: 1133–6.

Helle, W. & Sabelis, M. W. (eds) (1985) *Spider Mites. Their Biology, Natural Enemies and Control, Volume 1B*. Amsterdam: Elsevier.

Holling, C. S. (1959) Some characteristics of simple types of predation and parasitism. *Canadian Entomologist* **91**: 385–98.

Holt, R. D. (1985) Population dynamics in two-patch environments: some anomalous consequences of an optimal habitat distribution. *Theoretical Population Biology* **29**: 181–208.

Holt, R. D. & Hassell, M. P. (1993) Environmental heterogeneity and the stability of host–parasitoid interactions. *Journal of Animal Ecology* **62**: 89–100.

Huffaker, C. B. (1958) Experimental studies on predation: dispersion factors and predator–prey oscillations. *Hilgardia* **27**: 343–83.

Huffaker, C. B., Shea, K. P. & Herman, S. G. (1963) Experimental studies on predation: complex dispersion and levels of food in an acarine predator–prey interaction. *Hilgardia* **34**: 305–30

Huffaker, C. B., van de Vrie, M. & McMurtry, J. A. (1969) The ecology of tetranychid mites and their natural control. *Annual Review of Entomology* **14**: 125–74.

Hussey, N. W. & Parr, W. J. (1963a) Dispersal of the glasshouse red spider mite *Tetranychus urticae* Koch (Acarina: Tetranychidae). *Entomologia Experimentalis et Applicata* **6**: 207–14.

Hussey, N. W. & Parr, W. J. (1963b) The effect of glasshouse red spider mite (*Tetranychus urticae* Koch) on the yield of cucumbers. *Journal of Horticultural Science* **38**: 255–63.

Ives, A. R. (1992) Continuous-time models of host–parasitoid interactions. *American Naturalist* **140**: 1–29.

Jansen, V. A. A. & Sabelis, M. W. (1992) Prey dispersal and predator persistence. *Experimental and Applied Acarology* **14**: 215–31.

Johnson, D. L. (1983) Predation, Dispersal and Weather in an Orchard Mite System. PhD thesis: University of British Columbia, Vancouver.

Kareiva, P. (1990) Population dynamics in spatially complex environments: theory and data. *Philosophical Transactions of the Royal Society of London, Series B* **330**: 175–90.

Kennedy, G. G. & Smitley, D. R. (1985) Dispersal. In *Spider Mites. Their Biology, Natural Enemies and Control, Volume 1B*, ed. W. Helle & M. W. Sabelis, pp. 141–60. Amsterdam: Elsevier.

Kot, M. (1992) Discrete-time travelling waves: ecological examples. *Journal of Mathematical Biology* **30**: 413–36.

Lawson, A. B. & Walde, S. J. (1993) Comparison of the responses of two predaceous mites, *Typhlodromus pyri* and *Zetzellia mali*, to variation in prey density. *Experimental and Applied Acarology* **17**: 811–21.

Lawson, D. S., Nyrop, J. P. and Dennehy, T. J. (1996) Aerial dispersal of European red mites (Acari: Tetranychidae) in commercial apple orchards. *Experimental and Applied Acarology* **20**: 193–202.

Levin, S. A. & Segel, L. A. (1976) Hypothesis for origin of planktonic patchiness. *Nature* **259**: 659.

Levins, R. (1969) Some demographic and genetic consequences of environmental heterogeneity for biological control. *Bulletin of the Entomological Society of America* **15**: 237–40.

Liesering, R. (1960) Beitrag zum phytopathologischen Wirkungsmechanismus von *Tetranychus urticae* Koch (Tetranychidae, Acari). *Zeitschrift für Pflanzenkrankheiten* **67**: 524–42.

Margolies, D. C. (1993) Adaptation to spatial variation in habitat: spatial effects in agroecosystems. In *Evolution of Insect Pests / Patterns of Variation*, ed. K. C. Kim & B. A. McPheron, pp. 129–44. New York: John Wiley and Sons.

Margolies, D. C. (1995) Evidence of selection on spider mite dispersal rates in relation to habitat persistence in agroecosystems. *Entomologia Experimentalis et Applicata*, **76**: 105–8.

May, R. M. (1974) *Stability and Complexity in Model Ecosystems*. Princeton: Princeton University Press.

Maynard Smith, J. (1974) *Models in Ecology*. Cambridge: Cambridge University Press.

McArdle, B. H. & Gaston, K. J. (1992) Comparing population variabilities. *Oikos* **67**: 610–11.

McArdle, B. H. & Gaston, K. J. (1993) The temporal variability of populations. *Oikos* **67**: 187–91.

McLaughlin, J. F. & Roughgarden, J. (1991) Pattern and stability in predator–prey communities: how diffusion in spatially variable environments affects the Lotka–Volterra model. *Theoretical Population Biology* **40**: 148–72.

McLaughlin, J. F. & Roughgarden, J. (1992) Predation across spatial scales in heterogeneous environments. *Theoretical Population Biology* **41**: 277–99.

McMurtry, J. A. (1992) Dynamics and potential impact of 'generalist' phytoseiids in agroecosystems and possibilities for establishment of exotic species. *Experimental and Applied Acarology* **14**: 371–82.

Mobley, K. N. & Marini, R. P. (1990) Gas exchange characteristics of apple and peach leaves infested by European red mite and twospotted spider mite. *Journal of the American Society for Horticultural Science* 115: 757–61.

Murdoch, W. W. (1994) Population regulation in theory and practice. *Ecology* 75: 271–95.

Murdoch, W. W. & Oaten, A. (1975) Predation and population stability. *Advances in Ecological Research* 9: 1–131.

Murdoch, W. W., Briggs, C. J., Nisbet, R. M., Gurney, W. S. C. & Stewart-Oaten, A. (1992) Aggregation and stability in metapopulation models. *American Naturalist* 140: 41–58.

Nachman, G. (1981) Temporal and spatial dynamics of an acarine predator–prey system. *Journal of Animal Ecology* 50: 435–51.

Nachman, G. (1984) Estimates of mean population density and spatial distribution of *Tetranychus urticae* (Acarina: Tetranychidae) and *P. persimilis* (Acarina: Phytoseiidae) based upon the proportion of empty sampling units. *Journal of Applied Ecology* 21: 903–13.

Nachman, G. (1985) Sampling techniques. In *Spider Mites, Their Biology, Natural Enemies and Control, Volume 1B*, ed. W. Helle & M. W. Sabelis, pp. 175–82. Amsterdam: Elsevier.

Nachman, G. (1987a) Systems analysis of acarine predator–prey interactions. I. A stochastic simulation model of spatial processes. *Journal of Animal Ecology* 56: 247–65.

Nachman, G. (1987b) Systems analysis of acarine predator–prey interactions. II. The role of spatial processes in system stability. *Journal of Animal Ecology* 56: 267–81.

Nachman, G. (1988) Regional persistence of locally unstable predator/prey populations. *Experimental and Applied Acarology* 5: 293–318.

Nachman, G. (1991) An acarine predator–prey metapopulation system inhabiting greenhouse cucumbers. *Biological Journal of the Linnean Society* 42: 285–303.

Nachman, G. (1997) The effect of dispersal on the dynamics and persistence of an acarine predator–prey system in a patchy system. *Proceedings of the IX International Congress of Acarology*, Columbus, Ohio, 17–22 July, 1994. (In press).

Neubert, M. G., Kot, M. & Lewis, M. A. (1995) Dispersal and pattern formation in a discrete-time predator–prey model. *Theoretical Population Biology* 48: 7–43.

Nihoul, P. (1993) Asynchronous populations of *Phytoseiulus persimilis* Athias-Henriot and effective control of *Tetranychus urticae* Koch on tomatoes under glass. *Journal of Horticultural Science* 68: 581–8.

Nisbet, R. M. & Gurney, W. S. C. (1982) *Modelling Fluctuating Populations*. Chichester: John Wiley and Sons.

Nisbet, R. M., Briggs, C. J., Gurney, W. S. C., Murdoch, W. W. & Stewart-Oaten, A. (1993) Two-patch metapopulation dynamics. In *Patch Dynamics*, ed. S. A. Levin, T. M Powell & J. H. Steele, pp. 125–35. Berlin: Springer-Verlag.

Nyrop, J. N. (1988) Spatial dynamics of an acarine predator–prey system: *Typhlodromus pyri* (Acarina: Phytoseiidae) preying on *Panonychus ulmi* (Acarina: Tetranychidae). *Environmental Entomology* 17: 771–8.

Ohnesorge, B. (1981) Populations dynamische Untersuchungen in einem Räuber-Beutetier-System: *Phytoseiulus persimilis* A.-H- (Acarina; Phytoseiidae) und *Tetranychus urticae* Koch. *Zeitschrift für Angewandte Entomologie* 91: 25–49.

Reeve, J. D. (1988) Environmental variability, migration, and persistence in host–parasitoid systems. *American Naturalist* 132: 810–36.

Rohani, P. & Miramontes, O. (1995) Host–parasitoid metapopulations: the consequences of parasitoid aggregation on spatial dynamics and searching efficiency. *Proceedings of the Royal Society of London, Series B* 260: 335–42.

Ruxton, G. D. & Rohani, P. (1996) The consequences of stochasticity for self-organized spatial dynamics, persistence and coexistence in spatially extended host-parasitoid communities. *Proceedings of the Royal Society of London, Series B* 263: 625–31.

Sabelis, M. W. (1981) *Biological Control of Two-spotted Spider Mites using Phytoseiid Predators. Part 1. Modelling the Predator–prey Interaction at the Individual Level*. Agricultural Research Publication no. 910. Wageningen: Centre for Agricultural Publication and Documentation.

Sabelis, M. W. (1985a) Reproductive strategies. In *Spider Mites. Their Biology, Natural Enemies and Control, Volume 1A*, ed. W. Helle & M. W. Sabelis, pp. 265–78. Amsterdam: Elsevier.

Sabelis, M. W. (1985b) Predation on spider mites. In *Spider Mites. Their Biology, Natural Enemies and Control, Volume 1B*, ed. W. Helle & M. W. Sabelis, pp. 103–27. Amsterdam: Elsevier.

Sabelis, M. W. & van de Baan, H. E. (1983) Location of distant spider mite colonies by phytoseiid predators: demonstration of specific kairomones emitted by *Tetranychus urticae* and *Panonychus ulmi. Entomologia Experimentalis et Applicata* **33**: 303–14.

Sabelis, M. W., Vermaat, J. E. & Groeneveld, A. (1984) Arrestment responses of the predatory mite, *Phytoseiulus persimilis*, to steep odour gradients of a kairomone. *Physiological Entomology* **9**: 437–46.

Sabelis, M. W., Diekmann, O. & Jansen, V. A. A. (1991) Metapopulation persistence despite local extinction: predator–prey patch models of the Lotka–Volterra type. *Biological Journal of the Linnean Society* **42**: 267–83.

Schaeffer, M. & de Boer, R. J. (1995) Implications of spatial heterogeneity for the paradox of enrichment. *Ecology* **76**: 2270–7.

Segel, L. A. & Jackson, J. L. (1972) Dissipative structure: an explanation and an ecological example. *Journal of Theoretical Biology* **37**: 545–59.

Takafuji, A. & Chant, D. A. (1976) Comparative studies of two species of predacious phytoseiid mites (Acarina: Phytoseiidae), with special reference to their responses to the density of their prey. *Researches on Population Ecology* **17**: 255–310.

Takafuji, A., Tsuda, Y. & Miki, T. (1983) System behaviour in predator–prey interaction, with special reference to acarine predator–prey system. *Researches on Population Ecology* Supplement **3**: 75–92.

Taylor, L. R. (1961) Aggregation, variance, and mean. *Nature* **189**: 732–5.

Taylor, A. D. (1988) Large-scale spatial structure and population dynamics in arthropod predator–prey systems. *Annales Zoologici Fennici* **25**: 63–74.

Taylor, A. D. (1990) Metapopulations, dispersal, and predator–prey dynamics: an overview. *Ecology* **71**: 429–33.

Tomczyk, A. & Kropczynska, D. (1985) Effects on the host plant. In *Spider Mites. Their Biology, Natural Enemies and Control, Volume 1A*, ed. W. Helle & M. W. Sabelis, pp. 317–29. Amsterdam: Elsevier

Tomkiewicz, J., Skovgård, H., Nachman, G. & Münster-Swendsen, M. (1993) A rapid and non-destructive method to assess leaf injury caused by the cassava green mite, *Mononychellus tanajoa* (Bondar) (Acarina: Tetranychidae). *Experimental and Applied Acarology* **17**: 29–40.

Torii, T. (1971) The development of quantitative occurrence prediction of infestation by the rice stem-borer, *Chilo suppressalis* in Japan. *Entomophaga* **16**: 193–207.

van Baalen, M. & Sabelis, M. W. (1993) Coevolution of patch selection strategies of predator and prey and the consequences for ecological stability. *American Naturalist* **142**: 646–70.

van Baalen, M. & Sabelis, M. W. (1995) The milker–killer dilemma in spatially structured predator–prey interactions. *Oikos* **74**: 391–400.

van de Klashorst, G., Readshaw, J. L., Sabelis, M. W. & Lingeman, R. (1992) A demonstration of asynchronous local cycles in an acarine predator–prey system. *Experimental and Applied Acarology* **14**: 185–99.

van Lenteren, J. C. & Woets, J. (1988) Biological and integrated pest control in greenhouses. *Annual Review of Entomology* **33**: 239–69.

Walde, S. J. (1991) Patch dynamics of a phytophagous mite population. *Ecology* **72**: 1591–8.

Walde, S. J. (1994) Immigration and the dynamics of a predator–prey interaction in biological control. *Journal of Animal Ecology* **63**: 337–46.

Walde, S. J. (1995a) How quality of host plant affects a predator–prey interaction in biological control. *Ecology* **76**: 1206–19.

Walde, S. J. (1995b) Internal dynamics and metapopulations: experimental tests with predator–prey systems. In *Population Dynamics*, ed. N. Cappuccino & P. W. Price, pp.173–93. New York: Academic Press.

Walde, S. J., Nyrop, J. P. & Hardman, J. M. (1992) Dynamics of *Panonychus ulmi* and *Typhlodromus pyri*: factors contributing to persistence. *Experimental and Applied Acarology* **14**: 261–92.

Wanibuchi, K. & Saito, Y. (1983) The process of population increase and patterns of resource utilization of two spider mites, *Oligonycus ununguis* (Jacobi) and *Panonychus citri* (McGregor), under experimental conditions (Acarina: Tetranychidae). *Researches on Population Ecology* **25**: 185–98.

Weibull, W. (1951) A statistical distribution function of wide applicability. *Journal of Applied Mechanisms* **18**: 293–7.

Wennergren, U., Ruckelshaus, M. & Kareiva, P. (1995) The promise and limitations of spatial models in conservation biology. *Oikos* **74**: 349–56.

White, A., Begon, M. & Bowers, R. G. (1996) Host–pathogen systems in a spatially patchy environment. *Proceedings of the Royal Society of London, Series B* **263**: 325–32.

Woolhouse, M. E. J. & Harmsen, R. (1987) Prey-predator cycles and lags in space: descriptive models from a laboratory experiment. *Theoretical Population Biology* **32**: 366–82.

Zhang, Z.-Q. & Sanderson, J. P. (1993) Behavioural responses to prey density by three acarine predator–prey species with different degrees of polyphagy. *Oecologia* **96**: 147–56.

Zhang, Z.-Q. & Sanderson, J. P. (1995) Twospotted spider mite (Acari: Tetranychidae) and *Phytoseiulus persimilis* (Acari: Phytoseiidae) on greenhouse roses: spatial distribution and predator efficacy. *Journal of Economic Entomology* **88**: 352–7.

Zhang, Z.-Q., Sanderson, J. P. & Nyrop, J. P. (1992) Foraging time and spatial patterns of predation in experimental populations. *Oecologia* **90**: 185–96.

11

Habitat fragmentation and biological control

TEJA TSCHARNTKE AND ANDREAS KRUESS

Introduction

In the modern agricultural landscape, habitat destruction by agricultural activities is the most frequent cause of extinction in plants and animals. In addition to these deterministic causes of extinction, reductions in population size due to habitat fragmentation lead to further stochastic or random species losses (Shaffer, 1981; Saunders *et al.*, 1991; Tscharntke, 1992a; Baur & Erhardt, 1995). Habitat destruction reduces the total area available for certain organisms, while habitat fragmentation generates a spatial pattern of more-or-less isolated habitats or populations. This disruption to habitat continuity is accompanied by significant changes in habitat characteristics and community dynamics.

Habitat fragmentation is usually only considered in the context of conservation, but is also closely related to the efficiency of biological control in the agricultural landscape. First, the spatio-temporal mosaic of natural habitats can be manipulated in such a way that populations of beneficials increase relative to pests. The propagation of natural enemy populations is part of a conservation strategy to increase biodiversity by increasing connectivity of natural habitats in the landscape (see Gliessman, 1990; van Hook, 1994; Fry, 1995; Van Driesche & Bellows, 1996). Second, annual crop fields are like defaunated islands, since they are usually cleared each year during harvest; that is, human destruction of established plant–insect communities necessitates annual recolonization of crop fields with corresponding annual rejuvenation of communities. Conservation management to enhance biological control can be viewed as an effort to neutralize the island-status of crop fields with a mosaic of old and closely interconnected habitat patches.

Such considerations have been driven by changes in perspective and scale of agroecological analyses during the last decades. Acknowledgement of the need for environmental protection (such as nitrogen leaching or soil erosion) was supplemented with a focus on conservation and agroecology (see van Emden & Williams, 1974; Fry, 1995). In comparisons of diverse and uniform landscapes, insect species richness is positively related to the percentage of uncultivated area (Ryszkowski *et al.*, 1993; Kretschmer *et al.*, 1995), but little is known about diversity-enhanced pest control.

The classical species–area relationship

Based on empirical findings, the positive correlation between habitat area and species richness is, in most cases, a log-log relationship (discussed in detail by Rosenzweig, 1995; He & Legendre, 1996). The regression line $S = cA^z$ (McArthur & Wilson, 1967), where S is the species richness of the area, A the area, c the intercept (a constant), and the constant, z, a measurement of the slope, is based on a plot of untransformed values and typically resembles a saturation curve: changes in island size have a large effect on species richness only in small islands, and are much less pronounced in large islands. Essentially, two explanations for species–area relationships can be found in the literature:

(1) The area-*per se* hypothesis. The probability of finding a certain species is much lower in a small patch than in a large patch, regardless of whether these patches are islands. In addition, large islands support larger populations and, therefore, make random extinction events less probable (Shaffer, 1981).

(2) The habitat-heterogeneity hypothesis. Greater areas are more likely to include different habitat types, and different habitat types often have many unique species. Area and habitat heterogeneity often correlate so closely that either variable is equally as useful in predicting species richness (Rosenzweig, 1995).

Values of z, the slope of the regression line in the McArthur & Wilson (1967) species–area model, vary with the degree of habitat isolation and with the kind of species involved. Values for z are mostly between 0.25 and 0.33 in isolated oceanic or terrestrial habitat islands, but only between 0.13 and 0.18 when non-isolated terrestrial areas are compared (Rosenzweig, 1995). Such differences in species richness between contiguous areas of mainland should be due mainly to habitat diversity, whereas differences between isolated and non-isolated areas can be attributed to a separate area-*per se* effect. Furthermore, isolated areas have lower values for c than non-isolated areas, because extremely small patches depend almost entirely on immigration from nearby sources, and isolation reduces immigration rates. Monophagous herbivores have higher values for z than polyphagous species Zabel & Tscharntke (1998), and parasitoids have higher z values than herbivores (Kruess, 1996).

Disruption of species interactions in fragmented habitats

Some species suffer a disproportionate disadvantage on small or isolated habitats (see Pimm, 1991; Lawton, 1995), so interactions of species with their biotic or abiotic environment and thus community structure may be changed. These interactions are part of the functioning of the ecosystem and link population dynamics with ecosystem processes.

Losses of favorable interactions may handicap the target population in the community due to reduced pollination and seed set, seed dispersal, decomposition of dung or litter, or mutualistic mycorrhizal associations (Matthies *et al.*, 1995; Didham *et al.*, 1996; Steffan-Dewenter & Tscharntke, 1997). In contrast, disruption of antagonistic interactions may

favor target populations and enhance their persistence. For example, pathogenic fungal infections, seed predation, or mortality due to predation or parasitism may be reduced (Thomas, 1989; Matthies *et al.*, 1995; Hanski *et al.*, 1995).

Populations at the top of food chains are more likely to become extinct than those at lower trophic levels, so habitat fragmentation should generally affect specialized natural enemies more than their phytophagous victims, theoretically predicted by Pimm (1991), Lawton (1995), and Holt (1996). Plant populations are a more stable or predictable resource for herbivores than herbivore populations are for predators or parasitoids, so a greater susceptibility to extinction should make frequent colonization of habitat fragments more important for natural enemies than their victims. Isolated habitat fragments show reduced proportions of attack by a rust fungus on the legume *Lotus corniculatus*, reduced percentages of seed herbivory on *Bromus erectus* and *L. corniculatus* (Matthies *et al.*, 1995), and reduced percent parasitism on clover weevils (Kruess & Tscharntke, 1994). Furthermore, insects may be generally more sensitive to changes in cultivation practices than plants, because only plants have seeds and vegetative parts that survive through unfavorable conditions. The time between population crashes of plant or herbivore populations is often much shorter than the recovery time of populations of higher trophic levels. In such situations of local loss and recolonization, predator–prey interactions depend on highly vagile natural enemies (Bunce & Howard, 1990; de Vries *et al.*, 1996). Small populations may be 'rescued' by the conspecific immigrants from nearby source populations (Brown & Kodric-Brown, 1977).

Recent papers suggest that reduced rates of parasitism or predation are a common feature of fragmented populations of phytophagous or saprophagous insects. Frequent disturbances of phytotelmata (water-filled plant bodies) result in the loss of predators and the persistence of two-trophic level systems (Pimm & Kitching, 1987). The distance between colonies of the ant lion, *Euroleon nostras*, was negatively correlated with the probability of parasitism by a specialized chalcid wasp (Scherer &

Tscharntke, 1995). In beetles colonizing cow pats, the proportion of predatory species increases from 20% to 80% with age of the cow pat (Hanski, 1987). Predators in such ephemeral habitats tend to be generalists and aggregate in patches with the regionally most abundant prey, thereby contributing to stability.

Increasing patchiness in goldenrod fields leads to more frequent local outbreaks of aphid populations (Kareiva, 1987); both area and age of the patch contribute to low levels of aphid densities, presumably due to increased predation (Kareiva, 1990). Schoener *et al.* (1995) found that the hymenopteran parasitoid fraction was greatest in the largest and least isolated Bahamian islands. Faeth & Simberloff (1981) transplanted oak trees into an agricultural field and at the edge of the same field as a control, and the percentage parasitism of leaf-mining insects was lower in the field than at the field edge.

In the following parasitoid and predator case studies, the detailed effects of fragmentation on the impact of natural enemies will be shown.

Parasitoids

The size, isolation, and age of clover, vetch, and common reed stands were found to be important in predicting both loss of parasitoid species and reduction in parasitism rates. Manually established islands of red clover (*Trifolium pratense*) 100–500 m distant from the next large meadow were colonized by only a few species of the parasitoid community found on red clover in large meadows. In contrast, almost all of the herbivore species colonized these islands successfully (Kruess & Tscharntke, 1994; Kruess, 1996; Fig. 11.1). In a comparison of the insect community on red clover from large and small meadows, the numbers of parasitoid species decreased much more with meadow size than did the numbers of herbivore species (Kruess, 1996; Fig. 11.2). Thus, herbivores were released from heavy parasitism when they occured in isolated or small habitat fragments. Parasitism of stem-boring weevils decreased from about 80% of the maximum rate of parasitism on large meadows to 40% on small ones (Fig. 11.3(a)) and from 80% on non-isolated clover patches to 30%

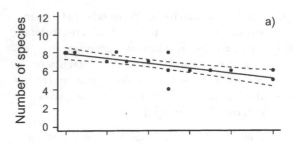

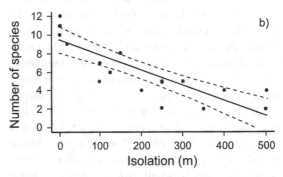

Figure 11.1. Effect of habitat isolation on species richness. Endophagous insects on manually established islands (1.2 m^2) of potted red clover (*Trifolium pratense*).
(a) Herbivores (Apionidae, Bruchidae, Cecidomyiidae): $y = 7.88 - 0.005\,x$; $F = 22.7$, $R^2 = 0.587$, $p < 0.001$, $n = 18$.
(b) Parasitoids centered upon Apionidae and Cecidomyiidae: $y = 9.38 - 0.016\,x$; $F = 39.33$, $R^2 = 0.71$, $p < 0.001$, $n = 18$.
Data from a field experiment with red clover islands isolated in the agricultural landscape of south-west Germany (Kruess & Tscharntke, 1994).

on isolated ones (Fig. 11.3(b)). Kruess (1996) found nearly identical effects of habitat fragmentation on the endophagous insects of *Vicia sepium*: small and isolated habitats had both reduced species richness and reduced rates of parasitism.

Parasitoid populations on both clover and vetches were not only affected by their trophic position, but also by other characteristics shared with extinction-prone species such as rarity and population variability. When all insects on the red clover or vetch islands were considered, absence of a species was negatively correlated with its abundance (Fig. 11.4(a)) and positively correlated with spatial population variability (Fig. 11.4(b)). Species with fluctuating populations

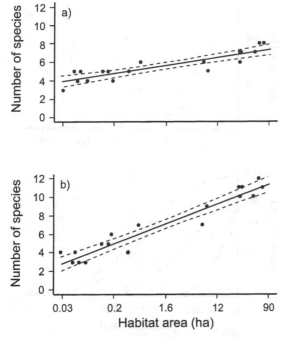

Figure 11.2. Effect of habitat area on species richness. Endophagous insects on naturally occuring red clover (*Trifolium pratense*) grown on differentially sized old meadows. (a) Herbivores: $y = 1.49 + 0.43 \times \ln x$; $F = 59.7$, $R^2 = 0.78$, $p < 0.001$, $n = 18$. (b) Parasitoids: $y = -3.35 + 1.07 \times \ln x$; $F = 180.1$, $R^2 = 0.914$, $p < 0.001$, $n = 18$. Data from a field study with red clover on old meadows in the agricultural landscape of south-west Germany (Kruess, 1996).

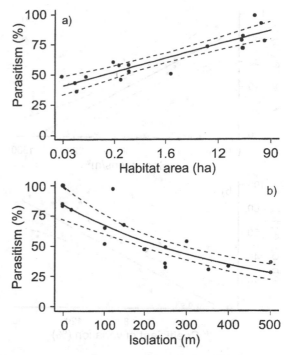

Figure 11.3. Effects of meadow area and isolation of habitat islands on the parasitism rate of stem-boring weevils (*Catapion seniculus, Ischnopterapion virens*) from red clover (*Trifolium pratense*). (a) Area effect: $y = 24.6 + 2.88 \times \ln x$, $F = 66.7$, $R^2 = 0.82$, $p < 0.001$, $n = 17$. (b) Isolation effect: $y = e^{(4.43 - 0.002 \times x)}$; $F = 54.6$, $R^2 = 0.773$, $p < 0.001$, $n = 18$. In each figure the greatest value of the parasitism rate is set to 100%. The range of parasitism rates is: (a) 11.9–44.3%, (b) 9.7–38.2%. Data from field studies in south-west Germany (see Figs 11.1 and 11.2: Kruess & Tscharntke, 1994; Kruess, 1996).

should generally be more susceptible to local extinction than species with stable populations (see den Boer, 1993).

Rarity is a well-known precursor to extinction (Gaston, 1994). Red Data Books mostly rely on information on the abundance of species as an estimate of their status. Local abundance and regional distribution are significantly correlated (Cornell & Lawton, 1992), i.e., most of the rare species suffer from a double jeopardy. This pattern suggests that dispersal between local populations is a frequent event. The hypothesis that species that are reduced to rarity by anthropogenic causes are at a greater disadvantage than species whose rarity is a long-term characteristic of the species needs to be tested. Rarity

and variability may be closely correlated (Fig. 11.4; Zabel & Tscharntke, 1998).

In a comparison of sewage treatment installations utilizing common reed (*Phragmites australis*), incidence curves showed that the two most abundant herbivores (gall makers) were much less affected by area loss or habitat age than their parasitoids (Fig. 11.5; Athen & Tscharntke, 1998; see also Tscharntke, 1992a; Tscharntke & Greiler, 1995). Parasitoid species differed in susceptibility to area reduction or age of their habitat: nine of the altogether ten species were significantly related to age and eight to area

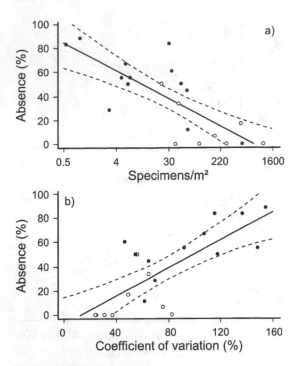

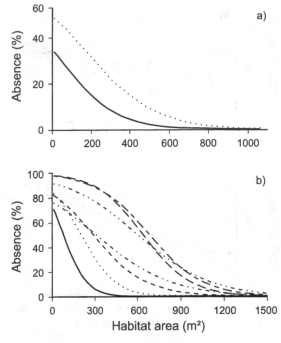

Figure 11.4. Percent absence of endophagous insects on red clover plots. (○, herbivores; ●, parasitoids.
(a) Correlation between the absence rate of each species and average abundance: $y = 77.43 - 11.66 \times \ln x$; $F = 28.35$, $R^2 = 0.60$, $p < 0.001$, $n = 21$. (b) Correlation between absence rate of each species and spatial variability (coefficient of variation, CV): $y = -7.02 + 0.58 \times x$; $F = 24.33$, $R^2 = 0.561$, $p = 0.0001$, $n = 21$. Data of a field experiment with 13 isolated clover plots and 5 control plots (as for Fig. 11.1). Absence rate is calculated using the data from all 18 plots; abundance and CV are calculated using data from the five control plots (Kruess & Tscharntke, 1994).

Figure 11.5. Percent absence of endophagous insects on common reed (*Phragmites australis*) in relation to area of sewage treatment plants. (a) —, chloropid fly, *Lipara pullitarsis* ($a = 0.113$, $b = -0.007$, $p = 0.039$), and ····, its parasitoid, *Stenomalina liparae* ($a = 1.068$, $b = -0.0058$, $p = 0.0138$); (b) —, Gall-midge, *Giraudiella inclusa* ($a = 1.035$, $b = -0.012$, $p = 0.0028$), and its parasitoids: ····, *Torymus arundinis* ($a = 1.725$, $b = -0.008$, $p = 0.002$, $\chi^2 = 9.241$) ---, *Platygaster szelenii* ($a = 1.618$, $b = -0.005$, $p = 0.008$, $\chi^2 = 6.85$); and ---, *Aprostocetus gratus* ($a = 1.618$, $b = -0.005$, $p = 0.008$, $\chi^2 = 6.85$); ·--·, *Aprostocetus orythyia* ($a = 1.209$, $b = -0.003$, $p = 0.03$, $\chi^2 = 4.704$); ——, *Eurytoma crassinervis* ($a = 3.918$, $b = -0.006$, $p = 0.003$, $\chi^2 = 8.746$); -—-, *Eudecatoma stagnalis* ($a = 3.704$, $b = -0.005$, $p = 0.005$, $\chi^2 = 7.948$); and ·---, *Aprostocetus calamarius* ($a = 2.424$, $b = -0.004$, $p = 0.014$, $\chi^2 = 6.044$). For all parasitoids except *E. stagnalis* logistic regression analyses also showed significant correlations between absence rate and habitat age, with species-specific differences in susceptibility to either age or size of habitats. Logistic regressions are based on $n = 28$ sewage treatment installations in northern Germany (Athen & Tscharntke, 1998). a, intercept; b, slope of regression line.

(Fig. 11.5). Furthermore, parasitism of two gall makers was significantly reduced on young treatment installations. For the chloropid fly *Lipara pullitarsis*, parasitism declined from about 30% to 0% (Fig. 11.6(a)), and for the gall midge *Giraudiella inclusa* from about 50% to 10% (Fig. 11.6(b)). Herbivores were thereby released from possible biocontrol.

These empirical studies can be related to the findings of Hawkins & Gross (1992), which showed that the proportion of parasitoid introductions in which some degree of control was achieved is correlated

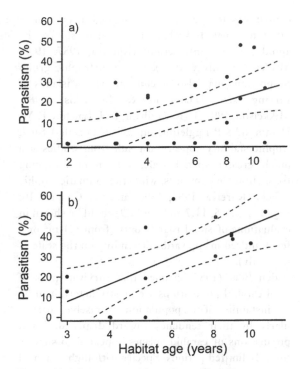

Figure 11.6. Percent parasitism of two phytophagous insects on sewage treatment installations of different age, using common reed (*Phragmites australis*). (a) Parasitism of *Lipara pullitarsis*: $y = 67.3 \times x - 28.3$; $F = 12.98$, $R^2 = 0.50$, $p = 0.0032$, $n = 15$; (b) Parasitism of *Giraudiella inclusa*: $y = 71.4 \times x - 26.4$; $F = 11.56$, $R^2 = 0.471$, $p = 0.0047$, $n = 15$. Percent parasitism was also positively correlated with host abundance: $y = 6.83 \times x - 3.87$; $F = 12.0$, $R^2 = 0.48$, $p = 0.004$, $n = 15$ (Athen & Tscharntke, 1998).

with the mean species richness of parasitoids centred upon the host species. Furthermore, the maximum susceptibility of a host to parasitoid attack, estimated by maximum parasitism rates, provides a good estimate of the probability that parasitoid introductions in classical biological control will reduce pest populations below some critical density defined as their economic threshold (Hawkins *et al.*, 1993). There is also a threshold for success: no control was achieved for any pest suffering less than 32–36% maximum parasitism (Hawkins & Cornell, 1994). Interestingly, in the field study shown in Figs 11.3(b) and 11.5(b), percent parasitism declined below this threshold due to reductions in age, size or isolation of habitat frag-

ments. Thus, these data suggest that habitat fragmentation may prevent success of a parasitoid imported for biological control.

Predators

Predators differ in many aspects from parasitoids, so fragmentation effects should also differ. Predators tend to be polyphagous, unlike most parasitoids, and they have more vagile, continuous populations, contrasting with the more or less isolated local populations of parasitoids. Dispersal ability and opportunistic resource use suggest that predators are the better agents for rapid colonization of crop fields and density-dependent regulation of the regionally most abundant prey.

Zabel & Tscharntke (1998) studied the effects of habitat fragmentation on the insect community of stinging nettle, *Urtica dioica*. In a comparison of 32 stands, species richness of herbivores correlated with habitat size, whereas the species richness of predators correlated only with habitat isolation. In logistic regressions, the probability that monophagous herbivores were absent could only be explained by habitat size (in 4 out of the 10 species: Fig. 11.7(a)), and in predators only by habitat isolation (in 3 out of the 12 species: Fig. 11.7(b)). These results show the differential susceptibility of herbivores and predators to habitat fragmentation. Predator and herbivore populations did not differ with respect to abundance or variability and, accordingly, absence of the three predators appeared to be only due to their trophic position (thereby contrasting with the above-mentioned parasitoids). These results on predator populations suggest that regular colonization of habitat fragments is of major importance, thus increased habitat connectivity in the agricultural landscape will disproportionally promote predator populations. Only predators with a quick response to decreasing local prey populations, in that they switch to another prey population, may be able to survive.

In a spatial model of *Tetranychus urticae–Phytoseiulus persimilis* dynamics, Walde & Nachman (Chapter 10) showed that as inter-plant distance decreased (resulting in enhanced dispersal), density

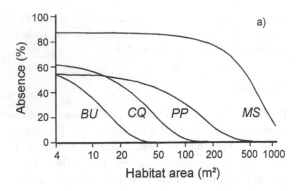

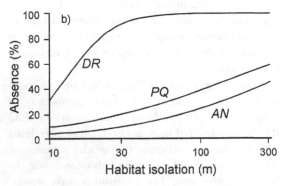

Figure 11.7. Percent absence of nettle insects on fragmented nettle patches. (a) Monophagous herbivores in relation to habitat area: *Brachypterus urticae* (*BU*): $a = 0.71$, $b = -0.135$, $\chi^2 = 9.2$, $p < 0.005$); *Cidnorhinus quadrimaculatus* (*CQ*): $a = 0.66$, $b = -0.042$, $\chi^2 = 11.2$, $p < 0.001$; *Phyllobius pomaceus* (*PP*): $a = 0.24$, $b = -0.011$, $\chi^2 = 4.9$, $p < 0.05$; *Macropsis scutellata* (*MS*): $a = 1.97$, $b = -0.004$, $\chi^2 = 3.7$, $p = 0.05$. (b) Predatory species in relation to habitat isolation (distance to the nearest *Urtica* habitat of at least 50 m²): *Deraeocoris ruber* (*DR*): $a = 7.36$, $b = -6.57$, $\chi^2 = 9.9$, $p < 0.0051$; *Propylaea 14-punctata* (*PQ*): $a = 4.0$, $b = -1.76$, $\chi^2 = 4.1$, $p < 0.05$; *Anthocoris nemorum* (*AN*): $a = 5.29$, $b = -2.06$, $\chi^2 = 3.8$, $P = 0.05$). Logistic regressions are based on $n = 32$ nettle patches in the agricultural landscape of northern Germany (Zabel & Tscharntke, 1998).

of prey decreased and density of predators increased. Predator–prey ratios were higher and predators tended to be more synchronized with prey, indicating better biological control at higher dispersal rates due to bridges connecting neighboring plants. Experimental results confirmed these models. The promotion of between-habitat dispersal obviously

contributes to the success of biological control at different spatial scales, as is also shown by aphid–predator interactions (Kareiva, 1987, 1990; Müller & Godfray, 1997). However, trophic group proportions may also be constant across islands of varying size (Hamilton & Stathakis, 1987; Mikkelson, 1993; R. K. Didham, J. H. Lawton, P. M. Hammond & P. Eggleton, unpublished results). Such trophic stability appears to result from studies at a much greater spatial scale, with islands ranging from 10 000 m² upwards, while the dramatic trophic effects (Kareiva, 1987, 1990; and see Chapter 10; Figs 11.3, 11.5, 11.7 and see 11.9) could be shown in evaluations of small islands only (from <1 m² or a few hundred m² upwards, depending on the scale of the study).

den Boer (1985) claimed that survival times of local carabid populations are mostly determined by the instability of the populations themselves, particularly by their tendency toward dispersal. Local populations of carabid species appeared to survive for only limited periods. Species with high dispersal powers persisted for about 10 years, whereas species with low dispersal powers persisted for about 40 years. Accordingly, local turnover in the field appeared to be high. In addition, fluctuations of local carabid subpopulations are significantly larger than the overall fluctuations, due to metapopulation dynamics of interconnected local populations (den Boer 1990, 1993).

Susceptibility of such polyphagous predators to fragmentation is ambigous: they should gain advantage by being generalists when they colonize islands (Becker, 1992), but they should also suffer disadvantage by their trophic-level position, in that they depend on prey availability. In general, habitat specialists are more susceptible to extinctions than generalists, because they depend on a blend of conditions that are less likely to occur on small patches. For monophagous or otherwise specialized insects the landscape produces a certain pattern of isolated habitat islands, whereas for polyphagous species, these islands may be connected by usable habitat patches, thereby producing a habitat continuum. With respect to the natural enemies of pest insects, the mostly specialized and monophagous parasitoids should be more affected than the generalist predators.

Abundance or species richness of natural enemies?

The goal of biological control is to enhance natural enemy reproduction relative to the pest in order to reduce pest densities. The question is whether agricultural practices directed toward the conservation of natural enemy populations should aim primarily at the enhancement of the abundance or species richness of beneficials. Species richness and abundance of natural enemies are often closely correlated, so the 'single-species' impact of a species-poor, but individual-rich community of beneficials cannot easily be separated from the 'multiple-species' impact of a species-rich community with only few individuals per species.

In biological control, there are two contrasting views on species-rich parasitoid assemblages. This disagreement may be dubbed the 'competition' *versus* 'complement' controversy. The competition view holds that competition among parasitoid species is most important (Zwölfer, 1970; Ehler & Hall, 1982, 1984). The argument is that endophagous, efficiently searching parasitoids may be suppressed by ectophagous, inefficient parasitoids when multiparasitism occurs. In classical biological control, single introductions of the very efficient, but poorly competitive species may therefore be superior to multiple species introductions. Theoretical studies have shown that invasion of less effective parasitoids may reduce the degree of control by the already existing agent (Hochberg, 1996; Murdoch & Briggs, 1996). In addition to competition, interference between predators and parasitoids may be caused by intraguild predation (Polis *et al.*, 1989). Ladybeetles (Coccinelidae) or syrphid flies may consume parasitized cereal aphids, and insectivorous birds such as yellowhammers and sparrows consume these syrphids and ladybeetles, but not aphids (Ehler, 1996; Tscharntke, 1997).

The complement view holds that species diverge in their host exploitation patterns as a result of individualistic responses to a heterogeneous host distribution (see Price, 1984; Tscharntke, 1992b; Hawkins, 1993; Chang, 1996). Each parasitoid species may concentrate on a subpopulation of the host that differs in space and time. As a result, both high parasitism and species coexistence are due to the complementary attack by many species. For example, the gall-inducing midge *Giraudiella inclusa* on the grass *Phragmites australis* is the host of 13 primary parasitoids that differ significantly in site of attack (Tscharntke, 1992b). The sites of successful parasitism are spatially or temporally separated, i.e., species concentrate on (i) galls of the 1st or the 2nd to 4th midge generation, (ii) galls on thin or thick shoots, (iii) galls on main shoots or side shoots, (iv) large or small gall clusters (i.e., they show positive or negative spatial density dependence), (v) certain sequence numbers of galled internode or (vi) combinations of these parameters (Tscharntke, 1992b).

Presumably, the two views focus on different aspects of a given parasitoid community, and their relative importance should change with the species involved. Competition is certainly important in many species-rich complexes, but coexistence of many species centred upon only one resource (one host) is the result of species-specific host exploitation patterns. In a review of 50 successful biological control programs, Myers *et al.* (1989) found that 32% of the successes were credited to a complex of introduced agents and 68% to only one of the several introduced species. Accordingly, in two out of three of the cases it was more important to get a single effective agent than to get all the agents.

With respect to landscape management in agricultural areas, strategies to enhance natural enemies should consider the species richness of parasitoid and predator species, either to simply increase the probability of inclusion of the most effective beneficials or to use their divergent exploitation strategies. Management can only rely on maximizing the abundance of one or a few species of natural enemies when their effectiveness has been well established. In addition, given an ever-changing environment, the presence of a great number of species provide the basis or capacity to adapt (Hawkins, 1993; Christensen *et al.*, 1996). Species-rich communities may support buffer populations that become important when the abundant populations decline.

This argument is akin to the SLOSS-debate (the 'single large or several small' conservation strategy) that is also about minimizing the extinction probability of selected species or maximizing species richness. Analyses of the parasitoid assemblages centered upon clover weevils from small and large meadows supported the findings of Quinn & Harrison (1988) in that 'several small' meadows supported more species than 'single large' ones (Kruess, 1996). However, in each of the small meadows both parasitoid species richness and percent parasitism were low. So in large refuge habitats, biological control agents had much greater impact in that both species richness and population densities of parasitoids were higher compared to any of the small habitats. In conclusion, conservation strategies should not only promote species richness with the help of a wide array of small patches, but also the persistence of large populations in large habitats. Both strategies have merit.

Similar questions arise with respect to other ecosystem functions. Allogamic plant populations visited by a wide array of pollinator species set more seed than little-visited populations (Steffan-Dewenter & Tscharntke, 1997). Each pollinator species seemingly concentrates on different flower subpopulations and may also differ in efficacy in time and space. Similarly, dung decomposition is faster when attack comes from a species-rich dung beetle community, since dung beetles differ in mode of decomposition (Didham *et al.* 1996).

Crop fields as islands

Until recently, relatively little fundamental research in population or community ecology has been directed at systems relevant to agriculture (Levins & Wilson, 1980). Consequently, ecological theory covers few aspects of agricultural systems, and extensive analyses of characteristic features of agricultural landscapes have not been performed. Agroecosystems may be analyzed according to the theory of island biogeography, because (i) the harvest of annual crops more or less completely eliminates populations and communities from crop fields (Levins & Wilson, 1980), (ii) crop fields are more or less isolated from refuge habitats with populations that colonize the defaunated crop fields (similar to colonizations of oceanic islands: Simberloff & Wilson, 1969), and (iii) the number of species is determined by a balance of immigration and extinction, according to the McArthur & Wilson (1967) 'equilibrium theory of island biography'. Communities of annual crops are dominated by species characterized by high dispersal abilities, lower trophic levels, little specialization and low dependence on other organisms. Crop fields harbor species typical of early successional stages such as annual, autogamic (self-pollinated) or apomictic weeds, and rapid colonizers that profit from the enemy-free and competition-free space.

Since crop fields are embedded in complex spatio-temporal landscape mosaics, the dynamics of non-agricultural habitat fragments may determine the local abundance of natural enemies of pests (Burel & Baudry, 1995; Fry, 1995; Holt *et al.*, 1995). Crop fields are usually surrounded by fragments of uncultivated habitats differing in size, isolation, vegetation or age, and understanding the dynamics of plant–insect communities on these successional habitat fragments is an integral part of the analysis of man-made agroecosystems.

The agroecological goal in biocontrol is the attempt to partially neutralize the island situation of crop fields. Crop fields should be closely embedded in the agricultural landscape, so populations may more easily switch between uncultivated and cultivated, or optimal and suboptimal, habitats. Enhancement of populations of natural enemies within crop fields comes from (i) population preservation due to extensive, rather than intensive, agricultural practices (e.g., reduced tillage, fertilizer or pesticide use), (ii) attraction of natural enemies by habitat manipulation within agroecosystems (e.g., vegetational diversification), and (iii) changing the spatio-temporal habitat mosaic in the agricultural landscape to increase favorable interactions.

'Clean', weed-free or crop residue-free farming often eliminates or disadvantages natural enemies. For example, conservation of bundles of straw has the potential to increase predator survival (Gilstrap, 1988; Shepard *et al.*, 1989), and retention of *Brassica* postharvest plants helps to conserve the parasitoid *Cotesia rubecula*, which attacks *Pieris rapae* larvae (van Driesche & Bellows, 1996). Nilsson (1985) found that ploughing reduced the emergence of parasitoids of rape pollen beetles from the soil by 50 to 100%. Vegetational diversification includes intercropping, that is the growing of two or more crop species, sowing of ground covers, or weedy culture. Reduction in pest populations due to increased vegetational diversity may be explained by the 'enemies hypothesis' that natural enemies will be augmented and better control herbivores when vegetational diversity is high. The 'resource concentration hypothesis' predicts that specialist herbivores will have higher immigration, lower emigration and higher reproduction rates in monocultures than in polycultures (Risch, 1987; Altieri, 1995; Vidal, 1995). Russell (1989) reviewed the effects of intercropping and found increased mortality due to natural enemies in 9 out of 13 cases, lowered levels in two cases and no effects in the remaining two cases. Vegetational diversity should enhance generalist species (predators) more than specialist species (parasitoids), since specialized species may behave like their host or prey and suffer from the reduced resource availability in diverse surroundings (Coll & Bottrell, 1996). Cabbage intersown with clover is known to support higher populations of polyphagous predators such as carabid and staphylinid beetles (O'Donnell & Croaker, 1975; Ryan *et al.*, 1980; Vidal, 1995).

Agricultural fields with persistent predator or parasitoid populations should be more common in perennial crop habitats than in annual crops, which are characterized by an annual time lag between the arrival of pest and enemy populations (Vandermeer, 1995). 'Resource concentration' effects are more common in annual crops. The increase in relative importance of natural enemies with age of succession is in conformity with general expectations from ecological theory (Southwood, 1988. Price, 1991). Due to the yearly turnover in annual crops, the significance of pest–enemy interactions should vary more than in perennial systems. For polyphagous

predators (beetles and spiders), the abundance is also affected by the type of crop and farming system (Booij & Noorlander, 1992).

Within-field and between-field structural diversity should equally increase beneficial/pest ratios. Patterns of annual crops may be designed in a way that enemy colonization is facilitated by coordinating plantings in space and time. For example, in winter rape (*Brassica napus*), bridges for natural enemies may come from (i) the temporal pattern of crops (continuous cultivation or closely neighboring crop plantings) or (ii) field margins and adjacent fallows, reducing spatial distance for crop colonization. Colonization of rape fields by parasitoids was found to be enhanced by locating them close to fields sown to rape the previous year (Hokkanen *et al.*, 1988).

Field margin strips may benefit natural enemies by (i) attraction of flower-visiting natural enemies and (ii) stabilization of enemy populations by providing alternative victims or shelter and nesting places. For example, uncultivated areas may be reservoirs of host plants belonging to families differing from that of the crop (e.g., nettle populations), thereby supporting non-pest populations of herbivores large enough to augment populations of polyphagous beneficials spreading into crop fields (Zabel & Tscharntke, 1998; for reviews, see Powell, 1986; Dennis & Fry, 1992; Nentwig, 1992). Parasitoid performance often depends on a spatial and temporal separation of host and nectar use. This is why strips inside cereal fields that have been sown with the intensely flowering and nectar-rich *Phacelia tanacetifolia* can attract parasitoids and enhance parasitism of pest insects (Wratten & van Emden, 1995; for a review, see Russell, 1989). Field margins obviously enhance the diversity of natural enemies in the agricultural landscape, but few studies prove that increases in numbers of predators result in decreases in numbers of pests (Ekbom & Wiktelius, 1985; Wratten & Powell, 1991; Ekbom *et al.*, 1992).

Six-year old field margin strips were found to significantly increase parasitism of the larvae of rape pollen beetle in the center of winter rape fields (Thies *et al.*, 1997. Fig. 11.8). When no strips or only 1-year old strips surround rape fields, the percentage of parasitism is significantly lower in the center than at

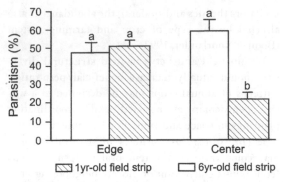

Figure 11.8. Effect of age of field margin strip on the parasitization of the rape pollen beetle, *Meligethes aeneus*, by the ichneumonid parasitoid, *Tersilochus heterocereus* in winter rape. Comparison of the effects of 1-year old strips (▧) and 6-year old strips (□). Edge, 1–2 m from field margin: $F = 0.1$, $p = 0.757$, $n = 20$. Center, 10–12 m from field margin: $F = 31.53$, $p < 0.001$, $n = 20$. Histograms show the mean and standard error. Analysis of variance is based on 20 field margins strips (3 m × 100 m) in northern Germany (Thies *et al.*, 1997). Homogenous groups separated by the Tukey test have been labelled by identical letters.

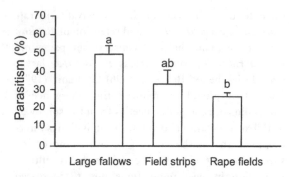

Figure 11.9. Effect of the surrounding habitat type on parasitization of the rape pollen beetle, *Meligethes aeneus*, by the ichneumonid parasitoid *Tersilochus heterocereus*, in manually created plots of summer rape. These summer rape plots were created in large fallows (about 1 ha, 6-year old naturally developed vegetation), and field margin strips (3 m × 100 m, 6-year old naturally developed vegetation) with winter rape fields. Data from a field experiment in northern Germany with manually created summer rape plots (2 m²) in each of the three habitat types ($F = 5.4$, $p = 0.034$, $n = 12$: Thies *et al.*, 1997).

the edge of the field. Only old strips support parasitoid populations large enough to spread into the field, increasing percent parasitism in the center from about 20% to 60% (Fig. 11.8). These results emphasize the need for reductions in field size or changes in field geometry as a prerequisite for enhancing mortality of pest insects. Square fields up to 1–3 ha may be most strongly affected by such 'edge effects', with beneficials spreading maximally 30–50 m into the field. However, former East Germany is characterized by crop fields covering on average 40.4 ha, and only southern and northern parts of former West Germany have on average 2.2 or 4.0 ha fields, respectively. Because of the time lag between the arrival of pests and natural enemies, the probability of insect outbreaks should increase with field size (Risch, 1987).

Furthermore, results on the parasitism of the rape pollen beetle emphasize the significance of habitat age (Fig. 11.8). In the agricultural landscape, disturbances such as cutting of vegetation or habitat destruction by ploughing affects small patches more than large ones. Accordingly, fragment age can be

important and is often positively correlated with fragment size.

Old fallows support even larger parasitoid populations attacking rape pollen beetles than do old field margins (Fig. 11.9). This result confirms expectations from the trophic-level hypothesis of island biogeography, in that the relative importance of higher trophic levels increases with area and decreases with isolation (see above, Fig. 11.3). In addition, field margins are characterized by a high proportion of edges. Life is different at fragment edges due to changes in microclimate or due to invasions by aerial plankton (seed rain, tiny arthropods) or other immigrants. The impact of agrochemicals such as fertilizers and pesticides applied in the crop fields is greatest on the edges of adjacent natural habitats. Lotka–Volterra equations predict that mortality factors affecting herbivore and predator populations equally, will disproportionally affect the recovery time of predators thereby increasing the probability of pest outbreaks (Wilson & Bossert, 1981).

In general, little is known of the effects of landscape structure on pest control in agroecosystems.

Both habitat protection and alternative agricultural practices are orientated towards promoting populations of natural enemies, so that they become sufficiently abundant, rapidly colonize crop fields and efficiently attack pests (van Driesche & Bellows, 1996). However, we are only at the threshold of understanding how to promote favorable biological interactions between natural habitats and crop fields and how to place this approach in the context of economically viable farming. Management practices should reduce isolation barriers and the adversity of the environment by preserving corridors or neighboring non-crop areas. Models of predator–prey metapopulation dynamics suggest that dispersal among local populations may allow the persistence of regional metapopulations, despite unstable fluctuations or local extinctions (Taylor, 1990). However, critical thresholds to habitat connectivity, i.e., the transition ranges across which small changes in spatial pattern produce abrupt shifts in ecological responses, are largely unknown, and time lags before species or interactions disappear make prospects even worse than they appear (Kareiva & Wennergren, 1995; With & Crist, 1995). Marino & Landis (1996) found that landscape complexity (but not distance to the nearest hedgerow) enhanced parasitization of armyworm in maize fields. The convergence of conservation and biological control in the evaluation of habitat fragmentation should result in further details of landscape design. Such analyses are a great challenge for the future and a contribution to the often cited sustainability of agriculture.

Summary

Efficiency of biological control in the agricultural landscape is closely related to effects of habitat fragmentation in that (i) the spatio-temporal mosaic of cultivated and uncultivated areas determines the efficiency of natural enemies, and (ii) annual crop fields are like defaunated islands with communities composed of rapid colonizers. In the agricultural landscape, the time between population crashes in isolated or disturbed environments is often shorter than the recovery time of populations of higher trophic levels. Recent field studies suggest that habitat fragments are characterized by reduced numbers of parasitoid and predator species and subsequently, reduced rates of parasitism or predation; that is, losses in species are paralleled by losses in species interactions. Due to their trophic position, natural enemies are less likely to be effective in very young, small or isolated habitat fragments. Such reductions in parasitism or predation releases potential or actual pests from possible biocontrol.

The agroecological goal in biocontrol is to partially neutralize the island situation of crop fields in that crop fields should be embedded in uncultivated parts of the landscape to enhance favorable interactions. Adjacent field margin strips or old fallows may increase the mortality of pests, examplified by a study on the parasitism of the rape pollen beetle but, in general, little is known of the effects of landscape structure on pest control.

Acknowledgements

Comments by Raphael K. Didham, Bradford A. Hawkins, Molly S. Hunter, and Michael Mühlenberg improved the chapter greatly. We also thank Olaf Athen, Carsten Thies, and Jörg Zabel for allowing us to use their results. Financial support came from the German Science Foundation (DFG), the German Ministry of the Environment, and the 'Forschungs- und Studienzentrum für Landwirtschaft und Umwelt' at the University of Göttingen.

References

Altieri, M. A. (1995) *Agroecology*. Boulder, CO: Westview Press.

Athen, O. & Tscharntke, T. (1998) Insect communities of *Phragmites* habitats used for sewage purification: effects of age and area on species richness and herbivore–parasitoid interactions. *Limnologica* (in press).

Baur, B. & Erhardt, A. (1995) Habitat fragmentation and habitat alterations: principal threats to most animal and plant species. *Gaia* 4: 221–6.

Becker, P. (1992) Colonization of islands by carnivorous and herbivorous Heteroptera and Coleoptera: effects of island area, plant species richness, and 'extinction' rates. *Journal of Biogeography* 19: 163–71.

Booij, C. J. H. & Noorlander, J. (1992) Farming systems and insect predators. In *Biotic Diversity in Agroecosystems*, ed. M. G. Paoletti & D. Pimentel, pp. 125–35. Amsterdam: Elsevier.

Brown, J. H. & Kodric-Brown, A. (1977) Turnover rates in insular biogeography: effects of immigration on extinction. *Ecology* 58: 445–9.

Bunce, R. G. H. & Howard, D. C. (1990) *Species Dispersal in Agricultural Habitats*. London: Belhaven Press.

Burel, F. & Baudry, J. (1995) Farming landscapes and insects. In *Ecology and Integrated Farming Systems*, ed. D. M. Glen, M. P. Greaves & H. M. Anderson, pp. 203–20. Chichester: John Wiley and Sons.

Chang, G. C. (1996) Comparison of single versus multiple species of generalist predators for biological control. *Environmental Entomology* 25: 207–12.

Christensen, N. L., Bartuska, A. M., Brown, J. H., Carpenter, S., D'Antonio, C., Francis, R., Franklin, J. F., MacMahon, J. A., Noss, R. F., Parsons, D. F., Petersen, C. H., Turner, M. G. & Woodmansee, R.G. (1996) The report of the Ecological Society of America committee on the scientific basis for ecosystem management. *Ecological Applications* 6: 665–91.

Coll, M. & Bottrell, D. G. (1996) Movement of an insect parasitoid in simple and diverse plant assemblages. *Ecological Entomology* 21: 141–9.

Cornell, H. V. & Lawton, J. H. (1992) Species interactions, local and regional processes, and limits to the richness of ecological communities – a theoretical perspective. *Journal of Animal Ecology* 61: 1–12.

den Boer, P. J. (1985) Fluctuations of density and survival of carabid populations. *Oecologia* 67: 322–30.

den Boer, P. J. (1990) The survival value of dispersal in terrestrial arthropods. *Biological Conservation* 54: 175–92.

den Boer, P. J. (1993) Are the fluctuations of animal numbers regulated or stabilized by spreading of risk? In *Dynamics of Populations*, ed. P.J. den Boer, P. J. M. Mols & H. Szyszko, pp. 23–8. Warsaw: Agricultural University.

Dennis, P. & Fry, G. L. A. (1992) Field margins: can they enhance natural enemy population densities and general arthropod diversity on farmland? *Agriculture, Ecosystems and Environment* 40: 95–115.

de Vries, H. H., den Boer, P. J. & van Dijk, T. S. (1996) Ground beetle species in heathland fragments in relation to survival, dispersal, and habitat preference. *Oecologia* 107: 332–42.

Didham, R. K., Ghazoul, J., Stork, N. E. & Davis, A. J. (1996) Insects in fragmented forests: a functional approach. *Trends in Ecology and Evolution* 11: 255–60.

Ehler, L. E. (1996) Structure and impact of natural enemy guilds in biological control of insect pests. In *Food Webs: Integration of Pattern and Dynamics*, ed. G.A. Polis & K.O. Winemiller, pp. 337–42. New York: Chapman and Hall.

Ehler, L. E. & Hall, R. W. (1982) Evidence for competitive exclusion of introduced natural enemies in biological control. *Environmental Entomology* 11: 1–4.

Ehler, L. E. & Hall, R. W. (1984) Evidence for competitive exclusion of introduced natural enemies in biological control: an addendum. *Environmental Entomology* 13: 5–7.

Ekbom, B. S. & Wiktelius, S. (1985) Polyphagous arthropod predators in cereal crops in central Sweden, 1979–1982. *Zeitschrift für Angewandte Entomologie* 99: 433–42.

Ekbom, B. S., Wiktelius, S. & Chiverton, P. A. (1992) Can polyphagous predators control the bird cherry-oat aphid (*Rhopalosiphum padi*) in spring cereals? *Entomologia Experimentalis et Applicata* 65: 215–223.

Faeth, S. H. & Simberloff, D. (1981) Experimental isolation of oak host plants: effects on mortality, survivorship, and abundance of leaf-mining insects. *Ecology* 62: 625–35.

Fry, G. (1995) Landscape ecology of insect movement in arable ecosystems. In *Ecology and Integrated Farming Systems*, ed. D. M. Glen, M. P. Greaves & H. M. Anderson, pp. 177–202. Chichester: John Wiley and Sons.

Gaston, K. J. (1994) *Rarity*. London: Chapman and Hall.

Gilstrap, F. E. (1988) Sorghum-corn-Johnsongrass and Banks grass mite: a model for biological control in field crops. In *The Entomology of Indigenous and Naturalized Systems in Agriculture*, ed. M. K. Harris & C. E. Rogers, pp.141–59. Boulder, CO: Westview Press.

Gliessmann, S. R. (1990) *Agroecology: Researching the Ecological Basis for Sustainable Agriculture*. New York: Springer-Verlag.

Hamilton, R. W. & Stathakis, D. G. (1987) Arthropod species diversity, composition and trophic structure at the soil level biotope of three northeastern Illinois prairie remnants. *Transactions of the Illinois State Academy of Science* 80: 273–98.

Hanski, I. (1987) Colonization of ephemeral habitats. In *Colonization, Succession, and Stability*, ed. A. J. Gray, M. J. Crawley & P. J. Edwards, pp. 155–85. Oxford: Blackwell Scientific.

Hanski, I., Pakkala, T., Kuussaari, M. & Lei, G. (1995) Metapopulation persistence of an endangered butterfly in a fragmented landscape. *Oikos* 72: 21–8.

Hawkins, B. A. (1993) Refuges, host population dynamics and the genesis of parasitoid diversity. In *Hymenoptera and Biodiversity*, ed. J. LaSalle & I. D. Gauld, pp. 235–56. Wallingford: CAB International.

Hawkins, B. A. & Cornell, H. V. (1994) Maximum parasitism rates and successful biological control. *Science* 266: 1886.

Hawkins, B. A. & Gross, P. (1992) Species richness and population limitation in insect parasitoid-host systems. *American Naturalist* 139: 417–23.

Hawkins, B. A., Thomas, M. B. & Hochberg, M. E. (1993) Refuge theory and biological control. *Science* 262: 1429–32.

He, F. & Legendre, P. (1996) On species-area relations. *American Naturalist* 148: 719–37.

Hochberg, M. (1996) Consequences for host population levels of increasing natural enemy species richness in classical biological control. *American Naturalist* 147: 307–18.

Hokkanen, H., Husberg, G. B. & Söderblom, M. (1988) Natural enemy conservation for the integrated control of the rape blossom beetle *Meligethes aeneus* F. *Annales Agricultura Fenniae* 27: 281–93.

Holt, R. D. (1996) Food webs in space: an island biogeographical perspective. In *Food Webs: Integration of Pattern and Dynamics*, ed. G. A. Polis & K. O. Winemiller, pp. 313–23. New York: Chapman and Hall.

Holt, R. D., Debinski, D. M., Diffendorfer, J. E., Gaines, M. S., Martinko, E. A., Robinson, G. R. & Ward, G. C. (1995) Perspectives from an experimental study of habitat fragmentation in an agroecosystem. In *Ecology and Integrated Farming Systems*, ed. D. M. Glen, M. P. Greaves & H. M. Anderson, pp. 147–75. Chichester: John Wiley and Sons.

Kareiva, P. (1987) Habitat-fragmentation and the stability of predator–prey interactions. *Nature* 326: 388–90.

Kareiva P. (1990) Population dynamics in spatially complex environments: theory and data. *Philosophical Transactions of the Royal Society of London, Series B* 330: 175–90.

Kareiva, P. & Wennergren, U. (1995) Connecting landscape patterns to ecosystem and population processes. *Nature* 373: 299–302.

Kretschmer, H., Pfeffer, H., Hoffman, J., Schrödl, G. & Fux, I. (1995) Strukturelemente in Agrarlandschaften Ostdeutschlands. *ZALF-Bericht (Zentrum für Agrarlandschafts- und Landnutzungsforschung Müncheberg)* 19: 1–164.

Kruess, A. (1996) *Folgen der Lebensraum-Fragmentierung für Pflanze–Herbivor–Parasitoid-Gesellschaften: Artendiversität und Interaktionen*. Bern: Paul Haupt.

Kruess, A. & Tscharntke, T. (1994) Habitat fragmentation, species loss, and biological control. *Science* 264: 1581–4.

Lawton, J. H. (1995) Population dynamic principles. In *Extinction Risks*, ed. J. H. Lawton & R. M. May, pp. 147–63. Oxford: Oxford University Press.

Levins, R. & Wilson, M. (1980) Ecological theory and pest managment. *Annual Review of Entomology* 25: 287–308.

Marino, P. C. & Landis, D. A. (1996) Effect of landscape structure on parasitoid diversity and parasitism in agroecosystems. *Ecological Applications* 6: 276–84.

Matthies, D., Schmid, B. & Schmid-Hempel, P. (1995) The importance of population processes for the maintenance of biological diversity. *Gaia* 4: 199–209.

McArthur, R. H. & Wilson, E. O. (1967) *The Theory of Island Biogeography*. Princeton: Princeton University Press.

Mikkelson, G. M. (1993) How do food webs fall apart? A study of changes in trophic structure during relaxation on habitat fragments. *Oikos* 67: 539–47.

Müller, C. B. & Godfray, H. C. J. (1997) Apparent competition between two aphid species. *Journal of Animal Ecology* 66: 57–64.

Murdoch, W. M. & Briggs, C. J. (1996) Theory for biological control. *Ecology* 77: 2001–13.

Myers, J. H., Higgins, C. & Kovacs, E. (1989) How many species are necessary for the biological control of insects? *Environmental Entomology* 18: 541–7.

Nentwig, W. (1992) Die nützlingsfördernde Wirkung von Unkräutern in angesäten Unkrautstreifen. *Zeitschrift für Pflanzenkrankheiten und Pflanzenschutz, Sonderheft* 13: 33–40.

Nilsson, C. (1985) Impact of ploughing on emergence of pollen beetle parasitoids after hibernation. *Zeitschrift für Angewandte Entomologie* 100: 302–8.

O'Donnell, M. S. & Croaker, T. H. (1975) Potential of intra-crop diversity in the control of *Brassica* pests. In *Proceedings of the 8th British Insecticide and Fungicide Conference U.K.*, pp. 101–7. Brighton.

Pimm, S. L. (1991) *The Balance of Nature? Ecological Issues in the Conservation of Species and Communities.* Chicago: University of Chicago Press.

Pimm, S. L. & Kitching, R. L. (1987) The determinants of food chain length. *Oikos* 50: 302–7.

Polis, G. A., Myers, C. A. & Holt, R. D. (1989) The ecology and evolution of intraguild predation: potential competitors that eat each other. *Annual Review of Ecology and Systematics* 20: 297–330.

Powell, W. (1986) Enhancing parasitoid activity in crops. In *Insect Parasitoids*, ed. J. K. Waage & D. J. Greathead, pp. 319–40. London: Academic Press.

Price, P. W. (1984) *Insect Ecology.* New York: Wiley and Sons.

Price, P. W. (1991) Evolutionary biology of parasites. *Biological Control* 1: 83–93.

Quinn, J. F. & Harrison, S.P. (1988) Effects of habitat fragmentation and isolation on species richness: evidence from biogeographic patterns. *Oecologia* 75: 132–40.

Risch, S. J. (1987) Agricultural ecology and insect outbreaks. In *Insect Outbreaks*, ed. P. Barbosa & J. C. Schultz, pp. 217–38. San Diego: Academic Press.

Rosenzweig, M. L. (1995) *Species Diversity in Space and Time.* Cambridge: Cambridge University Press.

Russell, E. P. (1989) Enemies hypothesis: a review of the effect of vegetational diversity on predatory insects and parasitoids. *Environmental Entomology* 18: 590–9.

Ryan, J., Ryan, M. F. & McNaeidhe, F. (1980) The effect of interrow plant cover on populations of the cabbage root fly *Delia brassicae* (Wied.). *Journal of Applied Ecology* 17: 31–40.

Ryszkowski, L., Karg, J., Margalit, G., Paoletti, M. G. & Zlotin, R. (1993) Above ground insect biomass in agricultural landscapes in Europe. In *Landscape Ecology and Agroecosystems*, ed. R. G. H. Bunce, L. Ryszkowski & M. G. Paoletti, pp. 71–82. Boca Raton: Lewis Publishers.

Saunders, D. A., Hobbs, R. J. & Margules, C. R. (1991) Biological consequences of ecosystem fragmentation: a review. *Conservation Biology* 5: 18–32.

Scherer, M. & Tscharntke, T. (1995) Habitat selection and dispersal of the ant lion *Euroleon nostras* (Fourcr.) (Planipennia, Myrmeleontidae). *Deutsche Gesellschaft für Allgemeine und Angewandte Entomologie* 10: 313–17.

Schoener, T. W., Spiller, D. A. & Morrison, L. W. (1995) Variation in the hymenopteran parasitoid fraction on Bahamian islands. *Acta Oecologia* 16: 103–21.

Shaffer, M. L. (1981) Minimum population sizes for species conservation. *Bioscience* 31: 131–4.

Shepard, M., Rapusas, H. R. & Estano, D. B. (1989) Using rice straw bundles to conserve beneficial arthropod communities in ricefields. *International Rice Research News* 14: 30–1.

Simberloff, D. S. & Wilson, E. O. (1969) Experimental zoogeography of islands. The colonization of empty islands. *Ecology* 50: 278–96.

Southwood, T. R. E. (1988) Tactics, strategies and templets. *Oikos* 52: 1–18.

Steffan-Dewenter, I. & Tscharntke, T. (1997) Bee diversity and seed set in fragmented habitats. *Acta Horticulturae* 437: 231–4.

Taylor, A. D. (1990) Metapopulations, dispersal, and predator–prey dynamics: an overview. *Ecology* 71: 429–33.

Thies, C., Denys, C., Ulber, B. & Tscharntke, T. (1997) The influence of field margins and fallowlands on pest–beneficial insects-interactions on oilseed rape (*Brassica napus*). *Mitteilungen der Gesellschaft für Ökologie* 27: 393–8.

Thomas, C. D. (1989) Predator–herbivore interactions and the escape of isolated plants from phytophagous insects. *Oikos* 55: 291–8.

Tscharntke, T. (1992a) Fragmentation of *Phragmites* habitats, minimum viable population size, habitat suitability, and local extinctions of moths, midges, flies, aphids, and birds. *Conservation Biology* 6: 530–6.

Tscharntke, T. (1992b) Coexistence, tritrophic interactions and density dependence in a species-rich parasitoid community. *Journal of Animal Ecology* 61: 59–67.

Tscharntke, T. (1997) Vertebrate effects on plant–invertebrate food webs. In *Multitrophic Interactions in Terrestrial Systems*, ed. A. C. Gange & V. K. Brown, pp. 277–97. Oxford: Blackwell Scientific.

Tscharntke, T. & Greiler, H.-J. (1995) Insect communities, grasses, and grassland. *Annual Review of Entomology* 40: 535–8.

Vandermeer, J. (1995) The ecological basis of alternative agriculture. *Annual Review of Ecology and Systematics* 26: 201–24.

van Driesche, R. G. & Bellows, T. S., Jr (1996) *Biological Control*. New York: Chapman and Hall.

van Emden, H. F. & Williams, G. F. (1974) Insect stability and diversity in agro-ecosystems. *Annual Review of Entomology* 19: 455–75.

van Hook, T. (1994) The conservation challenge in agriculture and the role of entomologists. *Florida Entomologist* 77: 42–73.

Vidal, S. (1995) Reduktion von Schadinsektenbefall im Freiland-Gemüsebau durch alternative Produktionsmethoden: Möglichkeiten und offene Fragen. *Mitteilungen der Deutschen Gesellschaft für allgemeine und angewandte Entomologie* 10: 237–42.

Wilson, E. O. & Bossert, W.H. (1981) *A Primer of Population Biology*. Sunderland: Sinauer Associates.

With, K. A. & Christ, T. O. (1995) Critical threshold in species responses to landscape structure. *Ecology* 76: 2446–59.

Wratten, S. D. & Powell, W. (1991) Cereal aphids and their natural enemies. In *The Ecology of Temperate Cereal Fields*, ed. L. G. Firbank, N. Carter, J. F. Darbyshire & G. R. Potts, pp. 233–57. Oxford: Blackwell Scientific.

Wratten, S. D. & van Emden, H. F. (1995) Habitat management for enhanced activity of natural enemies of insect pests. In *Ecology and Integrated Farming Systems*, ed. D. M. Glen, M. P. Greaves & H. M. Anderson, pp. 117–45. Chichester: John Wiley and Sons.

Zabel, J. & Tscharntke, T. (1998) Does fragmentation of *Urtica* habitats affect phytophagous and predatory insects differentially? *Oecologia* (in press).

Zwölfer, H. (1970) The structure and effect of parasite complexes attacking phytophagous host insects. In *Dynamics of Populations. Proceedings of the Advanced Study Institute on Dynamics of Numbers in Populations*, Netherlands, ed. P. J. den Boer & G. R. Gradwell, pp. 405–18. Wageningen: Centre for Agricultural Publishing and Documentation.

12

Outbreaks of insects: a dynamic approach

ALAN HASTINGS

Introduction

The basic premise of this chapter is that increased understanding of insect population dynamics will further our ability both to control and to predict the dynamics of insect pests (Metcalf & Luckmann, 1994). I will elaborate on this overall rationale below and will suggest approaches for achieving this goal. Mathematical models that can describe the dynamics of insects are vital for the design of effective control programs (Ruesink & Onstad, 1994). Yet, as I will argue here, despite much previous modeling of insect dynamics, further work is needed. Consequently, this chapter is much more prospective than retrospective.

As emphasized in Metcalf & Luckmann (1994), virtually all the theoretical underpinnings of insect pest control come from an equilibrium concept of insect population levels. In this vein, most analyses of host–parasitoid models used in biological control focus on the determination of the forces leading to a stable equilibrium, as an explanation for successful biological control. Similarly, the goal of biological control has often been stated to be the achievement of a stable equilibrium (Hassell, 1978). Yet, as many examples of forest insects (e.g., Berryman, 1988), plus essentially all insects affecting crops, show, equilibrium approaches are not at all appropriate when trying to understand the dynamics of many insect pests. Large departures from equilibrium, namely outbreaks, are the major features of interest in the dynamics of these insects. Thus, using an equilibrium concept as the central part of a theory to understand insect population dynamics is not appropriate.

There have been many studies that have developed models of insect outbreaks (see reviews by Berryman, 1987, and Cappuccino & Price, 1995), but these

models have rarely included the spatial component explicitly, and many of these models have in fact looked at the role of different equilibrium population levels. I will explicitly emphasize the transient behavior of the models, the spatial aspects and robust behavior that is not tied to particular models. The emphasis on transient dynamics, as opposed to equilibrium behavior, is necessary to reflect the fact that the evidence for equilibrium behavior in natural systems is weak (Connell & Sousa, 1983). The ultimate goal will be both basic and practical. Further understanding of the basic role of transient dynamics in understanding the dynamics of outbreaking insect pests should eventually lead to the design of approaches to control insect pests.

Insect outbreaks are receiving increasing attention through attempts to predict, understand and control insect populations (Barbosa & Schultz, 1987; Cappuccino & Price, 1995). I will review theoretical approaches that focus on dynamic explanations of insect outbreaks in both spatial and non-spatial contexts, drawing upon recent advances in understanding the dynamics of single and interacting ecological populations using the tools of modern dynamics, in particular recent results on the phenomenon of intermittency, or long transient behavior (Tél, 1990; Hastings & Higgins, 1994; and see Chapter 1). As emphasized in Berryman *et al.* (1990), I will stress the use of simple models drawing upon ecological theory, rather than detailed simulation models, as the best way to gain an understanding of the biology underlying insect outbreaks.

The specific questions I will ask fall under the general question, what are plausible dynamic explanations of insect outbreaks? More specific questions will fall into two general classes of explanations for outbreaks: those based on explicitly spatial considerations, and those based on local (in space) descriptions of population dynamics.

The biology of insect outbreaks

A good summary of insect outbreaks is the volume edited by Barbosa & Schultz (1987) (see also Cappuccino & Price, 1995 for more recent developments). In this volume, there is an overview of the classification of insect outbreaks, with numerous illustrative examples, by Berryman (1987). This classification provides a good starting point for a description of the underlying biological phenomena. Outbreaks, as I will use the term here, refer to insect populations reaching very high levels, far above normal. As characterized in Barbosa & Schultz (1987) and in Watt *et al.* (1990), outbreaks often are very irregular in both their levels and frequency, and any explanation must account for these overarching features.

There are likely to be multiple explanations for outbreaks, as insect outbreaks vary greatly in their dynamics. The major dichotomy outlined by Berryman (1987) is between insect outbreaks that spread from local epicenters, which are called *eruptive*, and those that do not spread from local epicenters, but remain local in space, which are called *gradient*. These two major classes of outbreaks will thus call for very different theoretical approaches. The eruptive outbreaks require an explicitly spatial model, while it may be possible to describe the dynamics of gradient outbreaks without an explicit spatial component.

Berryman further breaks up each class of outbreaks into those where the insects have a tendency to cycle around equilibrium, and those where the outbreaks appear to be tied explicitly to changes in environmental conditions.

Theories for outbreaks seek to explain several key observations:

- Why do some species exhibit outbreaks while others never appear to?
- What is the explanation for the temporal sequence of outbreak and non-outbreak behavior in a species that does exhibit outbreaks?

- A third question, which may have received less attention, is what causes an outbreak to end?

Possible explanations for insect outbreaks are also reviewed by Berryman (1987), and I briefly repeat his list of hypotheses here:

- Changes in the physical environment cause outbreaks.
- Genetic or physiological changes in the species cause outbreaks.
- Trophic interactions between insects and plants, or insects and their predators, cause outbreaks.
- Changes in host plants cause outbreaks.
- Outbreaks are common in some species because of their life histories.
- Overwhelming natural enemies causes outbreaks.
- Overwhelming plant defenses causes outbreaks.

What is intriguing about this list is that although dynamic models may underlie many of these explanations, only the suggestion that trophic interactions cause outbreaks really emphasizes the dynamic aspects of outbreaks. However, the dynamic approach to understanding tropic interactions has focused mainly on non-spatial models, with some notable exceptions such as Hassell *et al.* (1991). Thus, dynamic spatial models of insect outbreaks are in need of further study.

Lessons from some specific examples

Numerous examples of outbreaking insect species are reviewed in Barbosa & Schultz (1987) and in particular in Berryman (1987), in Watt *et al.* (1990) and in Cappuccino & Price (1995). Several other examples (Hanski, 1987; Hunter, 1991; Roland, 1993), in particular, emphasize the potential importance of a detailed dynamic spatial modeling approach for providing further insights into the dynamics of insect outbreaks. Hanski (1987) examined the dynamics of 11 species of pine sawflies in Northern Europe, of which approximately half are high outbreak species. He found a high degree of association between these outbreaking species and the presence of gregarious larvae. This implies that an explicit spatial model is needed. Hunter (1991) also distinguished between outbreaking and non-outbreaking species, among Macrolepidoptera feeding on northern hardwood trees. Once again, he found an association between gregarious feeding and the presence of outbreaks. In a somewhat different vein, the importance of spatial structure was emphasized by Roland (1993). Roland demonstrated that the best correlate with outbreaks of the forest tent caterpillar (*Malacosoma disstria*) in northern Ontario is the presence of forest edge.

One of the most extensive studies of insect outbreaks is the one by Legg (1981) analyzed by Bigger (1993a,b). In this study complete spatial data were available for occupancy of 238 trees by 23 species of insects (13 Homoptera, 1 Heteroptera and 9 Formicidae) based on 306 consecutive weekly observations in a cocoa plantation in Ghana. These observations correspond to between 6 and 50 generations of dynamics depending on the species. A preliminary examination of the data and the analyses by Bigger (1993a,b) demonstrated the important role played by spatial factors in determining the dynamics of the insect species. Moreover, almost all of the numerically important species showed very irregular patterns of dynamics through time, suggesting the importance of non-linear dynamic processes in determining the dynamic behavior and showing that equilibrium approaches cannot be used to understand these outbreaks.

On a longer time-scale, several recent studies (Swetnam & Lynch, 1993; Klimetzek, 1990) also provide clues into the regional behavior of outbreaks and the long time frequency and size of outbreaks. Klimetzek (1990) reviewed outbreak frequency and size of pine feeding insects in Germany over a 187 year period. Using tree ring data to extend their study to several centuries, Swetnam & Lynch (1993) showed that outbreaks of western spruce budworm in northern New Mexico are close to being periodic.

Using both the reviews in Barbosa & Schultz (1987) and Watt *et al.* (1990) and the specific examples mentioned here, one can conclude that spatially structured models that can describe the dynamics of insects will be required to understand the behavior of many insect outbreaks. However, some species also show gradient outbreaks. For these species, approaches that do not emphasize the spatial component may be appropriate.

Spatial modeling techniques

I reiterate that understanding the dynamics of eruptive outbreaking insects requires non-equilibrium approaches that explicitly include spatial structure. However, choice among different modeling approaches must be made with reference to biological requirements. Multiple modeling approaches should be used to provide further insight. I will begin by classifying approaches that include spatial structure into four broad categories to facilitate discussion of, and comparison between, different approaches. There is, of course, some overlap among the techniques. One method is based on reaction–diffusion models (Levin, 1974), which can be either discrete or continuous in time and space. A second method is based on looking at metapopulations by considering the number of patches that are in different states (Levins, 1969; Gilpin & Hanski, 1991; Hastings & Harrison, 1994). A third method is based on cellular automata and contact processes (Durrett & Levin, 1994). A final method is the use of integro-differential equation models, which essentially become coupled map lattices (Kaneko, 1993) when investigated numerically (Hastings & Higgins, 1994).

Reaction–diffusion models

These models can be traced back to the seminal papers of Fisher (1937), Skellam (1951) and Turing (1952), with a more recent review in Okubo (1980). In the simplest version, for one species in (one dimension in) continuous time and space, the model takes the form:

$$\frac{\partial N}{\partial t} = \frac{\partial}{\partial x}\left(D(x)\frac{\partial N}{\partial x}\right) + Nf(N,x) \qquad (12.1)$$

where $N(x,t)$ is the (density function for) the population size at x at time t; $D(x)$ is the diffusion coefficient, and the function $f(N,x)$ is the *per capita* rate of increase of the population at position x. The parameter D will be central to the approach, and has been estimated for many insects by Kareiva (1983).

Although it does not have a simple biological interpretation, it is the basic parameter describing movement in the 'null' model of movement, passive diffusion. The parameter D has units of length squared per time.

The model eqn (12.1) makes the assumption that individuals move randomly throughout their life (van den Bosch *et al.*, 1990, 1992). I will deal with modifications below. A discrete space analog for this model takes the form:

$$\frac{dN_i}{dt} = \sum_{j=1,n} D_{ij}(N_j - N_i) + Nf_i(N) \qquad (12.2)$$

where $N_i(t)$ is the population size at position i at time t; $D_{ij} = D_{ji}$ is the rate of exchange between patches i and j; and the function $f_i(N)$ is the *per capita* rate of increase of the population at position i.

These equations and their obvious analogs for more species have been used to approach several biological questions, in particular:

• What is the rate of spread of an invading organism?
• When do spatial patterns in population sizes occur, and is dispersal or spatial movement stabilizing or destabilizing?
• How large a suitable habitat is necessary to support a given species?
• How does a population respond to patches of different quality?

A number of studies have related these models to the spread of particular species (Skellam, 1951; Okubo, 1980; Kareiva, 1983; Lubina & Levin, 1988; Andow *et al.*, 1990). If all individuals start at a single point, and the *per capita* growth rate, f, is a constant, the solution to eqn (12.1) is a normal distribution with variance given by Dt. Surprisingly, the rate of spread is the same even for non-linear *per capita* growth rates (reviewed in Mollison, 1977; and van den Bosch *et al.*, 1990, 1992). These models are thus likely to be useful in the description of eruptive outbreaks, since at least some conclusions appear robust to changes in assumptions.

Dynamics that seem to mimic some of the behavior of insect outbreaks appear in models with age-dependent diffusion (Hastings, 1992), which indicate some of the potential changes to dynamics that arise when movement is age-dependent, in contrast to the classic diffusion model which has age-independent movement. The predictions of the former model, that age-dependent movement can be very destabilizing, suggest one mechanism for generating outbreaks, and one which is subject both to experimental test and perhaps control measures. Because typically only adults are highly mobile, it is extremely important to include age-dependent diffusion in models of insect populations.

Patch models

An alternate approach to including spatial structure begins with metapopulations (see reviews by Hanski, 1991; Hastings, 1991; Hastings & Harrison, 1994). The subpopulation, which, depending on the spatial scale, can be as small as a tree or larger, is the basic unit of description (Hastings, 1990). In the simplest model for a single species:

$$\frac{dp}{dt} = mp(1-p) - Ep \qquad (12.3)$$

where p is the fraction of patches occupied by the species, m is the rate at which empty patches are colonized per occupied patch, and E is the rate at which patches go extinct. Just as demography of single populations proceeds from a description that ignores size or other structuring variables to a description that includes size, one can move from a description of metapopulations that ignores size of subpopulations to one that includes size, going from ordinary differential equations to first order partial differential equations.

Several biological questions have been approached using these models:

• What determines whether a population persists?
• What are the dynamics of the metapopulation?
• What is the distribution of subpopulation sizes?
• What is the role of dispersal – is it stabilizing or destabilizing?

To describe insect outbreaks, one can use the extensions already developed (Metz & Diekmann, 1986; Hastings & Wolin, 1989; Hastings, 1991; Gyllenberg

& Hanski, 1992; Hastings & Harrison, 1994; Gyllenberg *et al.*, 1997), which include the number of individuals within a 'patch' to look at the role of gregarious feeding in the dynamics of insect outbreaks. In particular, these models show that there can be populations that show dramatic increases in population levels at a local scale, while maintaining stable dynamics at a global spatial scale. Thus, the metapopulation models provide one way of describing local outbreaks.

The natural extensions to two species of models of this kind have also been used to look at interactions between predator and prey (Hastings, 1977; Gurney & Nisbet, 1978) inspired in part by the classic experiments of Huffaker (1958). These models capture the essence of the experimental system: within each patch there are essentially local outbreaks of the prey, and then the prey are found by the predator. Thus these models predict only local outbreaks, which are unpredictable in space and time. Additionally, the total population size (over all patches) of the prey species can either remain constant or go through periodic fluctuations. Thus, in this model, local outbreaks are inevitable, and the end of local outbreaks is unpredictable. The likelihood of overall fluctuations (outbreaks that spread?) is increased as the predator (or parasitoid) becomes more efficient. Thus, perhaps not surprisingly, if one attempts to reduce the mean population level of the host by having a more efficient parasitoid, one risks introducing instabilities increasing the likelihood of global outbreaks.

Cellular automata approaches

The models of the previous section are unable to look at any issues relating to the spread of an outbreak. Much more specific biological information, and information about underlying habitat differences can be incorporated using these new computer intensive approaches (reviewed in Czaran & Bartha, 1992). A start towards applying these models to insects is found in Hassell *et al.* (1991) and Comins *et al.* (1992). These approaches represent each spatial location as being in one of a small number of states, such as high or low density of the species under considera-

tion. At each time-step, the state of each location changes according to rules that depend on the state of neighboring sites (the game of 'Life' is one simple, well studied example). These models have been used to approximate Nicholson–Bailey dynamics by Hassell *et al.* (1991) and Comins *et al.* (1992). These rules can include chance, and the effects of long-range dispersal. The models of Nicholson–Bailey dynamics show complex spatio-temporal patterns with large local fluctuations in population size.

A more specific study (Zhou *et al.*, 1995) has applied this approach to understanding the spread of gypsy moth outbreaks. The ability to relate these models to observed historical spread in determining parameters shows one advantage of this approach.

Coupled map lattice approaches

A final mathematical formalism has been called coupled map lattices by physicists (Kaneko, 1993). In this formalism one takes a discrete time, discrete space description of population dynamics. The local dynamics are described by density-dependent growth, with dispersal described by a function that gives the probability of an offspring ending up at a particular distance from its parent. These models can be developed directly, or arise as the result of numerically solving an equivalent description based on assuming that space is continuous, as in Hastings & Higgins (1994). What is most important about these models from the viewpoint of insect outbreaks is the tendency of these models to show the possibility of long transient behavior, or intermittency – sudden changes in the form of dynamics (Ott, 1993).

Hastings & Higgins (1994) studied the surprising behavior of the following simple model. They assumed that there was a finite length of favorable habitat, denoting position along the one-dimensional habitat by x, where $0 < x < L$. The number of larvae produced at position x in year t, $l(t,x)$, is given by the Ricker formulation:

$$l(t,x) = N(t,x)e^{r[1-N(t,x)]} \tag{12.4}$$

where they have scaled population size $N(t, x)$ so 1 is the equilibrium without spatial structure. Dispersal follows reproduction, so that the number of individuals at the location x is given by a summation of larvae released around the position x:

$$N(t+1,x) = \int_0^L l(t,y)g(y,x)dy \qquad (12.5)$$

where the probability that a larva released at y settles at x is:

$$g(y,x) = \frac{e^{-\delta(y-x)^2}}{\sqrt{\pi/\delta}} \qquad (12.6)$$

Here, δ is a parameter measuring the dispersal distance. The surprise is that in this model, the dynamics can exhibit very long transients, which mimics the concept of insect outbreaks (Hastings & Higgins, 1994). Moreover, the results were robust to changes in the local dynamics and changes in the dispersal distances over several orders of magnitude. What is also interesting is that the most dramatic fluctuations in population levels occurred at the edge of the suitable habitat, in concordance with the observation of Roland (1993) that outbreaks were correlated with 'edge' habitat. This question merits much further study, and requires incorporation of biological information about the behavior of hosts and parasitoids at the edge of their habitat.

Non-spatial modeling

Not all outbreaks require a spatial description, as noted above. Simple difference equation models, as introduced by May (1974, 1976) long ago showed the possibility of chaos in univoltine insects (recently reviewed by Logan & Allen, 1992; see also Logan & Hain, 1991; Hastings *et al.*, 1993). Difference equation models for hosts and parasitoids also have a long development (reviewed in Hassell, 1978). All these models of host–parasitoid, or predator–prey type, exhibit a form of dynamics that corresponds to several features of insect outbreaks – the host and parasitoid can be at low population levels, the host can increase in numbers, followed by an increase in numbers in the parasitoid, leading to a crash of both species. Here, I will not go into detail on these models, but instead describe some of the surprising behavior that has recently

been observed in analogous continuous-time models.

The presence of chaos in simple food web models in continuous-time was first suggested by the work of Gilpin (1979) who showed that chaos existed even in a Lotka–Volterra model for the one-predator two-prey web (with competition between the prey). For the three-species food chain model in continuous-time, the Lotka–Volterra model can be shown to be globally stable, but if non-linear functional responses are used chaos can be shown to exist, as demonstrated by Hastings & Powell (1991). In the simple three-species food chain, it is easy to demonstrate that there are multiple attractors – simultaneously stable solutions – with one attractor corresponding to elimination of one species, while another attractor corresponds to all species persisting (Abrams & Roth, 1994; McCann & Yodzis, 1994; Hastings, 1996). Thus initial conditions may play an important role (see Chapter 9).

The problem of multiple attractors can be solved by looking at asymptotic behavior, perhaps as it depends on initial conditions. However, there is a further problem – there can be very long transient behavior (McCann & Yodzis, 1994; Hastings, 1996). If the model studied by Hastings & Powell is allowed to reach its asymptotic, chaotic behavior, and then one parameter was changed slightly, the chaotic attractor can become unstable – no longer an attractor. However, the solution followed the 'ghost' of the chaotic attractor, the no longer stable chaotic solution, for a very long time before finally approaching the only stable solution remaining which corresponded to elimination of the top predator.

Thus concentrating on any asymptotic behavior, including chaos and not just equilibria, may be inappropriate (an excellent review of transients from a physicist's viewpoint is given in Tél, 1990). Transient behavior may be too long lasting to ever allow systems to approach their equilibrium – the ecological time-scale may be long relative to the time-scale of transient behavior. Or, systems may appear to have attained their asymptotic behavior, and then may suddenly shift dynamics. Sudden changes in dynamics of species or even community composition may not be reflecting changes in underlying dynamics, but may solely be the result of transient behavior. Once again, this kind of dynamical behavior corresponds exactly to the kinds of behavior observed in out-breaking insects.

In the framework of these models, the likelihood of outbreaks can be considered by using the well-developed theories looking at stability of host–parasitoid systems. In the non-chaotic models, the duration of any outbreak will be relatively predictable, and the same from outbreak to outbreak, but the time between outbreaks will be much more variable. The explanation for the variability in time between outbreaks is that this time corresponds to a time when both species are at low population levels, so small random fluctuations will lead to large changes in the time until the host population begins its explosive growth. For the chaotic models, the systems are even more unpredictable.

Conclusions

The approaches outlined here demonstrate the plausibility of describing insect outbreaks as transient dynamical features of models. If this suggestion indeed can be verified it will greatly change our approach to developing biological control measures. Verification of this suggestion will require both future modeling efforts and clever experimental tests coupled with observation of natural systems. Since the main thesis of this chapter is currently very preliminary, below I will outline required directions for future research.

What are the lessons from the models developed here for biological control? First, the suggestion is that outbreaks certainly correspond to species that do not exhibit equilibrium behavior. Thus, using a framework first summarized in Hassell (1978), one can begin to assess the likelihood of outbreaks by looking for species for which models predict instabilities – instabilities caused by features such as highly efficient parasitoids, very strong density dependence in univoltine systems, time delays, etc. One cautionary note is that apparent equilibrium behavior may not truly correspond to long-term equilibrium behavior, because of the possibility of transients. A second cautionary note is that features which may

appear to reduce the long-term equilibrium level of a species may in fact increase the likelihood of outbreaks.

In the models where one can look at questions of the sequence of outbreak and non-outbreak behavior, the time between outbreaks is typically not very predictable – although even in the models with transients the length of an outbreak may be more predictable. Thus, at this time, the only overall conclusion is that one should turn away from equilibrium concepts and focus on transients and dynamics. For those modeling insect outbreaks, this suggests more reliance on the use of numerical solutions of models, which are the only way to really look at dynamics, and less reliance on analytical approaches.

Since the conclusions here are so preliminary, what is the outlook for future developments? Building upon earlier work by Weinberger (1982) and van den Bosch *et al.* (1990, 1992) one can make a more general model of insect population dynamics that uses as its basis:

$$\sum_{\text{over all ages, } u, \text{ and locations, } y}$$

Probability of an egg at location x = Probability of an egg at y which hatched u weeks or months ago producing an egg at x

Changing this verbal formula into a mathematical one by putting the demography and dispersal into the words is straightforward. Using this formula requires making some simplifying assumptions, as is done by van den Bosch *et al.* (1990, 1992) for animals. One can make the appropriate simplifications to allow the application of this general formula to outbreaking insects, which will consist of assumptions about the spatial form of dispersal and about the demography. This model can then be investigated to see how likely are transient dynamics of the form observed by Hastings & Higgins (1994). One important kind of output from the modeling will be estimates of the frequency and size of outbreaks, which is the kind of typical behavior that can be determined from studies of intermittency (e.g., Ott, 1993).

Different modeling approaches, as outlined above, all need to be used because they make subtle different underlying assumptions. Although it appears that the metapopulation models and the diffusion models make incompatible predictions, an examination of non-equilibrium solutions of two patch models (Adler, 1993; Hastings, 1993, and unpublished results) show that some of the features of metapopulation models can be reproduced. I have recently completed (unpublished results) a numerical study of coupled, continuous-time, predator–prey models in two patches. The dynamics lead to a situation where the prey is essentially alternately found first in one patch, then in the other patch – essentially out-of-phase local outbreaks. Once again, the models will lead to predictions of the frequency and size of outbreaks. At a different spatial scale, the work of White *et al.* (1996) shows the potential importance of transients in the behavior of a host–pathogen system.

Turning to non-spatial approaches, the explanation of gradient outbreaks can take several forms (in addition to current approaches as outlined in Barbosa & Schultz, 1987). First, the metapopulation models discussed above may provide one way to understand this phenomenon. Second, the spatial models described above may be needed to understand why these outbreaks do not spread. However, some explanation may be possible in terms of local dynamics.

Intermittent behavior, or at least very long transient dynamics, has already been found in simple models of food webs, as noted above (see also Vandermeer, 1990). These models need to be modified to make them appropriate descriptions of insect population dynamics (see e.g., Beddington *et al.*, 1975). A focus on the dynamics of host–parasitoid systems within the context of food webs, rather than as individual interactions, may greatly increase our estimate of the likelihood and importance of transient dynamics in insect systems.

Also, the models need to be related both to previous and future experimental studies of the form described earlier. This effort presents a substantial challenge both to the theoretician, to develop testable predictions, and to the experimentalist, to develop observations on an appropriate spatial and temporal scale. Finally, the work will then need to be related to more practical issues from biological control.

Acknowledgements

I thank Peter Kareiva for helpful discussions. Research supported in part by a grant from the National Science Foundation.

References

Abrams, P. A. & Roth J. (1994) The effects of enrichment on three-species food chains with nonlinear functional responses. *Ecology* 5: 1118–30.

Adler, F. (1993) Migration alone can produce persistence of host-parasitoid models. *American Naturalist* 141: 642–50.

Andow, D. A., Kareiva, P. M., Levin, S. A. & Okubo, A. (1990) Spread of invading organisms. *Landscape Ecology* 4: 177–88.

Barbosa, P. & Schultz, J. C. (eds) (1987) *Insect Outbreaks*. New York: Academic Press.

Beddington, J. R., Free, C. A. & Lawton, J. H. (1975) Dynamic complexity in predator–prey models framed in difference equations. *Nature* 255: 58–60.

Berryman, A. A. (1987) The theory and classification of outbreaks. In *Insect Outbreaks*, ed. P. Barbosa & J. C. Schultz, pp. 3–30. New York: Academic Press.

Berryman, A. A. (ed.) (1988) *Dynamics of Forest Insect Populations: Patterns, Causes, Implications.* New York: Plenum Press.

Berryman, A. A., Millstein, J. A. & Mason, R. R. (1990) Modelling Douglas fir–Tussock moth population dynamics: the case for simple theoretical models. In *Population Dynamics of Forest Insects*, ed. A. D. Watt, S. R. Leather, M. D. Hunter & N. A. C. Kidd, pp. 369–80. Andover: Intercept.

Bigger, M. (1993a) Time series analysis of variation in abundance of selected cocoa insects and fitting of simple linear predictive models. *Bulletin of Entomological Research* 83: 153–69.

Bigger, M. (1993b) Ant–homopteran interactions in a tropical ecosystem. Description of an experiment on cocoa in Ghana. *Bulletin of Entomological Research* 83: 475–505.

Cappuccino, N. & Price, P. W. (eds) (1995) *Population Dynamics: New Approaches and Synthesis*. New York: Academic Press.

Comins, H. N., Hassell, M. P. & May, R. M. (1992) The spatial dynamics of host parasitoid systems. *Journal of Animal Ecology* 61: 735–48.

Connell, J. H. & Sousa, W. P. (1983) On the evidence needed to judge ecological stability or persistence. *American Naturalist* **121**: 789–824.

Czaran, T. & Bartha, S. (1992) Spatiotemporal dynamic models of plant populations and communities. *Trends in Ecology and Evolution* 7: 38–42.

Durrett, R. & Levin, S. A. (1994) Stochastic spatial models: a user's guide to ecological applications. *Philosophical Transactions of the Royal Society of London, Series B* **343**: 329–50.

Fisher, R. A. (1937) The wave of advance of advantageous genes. *Annals of Eugenics (London)* 7: 355–69.

Gilpin, M. E. (1979) Spiral chaos in a predator–prey model. *American Naturalist* **113**: 306–8.

Gilpin, M. E. & Hanski, I. A. (eds) (1991) *Metapopulation Dynamics: Empirical and Theoretical Investigations.* London: Academic Press.

Gurney, W. S. C. & Nisbet, R. M. (1978) Predator–prey fluctuations in patchy environments. *Journal of Animal Ecology* **47**: 85–102.

Gyllenberg, M. & Hanski, I. A. (1992) Single-species metapopulation dynamics – a structured model. *Theoretical Population Biology* **42**: 35–61.

Gyllenberg M., Hanski I. A. & Hastings A. (1997) Structured metapopulation models. In *Metapopulation Biology: Ecology, Genetics, and Evolution*, ed. I. A. Hanski & M. E. Gilpin, pp. 93–122. San Diego: Academic Press.

Hanski, I. A. (1987) Pine sawfly population dynamics: patterns, processes, problems. *Oikos* **50**: 327–35.

Hanski, I. (1991) Single-species metapopulation dynamics – concepts, models and observations. *Biological Journal of the Linnean Society* **42**: 17–38.

Hassell, M. P. (1978) *The Dynamics of Arthropod Predator–prey Systems.* Princeton: Princeton University Press.

Hassell, M. P., Comins, H. N. & May, R. M. (1991) Spatial structure and chaos in insect population dynamics. *Nature* **353**: 255–8.

Hastings, A. (1977) Spatial heterogeneity and the stability of predator–prey systems. *Theoretical Population Biology* **12**: 37–48.

Hastings, A. (1990) Spatial heterogeneity and ecological models. *Ecology* **71**: 426–8.

Hastings, A. (1991) Structured models of metapopulation dynamics. *Biological Journal of the Linnean Society* **42**: 57–71.

Hastings, A. (1992) Age dependent dispersal is not a simple process – density dependence, stability, and chaos. *Theoretical Population Biology* **41**: 388–400.

Hastings, A. (1993) Complex interactions between dispersal and dynamics – lessons from coupled logistic equations. *Ecology* **74**: 1362–72

Hastings, A. (1996) What equilibrium behavior of Lotka–Volterra models does not tell us about food webs. In *Food Webs: Intergration of Patterns and Dynamics*, ed. G. Polis & K. Winemiller, pp. 211–17. New York: Chapman and Hall.

Hastings, A. & Harrison, S. (1994) Metapopulation dynamics and genetics. *Annual Review of Ecology and Systematics* **25**: 167–88.

Hastings A. & Higgins, K. (1994) Persistence of transients in spatially structured ecological models. *Science* **263**: 1133–6.

Hastings, A. & Powell, T. (1991) Chaos in a three species food chain. *Ecology* **72**: 896–903.

Hastings, A. & Wolin, C. (1989) Within-patch dynamics in a metapopulation. *Ecology* **70**: 1261–6.

Hastings, A., Hom, C. L., Ellner, S., Turchin, P. & Godfray, H. C. J. (1993) Chaos in ecology: is mother nature a strange attractor? *Annual Review of Ecology and Systematics* **24**: 1–33.

Huffaker, C. B. (1958) Experimental studies on predation: dispersion factors and predator–prey oscillations. *Hilgardia* **27**: 343–83.

Hunter, A. F. (1991) Traits that distinguish outbreaking and nonoutbreaking Macrolepidoptera feeding on northern hardwood trees. *Oikos* **60**: 275–82.

Kaneko, K. (ed.) (1993) *Theory and Applications of Coupled Map Lattices.* New York: John Wiley and Sons.

Kareiva, P. (1983) Local movement in herbivorous insects: applying a passive diffusion model to mark-recapture field experiments. *Oecologia* **57**: 322–7.

Klimetzek, D. (1990) Population dynamics of pine-feeding insects: a historical study. In *Population Dynamics of Forest Insects*, ed. A. D. Watt, S. R. Leather, M. D. Hunter, & N. A. C. Kidd, pp. 3–10. Andover: Intercept.

Legg, J. T. (1981) *The Cocoa Swollen Shoot Research Project at the Cocoa Research Institute, Tafo, Ghana 1969–1978.* Technical Report, vol. 4. London: Overseas Development Administration.

Levin, S. A. (1974) Dispersion and population interactions. *American Naturalist*. **108**: 207–28.

Levins, R. (1969) Some demographic and genetic consequences of environmental heterogeneity for biological control. *Bulletin of the Entomological Society of America* **15**: 237–40.

Logan, J. A. & Allen, J. C. (1992) Nonlinear dynamics and chaos in insect populations. *Annual Review of Entomology* **37**: 455–77.

Logan, J. A. & Hain, F. (eds) (1991) *Chaos and Insect Ecology*. Blacksburg, VA: Virginia Agricultural Experiment Station, VPI and SU.

Lubina, J. & Levin, S. A. (1988) The spread of a reinvading organism: range expansion of the California sea otter. *American Naturalist* **131**: 526–43.

May, R. M. (1974) Biological populations with nonoverlapping generations: stable points, stable cycles and chaos. *Science* **186**: 645–7.

May, R. M. (1976) Simple mathematical models with very complicated dynamics. *Nature* **261**: 459–67.

McCann, K. & Yodzis, P. (1994) Biological conditions for chaos in a three-species food chain. *Ecology* **75**: 561–4.

Metcalf, R. L. & Luckmann, W. H. (eds) (1994) *Introduction to Insect Pest Management*, 3rd edn. New York: John Wiley and Sons.

Metz, J. A. J. & Diekmann, O. (1986) *The Dynamics of Physiologically Structured Populations*. New York: Springer-Verlag.

Mollison, D. (1977) Spatial contact models for ecological and epidemic spread. *Journal of the Royal Statistical Society* **B39**: 283–326.

Okubo, A. (1980) *Diffusion and Ecological Problems: Mathematical Models*. New York: Springer-Verlag.

Ott, E. (1993) *Chaos in Dynamical Systems*. Cambridge: Cambridge University Press.

Roland, J. (1993) Large-scale forest fragmentation increases the duration of tent caterpillar outbreak. *Oecologia* **93**: 25–30.

Ruesink, R. L. & Onstad, D. W. (1994) Systems analysis and modeling in pest management. In *Introduction to Insect Pest Management*, 3rd edn, ed. R. L. Metcalf & W. H. Luckmann, pp. 355–92. New York: John Wiley and Sons.

Skellam, J. G. (1951) Random dispersal in theoretical populations. *Biometrika* **38**: 196–218.

Swetnam, T. W. & Lynch, A. M. (1993) Multicentury, regional-scale patterns of western spruce budworm outbreaks. *Ecological Monographs* **63**: 399–424.

Tél, T. (1990) Transient chaos. In *Experimental Study and Characterization of Chaos*, ed. B.-L. Hao, pp. 149–211. Singapore: World Scientific Publishing.

Turing, A. M. (1952) The chemical basis of morphogenesis. *Philosophical Transactions of the Royal Society of London, Series B* **237**: 37–72.

van den Bosch, F., Metz, J. A. J. & Diekmann, O. (1990) The velocity of spatial population expansion. *Journal of Mathematical Biology* **28**: 529–65.

van den Bosch, F., Hengeveld, R. & Metz, J. A. J. (1992) Analysing the velocity of animal range expansion. *Journal of Biogeography* **19**: 135–50.

Vandermeer, J. (1990) Indirect and diffuse interactions: complicated cycles in a population embedded in a large community. *Journal of Theoretical Biology* **142**: 429–42.

Watt, A. D., Leather, S. R., Hunter, M. D. & Kidd, N. A. C. (eds) (1990) *Population Dynamics of Forest Insects*. Andover: Intercept.

Weinberger, H. F. (1982) Long time behavior of a class of biological models. *SIAM Journal of Mathematical Analysis* **13**: 353–96.

White, A., Begon, M. & Bowers, R. G. (1996) Host–pathogen systems in a spatially patchy environment. *Proceedings of the Royal Society of London, Series B* **263**: 325–32.

Zhou, G. F., Liebhold, A. M. & Zhou, G. F. (1995) Forecasting the spatial dynamics of gypsy moth outbreaks using cellular transition models. *Landscape Ecology* **10**: 177–89.

PART IV

Genetic/evolutionary considerations

Although it is safe to say that most biological control workers prefer to solve their pest problem in ecological time, the potential influence of evolutionary factors on long-term success has long been appreciated. Disadvantageous evolutionary changes in weed control programs have been a particular concern. For example, the agent may undergo a host–plant shift following introduction. Even so, formal theory has only recently addressed such concerns, and the application of theory to specific problems has been somewhat limited. This is likely to change in the near future, as the tools of molecular biology and the underlying genetic theory that supports them make the genetic manipulation of agents feasible and affordable.

A biological control agent achieving sustained control of a target potentially exerts strong selection pressures on that target to evolve resistance to the agent, analogous to the evolution of resistance to pesticides. This is a problem that is often mentioned in the biological control literature, but is not well studied. Holt *et al.* (Chapter 13) use a simple, population-genetic model to address it. Their focus is on spatial heterogeneity as measured by differences in the susceptibility of hosts across habitats of varying quality, echoing the theme of the previous section. They also stress the importance of non-equilibrium population dynamics in delaying or preventing the evolution of resistance to parasitoids. Although their discussion is theoretically driven, it sheds light on the applied problem of the breakdown of biological control over longer time periods.

Throughout the history of biological control, laboratory cultures of parasitoids have occasionally undergone drastic shifts in sex ratios, some of which have led to the loss of the culture. The coexistence of sexually and asexually reproducing strains of what appear to be the same parasitoid species has also been encountered numerous times, but it is only in this decade that 'selfish genetic elements' and the rickettsia *Wolbachia* have been identified as the likely reason for such heretofore unexplained phenomena. Hunter (Chapter 14) reviews the rapidly growing literature on these elements. She also makes clear that such elements can provide a tool that biological control workers can use to manipulate the genetic structure of natural enemies. Historically, these genetic elements and microorganisms have hindered biological control rather than benefited it, but theory suggests a number of ways they can be used to modify the attributes of potential agents to improve their control potential. It may only be a matter of time before the genetic manipulation of predators and parasitoids develops into a viable component of biological control practice.

The chapters in this book discuss the ecological effects of a diverse assortment of factors on the dynamics of biological control systems. Not surprisingly, most of these factors have important evolutionary consequences as well. In Chapter 15, Sabelis *et al.* place spatial and temporal heterogeneity, plant defenses, metapopulation structure, and 'bistable' dynamics in an evolutionary context. They discuss a tritrophic 'pancake predation' model developed specifically to understand evolution in spider mite systems under biological control. The theory has implications for integrated pest management, as well as biological control, including the selection of agents, the interplay between plant breeding and biological control, pesticide-induced disruption of biological control, and predator introduction strategies.

This section closes with a discussion of host selection by parasitoids by Luck & Nunney (Chapter 16). Host selection has been of long-standing practical importance to biological control, and it is now being fully integrated with evolutionary theory. Luck & Nunney describe the relationships between host selection and clutch size in *Trichogramma*, and between host selection and sex ratio in *Aphytis*. These provide striking examples of the role that theory can play in predicting the efficacy of prospective biological control agents.

13

Population dynamics and the evolutionary stability of biological control

ROBERT D. HOLT, MICHAEL E. HOCHBERG AND MICHAEL BARFIELD

Introduction

An appealing feature of classical biological control is that it has the potential to be self-sustaining, following initial establishment of the control agent, without the need for perpetual, large-scale human intervention. Most theoretical studies relevant to biological control (as in many chapters in this volume) explore population dynamics, making the quite reasonable assumption that basic biological properties of pest species and control agents are constant. Yet, over sufficiently long time-scales, neither target pests nor control agents are likely to have fixed properties. A truism of evolutionary biology is that all species harbor genetic variation for most characters (Lewontin, 1974) and so can respond *via* changes in genetic composition to environmental change. Introducing a biological control agent that limits pest numbers below economically significant levels surely counts as a significant environmental change, one that might be expected to have the potential to evoke an evolutionary response in the target pest species. This raises the question of what circumstances foster sustained biological control over evolutionary time-scales. Given strong selection, evolutionary time-scales permitting substantial changes in traits relevant to biological control could well be of the order of years or tens of years, not eons, given the high potential growth rates and population sizes, and short generation times, of many insect species. We suggest that the best biological control is that which can be evolutionarily stable, as well as ecologically persistent, and that a useful role of theory is to clarify when such evolutionary stability might be expected.

Recently (Holt & Hochberg, 1997) we have observed that while there are numerous instances of reduced pest control due to the evolution of resistance to chemical control agents, relatively few clear-cut examples exist for evolved breakdown in biological control. The few examples of evolved resistance that are reasonably unequivocal involve internal pathogens (e.g., myxomatosis, Dwyer *et al.*, 1990), rather than parasitoids or predators. Although for reasons discussed in detail in Holt & Hochberg (1997) it is difficult to quantify the seeming discrepancy in evolutionary responses to chemical *versus* biological control agents, there is little reason to doubt that it is real. This raises an interesting evolutionary puzzle.

Given that both chemical and biological control are likely to lead to novel selection pressures on target pest species, why is evolved resistance, leading to weakened control, not as obvious in biological as in chemical control systems? We (Holt & Hochberg, 1997) suggested that a number of complementary factors may contribute to this pattern. These include (i) the potential for greater costs of resistance to biological agents; (ii) the importance of behavioral plasticity in the control agent; (iii) the likelihood of coevolution, in which control agents evolve counter-adaptations to improved defenses in target species, and (iv) the role of spatial and temporal heterogeneity in weakening selection on target species. Scant data exist to assign relative importance to these factors; any may pertain to a given system.

Here, we use a simple theoretical model to examine the most novel of these four factors, the suggestion that spatial heterogeneity can weaken the evolutionary responses of a target pest species to an introduced control agent. The theoretical results reported below suggest that the constraining effects of spatial heterogeneity on host evolution can be greatly magnified by unstable population dynamics, which are likely to typify many biological control systems (see

Chapter 2). We should emphasize at the outset that our aim here is not to champion a particular model, but rather to use it to illustrate some of the dramatic effects spatial heterogeneity and population dynamics may have on evolutionary responses by host species to biological control agents. The particular conclusions we reach, drawn from a particular model, help highlight the need to examine the evolution of host resistance in a much wider range of models and ecological scenarios.

Sustained biological control requires that the population of the control agent persist, with oscillations in abundance bounded away from zero. Yet, an important aspect of natural enemy–victim interactions is their propensity towards instability when victim populations are limited to low levels by effective natural enemies (see Chapter 4). This observation has motivated the search for mechanisms stabilizing enemy–victim interactions, either at local equilibria or more broadly at landscape or regional scales (e.g., Hochberg, 1996). It is widely believed that spatial heterogeneity in one guise or another often helps to promote the persistence of strong natural enemy–victim interactions (e.g., Beddington *et al.*, 1978; Taylor, 1991). As we shall see, given that a host–parasitoid interaction persists because of spatial heterogeneity, the existence of unstable host–parasitoid population dynamics can have profound consequences for the rate of host evolution and the likely evolutionary stability of biological control.

A refuge model

We examine host evolution under biological control by parasitoids in situations where heterogeneity in parasitism is the ecological factor that permits persistence. Heterogeneity in parasitism may arise from a variety of sources, and can be broadly classified as involving individual-level phenomena (e.g., phenological and phenotypic variability in pests and/or natural enemies) or ensemble/population-level phenomena (e.g., density-dependent spatial variability). Such heterogeneities are generally recognized as stabilizers in host–parasitoid interactions, at least if the parasitoid limits host numbers to levels where direct density-dependence (namely self-limi-

tation) is weak (Beddington *et al.*, 1978; see Chapter 2 and Chapter 12).

The mathematical model we employ is the 'generalized proportional refuge model' (Holt & Hassell, 1993). Although the model was originally intended to reflect spatial protection for a fraction of the host population, the same model form can be interpreted as describing other host refuge/resistance mechanisms where host individuals are in two discrete classes – essentially either vulnerable to, or protected from, parasitism throughout their lives, and in which the density of hosts in either class plays no role in their vulnerability. Possible examples include the encapsulation of juvenile parasitoids by larval hosts, and the concealment of larval hosts within plant structures from adult parasitoid attack.

For the sake of clarity, however, we will refer to the refuge in spatial terms, such that a fraction of hosts reside in an 'exposed' habitat and the remaining hosts are in the 'refuge' habitat. Refuge and exposed habitats may arise because of barriers to parasitoid activity, or by behavioral preferences on the part of the adult parasitoid to search exclusively for hosts in certain areas. The fraction of the host population residing in the refuge reflects individual host behavior, e.g., the probability that a maternal host will lay an egg in one habitat, rather than another. This probability of refuge use can be viewed as a measure of habitat preference, a behavioral trait which can evolve if genetic variation exists in individual host preferences (Jaenike & Holt, 1991).

Several evolutionary questions naturally arise about this system. Under what circumstances does selection favor refuge use by the host? Given that a fixed fraction of hosts use a refuge, what is the strength of selection acting on other parameters (e.g., habitat-specific growth rates, or parasitoid attack rates) in the system? We show that the answers to the above evolutionary questions may be substantially influenced by dynamical properties of the system, suggesting an important function of population dynamics in modulating the rate of host evolution in biological control. Our ultimate goal is to assess the interplay of population dynamics, host evolution, and the effectiveness of biological control.

Elsewhere we (Hochberg & Holt, 1995, and

unpublished results) have examined a rather different and more complex model for coevolution in a host–parasitoid system with a refuge. Here, we deliberately examine a system with both a simple ecology (the parasitoid is assumed to be effective enough so that the host does not experience direct density-dependence), and a simple genetics (we assume clonal or haploid variation in the host species). Moreover, rather than examine the more complex scenario of coevolution between parasitoid and host, we concentrate on how population dynamics influence evolution in the host, with a genetically fixed parasitoid, and the consequences of such evolution for the effectiveness of biological control. An important task for future work will be to examine a broader range of host genetic architectures and permit coevolutionary responses by the natural enemy, in a wider range of ecological models for interacting hosts and parasitoids.

Each generation, a fraction, ε_i, of hosts in clone i are exposed to parasitism; the remaining fraction, $1 - \varepsilon_i$, are completely protected in the refuge habitat. In the absence of parasitism, hosts within and outside the refuge may have different expected net growth rates. *A priori*, it is reasonable to expect that host growth rates will tend to be smaller in the refuge than in exposed habitats; if this were not the case, then evolution would irrevocably be towards increased use of the refuge (barring other costs to protection from parasitism). Smaller refuge growth rates could be interpreted therefore as general 'costs' to protection from parasitism.

The model (from Holt & Hassell, 1993, generalized to multiple clones) is as follows:

$$N_i(t+1) = N_i(t)[\varepsilon_i \lambda_{1i} f_i(P(t)) + (1 - \varepsilon_i)\lambda_{2i}]$$
$$= N_i(t)W_i(t) \qquad (13.1)$$

$$P(t+1) = c\sum_i N_i(t)\varepsilon_i(1 - f_i(P(T)))$$

Here, t denotes generation t, $N_i(t)$ is the abundance of host clone i in generation t, and $P(t)$ is the abundance of the parasitoid in generation t. In effect, in each generation a given clone experiences two environments – one (a refuge) in which it has a constant growth rate, and the other in which it may have either a variable growth rate, varying with fluctua-

tions in parasitoid density, or a constant growth rate, if the system is at its demographic equilibrium and parasitoid numbers are constant.

In the remainder of this chapter, the quantity $f_i(P)$ is taken to be $e^{-a_i P}$, the usual Nicholson–Bailey form (e.g., Hassell, 1978). The quantities λ_1 and λ_2 are the growth rates of exposed and protected hosts, respectively. The quantity α_i is the attack rate, per parasitoid, on host clone i; c is the number of parasitoids emerging per parasitized host. For convenience, we set c equal to 1. The quantity $W_i(t)$ is the realized growth rate, or absolute fitness, of host clone i in generation t. As will be examined in more detail below, this fitness varies both with exposure to each habitat, and with parasitoid abundance. For notational clarity, in the following text the index i is suppressed except when referring explicitly to the dynamics of multiple clones.

As mentioned above, this model has previously been used to explore the importance of spatial heterogeneity in host growth rates for the persistence of the host–parasitoid interaction (Holt & Hassell, 1993). Here we recapitulate some key results. At equilibrium, parasitoid and host densities are, respectively:

$$P^* = \frac{1}{a}\ln\left[\frac{\varepsilon\lambda_1}{1 - (1-\varepsilon)\lambda_2}\right] \qquad (13.2)$$

$$N^* = \frac{1}{ac}\ln\left[\frac{\varepsilon\lambda_1}{1 - (1-\varepsilon)\lambda_2}\right]\left(\frac{\lambda_1}{\lambda_T - 1}\right)$$

where

$$\lambda_T = \varepsilon\lambda_1 + (1-\varepsilon)\lambda_2$$

is the host rate of increase at low densities, averaged over both habitats, in the absence of parasitism. For an equilibrium with positive densities to exist, two conditions must hold. First, $\lambda_T > 1$. Otherwise, the host goes extinct, dragging along with it the specialist parasitoid. Second, $(1 - \varepsilon)\lambda_2 < 1$. The quantity $(1 - \varepsilon)\lambda_2$ is the maximum rate of growth for those hosts residing in the refuge. If this quantity exceeds unity, these hosts (and hence the entire host population) grow geometrically, regardless of the fate of hosts exposed to parasitism. That is, the parasitoid cannot limit host numbers, and so there is no control.

The local stability analysis of this model is laid out fully in Holt & Hassell (1993), to which we refer the reader for details. Often, it is reasonable to assume that the growth rate of protected hosts differs from that of exposed hosts (e.g., because the two host classes experience different microclimates). The key factor determining stability in this model is the growth rate of hosts in the refuge, weighted by the fraction of hosts residing there. Even if hosts have a high rate of increase (averaged over both habitats), the system may be stable if $(1 - \varepsilon)\lambda_2 < 1$. However, if protected hosts have a very low rate of increase, or a very low fraction use the refuge (leaving most hosts exposed to parasitism), although the system may persist it is likely to exhibit cyclic or chaotic dynamics, often with large-amplitude fluctuations. Thus, the message for sustained biological control from the proportional refuge model is that protection from natural enemies should be costly for target host species – but not too much so – if control is to persist without large-scale outbreaks over ecological time-scales.

Host evolution in the proportional refuge model

Assume that a host is strongly limited by a parasitoid, and the interaction persists because the host uses a proportional refuge. How does host evolution modify the degree of control, and the importance of the refuge in allowing the interaction to persist, either at a stable point equilibrium or with bounded oscillations? There are two complementary techniques for examining such evolutionary questions. First, for multiple host clones, iterating the above equations allows one to analyze directly changes in the relative proportions of different clones in the total host population, and conditions for persistent polymorphism. Second, for any single clone the expression for W_i in eqn (13.1) defines a 'fitness generating function' (Brown & Vincent, 1992), which describes host fitness in generation t as a function of lower-level parameters, such as the attack rate. (Below, we will use 's' to denote an arbitrary parameter upon which W_i functionally depends). Manipulating the expression for W_i permits one to examine how small changes in parameter values translate into changes in host fitness. This analytic approach is typically difficult and indeed usually impossible when population dynamics are unstable (basically because one cannot solve for the time-series of population densities, given fluctuating populations, and so cannot analytically describe temporal variation in host fitness). In this chapter, we report the result of numerical studies of evolution in the system defined by eqn (13.1). (Similar results emerge from more detailed approximate, analytical studies of unstable systems: unpublished results).

We consider the direction of parameter evolution, first for stable population dynamics, and then for unstable population dynamics.

Stable population dynamics

Not surprisingly, the realized fitness of clone i declines with increasing parasitoid density. Because parasitoid density is determined by parasitoids emerging from all host clones, the fitness of one clone indirectly is depressed because other clones are present – an example of 'apparent competition' (Holt & Lawton, 1994). Holt & Lawton (1993) analyzed general models of apparent competition under the assumption that the system settles into a point equilibrium (i.e., with unvarying host and parasitoid densities). The stable coexistence of alternative clones sharing a parasitoid is usually impossible. The dominant clone out of any given clonal array is the one which sustains, and persists in the face of, the highest parasitoid density. This 'rule-of-thumb' provides a unifying principle for several of the following results.

Evolutionary biologists often assume that mutations causing a small change in a given character arise more frequently than mutations of large effect. Assume that fitness, W, is an arbitrary function of a parameter, s, and the initial population is fixed with $s = s'$. The fate of a mutation with a slightly different value of s, $s'' = s' + \delta s$ is determined by the difference between $W(s')$ and $W(s'')$; the greater this difference, the greater the selection favoring (or disfavoring) the new mutation. Using a Taylor expansion, we estimate the change in clonal fitness as $W(s'') = W(s') + \delta s(dW/ds|_{s=s'})$. The quantity dW/ds is the 'strength of selection' on a given parameter s; this quantity measures the marginal effect of a small increase in an arbitrary parameter s on fitness W. If the strength of selection is near zero, clones differing by a small amount in the value of s from s' are effectively selectively neutral. If, by contrast, the strength of selection is large in magnitude, selection more strongly favors (or disfavors) small deviations in s.

We consider the host parameters ε, a, λ_1, and λ_2 in turn, and assume that clonal variation exists in one parameter, while the other parameters remain fixed.

Evolution of host protection, $1-\varepsilon$

The strength of selection on use of the refuge is:

$$\frac{dW}{d(1-\varepsilon)} = -\lambda_1 e^{-aP} + \lambda_2 \qquad (13.3)$$

After substituting $P = P^*$, we find that $dW/d(1-\varepsilon) > 0$ if $\lambda_2 > 1$. As makes sense, hosts are selected towards increasing use of the refuge, where their positive growth rate permits them to grow, escaping control (until limited by other factors, beyond the scope of the model).

By contrast, $dW/d(1-\varepsilon) < 0$ if $\lambda_2 < 1$. In this case, hosts are selected to avoid the refuge, regardless of the presence of the parasitoid. As refuge use evolves toward lower values for ε, the inherent instability of the interaction can emerge, creating host outbreaks.

It would appear that in either case, host evolution tends toward lower levels of effective control (either failed limitation of the target host by the parasitoid, or episodic pest outbreaks). Below, we examine in more detail how selection acts on ε, given population dynamic instability.

Evolution in the attack parameter, a

In contrast to evolution in refuge use, which includes an explicit cost ($\lambda_1 > \lambda_2$), we can examine how exposed hosts may evolve cost-free resistance to parasitism *via* reductions in the parasitoid attack parameter, a. The strength of selection on the attack parameter is:

$$\frac{dW}{da} = -Pe^{-aP}\varepsilon\lambda_1 < 0 \qquad (13.4)$$

which is always negative. Therefore, as expected from the assumed lack of costs, the host should evolve so as to lower attack rates. By inspection of eqn (13.2), this implies higher host density, and hence lowered control. It is interesting that all else being equal, the more effective the parasitoid is at attacking exposed hosts (higher a), the *weaker* is selection on the host to evolve counter-adaptations. Moreover, the magnitude of eqn (13.4) declines with increasing density of parasitoids (as long as $aP > 1$). Basically, if total attacks are high enough, very few hosts escape parasitism at all, so small changes in the parameter a do not greatly reduce net parasitism rates and thus are not strongly favored.

Evolution in the basic growth rates, λ

Assume that use of the refuge is fixed, and host productivity within it is permitted to evolve. The strength of selection on growth in the refuge is simply:

$$\frac{dW}{d\lambda_2} = 1 - \varepsilon \qquad (13.5)$$

Thus, the strength of selection relevant to adaptation to the refuge is directly proportional to the magnitude of use of the refuge, but is independent of the absolute growth rates in either habitat.

Strength of selection on growth in the exposed habitat is:

$$\frac{dW}{d\lambda_1} = \varepsilon e^{-aP} = \frac{1 - (1-\varepsilon)\lambda_2}{\lambda_1} \qquad (13.6)$$

In contrast to eqn (13.5), larger growth rates in either habitat weaken selection on growth rate in the exposed habitat. The strength of selection for improved adaptation to the refuge exceeds that for improved adaptation to the exposed habitat (namely $dW/d\lambda_2 > dW/d\lambda_1$), provided that:

$$\lambda_1 + \lambda_2 > \frac{1}{1-\varepsilon}$$

The reason for the effect of host growth rate on the strength of selection on exposed hosts is that large host λ values tend to sustain large parasitoid populations, hence few hosts outside the refuge are likely to

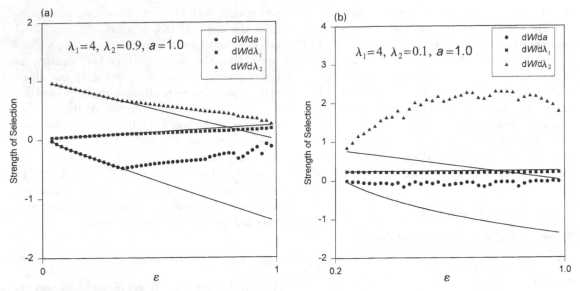

Figure 13.1. The strength of selection on three parameters (●, attack rate (*a*); ■, growth rate in exposed habitat (λ_1); ▲, growth rate in protected refuge (λ_2)) as a function of preference for the exposed habitat (ε). The parameters used are shown in the boxes. (a) moderate-quality refuge; (b) low-quality refuge. At low ε, the populations show stable dynamics. The continuous lines depict the expressions for strength of selection, assuming equilibrium, derived in the text. The divergence of the symbols from the lines indicates effects of population dynamics on the strength of selection. For unstable dynamics, the strength of selection was calculated as follows. First, a single clone is simulated using the parameters indicated in the graphs for 90 000 generations. For the next 10 000 generations, the values of $P(t)$ were recorded. This was then used to calculate the geometric mean fitness for a rare clone, over this same timespan. If sufficiently rare, the invading clone should have no effect on parasitoid dynamics. The rare clone was chosen to have parameters slightly different from the resident clone. This was done for two rare clones, each deviating by 1% in its parameter value (above and below, respectively) from the resident clone. To estimate dW_g/dq, the difference in these clones' realized value for W_g was divided by the difference in the parameter values. This process was repeated for ε in increments of 0.02. This numerical procedure agrees with the theoretical results, in the stable region. The parameters in (b) lead to much more unstable dynamics than those in (a) (see Holt & Hassell, 1993). The figure shows that there are correspondingly larger deviations in selection from equilibrial expectations.

escape parasitism and benefit from an increase in λ. This weakens selection for improved adaptation to the non-refuge habitat. It is useful to recall that the strength of selection is a measure of the net effect of a small change in a trait, averaged over all environments experienced by individuals expressing that trait. Because the model assumes that reproduction follows parasitism, high parasitism rates (which in this model are indirectly generated by high host growth rates) imply that most exposed individuals never live to realize a potential increase in fitness.

If potential pest populations have high intrinsic capacities for increase in the absence of biological

control, given that refuges stabilize the host–parasitoid interaction, there appears to be a bias in host adaptive evolution towards conditions within the refuge, relative to conditions outside.

Pulling these results together, the continuous lines of Fig. 13.1 depict for a particular combination of parameters how strength of selection on the various model parameters varies with ε.

In short, selection on the refuge use parameter of the host always tends to weaken biological control. By inspection of eqn (13.2), we see that host evolution on the attack parameter, *a*, always leads to higher host numbers, thus less effective control. One interesting result for stable systems is that control from very effective parasitoids (high a) might decay less rapidly than control from less-effective parasitoids (low a). There are two reasons for this: (i) as noted above, the strength of selection favoring a lower a, weakens with increasing a; and (ii) host abundance (and hence the likely pool of genetic variation) declines with increasing a. Finally, from eqn (13.2) it can be seen that host densities increase (albeit weakly) with increasing host growth rates, so selection on these parameters also implies somewhat lowered control.

Unstable population dynamics

As noted earlier and explored more fully in Chapter 4, a hallmark of host–parasitoid dynamics where hosts are limited to low densities by parasitoids is a propensity towards dramatic instability, with potentially damaging host outbreaks (see Chapter 12). We have seen that if a host–parasitoid interaction is stabilized because of a refuge, there is a tendency for parasitoid-dominated limitation to disappear over evolutionary time. In particular, if the refuge is a sink (i.e., $\lambda_2 < 1$, so that the host would go extinct were it restricted there), hosts are selected to avoid the refuge. Host evolution in general thus would appear to be inimical to long-term biological control. However, this particular conclusion assumes that the host–parasitoid system remains dynamically stable, as the host evolves.

We now turn our attention to evaluating selection when refuge use is low enough for parasitoid-driven instability. The basic question we ask is how such instability influences selection on the host. If dynamical instability weakens selection on hosts to escape parasitism, then systems that are to a degree unstable may evolve less rapidly than systems that are stable, and so retain control for longer spans of time. As before, we examine evolution in each parameter separately, but instead of reporting analytical results we now use numerical studies.

Evolution of habitat preference

We assume that the refuge is a sink habitat ($\lambda_2 < 1$), so that selection favors hosts that avoid the refuge if

populations attain constant levels. Our first procedure is to compete pairs of clones against each other, using numerical simulations of eqn (13.1). The clones have similar parameter values, except that they differ in their habitat preferences (measured by ε). Simulations were run until one clone dominated the host population (to within machine error), or for 100 000 generations. Figure 13.2 shows the outcome of these competitive trials for a number of parameter combinations.

It is useful to walk through the example of Fig. 13.2(a). To save space, this figure presents the results of two different scenarios in one frame: one of relatively high productivity of exposed hosts ($\lambda_1 = 10$, the upper triangle) and one of low productivity ($\lambda_1 = 4$, the lower triangle). In both cases the refuge habitat is assumed to be a slight sink, or $\lambda_2 = 0.9$. The two scenarios are illustrated on opposite signs of a diagonal line of slope 1. Because ε_i is bounded between 0 and 1, the upper and lower triangles can denote any possible pair of competing clones differing solely in habitat preference; above the line of slope 1, we call the clone with higher preference for the exposed habitat, clone 2 (namely $\varepsilon_2 > \varepsilon_1$), and below the line, the clone of higher preference for the exposed habitat is clone 1 (this is just a re-labeling). In zones denoted ε_j (where $j = 1$ or 2) that clone wins; in zones denoted C, there is clonal coexistence (= protected genetic polymorphism).

We demonstrated above that given stable dynamics, selection should favor avoiding the refuge if the absolute growth rate there is less than 1. Figure 13.2 shows that if one starts at high refuge use (low ε, implying stable dynamics), selection indeed favors higher exposure to the parasitoid; clones with higher values of ε replace clones with lower values of ε.

However, at higher exposure, the population dynamics become unstable. Figure 13.3 displays typical patterns in mean host abundance and coefficient of variation in abundance, as a function of ε. For sufficiently high ε, the dynamics are unstable and display increasingly large variance in numbers as ε increases. Figure 13.4 depicts an example of how the frequency of excursions to high levels (i.e., potentially worrisome pest outbreaks) tends to increase with ε.

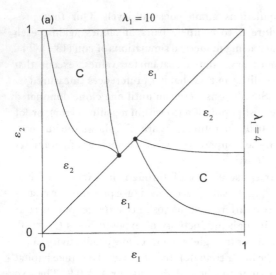

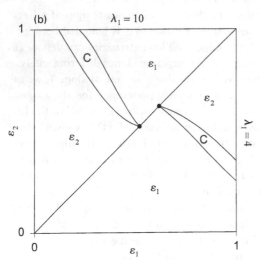

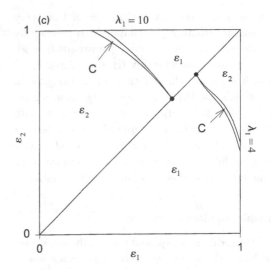

Figure 13.2. Domains of clonal dominance and coexistence. Equation (13.1) was iterated for pairs of clone, with each in turn initially rare while the other was present at its equilibrial density. In each case, $a = 1$. The subfigures correspond to different values for refuge growth, as follows: (a) $\lambda_2 = 0.9$; (b) $\lambda_2 = 0.5$; (c) $\lambda_2 = 0.1$. In the upper triangular domain of each figure, clone 2 has the higher preference for the exposed habitat; this is reversed in the lower triangular domain. The domains denoted ε_i indicate dominance by clone i (which can increase when rare, and excludes the alternative clone). The domains denoted C indicate pairwise clonal coexistence. The filled circles indicate the ultimate dominant clone (the evolutionarily stable strategy, ESS) when clones with intermediate values invade into systems with coexisting clonal pairs.

Population dynamics play a crucial role in determining the direction of evolution of refuge use. In particular, population instability tends to put a brake on the evolutionary decline in use of the refuge. From Fig. 13.2, we see that at high current values of exposure, *lower* values for exposure are favored. Thus, unstable population dynamics fosters use of an (on average) intrinsically costly, or low quality, refuge habitat.

If a pair of clones differ substantially in their use of the refuge, sustained coexistence can occur (area C of Fig. 13.2). However, if one starts with such a coexisting pair, and introduces a clone with an exposure level intermediate to those of the two resident clones (iterating the above model for three clones simultaneously), one of the original pair inevitably disappears. In our simulations of such three-way clonal competition, with repeated introductions of clones with intermediate habitat use (compared to a coexisting clonal pair) the system moves towards a single, intermediate value for ε (indicated by filled circles in the figure), a frequency of refuge use which

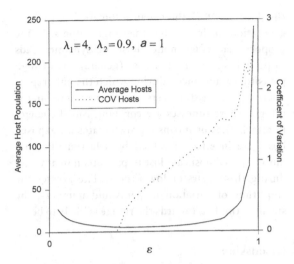

Figure 13.3. Average host abundance, and coefficient of variation in numbers, as a function of ε. —, average hosts; ·····, coefficient of variation in hosts.

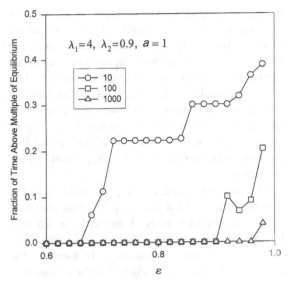

Figure 13.4. Fraction of time population trajectories reach a magnitude of $Y \times$ equilibrial values, where $Y = 10$ (circles), 100 (squares), and 1000 (triangles). The incidence of extreme population values increases with increasing ε.

defeats all lower or higher values for ε. This defines an evolutionarily stable strategy (ESS) refuge level. As the cost to being in the refuge increases (or quality there decreases), λ_2 declines, and the domain of parameter space permitting clonal coexistence decreases. Along with this shift in the domain of coexistence, the evolutionarily stable refuge level shifts towards lower levels (compare Figs 13.2(a), (b) and (c)).

The importance of initial conditions

The effect of refuge evolution upon the magnitude of realized control depends upon the starting conditions. If originally the refuge was not greatly utilized by the host, the dynamics are highly unstable. An evolutionary increase in refuge levels tends to greatly lower variability in host population abundance, measured for instance by the coefficient of variation or the frequency of excursions to a multiple of equilibrial host population size. In this case, host evolution can lead to more sustained control (namely fewer and smaller outbreaks). By contrast, if the initial population was heavily protected by the refuge, a decay in refuge level leads to increased mean abundance and an increase in variability, thus less effective control. The magnitude of this effect depends upon

the cost of being protected or alternatively by the inherent quality of the refuge; the higher the cost or lower the quality of the refuge, the more the magnitude of variability in abundance is observed to increase at the ESS (our unpublished results). Moreover, once the dynamics become unstable, selection favors continued refuge use, helping to prevent yet further decay in control.

Evolution of the attack parameter

As a second approach to analyzing parameter evolution, given population fluctuations, we empirically computed a strength of selection. In the numerical simulations we introduced clones differing by a small amount in a given parameter at very low densities into populations fluctuating around their equilibrium. The strength of selection was determined as the average change in geometric mean fitness over a specified time period, per unit change in the parameter (see the legend to Fig. 13.1 for more details on our protocol).

Figure 13.1 (filled circles) shows how population

fluctuations influence the strength of selection on the attack parameter, compared to expectations based upon equilibrial assumptions (when the symbols are coincident with the straight lines in the figure, the population dynamics are stable). The effect of instability is, in general, to weaken the strength of selection on a. Indeed, when refuge hosts have a low intrinsic growth rate (see Fig. 13.1(b)), the strength of selection on a is not markedly different from zero, regardless of the proportion of protected hosts.

This suggests that if a host–parasitoid interaction persists with bounded fluctuations in host numbers, because of the high costs of protection or use of a low-quality refuge, then this ecological mechanism fostering persistence simultaneously hampers the evolution of resistance by the host to the parasitoid.

Evolution of growth rates

Our numerical studies suggest that population instability may at times have a small effect on the strength of selection towards improved adaptation to the exposed state (e.g., Fig. 13.1, filled squares). By contrast, such instability increases the strength of selection towards improved adaptation to the refuge state (Fig. 13.1, filled triangles). This is particularly notable in Fig. 13.1(b) (where the parameter choices lead to large fluctuations in P), which shows that the strength of selection to the refuge actually *increases* with decreasing proportion of hosts protected by it – opposite the expectation derived given an assumption of equilibrial populations.

Our analysis of the proportional refuge model shows that combining spatial heterogeneity and unstable population dynamics can lead to what, at first glance, are counterintuitive effects on selection. To make the results more intuitive, it is helpful to recall some general effects of heterogeneous environments upon evolutionary dynamics. Considering spatial heterogeneity alone, one expects an automatic bias in selection towards habitats in which a species is most common, and in which they are most successful (see Holt, 1996). Considering temporal variability alone, because selection is governed by geometric mean fitness across generations (Seger &

Brockmann, 1987), there is an automatic bias towards generations of low, rather than high, fitness. In the proportional refuge model, unstable dynamics leads to periods of low host fitness (generations of high parasitoid abundance), during which the only reproductively successful hosts are those found in the refuge, where fitnesses are constant. Small changes in attack rates or intrinsic growth rates on exposed hosts are in effect devalued by selection, because most exposed hosts are lost to parasitism in any case during these years of low fitness. The greater the amplitude of variation in parasitoid numbers, the stronger this bias towards the refuge is likely to be.

Discussion

Effective biological control usually mandates that one must attempt to limit target pest species to densities at which the inherent instability of natural enemy–victim interactions is likely to be important. Often, spatial factors such as refuges may be required for such strong interactions to persist. Given that strong limitation of a target pest species by an introduced control agent may generate strong selection on the former species, potentially leading to the breakdown of control, it is of considerable interest to understand how spatial and temporal heterogeneities can modify such evolutionary dynamics. In Holt & Hochberg (1997) we examined a number of factors that influence the evolutionary stability of biological control. Here, we have demonstrated that the existence of unstable population dynamics can dramatically change the likely direction of evolution in a target host species, in a system which persists because the host can utilize a refuge. In particular, our theoretical results suggest that unstable dynamics can make it more difficult for pest species to evolve counter-adaptations to parasitism. Moreover, unstable dynamics foster the continued use of costly/low-quality refuges, which can be the key factor permitting the sustained persistence of inherently unstable host–parasitoid interactions. The general conclusion we draw is that for the purpose of sustained biological control, aiming for the tight regulation of the target pest by the control agent may in the end be self-defeating.

Holt (1997) explores why use of a suboptimal sink habitat (in this case, the refuge) can be favored by selection if populations have unstable dynamics. The basic idea is that unstable dynamics in one habitat necessarily mirrors temporal variation in fitness there. All that is needed for use of a second, low-quality habitat to be favored by selection is that in some generations, fitness there exceeds fitness in the (on average) higher-quality habitat. Utilization of low-quality habitats can be a form of 'bet-hedging' (Seger & Brockmann, 1987) fostered by variance in fitness in high-quality habitats. Weak use of the refuge leads to population fluctuations driven by the unstable host–parasitoid interaction, automatically paving the way for increased use of lower-quality refuge habitats lacking the parasitoid.

This suggests that the maintenance of biological control over evolutionary time-scales may reflect the interplay of a number of factors. Refuges are important ecologically in reducing the magnitude of population fluctuations, thus permitting the persistence of tightly-coupled host–parasitiod interactions. However, given stable population dynamics, the evolution of refuge use is in the direction of less effective limitation and regulation by the parasitoid (Hochberg & Holt, 1995). Indeed in some systems the very existence of unstable population dynamics may be needed for refuge levels to remain at an intermediate level (rather than all hosts being exposed, or protected) over evolutionary time-scales. Natural selection tends to ensure that some hosts will remain exposed to parasitoids, only if there are costs to protection in refuges (e.g., if refuges are intrinsically lower in quality than exposed habitats). Partial refuge use is evolutionarily stable, given unstable host–parasitoid dynamics. Highly effective control that leads to tight host regulation solely by parasitoids may thus be evanescent, given evolutionary responses to parasitism by hosts (Hochberg & Holt, 1995).

Our investigation considered the strength of selection, one parameter at a time. More generally, one could imagine that trait evolution involves trade-offs among parameters. Moreover, it is likely that there will be concurrent parameter evolution (i.e., coevolution) in the parasitoid, as well. It is also important to examine a much wider range of ecological models. A full examination of all these possibilities is beyond the scope of this chapter. However, the insights drawn from the proportional refuge model are likely to apply much more broadly. Evolution by natural selection tends to favor the optimization of any character that has a large strength of selection upon it, at the expense of characters with a low to zero strength of selection. Overall, even with trade-offs among parameters and parasitoid coevolution, unstable host–parasitoid dynamics should foster the employment of costly/low-quality refuges by hosts, and diminish the effectiveness of selection on the host to withstand parasitism by other mechanisms outside the refuge.

We suggest that a modicum of instability in biological control over ecological time-scales may in some circumstances facilitate the persistence of biological control over evolutionary time-scales. The conventional wisdom that one should seek natural enemies that lead to stable population dynamics in biological control may, in the end, lead to self-defeating efforts in control. This evolutionary perspective complements discussion of the utility of natural enemies that are local exterminators of target species – rather than regulators – in biological control programs (see Chapters 2 and 4). We caution that this potential long-term role of dynamic instability in facilitating long-term control has been demonstrated only for a rather simple model (the above proportional refuge model). We would like to leave the reader with two main messages. First, it is important that biological control theory add an evolutionary dimension to the existing rich, sophisticated literature on ecological mechanisms fostering persistence and stability of biological control systems. Second, one issue that warrants particular attention is the role of spatial heterogeneity and temporal variability in constraining evolutionary responses by target pest species to effective biological control agents. The long-term effectiveness of biological control may require either a finely balanced coevolutionary arms race, or a kind of conservatism in the basic ecological niche (broadly defined, Holt, 1996) of the pest. Spatial heterogeneity and unstable dynamics can foster such conservatism.

References

Beddington, J. R., Free, C. A. & Lawton, J. H. (1978) Modelling biological control: on the characteristics of successful natural enemies. *Nature* 273: 513–19.

Brown, J. S. & Vincent, T. L. (1992) Organization of predator–prey communities as an evolutionary game. *Evolution* 46: 1269–83.

Dwyer, G., Levin, S. A. & Buttel, L. (1990) A simulation model of the population dynamics and evolution of myxomatosis. *Ecological Monographs* 60: 423–47.

Hassell, M. P. (1978) *The Dynamics of Arthropod Predator–prey Systems.* Princeton: Princeton University Press.

Hochberg, M. E. (1996) An integrative paradigm for the dynamics of monophagous parasitoid–host interactions. *Oikos* 77: 556–61.

Hochberg, M. E. & Holt, R. D. (1995) Refuge evolution and the population dynamics of coupled host–parasitoid associations. *Evolutionary Ecology* 9: 633–61.

Holt, R. D. (1996) Demographic constraints in evolution: towards unifying the evolutionary theories of senescence and niche conservatism. *Evolutionary Ecology* 11: 1–10.

Holt, R. D. (1997) On the evolutionary stability of sink populations. *Evolutionary Ecology* 11: 723–32.

Holt, R. D. & Hassell, M. P. (1993) Environmental heterogeneity and the stability of host–parasitoid interactions. *Journal of Animal Ecology* 62: 89–100.

Holt, R. D. & Hochberg, M. E. (1997) When is biological control evolutionarily stable (or is it)? *Ecology.* 78: 1673–83.

Holt, R. D. & Lawton, J. H. (1993) Apparent competition and enemy-free space in insect host–parasitoid communities. *American Naturalist* 142: 623–45.

Holt, R. D. & Lawton, J. H. (1994) The ecological consequences of shared natural enemies. *Annual Review of Ecology and Systematics* 25: 495–520.

Jaenike, J. & Holt, R. D. (1991) Genetic variation for habitat preference: evidence and explanations. *American Naturalist* 137: S67–S90.

Lewontin, R. C. (1974) *The Genetic Basics of Evolutionary Change.* New York: Columbia University Press.

Seger, J. & H. J. Brockmann. (1987) What is bet-hedging? *Oxford Surveys in Evolutionary Biology* 4: 182–211.

Taylor, A. D. (1991) Studying metapopulation effects in predator–prey systems. *Biological Journal of the Linnaean Society* 42: 305–23.

14

Genetic conflict in natural enemies: a review, and consequences for the biological control of arthropods

MARTHA S. HUNTER

Introduction

We generally think of the DNA of an organism acting in a concerted manner to direct the development and behavior of that organism. However, a recent and rapidly growing body of literature suggests that the interests of different DNA lineages within organisms may sometimes be in conflict. Genetic conflict may arise between different genes or chromosomes within the nuclear genome, or between the nuclear genome and the DNA of cytoplasmic parasites such as viruses or bacteria. Although the origin of these elements spans a range of relatedness to the host, from within the same genetic population to unrelated micro-organisms, they function similarly to enhance their own fitness at the cost of other lineages; here, all will be categorized as 'selfish genetic elements'. The effects of these elements on the phenotype are varied, but they include sex ratio distortion, when the inheritance of the element is exclusively through one sex, and mating incompatibility, when the male of the pair carries a different strain of microbe than the female.

While the discovery of the existence of many of these types of elements is recent, and relatively few systems are well understood, preliminary evidence suggests that selfish genetic elements may be very widespread. For example, *Wolbachia*, bacteria that cause mating incompatibility in many insects and parthenogenesis in parasitic wasps, have been found in over 80 insect species, as well as in several isopods and a mite (Werren, 1997a). Parthenogenesis-inducing *Wolbachia* were unknown before the publication by Stouthamer *et al.* (1990a) yet they have since been found in several families of parasitic Hymenoptera, and appear to be well correlated with those families with the highest frequency of parthenogenetic, or thelytokous, reproduction. In general, the parasitic

Hymenoptera, which comprise a majority of agents used for biological control of arthropod pests, appear to be particularly susceptible to manipulation by selfish genetic elements. Not only parthenogenesis, but female-biasing and male-biasing sex ratio distortion, male-lethal infections and mating incompatibility have all been attributed to these elements. Male-lethal infections have been found with increasing frequency in the Coccinellidae as well.

In some cases, the success of a classical biological control program may rest on the early diagnosis of infection by one of these elements. Mass-rearing for an augmentative program may similarly depend on preventing the spread of a selfish genetic element in some cases, while in other instances the efficiency of the rearing program may be enhanced when the frequency of the element can be manipulated. The results of crossing tests between natural enemies may be incorrectly interpreted when mating incompatibility is caused by a microorganism. Lastly, some of these elements present potential opportunities for controlling or manipulating the sex ratio and genetic structure of populations of biological control organisms.

In the main section of this chapter, the types of selfish genetic elements found in natural enemies will be described. The elements to be discussed include cytoplasmic incompatibility *Wolbachia*, partheno-genesis-inducing *Wolbachia*, feminizing infections, male-lethal infections, and paternally inherited nuclear elements. The discussion will include their manifestation in populations and in laboratory cultures, how their fitness is enhanced at the expense of their carriers, the mechanism by which they influence the reproduction of their carriers, their effects on carrier fitness, and their transmission and spread in populations. A second section will present observations and some testable predictions about the distribution and frequency of these elements among species in nature, and also some thoughts on how the frequency of these elements may change in laboratory cultures. In the third section, the prospects for manipulating these elements will be presented. In most cases, these agents are clearly detrimental to natural enemy cultures, and should be eliminated if possible. In some instances, however, these agents may be beneficial. For example, a population of parasitic wasps infected with parthenogenesis microorganisms may, in some instances, be more efficiently mass-reared than a sexual population. Lastly the potential use of cytoplasmic incompatibility *Wolbachia* to serve as a driver to carry beneficial genes into natural enemy populations will be discussed.

Types of genetic conflict

Cytoplasmic incompatibility

Introduction to Wolbachia

Cytoplasmic incompatibility (CI) microorganisms are rickettsia in the genus *Wolbachia*. Recent evidence suggests *Wolbachia* is extremely widespread in arthropods (for a recent review of *Wolbachia* see Werren, 1997a). A survey in Panama estimated the rate of infection to be 16% of the species of insects sampled (Werren *et al.*, 1995a). Several mite species, including commercially available predatory species such as *Metaseiulus occidentalis* (Breeuwer & Jacobs, 1996; Johanowicz & Hoy, 1996) and *Phytoseiulus persimilis* (Breeuwer & Jacobs, 1996) have been found to harbor *Wolbachia* infections, and several isopods are infected with *Wolbachia* as well (Rousset *et al.*, 1992a). A *Wolbachia* relative has recently been found in a nematode (Sironi *et al.*, 1995). In the majority of species, the presence of *Wolbachia* is still known only from a molecular assay, and it is not clear what phenotypic effects these microbes cause, if any; some *Wolbachia* infections appear to be virtually asymptomatic (Giordano *et al.*, 1995; Hoffmann *et al.*, 1996). Molecular phylogenies of *Wolbachia* that have used either the 16S rDNA genes (Rousset *et al.*, 1992a;

Stouthamer *et al.*, 1993) or the bacterial cell cycle gene *ftsZ* (Werren *et al.*, 1995b) do not show any correspondence between particular lineages and the effect of these bacteria on their host. For example, *Wolbachia* that are known to cause parthenogenesis show up in both of the major subgroups of *Wolbachia* (Stouthamer *et al.*, 1993), suggesting parthenogenesis–induction has probably evolved more than once (Werren, 1997a). It appears, however, that *Wolbachia* that cause CI are by far the most common. These *Wolbachia* are found throughout the insects, while *Wolbachia* that cause parthenogenesis or feminization have not yet been found outside the Hymenoptera or Isopoda. Most of the studies cytoplasmic incompatibility have been done on a few model systems: mosquitoes and in particular *Culex pipiens* (Laven, 1967a; Yen & Barr, 1971, 1973; Subbarao, 1982), *Tribolium confusum* (Wade & Stevens, 1985, 1994; Stevens & Wade, 1990; Stevens & Wicklow, 1992; Stevens, 1993; Wade & Chang, 1995), *Drosophila* spp., and especially *D. simulans* (Hoffmann & Turelli, 1986, 1988; Hoffmann *et al.*, 1990, 1994, 1996; O'Neill & Karr, 1990; O'Neill, 1991; Turelli & Hoffmann, 1991; Turelli *et al.*, 1992; Turelli, 1994; Rousset & Solignac, 1995; Werren & Jaenike, 1995), and the pteromalid parasitoid *Nasonia vitripennis* (Ryan & Saul, 1968; Ryan *et al.*, 1985; Conner & Saul, 1986; Breeuwer & Werren, 1990, 1993a,b; Breeuwer *et al.*, 1992; Williams *et al.*, 1993; Perrot-Minnot *et al.*, 1996). In this review I will concentrate on CI in *Drosophila* and *Nasonia*. Although *Drosophila* is not likely to be a biological control agent in the near future, much of the recent research on CI has been done with *D. simulans*, and some of these studies provide the first real picture of the dynamics of a *Wolbachia* infection in the field. The research on *Nasonia* has more obvious relevance to biological control, and has also suggested how *Wolbachia* may be involved in speciation.

Infection by cytoplasmic incompatibility (CI) Wolbachia

Infection of arthropods by CI microorganisms may often go unrecognized. Most often, infection is suspected when crosses between populations are compatible in one direction, but in the reciprocal cross very few eggs hatch, or in the case of haplodiploid organisms such as parasitic wasps, haploid males are produced from fertilized eggs. Interestingly, the microbe is asymptomatic in females, but infected males will successfully fertilize eggs only of females with the same strain of microorganism. The action of this microorganism makes sense if one thinks in terms of clonal selection; males do not generally transmit the bacteria, but by preventing reproduction of uninfected females, or females with a different strain of the bacteria, the proportional representation of females infected with the same strain of bacteria increases in the population. Thus, although transmission of the bacteria is virtually exclusively through female oocytes, antibiotic treatment of the infected males restores compatibility. In CI, the conflict of interest between the genome of the bacteria and that of the male carrier is very clear. The inclusive fitness of the bacteria is increased by every mating of its male carrier, both those compatible matings that lead to the production of more infected individuals and those incompatible matings that prevent development of uninfected individuals or individuals infected with a different strain of *Wolbachia*. In contrast, the fitness of an infected male is decreased with each incompatible mating.

While in most cases, the manifestation of CI will be unidirectional incompatibility, the incompatibility effect of CI *Wolbachia* may be strain-specific, so that bidirectional incompatibility will result when the mating pair carry different strains (Yen & Barr, 1973; Breeuwer & Werren, 1990; O'Neill & Karr, 1990). It should be noted that if the populations used for a test cross are allopatric, and CI is not suspected, one might wrongly interpret the results of an incompatible cross as evidence that the test populations are different species. If CI is partly or wholly responsible for the reproductive incompatibility between populations, antibiotic treatment should allow the production of hybrid offspring.

Double infections of arthropods by CI *Wolbachia* also appear to be fairly common (Breeuwer *et al.*, 1992; Rousset & Solignac, 1995; Werren *et al.*, 1995a; Perrot-Minnot *et al.*, 1996). Males infected with two strains of *Wolbachia* are incompatible with singly infected and uninfected females while doubly infected females will produce normal numbers of offspring when mated with singly infected or uninfected males (Rousset & Solignac, 1995; Sinkins *et al.*, 1995a; Perrot-Minnot *et al.*, 1996). Double infections are thus predicted to spread in a population in an identical fashion to single infections.

It is important to remember, however, that incompatibility in a cross is not necessarily caused by *Wolbachia* (Stouthamer *et al.*, 1996), nor does the lack of incompatibility between populations mean CI *Wolbachia* are absent. One might not see any incompatibility if the microbe has spread to fixation across the entire range of a species, or if one is performing crosses within the geographic area where one strain has become fixed. Also, recent surveys of insects for *Wolbachia* have turned up a few instances where the infection is asymptomatic; no incompatibility is detected in crosses between infected and uninfected individuals, and no other effects have been associated with the microbe (Giordano *et al.*, 1995; Hoffmann *et al.*, 1996). The asymptomatic infection in Australian *D. simulans* has been proposed to persist as a neutral trait due to almost perfect maternal transmission and a lack of deleterious effects on host fitness (Hoffmann *et al.*, 1996). Theory suggests that strains that produce lower levels of incompatibility may spread in competition with CI strains if the rate of transmission is higher and there are fewer detrimental effects on the fitness of their hosts (Turelli, 1994).

Effects of CI Wolbachia *on carrier fitness*

The degree of incompatibility, that is, the degree to which normal progeny production is suppressed in incompatible crosses, appears to be determined by one or more of the following factors: the strain of *Wolbachia* (Breeuwer & Werren, 1993b; Giordano *et al.*, 1995), the genotype of the host (Boyle *et al.*, 1993), and the density of bacteria in the eggs (Binnington & Hoffmann, 1989; Boyle *et al.*, 1993; Breeuwer & Werren, 1993a; Bressac & Rousset, 1993; Sinkins *et al.*, 1995b). In an experiment in which asymptomatic *Wolbachia* from *D. mauritiana* was injected into eggs of *D. simulans*, *D. simulans* became infected but did not show incompatibility, indicating that the *Wolbachia* strain had a similar effect on two different hosts (Giordano *et al.*, 1995). An interaction between the nuclear genome and the *Wolbachia* strains also appears not to be a factor in the expression of bidirectional incompatibility (Breeuwer & Werren, 1993b). A series of backcrosses was performed causing introgression of the nuclear genotype of one species of *Nasonia* infected with *Wolbachia* into another infected species. Even though the individuals then had virtually the same genetic background, the crosses were no more compatible than when the *Wolbachia* strains were in different genetic backgrounds (Breeuwer & Werren, 1993b). The work of Boyle *et al.* (1993) suggested host factors may sometimes be important, however. When *Wolbachia*-bearing egg cytoplasm was transferred from *D. simulans* to *D. melanogaster*, lower levels of incompatibility were observed in the new host, but when the *Wolbachia* was then transferred back into an uninfected strain of *D. simulans*, the levels of incompatibility were restored. Lastly the density of *Wolbachia* appears to be important, and may vary in male and female reproductive tissue with strain, age of the insect, or time since the infection was introduced (Hoffmann *et al.*, 1990, 1996; Boyle *et al.*, 1993; Breeuwer & Werren, 1993a). While there is not always a correlation between density and the degree of incompatibility among strains (Hoffmann *et al.*, 1996), the correlation appears stronger within strains. For example, in *D. simulans* males there is a drop in the relative number of spermatocysts that carry bacteria with age (Binnington & Hoffmann, 1989; Bressac & Rousset, 1993), and this drop is correlated with a decline in the expression of incompatibility (Hoffmann *et al.*, 1990).

The fitness cost to females of carrying the *Wolbachia* infection is variable. In *D. simulans* where this question has been addressed most exhaustively, infected flies were found to have a reduction in fecundity of approximately 22% in the laboratory, but estimates from field-collected flies give a lower estimate of the cost, approximately 10% (Hoffmann *et al.*, 1990). No evidence was found for lower viability of immatures throughout development or for dif-

ferential mating success of adults (Hoffmann *et al.*, 1990). Similarly, no evidence was found for differences in sperm competition associated with infection (Hoffmann *et al.*, 1990).

Mechanism

The mechanism by which the bacteria in the male prevents reproduction of his uninfected mate is still unknown, although the evidence suggests the male chromosomes are modified in such a way that only eggs carrying the same strain of microbe as the male allow karyogamy, the normal fusion of the egg and sperm pronucleus, to proceed (Werren, 1997a). A current model suggests that the *Wolbachia* in the male serves as a sink, binding a host protein or proteins. Sperm from infected males may then be rescued when there is *Wolbachia* in the female dragging the host protein from the ovaries into the egg to rescue fertilization (T. Karr, pers. comm.). This model remains tentative, and no hypotheses have yet been advanced to explain how the action of *Wolbachia* can be specific to different strains.

The mechanism is likely to be the same in the haplodiploid Hymenoptera, but the result is different; following destruction of the male set of chromosomes, male offspring are produced. Cytogenetic studies of eggs fertilized by infected males in the hymenopteran parasitoid *Nasonia vitripennis* (Walker) have shown the male chromosomes are improperly condensed and may appear shredded during the first cleavage division (Ryan & Saul, 1968; Ryan *et al.*, 1985; Breeuwer & Werren, 1990). The paternal chromosomes are generally completely lost, but fragments may be pulled into association with the dividing daughter nuclei (Reed & Werren, 1995), and occasionally fragments with centromeres may be incorporated into the genome (Ryan *et al.*, 1985, 1987). Aneuploidy (when the number of chromosomes is not a multiple of the haploid number) may cause death of the zygote, so both a male-biased sex ratio and a depression of offspring production may occur in an incompatible cross (Breeuwer & Werren, 1993a).

In the haplodiploid spider mites, CI generally results in the death of fertilized eggs (Breeuwer,

1997) most likely because of aneuploidy. Spider mites have holokinetic chromosomes; the spindle attaches not at a centromere as in most genetic systems but anywhere along their length. This is likely to increase the probability that fragments of paternal chromosomes end up in the zygote, causing zygote death (Breeuwer, 1997).

Transmission and spread in populations

The evidence from laboratory experiments and field studies indicates that *Wolbachia* is almost entirely transmitted vertically through the egg cytoplasm. A laboratory study of *D. simulans* found a low (approximately 1%) rate of paternal transmission (Hoffmann & Turelli, 1988), but field studies of the same insect found that the mitochondrial restriction pattern and infection status were highly correlated, suggesting paternal transmission does not occur with detectable frequency in the field (Turelli *et al.*, 1992).

In spite of the lack of experimental evidence for horizontal transmission within species, however, molecular phylogenies of *Wolbachia* show almost no evidence for coevolution of the microbes and their hosts, making infrequent horizontal transmission the most likely explanation for the origin of *Wolbachia* in most species (O'Neill *et al.*, 1992; Rousset *et al.*, 1992b; Stouthamer *et al.*, 1993; Moran & Baumann, 1994; Werren, *et al.* 1995b). The means of horizontal transfer is still very obscure; recent studies of the incidence of *Wolbachia* in particular food webs have not turned up any obvious patterns (Werren & Jaenike, 1995; Y. Gottlieb & E. Zchori-Fein, unpublished results). Some spider mites carry *Wolbachia* and are possible sources of the *Wolbachia* found in predatory mites (Breeuwer & Jacobs, 1996). The possibility that parasitoids may serve as vectors for *Wolbachia* is supported by a close phylogenetic relationship between strains of *Wolbachia* found in the pteromalid parasitoid *Nasonia giraulti* and in its blowfly host (Werren *et al.*, 1995b). No other obvious associations emerge from the phylogeny.

The factors influencing the change of frequency of CI *Wolbachia* within populations appears complex. They include the degree of incompatibility, the fitness costs to infected individuals, and the maternal transmission rate (Werren, 1997a). In uninfected populations where there are no deleterious fitness effects on infected females and where maternal transmission is 100%, theory predicts that CI *Wolbachia* should increase in frequency until fixation (Caspari & Watson, 1959). A reduction in the fecundity of infected females is predicted to cause an unstable equilibrium equal to the reduction in fecundity of infected females divided by the reduction in hatch (or reduction in female production for haplodiploids) of incompatible crosses. Below this equilibrium the *Wolbachia* should decrease in frequency, and above it the *Wolbachia* should spread to fixation (Caspari & Watson, 1959; Hoffmann *et al.*, 1990).

The maternal transmission rate of CI *Wolbachia* appears generally high, but probably is rarely 100%, and may vary according to the density of *Wolbachia* in the host, and the degree to which the host has been exposed to environmental agents which 'cure' the host. Environmental curing may take the form of exposure to high temperatures (Hoffmann *et al.*, 1990) or to naturally occurring antibiotics, as can be experienced by *Tribolium* in flour contaminated by antibiotic-producing bacteria or fungi (Stevens & Wicklow, 1992). Perrot-Minnot *et al.* (1996) found that a prolonged diapause in *Nasonia* larvae in the laboratory could also cause the loss of a *Wolbachia* infection. Imperfect maternal transmission, possibly as a result of environmental 'curing' has been considered the most likely reason that in mostly infected populations of *D. simulans* in California, about 6% of the individuals remain uninfected (Hoffmann *et al.*, 1990).

One surprising factor that may influence CI *Wolbachia* spread is sperm competition. Wade & Chang (1995) found that sperm from infected *Tribolium* males may out-compete sperm from uninfected males, although similar experiments in *Drosophila* failed to find evidence for differential sperm competition (Hoffmann *et al.*, 1990).

The role of CI Wolbachia in speciation

It seems likely that CI *Wolbachia* play an important role in causing and/or maintaining reproductive isolation of some populations, and may thus be involved in speciation. Theory suggests that a single strain of *Wolbachia* causing unidirectional incompatibility is not likely to lead to speciation because it is predicted to spread rapidly throughout a population (Caspari & Watson, 1959). However, bidirectional incompatibility may prevent gene flow between carriers of two *Wolbachia* incompatibility types. Recent studies suggest that *Wolbachia* may have been involved in speciation of *Nasonia*, a genus of pteromalid parasitoids (Breeuwer & Werren, 1990).

Three sibling species of *Nasonia* occur in North America. *Nasonia vitripennis* is used commercially for biological control of muscoid flies in stables and poultry houses (Rueda & Axtell, 1985), and may be found in nature parasitizing flies in carrion and birds nests (Darling & Werren, 1990). *Nasonia vitripennis* is cosmopolitan, and is a generalist with respect to habitat while the other two species, *N. giraulti* and *N. longicornis* are geographically separated, restricted to birds nests, and sympatric with *N. vitripennis* throughout much of their ranges (Darling & Werren, 1990; Werren, 1997a). All three species are reproductively isolated, but the isolation is largely due to CI *Wolbachia* in each species causing bidirectional incompatibility when species are crossed (Breeuwer & Werren, 1990; Werren, 1997a). When treated with antibiotics, fertile hybrid offspring are produced, although relatively high mortality of the F2 generation indicates that other post-zygotic isolating mechanisms are also involved in maintaining isolation (Breeuwer & Werren, 1990, 1995). What cannot yet be determined is whether CI *Wolbachia* was important in causing reproductive isolation between populations of *Nasonia* within a species, or whether it was more simply involved with maintaining isolation of new species (Werren, 1997a). The problem with the former idea is that theory predicts it will be difficult for a new incompatibility type of *Wolbachia* to invade a population where one is established (Turelli, 1994), thus the presumed starting condition of an interbreeding population with stable frequencies of two incompatibility types is expected to be rare. However, if two incompatibility types arise in allopatric populations, *Wolbachia* may well be influential in maintaining reproductive isolation when

sympatry is re-established (Werren, 1997b). Furthermore, CI *Wolbachia* need not be the only mechanism promoting isolation to be important in speciation. Combinations of factors, including unidirectional incompatibility in one direction with hybrid sterility in the other, could result in bidirectional isolation and speciation (Werren, 1997a,b).

The recent survey indicating that *Wolbachia* can be found in approximately 16% of Panamanian insects (Werren *et al.*, 1995a) attests to the real potential for CI *Wolbachia* to be involved in reproductive isolation, and the importance of considering this possibility when conducting crossing studies. The new evidence that *Wolbachia* is present in several predatory mite species is a case in point (Breeuwer & Jacobs, 1996; Johanowicz & Hoy, 1996). At least some of the unidirectional incompatibility observed among allopatric populations of *Metaseiulus occidentalis* (Hoy & Cave, 1988) is likely to be a result of infection by CI *Wolbachia*. As mentioned above, reciprocal incompatibility may also be due to CI *Wolbachia* when both populations are infected with a different strain. Bidirectional incompatibility has been found in *D. simulans* (O'Neill & Karr, 1990), two mosquito species (Werren, 1997a), and bidirectional incompatibility associated with CI *Wolbachia* has recently been found among several populations of the diapriid parasitoid *Trichopria drosophilae* (J. J. M. van Alphen & J. H. Werren, unpublished results).

Parthenogenesis-inducing (PI) *Wolbachia*

Introduction to parthenogenesis in parasitoids

Thelytokous reproduction or parthenogenesis (see Definitions, below) is relatively common in the Hymenoptera; 270 parthenogenetic species have been documented (Luck *et al.*, 1993), and rough proportional estimates vary by family from 2% in the Braconidae to 35% and 40% in the chalcidoid families Aphelinidae and Signiphoridae, respectively (DeBach, 1969; Luck *et al.*, 1993). It appears that parthenogenesis is more common in chalcidoids than in ichneumonoids (Luck *et al.*, 1993). Many species reproduce exclusively by parthenogenesis, but other species are comprised of both parthenogenetic and sexual populations that may be allopatric or sympatric (Flanders, 1945, 1950; Rössler & DeBach, 1972; Hung *et al.*, 1988; Aeschlimann, 1990; Stouthamer *et al.*, 1990a,b; Zchori-Fein *et al.*, 1994). Morphologically identical parasitoid forms with different reproductive modes have long posed a number of questions for biological control workers (DeBach, 1969; Rössler & DeBach, 1972). Are sexual and parthenogenetic forms genetically distinct? Should they be considered the same or different species? Lastly, are parthenogenetic forms less genetically variable and if so, does that influence their ability to adapt to new habitats (Rössler & DeBach, 1973; Hung *et al.*, 1988)?

Definitions

The term 'parthenogenesis' may refer to the mode of reproduction of all haplodiploid organisms like the Hymenoptera, because females are able to produce haploid male progeny without mating (Whiting, 1945). 'Arrhenotokous reproduction' has been used to refer to sexual reproduction of haplodiploids, that is when females produce diploid female progeny from fertilized eggs and males from unfertilized eggs. 'Thelytokous reproduction' refers to the production of daughters from unfertilized eggs, by diploid parthenogenesis (White, 1954). Here I will use 'parthenogenesis' as equivalent to 'thelytoky' because its meaning is more generally understood. 'Deuterotoky' refers to parthenogenetic reproduction of both sexes (Whiting, 1945). With the hindsight that our current knowledge of PI *Wolbachia* affords, it seems that many of the observations of 'deuterotoky' may simply refer to populations with imperfect transmission of the microbe. In some species of parasitoids, however, this term may have more biological significance. In the aphelinid parasitoid of white peach scale, *Ptertoptrix orientalis*, both unmated and mated females regularly produce males and females from differentiated ovarioles (Garonna & Viggiani, 1988; Viggiani & Garonna, 1993). It is as yet unknown whether microorganisms play a role in the evolution of this extremely unusual life history.

Infection by PI Wolbachia

Microorganisms that cause thelytokous parthenogenesis in parasitic wasps were discovered relatively recently. Stouthamer *et al.* (1990a) studied the inheritance of parthenogenesis in mixed populations of sexual and parthenogenetic *Trichogramma*. They crossed sexual females to males derived from antibiotic treatment of parthenogenetic females, and then in succeeding generations backcrossed the progeny to males from the parthenogenetic lines until the nuclear genome was almost entirely from the parthenogenetic line. Virgin females of this hybrid line produced only male offspring, however, suggesting the factor causing parthenogenesis was inherited through the cytoplasm. Antibiotic or heat treatment of parthenogenetic females caused them to behave as unmated sexual females, producing only male offspring. Subsequent studies of the oocytes showed microorganisms in the oocytes of parthenogenetic females but not in sexual females or antibiotic-cured parthenogens (Stouthamer & Werren, 1993), and sequencing of the 16S rRNA region showed these bacteria to be *Wolbachia* (Rousset *et al.*, 1992a; Stouthamer *et al.*, 1993). Phylogenies based on 16S rRNA sequences and sequences of *ftsZ*, a cell cycle gene, furthermore indicate that PI *Wolbachia* have arisen independently several times within lineages that cause cytoplasmic incompatibility (Stouthamer *et al.*, 1993; Werren *et al.*, 1995b).

The fitness advantage to a cytoplasmically inherited bacteria causing parthenogenesis is straightforward. A cytoplasmic element will increase in frequency in a population if infected females produce more infected daughters than uninfected females produce healthy daughters (Bull, 1983; Werren, 1987). By converting unfertilized, incipient male eggs to female eggs, PI *Wolbachia* make carriers out of non-carriers. In a population that would ordinarily have equal numbers of males and females, infected females may produce as much as twice as many daughters as uninfected females. Furthermore, PI *Wolbachia* may easily spread to fixation because males are not required to produce female offspring.

Wolbachia that cause parthenogenesis have been found only in the Hymenoptera to date, and most of the reports have been in the Chalcidoidea (Table 14.1).

Table 14.1. *Research on selfish genetic elements in natural enemies of arthropods*

Parasitoid or predator	Effect	Probable agent[†]	Research[‡]	Selected reference(s)
Hymenoptera:				
Aphelinidae				
Encarsia formosa	P	1	B,C,D	Zchori-Fein *et al.* (1992), Werren *et al.* (1995b)
E. hispida	P	1	A,C	M. S. Hunter, unpublished results
Aphytis lingnanensis	P	1	A,B,C,D	Zchori-Fein *et al.* (1994, 1995)
A. diaspidis	P	1	A,B,C,	Zchori-Fein *et al.* (1995)
Aphytis yanonensis	P	1	B,D	Werren *et al.* (1995b), R. Roush, unpublished results
Eretmocerus staufferi	P	1	C	M. S. Hunter, unpublished results
Encarsia pergandiella	MBE	?	A, E, G, J	Hunter *et al.* (1993)
Braconidae				
Asobara tabida	CI	1	B	Werren *et al.* (1995b)
Diapriidae				
Trichopria drosophilae	CI	1	A,B?	J. J. M. van Alphen and J. H. Werren, unpublished results cited in Werren (1997)
Encyrtidae				
Apoanagyrus diversicornis	P	1	A,B,C,I	Pijls *et al.* (1996), J. H. Werren, unpublished results
Trechnites psyllae	P	1	C	T. Unruh, unpublished results
Eucoilidae				
Leptopilina australis	P	1	B	Werren *et al.* (1995b)
Leptopilina clavipes	P	1		Werren *et al.* (1995b)
Mymaridae				
Caraphractus cinctus	MLI	?	A	Jackson (1958)
Pteromalidae				
Nasonia vitripennis	CI	1	A,B,C,D,E,	Saul (1961), Ryan & Saul (1968), Breeuwer & Werren (1990), Breeuwer *et al.* (1992)
	FI	?	A,J	Skinner (1982)
	MLI	2	A,C,D,J	Skinner (1985), Huger *et al.* (1985), *Werren et al.* (1986), Gherna *et al.* (1991), Balas *et al.* (1996)
	MBE	3	A,B,D,E,F,G,H,J	Werren *et al.* (1981), Werren & van den Assem (1986), Skinner (1987), Nur *et al.* (1988), Beukeboom & Werren (1993a,b), Beukeboom (1994)
N. giraulti	CI	1	A,B,C	Breeuwer & Werren (1990), Breeuwer *et al.* (1992), Werren (1997a)
N. longicornis	CI	1	A,B,C	Breeuwer *et al.* (1992), Werren (1997a,b)
Muscidifurax uniraptor	P	1	B,E	Legner (1985), Stouthamer *et al.* (1993), Werren (1995b)
Spalangia fuscipes	P	1	B	Werren (1995b)
Trichogrammatidae				
Trichogramma deion	P	1	A,B,C,D,E,H	Stouthamer *et al.* (1990a,b), Stouthamer & Werren (1993), Stouthamer *et al.* (1993), Stouthamer & Kazmer (1994)
T. cordubensis	P	1	B,D	Stouthamer & Werren (1993), Stouthamer *et al.* (1993)
T. pretiosum	P	1	A,B,C,D	Stouthamer et al. (1990a,b), Stouthamer & Werren (1993), Stouthamer *et al.* (1993)

Table 14.1. (*cont.*)

Parasitoid or predator	Effect	Probable agent[†]	Research[‡]	Selected reference(s)
T. chilonis	P	1	A,C	Stouthamer et al. (1990a,b),
T. platneri	P	1	A,C	Stouthamer et al. (1990a,b), Stouthamer & Werren (1993)
T. kaykai (= *T.* nr. *deion*)	P	1	A,H,J	Stouthamer & Kazmer (1994)
T. brevicapillum	P	1	C,D	Stouthamer & Werren (1993)
T. oleae	P	1	C,D,	Stouthamer & Werren (1993)
T. rhenana	P	1	C,D	Stouthamer & Werren (1993)
T. sibericum	P	1	B,C	Schilthuizen & Stouthamer (1997)
T. nubilale	P	1	B,C	Schilthuizen & Stouthamer (1997)
Coleoptera:				
Coccinellidae				
Adalia bipunctata	MLI	4	A,B,C,J	Hurst *et al.* (1992, 1993), Werren *et al.* (1994)
Coleomegilla maculata	MLI	?	A,C	Hurst *et al.* (1996)
Harmonia axyridis	MLI	?	A	Matsuka *et al.* (1975)
Hippodamia convergens	MLI	?	A	Shull (1948)
Menochilus sexmaculatus	MLI	?	A,C	Gotoh & Niijima (1986)
Acarina:				
Phytoseiidae				
Galendromus occidentalis	CI?	1	A,B	Hoy & Cave (1988), Johanowicz & Hoy (1996), Breeuwer & Jacobs (1996)
Phytoseiulus persimilis	?	1	B	Breeuwer & Jacobs (1996)
Neoseiulus barkeri	?	1	B	Breeuwer & Jacobs (1996)
N. bibens	?	1	B	Breeuwer & Jacobs (1996)

Notes:
Abbreviations: P, parthenogenesis; MBE, male-biasing element; CI, cytoplasmic incompatibility; FI, feminizing infection; MLI, male lethal infection.
[†] 1, Bacteria, *Wolbachia* sp.; 2, Bacteria, *Arsenophonus nasoniae*; 3, supernumerary chromosome; 4, bacteria, *Rickettsia* sp.; ?, unknown.
[‡] A, controlled matings and brood and/or sex ratio analysis; B, PCR probe with primers specific to the agent and sequencing; C, antibiotic treatment; D, microscopic observation of agent; E, cytogenetics; F, selection experiments; G, radiation analysis; H, electrophoretic and/or visible marker studies; I, heat treatment; J, field studies.

In addition to the *Trichogramma* spp. studied by Stouthamer *et al.* (1990a), and others (Chen *et al.*, 1992), parthenogenetic wasps in the following families have been 'cured' with antibiotic treatment; Encyrtidae (Pijls *et al.*, 1996; T. R Unruh, unpublished results), Aphelinidae (Zchori-Fein *et al.*, 1992, 1995; Werren *et al.*, 1995b; my unpublished results), and Pteromalidae (Stouthamer *et al.*, 1993). The only other superfamily in which PI *Wolbachia* have been found is the Cynipoidea, in two eucoilid parasitoids of *Drosophila* in the genus *Leptopilina* (Werren *et al.*, 1995b). Not all types of antibiotics are effective in curing parthenogens infected with *Wolbachia* (Stouthamer *et al.*, 1990a; Kajita, 1993; Pijls *et al.*, 1996), and not all wasps are cured in the first generation. In some species, treated mothers produce all female offspring, but F1 daughters will produce mostly male offspring without further treatment with antibiotics (Zchori-Fein *et al.*, 1995; Pijls *et al.*, 1996).

It is not yet clear what proportion of all

parthenogenetic species exhibit this mode of reproduction as a result of a *Wolbachia* infection. While still only a few parthenogenetic species are known that have been cured with antibiotics, an earlier literature describes parthenogenetic populations that may be cured with heat treatment of the adult females (Flanders, 1945, 1965; Wilson & Woolcock, 1960; Bowen & Stern, 1966; Eskafi & Legner, 1974; Gordh & Lacey, 1976; Laraichi, 1978; Legner, 1985). These species are also likely to be infected with PI *Wolbachia* (Stouthamer *et al.*, 1990a). However, not all parthenogenetic parasitoids are infected with *Wolbachia* (Stouthamer *et al.*, 1990a). Parthenogenesis may have a simple genetic basis or be a result of gene interactions following hybridization (White, 1964; Nagarkatti, 1970) and we expect that PI *Wolbachia* will be restricted to those haplodiploid organisms that do not have single locus sex determination (Werren, 1997a). Unfortunately the instances where parthenogenesis appears not to be caused by *Wolbachia* have rarely been reported in the literature. Reports include some trichogrammatids (Stouthamer *et al.*, 1990a), a mymarid (J. T. Cronin & D. R. Strong, unpublished results cited in Cronin & Strong, 1996), and *Venturia canescens*, an ichneumonid (R. Butcher, S. Hubbard & W. Whitfield, unpublished results).

Effects of PI Wolbachia *on carrier fitness*

We as yet know relatively little about the fitness costs to infection with PI *Wolbachia*. In *Trichogramma deion* and *T. pretiosum*, two species with sympatric sexual and parthenogenetic forms, Stouthamer & Luck (1993) demonstrated that the fecundity of infected females was substantially lower than that of sexual females derived from cured parthenogens. In fact when hosts were not limited, sexual females produced the same number or more daughters than the parthenogenetic forms. It appears that host-limitation may be necessary to maintain parthenogens in the population (Stouthamer & Luck, 1993). In contrast, experiments on the aphelinid whitefly parasitoid *Encarsia formosa* showed no significant increase in fecundity following treatment with antibiotics (Stouthamer *et al.*, 1994). One important

difference between these systems is that *E. formosa* is fixed for PI *Wolbachia*. One might expect that a lower fitness cost to infection would more readily lead to fixation in a population, and furthermore that negative effects may be attenuated as a result of the evolution of the bacteria over a long association with the host (Stouthamer *et al.*, 1994).

The effects of a PI *Wolbachia* infection on parasitoid behavior have not been well explored, yet when *Wolbachia* infection has become fixed in a population, one might expect evolutionary changes of the wasp behavior. Indirect evidence that this has occurred can be found in parthenogenetic populations of whitefly parasitoids in the genus *Encarsia*. In almost all sexual species in this genus, females are produced as primary parasitoids of armored scale or whiteflies, and males are produced as hyperparasitoids, developing on females of their own species or on other primary parasitoids. The host relationships in this group are obligate; virgin females typically refrain from ovipositing in homopteran hosts, and males do not ordinarily complete development in homopterans (Gerling *et al.*, 1987; Hunter, 1991). In parthenogenetic *Encarsia*, females are produced in homopteran hosts. This is interesting because in populations fixed for PI *Wolbachia*, females are behaving differently than their sexual relatives by laying unfertilized eggs in homopterans. In two populations of parthenogenetic *Encarsia* that could be cured with antibiotics, *E. formosa* and *E. hispida*, the oviposition behavior of antibiotic-treated females was compared with antibiotic-treated virgin sexual females of *Encarsia pergandiella* in no-choice exposures to either greenhouse whitefly nymphs or *E. formosa* pupae (my unpublished results). *Encarsia pergandiella* oviposited almost exclusively in *E. formosa* pupae, while *E. formosa* oviposited almost exclusively in whitefly nymphs. In each case, males emerged only from the host where most of the oviposition had occurred. Intermediate behavior was seen in *E. hispida*; which oviposited at similar rates in both whitefly nymphs and *E formosa* pupae, but produced few males from *E. formosa* pupae (my unpublished results). This experiment suggests that infection with *Wolbachia* may lead to evolutionary changes both in the behavior of females and in the host-specific developmental requirements of larval *Encarsia*.

Mechanism

Early development of haploid eggs from females infected with PI *Wolbachia* has been observed in *Trichogramma* (Stouthamer & Kazmer, 1994). Meioisis proceeds normally, and the egg pronucleus starts to divide mitotically, but anaphase is incomplete, and both sets of chromosomes remain in the same nucleus. The consequence of this mechanism of diploid restoration is that the resulting female is homozygous at all loci. In a study of three species of *Trichogramma* with sympatric populations of parthenogenetic and sexual forms, mating of a parthenogenetic female with a sexual male results in normal fertilization and development of some of her eggs (those she allocates to daughters) and doubling of the chromosome complement of the unfertilized eggs (those she allocates to sons). These females thus produce all infected female progeny, but the fertilized eggs give rise to heterozygous daughters and the unfertilized eggs give rise to homozygous daughters (Stouthamer & Kazmer, 1994). Little is known about the biochemical mechanism by which PI *Wolbachia* manipulate the reproduction of their carriers.

Transmission and spread in populations

Like CI *Wolbachia* the current evidence indicates that within populations PI *Wolbachia* are transmitted only vertically, although molecular phylogenies suggest that horizontal transmission has occurred among species (Stouthamer *et al.*, 1993; Werren *et al.*, 1995b; Schilthuizen & Stouthamer, 1997). The phylogenies further suggest that parthenogenesis induction has arisen several times independently within CI lineages of *Wolbachia*. However, it is possible, if unlikely, that no functional change in the *Wolbachia* has occurred to cause parthenogenesis, rather the host environment determines the symptoms of an infection (Werren *et al.*, 1995b). Experimental transfer of one type of *Wolbachia* to a cured carrier of another type could address this question.

The rate of spread (or decline) of PI *Wolbachia* in a population of primary parasitoids should depend on the rate of maternal transmission of the microbe to daughters, the proportion of daughters produced by infected females, the fitness costs to infection (if any) and the sex ratio of uninfected females (Werren, 1987). Both the maternal transmission rate and the proportion of daughters produced by infected females appear to be very close to 1.0 in the systems studied to date. The fitness costs to infection appear to vary between species as discussed above. The typical sex ratio of uninfected females is important because the difference between the number of daughters produced by uninfected females and the number of infected daughters produced by infected females will determine the rate of spread of *Wolbachia* in the population.

Natural populations of parasitoids infected with PI *Wolbachia* range from those that are fixed for the microbe to populations with a very low frequency of infected individuals. The latter category has been best documented by Stouthamer & Kazmer (1994) in *Trichogramma* spp. which may have infection frequencies of less than 5%. In *T. kaykai* (= *T.* nr. *deion*), field studies suggest that the frequency of parthenogenetic females ranges from 6 to 26% (Stouthamer & Kazmer, 1994). Furthermore, many of these females (a minimum of 32% at one site and 81% at another site) are heterozygous, indicating a high rate of mating with sexual males and gene flow from the sexual to the parthenogenetic forms in the field (Stouthamer & Kazmer, 1994).

Indirect evidence of polymorphism for reproductive mode is provided by the fairly common observation of laboratory cultures 'turning' parthenogenetic (Flanders, 1945), or of parasitoid species used in biological control that are sexual in their country of origin but parthenogenetic where they have been introduced (Hung *et al.*, 1988). The majority of such cases can parsimoniously be explained by competitive exclusion of the sexual form by the parthenogenetic form (DeBach, 1969; Stouthamer & Luck, 1991). These observations are not restricted to parthenogenesis caused by *Wolbachia*, however, nor do most of them indicate whether gene flow between the sexual and parthenogenetic forms occurred.

Gene flow between rare males (or males produced by heat or antibiotic treatment) of parthenogenetic forms and uninfected sexual females has been documented in a number of instances (Orphanides & Gonzales, 1970; Legner, 1987; Stouthamer *et al.*, 1990a), as the presence of females in this cross is an indication of successful egg fertilization. Only a couple of studies have been able to demonstrate gene flow in the reciprocal cross, because regardless of whether eggs laid by parthenogenetic females are fertilized, they will all be females. Rössler & DeBach (1972) used a visible eye-color mutant to detect gene flow in *Aphytis mytilaspidis* while Stouthamer & Kazmer (1994) used both a visible marker and allozymes to detect gene flow in *Trichogramma*. The nature of gene flow in mixed populations of wasps infected and uninfected with PI *Wolbachia* is important for understanding the evolution of both sub-populations. Gene flow from the uninfected to the infected forms, for example, is important for maintaining sexual competence in infected forms (Stouthamer *et al.*, 1990b). Gene flow from the infected to the uninfected forms is dependent on incomplete maternal transmission of the microbe. Gene flow in this direction will influence the degree to which the uninfected subpopulation is able to evolve in response to the infection, for example by biasing their sex ratio towards males (Werren, 1987).

Theory predicts that polymorphism in infection status with PI *Wolbachia* may be maintained by imperfect maternal transmission, by resistance genes, or by a heterogeneous environment. Stouthamer & Luck (1993) found that when given unlimited hosts, sexual females may actually produce more daughters than infected females, while the reverse is true when hosts are limited. This led to the hypothesis that in an environment that is heterogeneous with respect to host density, the polymorphism may be maintained by varying selection for either infected or uninfected females (Stouthamer & Luck, 1993; Werren, 1997a).

Sexual function in cured parthenogens

In populations of wasps fixed for PI *Wolbachia*, it may be difficult to restore sexual reproduction using males produced by antibiotic treatment of infected females. The loss of sexual function has been most closely studied in the Aphelinidae. In *Encarsia formosa*, males court and copulate with females, but sperm is not transferred (Zchori-Fein *et al.*, 1992) The seminal vesicles of males in this species appear empty of mature sperm (my unpublished results) In *Encarsia hispida* sperm production appears very low and variable relative to sexual congeners (my unpublished results). In contrast, in *Aphytis lingnanensis* males produced mature sperm that was transferred to the female spermatheca about half the time, but mated, 'cured' females did not produce female offspring, suggesting the sperm was not used or was not able to fertilize the eggs (Zchori-Fein *et al.*, 1995) Similarly, observations of *Eretmocerus staufferi* showed that males produced mature sperm, and sperm was transferred at low frequency, but antibiotic-treated females confined with males did not produce any more female offspring than unmated females (my unpublished results).

Another barrier to sexual reproduction was found in a study of allopatric sexual and parthenogenetic populations of the encyrtid mealybug parasitoid *Apoanagyrus diversicornis*. Pijls *et al.* (1996) found that while males of the sexual population and males produced from antibiotic treatment of the parthenogenetic population were equally successful at inseminating females of the sexual population, parthenogenetic females were more likely to resist mating attempts by either types of males. Hybrids of the two forms produced viable F2 offspring, suggesting the absence of any post-zygotic barriers to interbreeding of these populations (Pijls *et al.*, 1996).

Several of the studies of parthenogenesis cured by heat treatment also document barriers to reproduction. These include failure to copulate (Wilson & Woolcock, 1960; Gordh & Lacey, 1976) and other pre-zygotic barriers (Bowen & Stern, 1966). In contrast, one study shows successful sexual reproduction of heat-cured *Trichogramma cordubensis* after at least 41 generations of parthenogenetic reproduction (Cabello-García & Vargas-Piqueras, 1985).

Most of the examples above suggest that decay of both physiological and behavioral aspects of sexual reproduction may occur by random mutation when they are not maintained by sexual reproduction (Carson *et al.*, 1982; Stouthamer *et al.*, 1990b).

Direct interference of the microbes in sexual reproduction cannot be ruled out; one might expect then to see barriers to sex to be correlated with infection status. Instead, in mixed populations, gene flow from the sexual forms to the parthenogenetic forms is associated with the maintenance of sexual reproduction in cured parthenogens (Stouthamer & Kazmer, 1994), and isolated populations with PI *Wolbachia* may still reproduce sexually when cured (Cabello-García & Vargas-Piqueras, 1985; Pijls *et al.*, 1996). We have yet to understand the rate of decay of sexuality in populations fixed for PI *Wolbachia*, and whether there are homologous genes or gene complexes across parasitoid species most likely to be affected by the cessation of sex.

Femininizing infections

Femininizing infections are caused by cytoplasmically inherited agents that induce eggs that would normally develop as males to develop as females. The fitness advantage to feminizing elements is very similar to that accruing to PI *Wolbachia*; by increasing the proportion of eggs that develop as females, the element increases its transmission rate. Feminizing infections differ from parthenogenesis-inducing infections in that populations with feminizing infections require males for egg fertilization. Fixation of a feminizing infection thus causes population extinction.

Feminizing infections are best known in crustaceans, in amphipods and isopods in particular (for a thorough review, see Hurst, 1993). In the marine shrimp *Gammarus*, females infected with one of a couple of species of microsporidian produce all-female broods (Dunn *et al.*, 1993). In an extremely well-studied isopod system, *Armidillarium vulgare*, sex is determined genetically, but genetic males become females when they are infected by *Wolbachia* (Rigaud & Juchault, 1992, 1993; Rousset *et al.*, 1992a).

In insects, a feminizing factor called 'maternal sex ratio' or MSR has been found in the parasitoid *Nasonia vitripennis*. MSR causes females to fertilize approximately 95% of their eggs (Skinner, 1982, 1987; Werren, 1987). This element exists in field populations in Utah at a rate of approximately 17% (Skinner, 1983). The feminizing factor is maternally inherited, with apparently perfect transmission and so is likely to be a cytoplasmic element such as a bacteria, but the identity of the factor, as well as its mechanism, fidelity of transmission, and spread in populations, is not yet known (Skinner, 1982; Werren, 1987).

Male-lethal infections (MLIs)

Male-lethal infections (MLIs) are cytoplasmic agents that are generally asymptomatic in females but cause death in males, usually very early in development. Infected females produce highly female-biased or exclusively female progeny, but the number of progeny produced is decreased by the proportion of offspring that would ordinarily be male. The frequencies of MLIs are characteristically low in populations (Ebbert, 1993). In a mass culture of an insect that typically has equal numbers of males and females, the only clue to the presence of a male-lethal infection may be a slightly female-biased sex ratio, while in a species with variable sex ratios, the infection may be cryptic, and revealed only when the progeny sex ratios of large numbers of individual females are examined.

MLIs are spread widely in insects (for reviews, see Ebbert, 1993; Hurst, 1993; and Hurst & Majerus, 1993); they have been found in parasitic Hymenoptera (Jackson, 1958; Skinner, 1985), Coleoptera and in particular the Coccinellidae (Shull, 1948; Matsuka *et al.*, 1975; Hurst *et al.*, 1992, 1996), several Lepidoptera (Clarke & Sheppard, 1975), Heteroptera (Leslie, 1984; Groeters, 1996) and in several species of *Drosophila* (Williamson & Poulson, 1979; Ebbert, 1991). The cytoplasmic agents that cause male death are similarly diverse and include bacteria (*Rickettsia* and *Arsenophonus nasoniae*: Gherna *et al.*, 1991; Werren *et al.*, 1994; Hurst *et al.*, 1996) and spiroplasmas (Williamson & Poulson, 1979). Within the Coccinellidae alone, there are two distinct agents that cause male death (Hurst *et al.*, 1996).

While it is clear that a cytoplasmically-inherited microorganism would be indifferent to the fate of males, it is less obvious how it may be benefited by male death. Several hypotheses have been put forward to explain the adaptive evolution of the male-killing trait, and different explanations may apply to different systems. In some cases, the clonal relatives of the microbe may benefit when resources are released to infected females as a result of male death (Skinner, 1985). Resource release may be direct, when females actually cannibalize their brothers, or indirect, when the absence of males leads to reduced competition on a food source (Hurst & Majerus, 1993). Perhaps the most persuasive examples of fitness benefit to infected females are in systems such as the aphidophagous Coccinellidae where egg cannibalism is common, and first instar females have been shown to have greater survivorship as a result of an egg meal (Hurst & Majerus, 1993). Theory suggests that infected females may also benefit from the death of male siblings when this results in an avoidance of inbreeding (Werren, 1987). This hypothesis is more difficult to test as it requires detailed knowledge of the mating structure of natural populations as well as the degree of inbreeding depression. However, Werren (1987) has demonstrated theoretically that even relatively low levels of inbreeding can lead to the maintenance of male-killing. In addition to the adaptive hypotheses there is one non-adaptive explanation; in some cases the invasion of the microbe may be partially or wholly due to horizontal transmission. When the microbe is horizontally transmitted, male death may contribute to the invasion of a MLI if it increases the fitness of infected females, or it may be a neutral trait.

Transmission and spread of MLIs in populations

MLIs may be exclusively vertically transmitted, or may be both vertically and horizontally transmitted. Horizontal transmission of the spiroplasma that causes a MLI was looked for but not found in *Drosophila willistoni* (Ebbert, 1991). The bacteria that causes male death in the parasitoid *Nasonia vitripennis* is actually injected into the hemolymph of the host at oviposition and acquired by both offspring and unrelated larvae feeding on the same host

(Werren *et al.*, 1986). In addition, male killing by microsporidians infecting mosquitoes may be maintained by horizontal transmission (Hurst, 1993). Unlike the other systems discussed, male killing by microsporidians occurs relatively late in development, following the build up of microsporidian populations, and contagious infection may occur following the death of male larvae (Hurst, 1993).

The frequency of MLIs in natural populations is invariably quite low. In many samples of the *D. willistoni* group in the Carribean, the frequency of the MLI was generally found to be less than 5%. In a natural population of the coccinellid *Adalia bipunctata* Hurst *et al.* (1993) found that 6 out of 82 females, about 7%, carried the infection. The male-killing bacteria in *Nasonia* was found at a frequency of 4–5% in a natural population in Utah (Skinner, 1983; Balas *et al.*, 1996).

The limited frequency of MLIs may in part stem from incomplete maternal transmission. The rate of maternal transmission is generally between 85 and 99% (Hurst & Majerus, 1993). As well as imperfect transmission in a majority of females, in a minority of cases in *D. willistoni* and in the coccinellids *Harmonia axyridis* and *Coleomegilla maculata*, the infection appears to take time to develop (Matsuka *et al.*, 1975; Ebbert, 1991; Hurst *et al.*, 1996). Females with this condition, called 'progressive sex ratio' by Hurst *et al.* (1996) produce males and uninfected females shortly after emergence, but male egg hatch decreases and the infection rate of females increases as females age. It appears that in these females the bacteria take a longer time to build up to the numbers required to kill males. Whether this is evidence of heritable resistance is unclear, but it has been postulated that this type of resistance could, in turn, select for earlier reproduction of infected females (Hurst *et al.*, 1996).

Another explanation of the low frequency of MLIs in natural populations may be due to their negative effects on the fecundity of female carriers. In a review of 15 species with MLIs, no examples of the infection having a positive influence on fecundity were found, while in five cases, females had reduced fecundity, and in two cases, workers had shown no influence on fecundity (Hurst & Majerus, 1993).

MLIs in coccinellids

Aphidophagous coccinellids may be particularly susceptible to MLIs. They have been found in five species (Table 14.1: Shull, 1948; Matsuka *et al.*, 1975; Gotoh & Niijima, 1986; Hurst *et al.*, 1992, 1996), and are caused by at least two different infectious agents (Hurst *et al.* 1996). That MLIs are still being found in relatively well-studied species like *Coleomegilla maculata* suggests that careful observation may turn up more of these infections.

The life-history attribute that may make aphidophagous coccinellids prone to MLI is cannibalism of unhatched eggs by neonate larvae. Generally eggs hatch over a few hours and first-emerging offspring are likely to consume unhatched eggs. Five to ten percent of ladybirds may die from cannibalism (Hurst & Majerus, 1994). First instar larvae must search for aphid prey and are often unsuccessful in early encounters (Banks, 1955); those that have had an egg meal have increased survival time (Hurst & Majerus, 1993). Osawa (1992) found that first instar *Harmonia axyridis* fed eggs in the laboratory were more likely to reach the second instar when released in the field compared to unfed individuals. The density of aphids determined the degree of difference in survivorship (Osawa, 1992). The presence of non-viable male eggs then means that females infected with MLI are more likely to be able to benefit from an egg meal then females from uninfected clutches. Hurst *et al.* (1992) point out two other consequences of the presence of non-viable male eggs that will lead to an increase in female fitness. The availability of these eggs in a clutch simply reduces the probability that late-hatching viable female eggs will be cannibalized. It also protects female eggs by reducing by approximately half the number of potential cannibals (Hurst *et al.*, 1992).

MLIs in parasitic Hymenoptera

MLIs have also been found in parasitic Hymenoptera. A MLI is the probable cause of male death in a mymarid, *Caraphractus cinctus*, that attacks the eggs of Dystiscidae (Jackson, 1958). Jackson (1958) found that unmated female progeny of one particular female produced only a small fraction of the number of males that other unmated females produced, while mated females from this isofemale line produced comparable numbers of females but few males.

A better studied example is 'son-killer', a bacteria, *Arsenophonus nasoniae* (Gherna *et al.*, 1991), that has been found at a frequency of 4–5% in a population of *Nasonia vitripennis* (Skinner, 1983; Balas *et al.*, 1996). While *Nasonia* females typically produce sex ratios of 11–15% males, infected females produce sex ratios of 0–5% males (Skinner, 1985). Skinner (1985) found that the rate of both vertical and horizontal transmission (to unrelated wasps developing on the same host) is approximately 97%. He speculated that all developing wasps on a host acquire the bacteria from the host hemolymph. This idea was supported by histopathological studies, which indicated that females probably transfer the bacteria to the host hemolymph during parasitism where it is ingested by the developing wasps (Huger *et al.*, 1985). The bacteria does not kill males that feed on it, but rather, male eggs laid by females that fed on the bacteria as larvae do not develop (Werren *et al.*, 1986). While males are killed, female offspring production is not significantly different between infected and uninfected individuals (Skinner, 1985).

The high degree of potential horizontal transmission and the apparent lack of transmission within eggs (Huger *et al.*, 1985) distinguishes son-killer from other MLIs. While Skinner (1985) suggested that the maintenance of the infection may be due to the release of host resources to females developing on a host with fewer siblings, an examination of the size of females coming from infected and uninfected broods in the field showed no significant difference (Balas *et al.*, 1996). However, Balas *et al.* (1996) point out that only a very subtle improvement in fitness of approximately 5% is necessary for maintaining the MLI at a low frequency. Furthermore, horizontal transmission alone may be the major force maintaining this infection, and male killing a less important factor, perhaps just beneficial enough to the fitness of infected females to allow the bacteria to out-compete a non-lethal strain (Balas *et al.*,

1996). Alternatively, male-killing may be a neutral trait.

Paternally-inherited nuclear elements

The 'paternal sex ratio' of Nasonia vitripennis

The 'paternal sex ratio' factor (PSR) is a paternally-inherited element in the pteromalid *Nasonia vitripennis* that converts fertilized, incipient female eggs to males. PSR was discovered when selection experiments for male- and female-biased sex ratios revealed individuals that produced only male offspring (Werren *et al.*, 1981). Because of haplodiploidy, females that produce only sons are relatively common in laboratory cultures of Hymenoptera; these are usually just females that have failed to become inseminated. However, in this instance the male-bias was stably inherited over several generations. Crosses further demonstrated that the male-bias was due to the genetic constitution of the male parent (Werren *et al.*, 1981). This result was extraordinary because in a haplodiploid system, males do not usually contribute genetically to male offspring. Crosses using visible markers suggested that the trait was not part of the normal chromosomal complement of the male, and Werren *et al.* (1981) concluded that the trait was extra-chromosomal. Later studies showed that in fact the PSR trait is due to the presence of a supernumerary, or 'B' chromosome (Nur *et al.*, 1988). This chromosome is part of the paternal genome until an egg is fertilized with PSR-bearing sperm. At this point the PSR chromosome disassociates from the rest of the paternal set of chromosomes, which become overcondensed and eventually disintegrate (Werren *et al.*, 1987). The resulting fertilized egg is then haploid and develops as a male bearing PSR.

PSR is one of the most extreme examples of a selfish genetic element and, because it is paternally inherited, the exception that proves the rule for transmission of these elements. All of the discussion up to this point has focused on elements that are maternally inherited and so favor feminization, parthenogenesis, male-killing, or differential fertilization of infected eggs. One would not expect elements that are exclusively transmitted in egg cytoplasm such as bacteria or protists to cause masculinization; even viruses that may be transmitted by sperm (Mulcahey & Pascho, 1984), are more easily transmitted in eggs. However, unpaired chromosomes such as PSR are better transmitted by males in a haplodiploid system because in females, the meiotic reduction division is more likely to cause their elimination (Werren, 1991). This was demonstrated experimentally in *Nasonia*. Partially deleted PSR chromosomes that had lost their ability to destroy the paternal chromosomes and so ended up in both males and females were transmitted at a high rate through males but at a low rate through females (Beukeboom & Werren, 1993a).

Mechanism

While the origin of the PSR element is still unclear, Werren (1991) suggests that it may have originated from cytoplasmic incompatibility between *N. vitripennis* and a sympatric species, *N. longicornis*. Both of these species have a strain of CI *Wolbachia* that causes bidirectional incompatibility (see The role of CI *Wolbachia* in speciation, above), and fragments of the paternal chromosomes may be inherited by the male offspring that result from an incompatible cross (Ryan & Saul, 1968; Breeuwer & Werren, 1990). Further support for this idea comes from sequences of repetitive DNA on the PSR chromosome that show some similarities to sequences on the autosomes of not only *N. vitripennis* but also *N. longicornis* and *N. giraulti* (Eickbush *et al.*, 1992). The way that PSR functions to cause destruction of the rest of the paternal genome is also as yet unknown, but studies of partially deleted PSR chromosomes suggest that rather than the expression of a specific gene product, PSR may act as a sink for a protein needed for proper processing and condensation of male chromosomes (Eickbush *et al.*, 1992). The ability of partially deleted PSR chromosomes to cause genome loss appears not to depend on the areas deleted but rather the overall size of the chromosome. Chromosomes below a minimum size do not cause genome loss (Beukeboom *et al.*, 1992; Beukeboom & Werren, 1993a).

Transmission and spread in populations

The invasion criteria and equilibrium frequency for PSR have been modeled in panmictic (Skinner, 1987; Werren, 1987) and subdivided (Beukeboom & Werren, 1992) populations. The three factors that are important are the transmission rate, the fertilization rate (or proportion of female eggs among non-carriers of PSR), and the deme size. Given perfect transmission, the fertilization rate needs to be greater than 50% for PSR to invade. The transmission rate of PSR is generally between 94 and 100% (Beukeboom & Werren, 1993b). Thus, PSR depends for its invasion on a system with characteristically female-biased sex ratios. *Nasonia vitripennis* has a demic population structure, and often sex ratios are female-biased as a result of inbreeding and local mate competition. In a population cage experiment looking at the effect of deme size on the maintenance of PSR, Beukeboom & Werren (1992) found that PSR frequency was very low in small demes because of the sex ratio variance among demes. The demes with PSR produced few females to mate with the PSR males. Only at a larger deme size and in a panmictic population (in which the fertilization frequency ranged from 0.7 to 0.9) was PSR maintained. Furthermore, the presence of the feminizing factor MSR (maternally inherited sex ratio) increased the equilibrium frequency of PSR by causing most eggs to be fertilized (Beukeboom & Werren, 1992). PSR and MSR are found in the same populations in Utah (Skinner, 1983; Werren, 1987) and it has been suggested that the frequency of PSR in the field may be partly determined by the frequency of MSR (Beukeboom & Werren, 1992).

The 'primary male' factor in Encarsia pergandiella

A second paternally-inherited element was subsequently discovered in an unrelated whitefly parasitoid in the Aphelinidae, *Encarsia pergandiella* (Hunter *et al.*, 1993). This species is an autoparasitoid, thus females are primary parasitoids and males are normally obligate hyperparasitoids as described earlier (see Infection by PI *Wolbachia*, above). The first sign of something unusual in this

population was the observation of males developing as primary parasitoids on the whitefly hosts. These males, called *primary males* were found to be haploid but develop from fertilized eggs. Like males carrying the PSR chromosome, a radiation experiment in *E. pergandiella* indicated that it was the paternal genome that was lost in primary males, and that the factor was inherited from father to son. While the obvious candidate for such an element was a supernumerary chromosome, karyotypes of primary males in comparison with normal hyperparasitic males did not show a supernumerary chromosome. The causal agent of primary males in *E. pergandiella* is as yet unknown.

The transmission dynamics of the primary male factor in *E. pergandiella* is worth noting because of its difference from PSR. While PSR appears to have close to perfect transmission, only about 28% of eggs fertilized by sperm from primary males develop into primary males (Hunter *et al.*, 1993). The probable reason this element can be maintained in this population with such a low transmission rate is because of the extremely high rate of egg fertilization, on average 88% (Hunter *et al.*, 1993). In turn, the high rate of egg fertilization is caused by the peculiar biology of the wasp and the parasitoid–whitefly dynamics. The density of wasp hosts is usually an order of magnitude lower than that of whitefly hosts, so most of the hosts encountered by females are whiteflies and are suitable only for oviposition of a fertilized egg (Hunter, 1993).

Genetic conflict and biological control

Distribution of selfish genetic elements – observation and predictions

While the incidence of most of the selfish genetic elements described above appears relatively low, the current accelerating rate of their discovery suggests that many more are yet to be discovered. As an example, cytoplasmic incompatibility (CI) was known only from mosquitoes and *Nasonia vitripennis* during the period from the late 1950s until the mid 1970s (Caspari & Watson, 1959; Saul, 1961; Laven, 1967a; Yen & Barr, 1971, 1973), when it was found in a moth, *Ephestia cautella* (Brower, 1976) and a tephri-tid fruit fly, *Rhagoletis cerasi* (Boller *et al.*, 1976). In the 1980s CI was found in the alfalfa weevil, *Hypera postica* (Hsiao & Hsiao, 1985), the flour beetle, *Tribolium confusum* (Wade & Stevens, 1985), the small brown planthopper, *Laodelphax striatellus* (Noda, 1984), and *Drosophila simulans* (Hoffmann & Turelli, 1986). The tally is now more than 80 species of insects that have been found to carry *Wolbachia* (Werren, 1997a). Probably most of these infections cause cytoplasmic incompatibility (Werren *et al.*, 1995a). Another indication that selfish genetic elements may be more widely distributed than is cur-rently documented in natural enemies is that intense study has revealed several elements in a single species, *Nasonia vitripennis* (see Table 14.1). *Nasonia* is one of only two species of parasitoids that has been the subject of long-term genetic study, the other being *Bracon hebetor*. While sex ratio distorters and CI have not been discovered in *Bracon*, four have been found in *Nasonia*. The sample size is obviously inadequate, but it seems unlikely that *Nasonia* is unique among parasitic Hymenoptera.

In this section, the expectation of discovery of the various types of genetic element in natural enemies will be discussed. In most cases, the observations are few, and we have only a few general theoretical pre-dictions to limit our search. These predictions are entirely untested, and should be viewed as a first approach to the problem. We are also limited in our understanding by our ignorance of how most of these elements arise in species; the predictions then apply exclusively to establishment within a population, and naively assume that all species may acquire any selfish genetic element.

There are no observed or theoretical limits to the occurrence of CI *Wolbachia*. Infection by *Wolbachia* is spread across several orders of insects, and may only be limited by what appears to be relatively rare events of horizontal transmission and establishment in new species (Sinkins *et al.*, 1997). In the parasitic Hymenoptera, CI has been recorded only twice (Werren, 1997a) and it has not been recorded in other parasitoids or predatory beetles. However, Werren *et al.*'s (1995a) survey of 154 Panamanian insects sug-gested that the Diptera and Hymenoptera may have a slightly higher rate of infection with *Wolbachia* (of unknown phenotypic effects) than other orders of insects.

Parthenogenesis-inducing (PI) *Wolbachia* have been found only in parasitic Hymenoptera to date (see Table 14.1). To cause parthenogenetic develop-ment in the haplodiploid Hymenoptera, the *Wolbachia* prevents the completion of the first mitotic division, causing a haploid genome to double, and an incipient male to develop as a female (Stouthamer & Kazmer, 1994). PI *Wolbachia* is likely therefore to be limited at least to haplodiploid systems, where the specific and relatively simple manipulation of haploid embryogenesis by the microbe leads to a change in sex, and eggs do not need to be fertilized to develop (Werren, 1997a).

Another restriction to invasion of parasitoids by PI *Wolbachia* is likely to be single locus sex determina-tion (Stouthamer & Kazmer, 1994). In many Hymenoptera, female sex appears to be determined by heterozygosity at a single variable locus. Hemizygous or homozygous individuals are male. Diploid males may appear in laboratory cultures as a result of inbreeding, and are generally sterile. Although the distribution of single locus (or multiple locus) sex determination in the Hymenoptera is not entirely understood, all of the documented cases have been in the Symphyta (sawflies), the Aculeata (ants, bees, and stinging wasps), and among the Parasitica, in the Ichneumonoidea (Cook, 1993). No instances of this sex determination system have been found in the Chalcidoidea, despite a few intensive studies (Cook, 1993). Because the cytogenetic mechanism involved in doubling the chromosome complement in at least *Trichogramma* infected with PI *Wolbachia* causes complete homozygosity at all loci, PI *Wolbachia* is not expected to be able to spread in species with single or even multiple locus sex determination (Stouthamer & Kazmer, 1994).

Male-lethal infections (MLIs) have been found in a couple of species of parasitic Hymenoptera, and in five species of Coccinellidae, as well as in Lepidoptera, and Heteroptera (see Table 14.1: Ebbert, 1993; Hurst, 1993; Hurst & Majerus, 1993). These infections are invariably at low frequency in natural populations and may be cryptic even in laboratory populations. The recent discovery of a MLI in the well-studied coccinellid *Coleomegilla maculata* emphasizes this point (Hurst *et al.*, 1996). Alternative and non-exclusive hypotheses about the benefit to female fitness of male-killing make it difficult to predict the distribution of vertically-transmitted MLIs. Obvious candidates for MLIs are species in which egg cannibalism of siblings results in greater survivorship. In species such as the aphidophagous coccinellids, dead male eggs are likely to increase the fitness of sisters that consume them as well as decreasing the possibility that they themselves are consumed (Hurst & Majerus, 1993). Potential other vulnerable species are those in which siblings develop on a discrete resource, and females are larger and more fecund when released from competition with their brothers. While many parasitoids would seem to fall into this category, the one test of this hypothesis in *Nasonia vitripennis* failed to find any effect of competitive release on female size (Balas *et al.*, 1996). Theory also predicts that an avoidance of inbreeding can lead to maintenance of a MLI (Werren, 1987). In these species, sibling mating must occur with some frequency, but so must inbreeding depression. Since these are generally inversely related, it may be difficult to determine *a priori* species in which the conditions for this hypothesis hold (Hurst & Majerus, 1993). Lastly, much less restrictive conditions apply to MLIs with horizontal transmission. In these types of infections, such as 'son-killer' in *Nasonia vitripennis*, male killing may provide an 'incremental gain' in fitness to females over an infection that does not kill males (Balas *et al.*, 1996), or it may be a neutral trait.

Paternally-inherited elements that cause genome loss have been found only in two unrelated species of parasitic Hymenoptera (see Table 14.1: Werren, 1991; Hunter *et al.*, 1993), and they may be very rare, as would be suggested by their extraordinary mode of action, and the probable origin of PSR in *Nasonia* (Werren, 1991). It is important to note, however, that this type of element is likely to be the most difficult to detect. A relatively few parasitoid females that produce exclusively males would generally be assumed to be uninseminated. It was only because of a selection experiment for male-biased sex ratios in

Nasonia that PSR was found (Werren *et al.*, 1981), and only because 'primary males' in *Encarsia pergandiella* developed on a host usually reserved for females that the (as yet unknown) factor causing genome loss was discovered in this species (Hunter *et al.*, 1993). The conditions for invasion of these elements are female-biased sex ratios and, ordinarily, a high transmission fidelity between father and son (Skinner, 1987; Werren, 1987). Sex ratio and transmission rate are related in such a way that an element with a low transmission rate can persist only when the sex ratio is extremely female-biased (Skinner, 1987; Werren 1987).

Changes in frequency of selfish genetic elements in laboratory cultures

Many of the elements discussed are at relatively low frequency in natural populations of natural enemies, and may not have a great impact on a biological control program if the frequencies remain low. However, bringing natural enemies into culture in the laboratory generally involves making a number of quite radical changes in their environment and may lead to changes in selection on the selfish genetic elements for which they serve as carriers. An anecdotal example mentioned earlier is the appearance of parthenogenetic reproduction in laboratory cultures, presumably due to an increase in the relative frequency of parthenogens in a mixed culture (Flanders, 1945; DeBach, 1969; Stouthamer & Luck, 1991). One could suggest reasons why parthenogenetic females may have increased fitness in laboratory cultures, for example because of the lack of appropriate mating cues for the sexual individuals, or because of selection among genotypes for relatively constant laboratory conditions followed by a lack of recombination in successful parthenogens. However, because a low frequency of parthenogens would ordinarily go unnoticed, decreases in frequency leading to extinction of the parthenogenetic form, or a lack of change in frequency may be as common in laboratory cultures as an increase in frequency. It is only in the last instance that parthenogenesis will be noticed.

We know very little about how various types of element respond to laboratory selection, drift, and inbreeding, but an extremely general set of testable predictions can be made. (i) If a culture is being established from few individuals one might expect that the frequency of a particular element may be different from the field simply due to sampling error. (ii) Cytoplasmic incompatibility is most likely to appear when establishing a culture from several geographic locations and will result in decreased egg hatch in diploid species or a male-biased sex ratio in haplodiploid species. (iii) Any element that is transmitted horizontally as well as vertically may increase under the typically high density culture conditions. (iv) Lastly, one might expect that selection in the laboratory might act to decrease the frequency of a MLI in a coccinellid if, when hosts are provided at high densities, there is little gain in survivorship of females that have an egg meal of their dead brothers relative to those that do not. However, females would still be expected to gain from a lower probability of being eaten themselves when male killing occurs.

Manipulation

The presence of selfish genetic elements in a natural enemy population will in most instances be detrimental to their use in biological control. In some cases they may be relatively easy to remove from a laboratory culture. Male-lethal bacterial infections in coccinellids, for example, may be permanently cured by the administration of antibiotics (Gotoh & Niijima, 1986; Hurst *et al.*, 1992; 1996). The male-lethal spiroplasma in *Drosophila* may also be cured by heat treatment (Williamson & Poulson, 1979). Antibiotic treatment would ordinarily be expected to cure cultures of CI *Wolbachia* as well. PI *Wolbachia* may not always be detrimental to the use of a parasitoid in biological control (see below), and, as discussed earlier, it may be impossible to restore sexual reproduction in populations that have become fixed for the microbe. In mixed populations, however, antibiotic treatment will permanently restore sexual reproduction to the parthenogenetic forms (Stouthamer *et al.*, 1990a). As mentioned above, the challenge in dealing with a paternally-inherited male-biasing nuclear element such as PSR is identifying it. If such an element is identified it may readily

be removed from a culture by rearing females individually and eliminating the male-biased families.

PI *Wolbachia* may in some instances be usefully manipulated to enhance biological control. In species with mixed populations of sexuals and parthenogens, investigators have a choice of which form to work with, or may use both at different phases of a project (Stouthamer, 1993). Mass-rearing for augmentative release may be more cost effective when a parthenogenetic form is used because none of the hosts provided will produce males. Parthenogenetic forms may also have a greater rate of increase, but not necessarily so. The cost in fecundity to bearing the infection may be relatively severe, such that parthenogenetic females given unlimited hosts produce fewer female offspring than sexual females (Stouthamer & Luck, 1993). In other species, there may be no measurable fecundity cost (Stouthamer *et al.*, 1994). Clearly, knowing something about the fitness cost of the infection is critical for making a choice between sexual and parthenogenetic forms for mass-rearing. For inoculative biological control, one would expect that the higher rates of recombination in sexual populations may better enable them to adapt to a new habitat, but sexual forms may have difficulty finding a mate at the low densities expected following release (Stouthamer, 1993). Stouthamer (1993) suggests that when a population consists of interbreeding sexual and parthenogenetic forms, it should be easier for parthenogens to invade a sexual population newly released in the field than the reverse. This is because matings between asexual females and sexual males produce asexual offspring and, probably more importantly, sexuals may be limited by the availability of mates at low densities. For this reason Stouthamer (1993) recommends an initial release of sexual forms followed by a release of parthenogens when host densities are reduced. The rationale is that sexual forms will adapt to local conditions, and when parthenogens are released, introgression of sexual genes into the parthenogenetic subpopulation will lead to adaptation of the parthenogenetic forms.

Stouthamer's (1993) predictions form a set of testable hypotheses for a release strategy for a parasitoid with two reproductive forms. Perhaps uniquely in biological control, such tests would enable us to investigate the effect of one life-history characteristic in isolation, reproductive mode, on the effectiveness of biological control in different circumstances. At this time, however, we are still limited in the systems in which we can test these ideas to naturally infected species. There has not yet been a successful transfer of PI *Wolbachia* between species (Werren, 1997a).

The potential for manipulation of CI for control of its pestiferous insect hosts, in particular mosquitoes, has long been realized (Laven, 1967b). In one approach, males bearing CI *Wolbachia* are used essentially as sterile males, causing incompatible matings with uninfected females in the field. Laven (1967b) used this approach successfully to eradicate the mosquito *Culex pipiens fatigans* in a small isolated village in Burma. The limitations with this approach are that large numbers of males must be introduced, and eradication is only likely to be temporary where migration is possible (Curtis *et al.*, 1982). A second approach, which may have potential for transforming natural enemies, involves using unidirectional incompatibility simply as a mechanism to drive beneficial genes into a population. The dramatic sweep of *Wolbachia* and an associated mitochondrial variant into *Drosophila simulans* populations in California demonstrates the potential power of this technique (Turelli & Hoffman, 1991; Turelli *et al.*, 1992). Because of the necessity that the *Wolbachia* and the beneficial genes do not become uncoupled, the genes of interest must be carried on a maternally-inherited expression vector (Sinkins *et al.*, 1997). Candidates for such a vector include a transformed *Wolbachia* strain, or other cytoplasmic elements such as mitochondria, nutritive symbionts, or transovarially transmitted viruses (Sinkins *et al.*, 1997).

In the future, it may be possible to drive nuclear genes into a population if the genes that cause incompatibility could be isolated and used to transform a nuclear genome (Sinkins *et al.*, 1997). Theory suggests that there would be an unstable equilibrium in such a population, and one would need to release a larger number of transformed individuals to prevent extinction at low frequencies (Sinkins *et al.*, 1997), but this approach would have the advantage of not requiring a maternally inherited expression vector to drive the gene of interest into the population.

There are still formidable challenges to the effective use of CI for control of pests or for transforming natural enemies of pests. Experimental transfer of *Wolbachia* between *Drosophila* species (Boyle *et al.*, 1993), and between a mosquito and *Drosophila* (Braig *et al.*, 1994) has been performed, but infection of species other than *Drosophila* has not yet been performed (Sinkins *et al.*, 1997). Using *Wolbachia* as an expression vector may have the drawback that the gene of interest may be expressed only in the tissues where *Wolbachia* is found, in the reproductive system in particular (Sinkins *et al.*, 1997). An equally critical challenge for the use of this technique to improve biological control is to find and then isolate single locus traits that will improve the effectiveness of a natural enemy (Hoy, 1994). Insecticide resistance has been shown to be beneficial in some instances (Hoy, 1990), but other discrete characters have rarely been identified, and many of the life history and behavioral traits that might be of interest are likely to be polygenically inherited (Stearns, 1992).

Conclusions

It should be apparent from the preceding that our understanding of selfish genetic elements in arthropod natural enemies is nascent. There is a tremendous growth in the literature describing these elements, but relatively few systems have been studied in any detail in the laboratory, and even fewer have been investigated in the field. There are obvious gaps in our knowledge that make deciding what types of elements one would expect to find, and with what likelihood they will occur, difficult. We need to know the frequency of selfish genetic elements among and within populations of natural enemies. For example, what proportion of the parthenogenetic chalcidoid species carry PI *Wolbachia*? And what proportion of predominantly sexual populations have a low frequency of females infected with PI *Wolbachia*? The relatively simple procedures required to treat parthenogenetic females with antibiotics and observe the sex ratio of their offspring for one or two generations would seem to make it possible for workers to investigate at least the first question as a matter of routine. As polymerase chain reaction (PCR) techniques become more available, probing collections of field-collected individuals for the presence of *Wolbachia* may also be feasible for many research teams.

PI *Wolbachia* are special among selfish genetic elements in that they may sometimes be beneficial for biological control efforts. In order to work with natural infections we need surveys for parthenogenesis microorganisms, as mentioned above. Also, the potential application of these microbes will be greatly expanded if we can develop a technique for artificially infecting uninfected populations. Transfer experiments would permit the study of the role of genetic variation in the host and microorganism in the transmission and expression of the microbe, and permit a comparison of the relative fitness of uninfected and infected carriers. Most importantly, however, sexuals and parthenogens should be compared for biological control efficacy (Stouthamer, 1993).

Lastly, common reports of sex ratio shifts, loss of cultures, and failure of establishment of natural enemies may sometimes be explained by changing frequencies of selfish genetic elements when brought into culture or released in a new habitat. A useful avenue of research would be to further develop and test predictions of how the frequency of particular elements will change when brought into the laboratory, and when released with their carriers in inoculative or augmentative biological control programs. We do not know how often these elements play a role in the success or failure of a natural enemy in biological control, but it is clear that they are more common and more influential than their first discoverers ever imagined, and careful consideration of their potential involvement can only improve our chances of success.

Acknowledgements

Many thanks to Richard Stouthamer and Jack Werren for interesting and constructive discussions on several topics, and to Jack Werren, Richard Stouthamer, Scott O'Neill, and Hans Breeuwer for providing me with manuscripts in review or in press. I would also like to thank Robert Butcher, Yuval Gottlieb, Tim Karr, Scott O'Neill, and Einat Zchori-Fein for sharing unpublished data, and Tim Collier, Buck Cornell, Charles Godfray and Brad Hawkins for comments on a previous version of the manuscript.

References

Aeschlimann, J. P. (1990) Simultaneous occurence of thelytoky and bisexuality in Hymenopteran species, and its implications for the biological control of pests. *Entomophaga* **35**: 3–5.

Balas, M. T., Lee, M. H. & Werren, J. H. (1996) Distribution and fitness effects of the son-killer bacterium in *Nasonia*. *Evolutionary Ecology* **10**: 593–607.

Banks, C. J. (1955). An ecological study of Coccinellidae (Col.) associated with *Aphis fabae* Scop. on *Vicia fabia*. *Bulletin of Entomological Research* **46**: 561–87.

Beukeboom, L. W. (1994) Phenotypic fitness effects of the selfish B chromosome, paternal sex ratio (PSR) in the parasitic wasp *Nasonia vitripennis*. *Evolutionary Ecology*, **8**: 1–24.

Beukeboom, L. W. & Werren, J. H. (1992) Population genetics of a parasitic chromosome: experimental analysis of *PSR* in subdivided populations. *Evolution* **46**: 1257–68.

Beukeboom, L. W. & Werren, J. H. (1993a) Deletion analysis of the selfish B chromosome, paternal sex ratio (PSR), in the parasitic wasp *Nasonia vitripennis*. *Genetics* **133**: 637–48.

Beukeboom, L. W. & Werren, J. H. (1993b) Transmission and expression of the parasitic paternal sex ratio (PSR) chromosome. *Heredity* **70**: 437–43.

Beukeboom, L. W., Reed, K. M. & Werren, J. H. (1992) Effects of deletions on mitotic stability of the paternal-sex-ratio (PSR) chromosome from *Nasonia*. *Chromosoma* **102**: 20–6.

Binnington, K. C. & Hoffmann, A. A. (1989) *Wolbachia*-like organisms and cytoplasmic incomptibility in *Drosophila simulans*. *Journal of Invertebrate Pathology* **54**: 344–52.

Boller, E. F., Russ, K., Vallo, V. & Bush, G. L. (1976) Incompatible races of European cherry fruit fly, *Rhagoletis cerasi* (Diptera: Tephritidae), their origin and potential use in biological control. *Entomologia Experimentalis et Applicata* **20**: 237–47.

Bowen, W. R. & Stern, V. M. (1966) Effect of temperature on the production of males and sexual mosaics in a uniparental race of *Trichogramma semifumatum* (Hymenoptera: Trichogrammatidae). *Annals of the Entomological Society of America* **59**: 823–34.

Boyle, L., O'Neill, S. L., Robertson, H. M. & Karr, T. L. (1993) Interspecific and intraspecific horizontal transfer of *Wolbachia* in *Drosophila*. *Science* **260**: 1796–9.

Braig, H. R., Guzman, H., Tesh, R. B. & O'Neill, S. L. (1994) Replacement of a natural *Wolbachia* symbiont of *Drosophila simulans* with a mosquito counterpart. *Nature* **367**: 453–5.

Breeuwer, J. A. J. (1997) *Wolbachia* and cytoplasmic incompatibility in the spider mites *Tetranychus urticae* and *T. turkestani*. *Heredity* **79**: 41–7.

Breeuwer, J. A. J. & Jacobs, G. (1996) *Wolbachia*: intracellular manipulators of mite reproduction. *Experimental and Applied Acarology* **20**: 421–34.

Breeuwer, J. A. J. & Werren, J. H. (1990) Microorganisms associated with chromosome destruction and reproductive isolation between two insect species. *Nature* **346**: 558–60.

Breeuwer, J. A. J. & Werren, J. H. (1993a) Cytoplasmic incompatibility and bacterial density in *Nasonia vitripennis*. *Genetics* **135**: 565–74.

Breeuwer, J. A. J. & Werren, J. H. (1993b) Effect of genotype on cytoplasmic incompatibility between two species of *Nasonia*. *Heredity* **70**: 428–36.

Breeuwer, J. A. J. & Werren, J. H. (1995) Hybrid breakdown between two haplodiploid species: the role of nuclear and cytoplasmic genes. *Evolution* **49**: 705–17.

Breeuwer, J. A. J., Stouthamer, R., Burns, D. A., Pelletier, D. A., Weisburg, W. G. & Werren, J. H. (1992) Phylogeny of cytoplasmic incompatibility microorganisms in the parasitoid wasp *Nasonia* (Hymenoptera: Pteromalidae) based on 16S ribosomal DNA sequences. *Insect Molecular Biology* **1**: 25–36.

Bressac, C. & Rousset, F. (1993) The reproductive incompatibility system in *Drosophila simulans*: DAPI-staining analysis of *Wolbachia* symbionts in sperm cysts. *Journal of Invertebrate Pathology* **61**: 226–30.

Brower, J. H. (1976) Cytoplasmic incompatibility: occurrence in a stored-product pest *Ephestia cautella*. *Annals of the Entomological Society of America* **69**: 1011–15.

Bull, J. J. (1983) *Evolution of Sex Determining Mechanisms*. Menlo Park, CA: Benjamin/Cummings Publishing Co.

Cabello-García, T. & Vargas-Piqueras, P. (1985) Temperature as a factor influencing the form of reproduction of *Trichogramma cordubensis* Vargas and Cabello (Hym., Trichogrammatidae). *Zeitschrift für Angewandte Entomologie* **100**: 434–41.

Carson, H. L., Chang, L. S. & Lyttle, T. W. (1982) Decay of female sexual behavior under parthenogenesis. *Science* **218**: 68–70.

Caspari, E. & Watson, G. S. (1959) On the evolutionary importance of cytoplasmic sterility in mosquitoes. *Evolution* **13**: 568–70.

Chen, B.-H., Kfir, R. & Chen, C. N. (1992) The thelytokous *Trichogramma chilonis* in Taiwan. *Entomologia Experimentalis et Applicata* **65**: 187–94.

Clarke, C. & Sheppard, P. M. (1975) All-female broods in the butterfly *Hypolimnas bolina* (L.). *Proceedings of the Royal Society of London, Series B* **189**: 29–37.

Conner, G. W. & Saul, G. B. (1986) Acquisition of incompatibility by inbred wild-type stocks of *Mormoniella*. *Journal of Heredity* **77**: 211–13.

Cook, J. M. (1993) Sex determination in the Hymenoptera: a review of models and evidence. *Heredity* **71**: 421–35.

Cronin, J. T. & Strong, D. R. (1996) Genetics of oviposition success of a thelytokous fairyfly parasitoid, *Anagrus delicatus*. *Heredity* **76**: 43–54.

Curtis, C. F., Brooks, G. D., Ansari, M. A., Grover, K. K., Krishnamurthy, B. S., Rajagopalan, P. K., Sharma, L. S., Sharma V. P., Singh, D. S., Singh, K. R. P. & Yasuno, M. (1982) A field trial on control of *Culex quinquefasciatus* by release of males of a strain integrating cytoplasmic incompatibility and a translocation. *Entomologia Experimentalis et Applicata* **31**: 181–90.

Darling, D. C. & Werren, J. H. (1990) Biosystematics of *Nasonia* (Hymenoptera: Pteromalidae): two new species reared from birds' nests in North America. *Annals of the Entomological Society of America* **83**: 352–70.

DeBach, P. (1969) Uniparental, sibling and semi-species in relation to taxonomy and biological control. *Israel Journal of Entomology* **4**: 11–28.

Dunn, A. M., Adams, J. & Smith, J. E. (1993) Transovarial transmission and sex ratio distortion by a microsporidian parasite in a shrimp. *Journal of Invertebrate Pathology* **61**: 248–52.

Ebbert, M. A. (1991) The interaction phenotype in the *Drosophila willistoni*-spiroplasma symbiosis. *Evolution* **45**: 971–88.

Ebbert, M. A. (1993) Endosymbiotic sex ratio distorters in insects and mites. In *Evolution and Diversity of Sex Ratio in Insects and Mites*, ed. D. L. Wrensch & M. A. Ebbert, pp. 150–91. New York: Chapman and Hall.

Eickbush, D. G., Eickbush, T. H. & Werren, J. H. (1992) Molecular characterization of repetitive DNA sequences from a B chromosome. *Chromosoma* **101**: 575–83.

Eskafi, F. M. & Legner, E. F. (1974) Parthenogenetic reproduction in *Hexacola* sp. near *websteri* (Hym., Cynipidae) a parasite of *Hippelates* eye gnats (Diptera, Chloropidae). *Annals of the Entomological Society of America* **67**: 767–8.

Flanders, S. E. (1945) The bisexuality of uniparental Hymenptera, a function of the environment. *American Naturalist* **79**: 122–41.

Flanders, S. E. (1950) Races of apomictic parasitic Hymenoptera introduced into California. *Journal of Economic Entomology* **43**: 719–20.

Flanders, S. E. (1965) On the sexuality and sex ratios of hymenopterous populations. *American Naturalist* **49**: 489–94.

Garonna, A. P. & Viggiani, G. (1988) Studi biologici sull'*Archenomus orientalis* Silvestri: 1. Longevità degli adulti, fecondità, fertilità e progenie. *Atti XV Congresso Nazionale Entomologia, L'Aquila, Italy*, pp. 859–66.

Gerling, D., Spivak, D. & Vinson, S. B. (1987) Life history and host discrimination of *Encarsia deserti* (Hymenoptera: Aphelinidae), a parasitoid of *Bemisia tabaci* (Homoptera: Aleyrodidae). *Annals of the Entomological Society of America* **80**: 224–9.

Gherna, R. L., Werren, J. H., Weisburg, W., Cote, R., Woese, C. R., Mandelco, L. & Brenner, D. J. (1991) *Arsenophonus nasoniae* gen. nov., sp. nov., the causative agent of the son-killer trait in the parasitic wasp *Nasonia vitripennis*. *International Journal of Systematic Bacteriology* **41**: 563–5.

Giordano, R., O'Neill, S. L. & Robertson, H. M. (1995) *Wolbachia* infections and the expression of cytoplasmic incompatibility in *Drosophila sechellia* and *D. mauritiana*. *Genetics* **140**: 1307–17.

Gordh, G. & Lacey, L. (1976) Biological studies of *Plagiomerus diaspidus* Crawford, a primary internal parasite of diaspid scale insects (Hymenoptera: Encyrtidae; Homoptera: Diaspididae). *Proceedings of the Entomological Society of Washington* **78**: 132–44.

Gotoh, T. & Niijima, K. (1986) Characteristics and agents of abnormal sex ratio in two aphidophagous coccinellid species. In *Ecology of the Aphidophaga*, ed. I. Hodek, pp. 545–50. Prague: Academia.

Groeters, F. R. (1996) Maternally inherited sex ratio distortion as a result of a male-killing agent in *Spilostethus hospes* (Hemiptera: Lygaeidae). *Heredity* **76**: 201–8.

Hoffmann, A. A. & Turelli, M. (1986) Unidirectional incompatibility between populations of *Drosophila simulans*. *Evolution* **40**: 692–701.

Hoffmann, A. A. & Turelli, M. (1988) Unidirectional incompatibility in *Drosophila simulans:* inheritance, geographic variation and fitness effects. *Genetics* **119**: 435–44.

Hoffmann, A. A., Turelli, M. & Harshman, L. G. (1990) Factors affecting the distribution of cytoplasmic incompatibility in *Drosophila simulans*. *Genetics* **126**: 933–48.

Hoffmann, A. A., Clancy, D. J. & Merton, E. (1994) Cytoplasmic incompatibility in Australian populations of *Drosophila melanogaster*. *Genetics* **136**: 993–9.

Hoffmann, A. A., Clancy, D. J. & Duncan, J. (1996) Naturally-occurring *Wolbachia* infection in *Drosophila simulans* that does not cause cytoplasmic incompatibility. *Heredity* **76**: 1–8.

Hoy, M. A. (1990) Pesticide resistance in arthropod natural enemies: variability and selection responses. In *Pesticide Resistance in Arthropods*, ed. R. T. Roush & B. E. Tabashnik, pp. 203–36. New York: Chapman and Hall.

Hoy, M. A. (1994) *Insect Molecular Genetics*. New York: Academic Press.

Hoy, M. A. & Cave, F. E. (1988) Premating and postmating isolation among populations of *Metaseiulus occidentalis* (Nesbitt) (Acarina: Phytoseiidae). *Hilgardia* **56**: 1–20.

Hsiao, C. & Hsiao, T. H. (1985) Rickettsia as the cause of cytoplasmic incompatibility in the alfalfa weevil *Hypera postica*. *Journal of Invertebrate Pathology* **45**: 244–6.

Huger, A. M., Skinner, S. W. & Werren, J. H. (1985) Bacterial infections associated with the son-killer trait in the parasitoid wasp *Nasonia* (= *Mormoniella*) *vitripennis* (Hymenoptera: Pteromalidae). *Journal of Invertebrate Pathology* **46**: 272–80.

Hung, A. C. F., Day, W. H. & Hedlund, R. C. (1988) Genetic variability in arrhenotokous and thelytokous forms of *Mesochorus nigripes* [Hym.: Ichneumonidae]. *Entomophaga* **33**: 7–15.

Hunter, M. S. (1991) Sex Ratio in an Autoparasitoid *Encarsia pergandiella* Howard (Hymenoptera: Aphelinidae). PhD Thesis: Cornell University.

Hunter, M. S. (1993) Sex allocation in a field population of an autoparasitoid. *Oecologia* **93**: 421–8.

Hunter, M. S., Nur, U. & Werren, J. H. (1993) Origin of males by genome loss in an autoparasitoid wasp. *Heredity* **70**: 162–71.

Hurst, G. D. D. & Majerus, M. E. N. (1993) Why do maternally inherited microorganisms kill males? *Heredity* **71**: 81–95.

Hurst, G. D. D. & Majerus, M. E. N. (1994) Feminist bacteria of ladybird beetles. *Natural History* **103**: 32–4.

Hurst, G. D. D., Majerus, M. E. N. & Walker, L. E. (1992) Cytoplasmic male killing elements in *Adalia bipunctata* (Linnaeus) (Coleoptera : Coccinellidae). *Heredity* **69**: 84–91.

Hurst, G. D. D., Majerus, M. E. N. & Walker, L. E. (1993) The importance of cytoplasmic male killing elements in natural populations of the two spot ladybird, *Adalia bipunctata* (Linnaeus) (Coleoptera: Coccinellidae). *Biological Journal of the Linnean Society* **49**: 195–202.

Hurst, G. D. D., Hammarton, T. C., Obrycki, J. J., Majerus, T. M. O., Walker, L. E., Bertrand, D. and Majerus, M. E. N. (1996) Male-killing bacterium in a fifth ladybird beetle, *Coleomegilla maculata* (Coleoptera: Coccinellidae). *Heredity* **77**: 177–85.

Hurst, L. D. (1993) The incidences, mechanisms and evolution of cytoplasmic sex ratio distorters in animals. *Biological Reviews* **68**: 121–93.

Jackson, D. J. (1958) Observations on the biology of *Caraphractus cinctus* Walker (Hymenoptera: Mymaridae), a parasitoid of the eggs of Dytiscidae. I. Methods of rearing and numbers bred on different host eggs. *Transactions of the Royal Entomological Society of London* **110**: 533–54.

Johanowicz, D. L. & Hoy, M. A. (1996) *Wolbachia* in a predator–prey system: 16S ribosomal DNA analysis of two phytoseiids (Acari: Phytoseiidae) and their prey (Acari: Tetranychidae). *Annals of the Entomological Society of America* **89**: 435–41.

Kajita, H. (1993) Induction of males in the thelytokous wasp *Encarsia formosa* Gahan (Hymenoptera: Aphelinidae). *Applied Entomology and Zoology* **28**: 115–17.

Laraichi, M. (1978) L'effet de hautes températures sur le taux sexuel de *Oencyrtus fecundus* Hymenoptera: Encyrtidae. *Entomologia Experimentalis et Applicata* **23**: 237–42.

Laven, H. (1967a) Speciation and evolution in *Culex pipiens*. In *Genetics of Insect Vectors of Disease*, ed. J. W. Wright & R. Pal, pp. 251–75. Amsterdam: Elsevier.

Laven, H. (1967b) Eradication of *Culex pipiens fatigans* through cytoplasmic incompatibility. *Nature* **216**: 383–4.

Legner, E. F. (1985) Effects of scheduled high temperature on male production in thelytokous *Muscidifurax uniraptor* (Hymenoptera: Pteromalidae). *Canadian Entomologist* **117**: 383–9.

Legner, E. F. (1987) Transfer of thelytoky to arrhenotokous *Muscidifurax raptor* Girault and Sanders (Hymenoptera: Pteromalidae). *Canadian Entomologist* **119**: 265–71.

Leslie, T. F. (1984) A sex ratio condition in *Oncopeltus fasciatus*. *Journal of Heredity* **75**: 260–4.

Luck, R. F., Stouthamer, R. & Nunney, L. P. (1993) Sex determination and sex ratio patterns in parasitic Hymenoptera. In *Evolution and Diversity of Sex Ratio of Insects and Mites*, ed. D. L. Wrensch & M. A. Ebbert, pp. 442–76. New York: Chapman and Hall.

Matsuka, M., Hashi, H. & Okada, I. (1975) Abnormal sex-ratio found in the lady beetle *Harmonia axyridis* Palla (Coleoptera: Coccinellidae). *Applied Entomology and Zoology* **10**: 84–9.

Moran, N. & Baumann, P. (1994) Phylogenetics of cytoplasmically inherited microorganisms of arthropods. *Trends in Ecology and Evolution* **9**: 15–20.

Mulcahey, D. & Pascho, R. J. (1984) Adsorption to fish sperm of vertically transmitted fish viruses. *Science* **225**: 333–5.

Nagarkatti, S. (1970) The production of a thelytokous hybrid in an interspecific cross between two species of *Trichogramma* (Hym. Trichogrammatidae). *Current Science* **39**: 76–8.

Noda, H. (1984) Cytoplasmic incompatibility in allopatric field populations of the small brown planthopper, *Laodelphax striatellus*, in Japan. *Entomologia Experimentalis et Applicata* **35**: 263–7.

Nur, U., Werren, J. H., Eickbush, D. G., Burke, W. D. & Eickbush, T. H. (1988) A 'selfish' B chromosome that enhances its transmission by eliminating the paternal genome. *Science* **240**: 512–4.

O'Neill, S. L. (1991) Cytoplasmic incompatibility in *Drosophila* populations: influence of assortative mating on symbiont distribution. *Journal of Invertebrate Pathology* **58**: 436–43.

O'Neill, S. L. & Karr, T. L. (1990) Bidirectional incompatibility between conspecific populations of *Drosophila simulans*. *Nature* **348**: 178–80.

O'Neill, S. L., Giordano, R., Colbert, A. M. E. & Karr, T. L. (1992) 16S rRNA phylogenetic anaylsis of the bacterial endosymbionts associated with cytoplasmic incompatibility in insects. *Proceedings of the National Academy of Science of the USA* **89**: 2699–702.

Orphanides, G. M. & Gonzales, D. (1970) Identity of a uniparental race of *Trichogramma pretiosum* (Hymenoptera: Trichogrammatidae). *Annals of the Entomological Society of America* **63**: 1784–6.

Osawa, N. (1992) Sibling cannibalism in the ladybird beetle *Harmonia axyridis*: fitness consequences for mother and offspring. *Researches on Population Ecology* **31**: 153–60.

Perrot-Minnot, M., Guo, L. R. & Werren, J. H. (1996) Single and double infections with *Wolbachia* in the parasitic wasp *Nasonia vitripennis*: effects on compatibility. *Genetics* **143**: 961–72.

Pijls, J. W. A. M., van Steenbergen, J. J. & van Alphen, J. J. M. (1996) Asexuality cured: the relations and differences between sexual and asexual *Apoanagyrus diversicornis*. *Heredity* **76**: 506–13.

Reed, K. M. & Werren, J. H. (1995) Induction of paternal genome loss by the paternal-sex ratio chromosome and cytoplasmic incompatibility bacteria (*Wolbachia*): a comparative study of early embryonic events. *Molecular Reproduction and Development* **40**: 408–18.

Rigaud, T. & Juchault, P. (1992) Genetic control of the vertical transmission of a cytoplasmic sex factor in *Armadillidium vulgare* Latr. (Crustacea, Oniscidea). *Heredity* **68**: 47–52.

Rigaud, T. & Juchault, P. (1993) Conflict between feminizing sex ratio distorters and an autosomal masculinizing gene in the terrestrial isopod *Armadillidium vulgare* Latr. *Genetics* **133**: 247–52.

Rössler, Y. & DeBach, P. (1972) The biosystematic relations between a thelytokous and an arrhenotokous form of *Aphytis mytilaspidis* (Le Baron) [Hymenoptera: Aphelinidae] I. The reproductive relations. *Entomophaga* **17**: 391–423.

Rössler, Y. & DeBach, P. (1973) Genetic variability in a thelytokous form of *Aphytis mytilaspidis* (Le Baron) (Hymenoptera: Aphelinidae). *Hilgardia* **42**: 149–75.

Rousset, F. & Solignac, M. (1995) Evolution of single and double *Wolbachia* symbioses during speciation in the *Drosophila simulans* complex. *Proceedings of the National Academy of Science of the USA* **92**: 6389–93.

Rousset, F., Bouchon, D., Pintureau, B., Juchault, P. & Solignac, M. (1992a) *Wolbachia* endosymbionts responsible for various alterations of sexuality in arthropods. *Proceedings of the Royal Society of London, Series B* **250**: 91–8.

Rousset, F., Vautrin, D. & Solignac, M. (1992b) Molecular identification of *Wolbachia*, the agent of cytoplasmic incompatibility in *Drosophila simulans*, and variability in relation with host mitochondrial types. *Proceedings of the Royal Society of London, Series B* **247**: 163–8.

Rueda, L. M. & Axtell, R. C. (1985) Guide to common species of pupal parasites (Hymenoptera: Pteromalidae) of the house fly and other muscoid flies associated with poultry and livestock manure. *North Carolina Agricultural Research Service Technical Bulletin* **278**: 1–88.

Ryan, S. L. & Saul, G. B. (1968) Post-fertilization effect of incompatibility factors in *Mormoniella*. *Molecular and General Genetics* **103**: 29–36.

Ryan, S. L., Saul, G. B. & Conner, G. W. (1985) Aberrant segregation of R-locus genes in male progeny from incompatible crosses in *Mormoniella*. *Journal of Heredity* **76**: 21–6.

Ryan, S. L., Saul, G. B. & Conner, G. W. (1987) Separation of factors containing R-locus genes in *Mormoniella* stocks derived from aberrant segregation following incompatible crosses. *Journal of Heredity* **78**: 273–5.

Saul, G. B. (1961) An analysis of non reciprocal cross incompatibility in *Mormoniella vitripennis* (Walker). *Zeitschrift für Vererbungslehre* **92**: 28–33.

Schilthuizen, M. & Stouthamer, R. (1997) Horizontal transmission of parthenogenesis-inducing microbes in *Trichogramma* wasps. *Proceedings of the Royal Society of London, Series B* **264**: 361–6.

Shull, A. F. (1948) An all-female strain of lady beetles with reversions to normal sex ratios. *American Naturalist* **82**: 241–51.

Sinkins, S. P., Braig, H. R. & O'Neill, S. L. (1995a) *Wolbachia* superinfections and the expression of cytoplasmic incompatibility. *Proceedings of the Royal Society of London, Series B* **261**: 325–30.

Sinkins, S. P., Braig, H. R. & O'Neill, S. L. (1995b) *Wolbachia pipientis*: bacterial density and unidirectional cytoplasmic incompatibility between infected populations of *Aedes albopictus*. *Experimental Parasitology* 81: 284–91.

Sinkins, S. P., Curtis, C. F. & O'Neill, S. L. (1997) The potential application of inherited symbiont systems to pest control. In *Influential passengers: Inherited Microorganisms and Invertebrate Reproduction*, ed. S. L. O'Neill, A. A. Hoffmann & J. H. Werren, pp. 155–75. Oxford: Oxford University Press.

Sironi, M., Bandi, C., Sacchi, L., Sacco, B. D., Damiani, G. & Genchi, C. (1995) Molecular evidence for a close relative of the arthropod endosymbiont *Wolbachia* in a filarial worm. *Molecular and Biochemical Parasitology* 74: 223–7.

Skinner, S. W. (1982) Maternally inherited sex ratio in the parasitoid wasp *Nasonia vitripennis*. *Science* 215: 1133–4.

Skinner, S. W. (1983) Extrachromosomal Sex Ratio Factors in the Parasitoid Wasp, *Nasonia* (= *Mormoniella*) *vitripennis*. PhD Thesis: University of Utah.

Skinner, S. W. (1985) Son-killer: a third extrachromosomal factor affecting the sex ratio in the parasitoid wasp, *Nasonia* (= *Mormoniella*) *vitripennis*. *Genetics* 109: 745–59.

Skinner, S. W. (1987) Paternal transmission of an extrachromosomal factor in a wasp: evolutionary implications. *Heredity* 59: 47–53.

Stearns, S. C. (1992) *The Evolution of Life Histories*. Oxford: Oxford University Press.

Stevens, L. (1993) Cytoplasmically inherited parasites and reproductive success in *Tribolium* flour beetles. *Animal Behaviour* 46: 305–10.

Stevens, L. & Wade, M. J. (1990) Cytoplasmically inherited reproductive incompatibility in *Tribolium* flour beetles: the rate of spread and effect on population size. *Genetics* 124: 367–72.

Stevens, L. & Wicklow, D. T. (1992) Multispecies interactions affect cytoplasmic incompatibility in *Tribolium* flour beetles. *American Naturalist* 140: 642–53.

Stouthamer, R. (1993) The use of sexual versus asexual wasps in biological control. *Entomophaga* 38: 3–6.

Stouthamer, R. & Kazmer, D. (1994) Cytogenetics of microbe-associated parthenogenesis and its consequences for gene flow in *Trichogramma* wasps. *Heredity* 73: 317–27.

Stouthamer, R. & Luck, R. F. (1991) Transition from bisexual to unisexual cultures in *Encarsia perniciosi* Tower (Hymenoptera: Aphelinidae): new data and a reinterpretation. *Annals of the Entomological Society of America* 84: 150–7.

Stouthamer, R. & Luck, R. F. (1993) Influence of microbe-associated parthenogenesis on the fecundity of *Trichogramma deion* and *T. pretiosum*. *Entomologia Experimentalis et Applicata* 67: 183–92.

Stouthamer, R. & Werren, J. H. (1993) Microbes associated with parthenogenesis in wasps of the genus *Trichogramma*. *Journal of Invertebrate Pathology* 61: 6–9.

Stouthamer, R., Luck, R. F. & Hamilton, W. D. (1990a) Antibiotics cause parthenogenetic *Trichogramma* (Hymenoptera/Trichogrammatidae) to revert to sex. *Proceedings of the National Academy of Science of the USA* 87: 2424–7.

Stouthamer, R., Pinto, J. D., Platner, G. R. & Luck, R. F. (1990b) Taxonomic status of thelytokous forms of *Trichogramma* (Hymenoptera: Trichogrammatidae). *Annals of the Entomological Society of America* 83: 475–581.

Stouthamer, R., Breeuwer, J. A. J., Luck, R. F. & Werren, J. H. (1993) Molecular identification of microorganisms associated with parthenogenesis. *Nature* 361: 66–8.

Stouthamer, R., Luko, S. & Mak, F. (1994) Influence of parthenogenesis *Wolbachia* on host fitness. *Norwegian Journal of Agricultural Sciences* Supplement no. 16: 117–22.

Stouthamer, R., Luck, R. F., Pinto, J. D., Platner, G. R. & Stephens, B. (1996) Non-reciprocal cross-incompatibility in *Trichogramma deion*. *Entomologia Experimentalis et Applicata* 80: 481–9.

Subbarao, S. K. (1982) Cytoplasmic incompatibility in mosquitoes. In *Recent Developments in the Genetics of Insect Disease Vectors*, pp. 313–41. Champaign, IL: Stipes Publishing.

Turelli, M. (1994) Evolution of incompatibility-inducing microbes and their hosts. *Evolution* 48: 1500–13.

Turelli, M. & Hoffman, A. A. (1991) Rapid spread of an inherited incompatibility factor in California *Drosophila*. *Nature* 353: 440–2.

Turelli, M., Hoffman, A. A. & McKechnie, S. W. (1992) Dynamics of cytoplasmic incompatibility and mtDNA variation in natural *Drosophila simulans* populations. *Genetics* 132: 713–23.

Viggiani, G. & Garonna, A. P. (1993) Le specie italiane del complesso *Archenomus* Howard, *Archenomiscus* Nikolskaja, *Hispaniella* Mercet e *Pteroptrix* Westwood, con nuove combinazioni generiche (Hymenoptera: Aphelinidae). *Bollettino del Laboratorio Entomologia agraria Filippo Silvestri* **48 (1991)**: 57–88.

Wade, M. J. & Chang, N. W. (1995) Increased male fertility in *Tribolium confusum* beetles after infection with the intracellular parasite *Wolbachia*. *Nature* **373**: 72–4.

Wade, M. J. & Stevens, L. (1985) Microorganism mediated reproductive isolation in flour beetles (Genus *Tribolium*). *Science* **227**: 527–8.

Wade, M. J. & Stevens, L. (1994) The effect of population subdivision on the rate of spread of parasite-mediated cytoplasmic incompatibility. *Journal of Theoretical Biology* **167**: 81–7.

Werren, J. H. (1987) The coevolution of autosomal and cytoplasmic sex ratio factors. *Journal of Theoretical Biology* **124**: 317–34.

Werren, J. H. (1991) The paternal-sex-ratio chromosome of *Nasonia*. *American Naturalist* **137**: 392–402.

Werren, J. H. (1997a) Biology of *Wolbachia*. *Annual Review of Entomology* **42**: 587–609.

Werren, J.H. (1997b) *Wolbachia* and speciation. In *Endless Forms: Species and Speciation*, ed. D. Howard & S. Berlocher. Oxford: Oxford University Press.

Werren, J. H. & Jaenike, J. (1995) *Wolbachia* and cytoplasmic incompatibility in mycophagous *Drosophila* and their relatives. *Heredity* **75**: 320–6.

Werren, J. H. & van den Assem, J. (1986) Experimental analysis of a paternally inherited extrachromosomal factor. *Genetics* **114**: 217–33.

Werren, J. H., Skinner, S. W. & Charnov, E. L. (1981) Paternal inheritance of a daughterless sex ratio factor. *Nature* **293**: 467–8.

Werren, J. H., Skinner, S. W. & Huger, A. M. (1986) Male-killing bacteria in a parasitic wasp. *Science* **231**: 990–2.

Werren, J. H., Nur, U. & Eickbush, D. (1987) An extrachromosomal factor causing loss of paternal chromosomes. *Nature* **327**: 75–6.

Werren, J. H., Hurst, G. D. D., Zhang, W., Breeuwer, J. A. J., Stouthamer, R. & Majerus, M. E. N. (1994) Rickettsial relative associated with male killing in the ladybird beetle (*Adalia bipunctata*). *Journal of Bacteriology* **176**: 388–94.

Werren, J. H., Guo, L. & Windsor, D. W. (1995a) Distribution of *Wolbachia* in neotropical arthropods. *Proceedings of the Royal Society of London, Series B* **262**: 147–204.

Werren, J. H., Zhang, W. & Guo, L. R. (1995b) Evolution and phylogeny of *Wolbachia*: reproductive parasites of arthropods. *Proceedings of the Royal Society of London, Series B* **261**: 55–71.

White, M. J. D. (1954) *Animal Cytology and Evolution*. Cambridge: Cambridge University Press.

White, M. J. D. (1964) Cytogenetic mechanisms in insect reproduction. In *Insect Reproduction* 2nd edn, ed. K. C. Highnam, pp. 1–12. London: Royal Entomological Society.

Whiting, P. W. (1945) The evolution of male haploidy. *Quarterly Review of Biology* **20**: 231–60.

Williams, E. H., Fields, S. & Saul, G. B. (1993) Transfer of incompatibility factors between stocks of *Nasonia* (=*Mormoniella*) *vitripennis*. *Journal of Invertebrate Pathology* **61**: 206–10.

Williamson, D. L. & Poulson, D. F. (1979) Sex ratio organisms (spiroplasmas) of *Drosophila*. *The Mycoplasmas* **3**: 175–208.

Wilson, F. & Woolcock, L. T. (1960) Temperature determination of sex in a parthenogenetic parasite, *Ooencyrtus submetallicus* (Howard) (Hymenoptera: Encyrtidae). *Australian Journal of Zoology* **8**: 153–69.

Yen, J. H. & Barr, A. R. (1971) New hypothesis of the cause of cytoplasmic incompatibility in *Culex pipiens* L. *Nature* **232**: 657–8.

Yen, J. H. & Barr, A. R. (1973) The etiological agent of cytoplasmic incompatibility in *Culex pipiens*. *Journal of Invertebrate Pathology* **22**: 242–50.

Zchori-Fein, E., Roush, R. T. & Hunter, M. S. (1992) Male production induced by antibiotic treatment in *Encarsia formosa* (Hymenoptera: Aphelinidae), an asexual species. *Experientia* **48**: 102–5.

Zchori-Fein, E., Rosen, D. & Roush, R. T. (1994) Microorganisms associated with thelytoky in *Aphytis lingnanensis* Compere (Hymenoptera: Aphelinidae). *International Journal of Morphology and Embryology* **23**: 169–72.

Zchori-Fein, E., Faktor, O., Zeidan, M., Gottlieb, Y., Czosnek, H. & Rosen, D. (1995) Parthenogenesis-inducing microorganisms in *Aphytis* (Hymenoptera: Aphelinidae). *Insect Molecular Biology* **4**: 173–8.

15

The evolution of overexploitation and mutualism in plant–herbivore–predator interactions and its impact on population dynamics

MAURICE W. SABELIS, MINUS VAN BAALEN, JAN BRUIN, MARTIJN EGAS,
VINCENT A. A. JANSEN, ARNE JANSSEN AND BAS PELS

Introduction

Populations of arthropod herbivores may show periodical outbreaks, large amplitude cycles, strongly bounded fluctuations or stable equilibria. At small spatial scales they may also show dynamics different from those at large spatial scales, and they may even go extinct. There is a need to experimentally test models predicting the dynamical consequences of the mechanisms underlying these patterns. In particular, what needs an explanation is the observation that plants retain a green appearance despite attacks by herbivores (Hairston et al., 1960; Strong et al., 1984). There are two possible explanations: (i) plants defend themselves effectively against herbivores, (ii) predators suppress herbivore populations to very low levels. Hairston et al. (1960) tacitly ignored the first and emphasized the latter in formulating their so-called 'the world is green' hypothesis. The two explanatory mechanisms, however, are not mutually exclusive; the plant may promote predator foraging success and the herbivore's enemies may use the facilities offered by the plant.

Price et al. (1980) stimulated a growing awareness that plants may defend themselves both directly against herbivorous arthropods and indirectly by promoting natural enemies, for example, by providing protection, food, and alarm signals to the enemies of the herbivore. Direct defenses include plant structures that hinder feeding by the herbivore, and secondary plant compounds that inhibit digestion, intoxicate the herbivore or deter feeding. Before Price et al. (1980) appeared, examples of indirect defenses were already well-known from ant–plant interactions (Janzen, 1966; Bentley, 1977; see also historical reviews by Beattie, 1985, and Jolivet, 1996). Eighteen years later, it is clear that mutualistic interactions between plants and predatory arthropods are widespread (Dicke & Sabelis, 1988, 1989, 1992; Drukker et al., 1995; Turlings et al., 1995; Walter, 1996).

The aim of this chapter is to provide a rationale for the evolution of mutualisms between plants and predatory arthropods based on natural selection theory. To understand the evolution of such mutualisms, one needs to identify the advantage to the plant, predict the consequences for the population dynamics of herbivorous and predatory arthropods, and elucidate how the dynamics in turn may affect the evolution of the mutualistic interaction. Understanding the evolution of plant defenses and their consequences for population dynamics at all three trophic levels will help us understand long-term changes in the effectiveness of pest control strategies, and thus to design long-term pest management strategies.

The structure of the chapter is as follows. First, we emphasize the complexities arising from the fact that the population dynamics of the plant, herbivorous, and predatory arthropods operate on very different spatial and temporal scales. Second, we introduce a simple predator–prey model that assumes that the spatial scale of the dynamics is identical to that of an individual plant and the temporal scale is shorter than the generation time of the plant. Based on this model we identify the main categories of defensive strategies of a plant. Third, we summarize the various ways in which plants promote the effectiveness of predatory arthropods. Fourth, we analyze the tritrophic game, assuming equilibrium dynamics, and consider the case where plant defense strategies depend on the defense of neighboring plants. We then freeze the evolutionary process and consider the (meta)population dynamics of tritrophic systems. We also consider what happens when evolution and population dynamics are both in action and provide a hypothesis on optimal investment of plants in direct and indirect defense. Finally, we discuss the implications for plant resistance breeding, evolution of differential pesticide resistance in predators *versus* prey, and the effectiveness of biological control.

On scales of time and space

To study dynamics in tritrophic systems, it is not always necessary to model changes in populations at each trophic level. Arthropod predator–herbivore systems on plants have usually been modeled as ditrophic systems with the plant as an environmental variable (e.g., Hassell, 1978). The assumptions underlying this simplification follow from the ecological time-scale arguments reviewed below. Another assumption underlying classic models of predator–prey interactions on plants is a high degree of population mixing at the level of predators and herbivores. However, later models consider that predators have to cope with clumped herbivore distributions (e.g., Hassell, 1978). Clumped distributions reflect population viscosity and impose a spatial scale that may be relevant to the role of plant defenses. One may ask whether the defense of an individual plant initially affects only the herbivore population it harbors (and much later the herbivore population as a whole) or whether it permeates population-wide and without delay (as a consequence of high herbivore mobility). We hope to convince the reader that the evolution of direct and indirect plant defenses cannot be properly understood without considering the dynamical and evolutionary responses at higher trophic levels.

Ecological time-scales

Interactions among populations of plants, herbivores, and predators are characterized by quite different time-scales at each trophic level. Typically, plants have much longer generation times than arthropods, and generation times of predator and prey depend largely on body size (Sabelis, 1992). True predators are usually as big as, or larger than, their prey, and, if larger, their generation times are

also usually longer, although usually not as long as the generation time of the plants. These time-scale differences are of crucial importance. When host–plant populations change over much longer time-scales than predatory and herbivorous arthropods the plant population may often be considered to be constant. Similarly, when the predator is *much* larger than its prey, changes in the predator population are irrelevant to the population fluctuations of the herbivores. This may apply to insectivorous mammals and birds, but not to predatory arthropods. However, if the predatory arthropods feed on many different prey, their population changes are less dependent on abundance of the herbivore under consideration and so predator populations may be considered constant.

Time-scale and dietary-scale considerations are pivotal in deciding whether to model the dynamics of the herbivore alone, the herbivore and the predator, or all three interacting populations. Because the generation time of plants is an order of magnitude longer than the generation times of the arthropods, arthropod predator–herbivore systems are commonly modeled as two-species interactions. However, if leaf lifespans are comparable to arthropod lifespans, and leaf abundance rather than numbers of plants determine herbivore food supply, it is necessary to model all three trophic levels (Sabelis *et al.*, 1991; Jansen & Sabelis, 1995). This is not the only reason why it may be more appropriate to consider the full tritrophic system; the spatial distributions of herbivores and predators provide another.

Hierarchical spatial scales

Many herbivorous arthropods form patchy infestations on their host plants. The causes are varied. Weakened hosts or hosts whose defenses are overwhelmed by pioneer attacks become attractive to con- and heterospecific herbivores, thereby creating local population aggregations, as in the case of bark beetles (Berryman *et al.*, 1985; De Jong & Sabelis, 1988, 1989). In other cases aggregations result from large egg clutches of the founder(s), as in various species of moths (larch bud moths, gypsy moths, ermine moths), or they result from founders with a high intrinsic capacity of population increase (relative to the rate of dispersal), as in many smaller herbivorous arthropods. Such aggregations may have a lifespan close to that of the arthropod, as in many moths and bark beetles, but they may also persist for several generations. The latter is the case for smaller arthropods (scale insects, mealybugs, aphids, leafhoppers, whiteflies, thrips, spider mites, and rust mites) because of their short generation time, low *per capita* food demands and their tendency to suppress long-range dispersal, as long as food is nearby. The major point is that aggregations represent a hierarchical level between that of the individual and the population. Multi-generation aggregations have a lifespan approaching that of annual plants (or the lifetime of leaves from perennials), and these systems should be modeled as tritrophic (Sabelis *et al.*, 1991; Jansen & Sabelis, 1995). Individual plants (or leaves) are then the units at the lowest trophic level and in the simplest case each plant harbors a separate multi-generation population of predators and herbivores. In that case, the spatial scales of individual plants and local arthropod populations overlap. If, however, the local arthropod population covers several individual plants, then modeling requires specification of plant neighborhoods or patches of plants subject to attack by the same herbivore population. This obviously complicates modeling population dynamics and, to an even greater extent, evolutionary dynamics of tritrophic interactions.

Metapopulation dynamics and evolutionary time scales

One reason is that the unit of selection is the individual plant and the selective advantage of defense depends on the extent to which a plant influences local dynamics of herbivores and predators by direct and/or indirect defenses. Another is that predators and herbivores are independent players in the evolutionary game. Whether their food-exploitation strategies are in the interest of the plant remains to be seen. Moreover, individual herbivores may differ in the exploitation strategy they employ, and the success of each strategy does not merely depend on

reproductive performance when exploiting a plant alone, but also on how it is affected by competition with other exploiters on the plant. The composition of exploitation strategies within each local population depends on the overall dynamics of the ensemble of local populations in the tritrophic system. This so-called metapopulation dynamics depends in turn on the relative success of the strategies competing within local populations. Thus, metapopulation dynamics, the evolution of food source exploitation strategies, and the evolution of plant defense strategies are intertwined.

We present some early results of analyses of this complex problem, but not without alerting the reader that evolution in tritrophic systems with a patchy structure is still largely an open problem, awaiting a more comprehensive analysis.

Predator–herbivore dynamics on the plant

To understand the range of possible plant defense strategies it is instructive to model the dynamics of herbivores assuming that the spatial scale covered by a local predator and prey population corresponds to that of an individual plant (or group of kin-related plants). We want the simplest model of herbivore dynamics directly after discovery by predators and then ask how an individual plant can influence the dynamics so as to decrease damage by the herbivores.

We assume that herbivores form patchy aggregations on clusters of leaves, and that predators entering such patches can move around freely and spend little time in moving between patches. Thus, the leaf cluster is a coherent and homogeneous arena for strongly coupled predator–prey interactions (Fig. 15.1).

We further assume that the predators search for the highest prey densities and minimize interference with competitors. Finally, we assume that herbivore density within newly infested leaf areas is constant (Sabelis, 1991; Sabelis & Janssen, 1994). Thus, herbivores raise a fixed number of offspring per unit of plant area and predators reaching a freshly colonized leaf site continue to eat prey until they do better by moving to another site nearby. These assumptions lead to a constant rate of predation, which is much

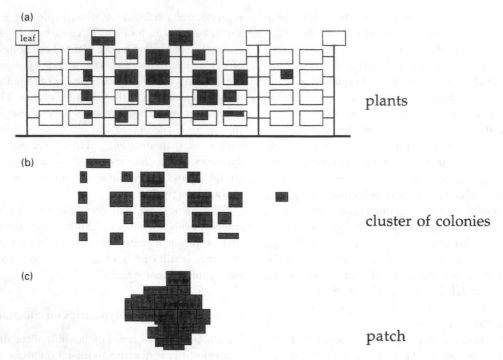

(a)

leaf

plants

(b)

cluster of colonies

(c)

patch

Figure 15.1. A patchy infestation of small herbivorous arthropods in a row of plants (a). It is inspired by observations on spider mites, but in essence applies to many other herbivorous arthropods. The infested leaf parts are excised (b) and then put together as a jig-saw puzzle (c). The latter forms the patch or arena within which the interaction between predator and prey takes place (assuming negligible time spent in moving between infested leaves).

like 'eating a pancake': a constant amount of food at each bite until there is nothing left of the pancake.

As long as the pancake is not completely eaten, predators can maximize the rate of predation, development, and reproduction. Hence, under conditions of a stable age distribution they will reach their intrinsic capacity for population increase. Similar assumptions can be made with respect to herbivore growth capacity in the absence of the predators. Let us for simplicity assume that prey supply is sufficient to make predators stay until virtually all prey are eaten. The dynamics of predator and herbivore numbers can be described by two linear differential equations, which differ from the classic Lotka–Volterra models in that the predation term now only depends on the number of predators (Sabelis, 1992; Janssen & Sabelis, 1992):

$$\frac{dN}{dt} = \alpha N - \beta P$$

$$\frac{dP}{dt} = \gamma P \tag{15.1}$$

This is the so-called 'pancake predation' model with t being the time since initiation of the predator–prey interaction; N_t is the number of prey at time t; P_t is the number of predators at time t; α is the rate of prey population growth; β is the maximum predation rate; and γ is the rate of predator population growth. Analytical solutions for the number of predators and prey are readily obtained:

$$N_t = N_0 e^{\alpha t} - P_0 \frac{\beta}{\gamma - \alpha}(e^{\gamma t} - e^{\alpha t})$$

$$P_t = P_0 e^{\gamma t} \tag{15.2}$$

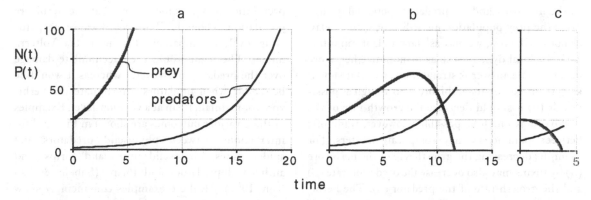

Figure 15.2. Three types of local predator–prey dynamics according to the pancake predation model (with parameters $\alpha=0.3$, $\beta=1$, $\gamma=0.25$, $\mu=0$, and $N_0=25$): (a) prey increase ($P_0=8$), $N_0/P_0=25/8$; (b) delayed prey decline ($P_0=3$), $N_0/P_0=25/3$; and (c) immediate prey decline ($P_0=1$) $N_0/P_0=25/1$. The general conditions for each of these types of dynamics are discussed in the text.

Predators grow exponentially until a time, τ, when all prey are eaten (and the predators move on to another food source). The time from predator invasion to prey extinction can be expressed as function of α, β, and γ and the initial numbers of predator and prey, N_0 and P_0:

$$\tau = \tau(\alpha, \beta, \gamma, N_0, P_0)$$

$$\tau = \frac{1}{\gamma - \alpha} \ln\left(1 + \frac{\gamma - \alpha}{\beta}\frac{N_0}{P_0}\right) \qquad (15.3)$$

Three types of dynamical behavior of the prey population may occur (Fig. 15.2): (i) continuous increase (but at a pace slower than the intrinsic rate of prey population growth: *continued increase; predator surfing on a prey wave*), (ii) initial increase, followed by decrease until extinction (*delayed decline*), and (iii) continuous decay until extinction (*immediate decline*). In the first case, predatory arthropods cannot prevent local prey population outbreaks and in the real world this would end in overexploitation of the plant. Immediate decline of the herbivores occurs when the differential equation for herbivore dynamics is negative:

$$\alpha N_0 < \beta P_0$$

or:

$$\frac{P_0}{N_0} > \frac{\alpha}{\beta}$$

In words, the ratio of predators to herbivores should exceed the ratio of the *per capita* population growth rate of the herbivore and the maximum *per capita* predation rate.

The conditions for immediate or delayed herbivore decline are found where the time to prey extinction has a finite value:

$$\frac{P_0}{N_0} > \frac{\alpha - \gamma}{\beta}$$

If this condition is not met, herbivores continue to increase and predators 'surf' on the 'population growth wave' of the herbivore, resulting in the overexploitation of the plant.

The overall damage incurred by the plant over the interaction period can be expressed in the number of herbivore-days, D_r, or the area under the curve in Fig. 15.2 expressing the temporal changes in the size of the herbivore population:

$$D(\alpha, \beta, \gamma, \tau, N_0, P_0) = \frac{1}{\alpha}\left[P_0\frac{\beta}{\gamma}(e^{\gamma\tau} - 1) - N_0\right] \qquad (15.4)$$

Note that this measure of damage strongly depends on the exponential term and thus on the time to prey extinction (τ) and the *per capita* growth rate of the predator population (γ). Thus, given the initial numbers (N_0 and P_0) and estimates of the parameters (α, β, and γ) it is possible to assess the overall damage by the herbivore and the predator's potential to suppress the prey population immediately, with a delay, or not at all. Now, we may ask how a plant can influence the local dynamics so as to minimize herbivore damage. The answer is straightforward: (i) it should keep the predator-to-herbivore ratio as high as possible, or (ii) it should decrease the growth rate of the herbivore as much as possible. However, trade-offs between parameters may complicate matters. For example, decreasing the growth rate of the herbivore (α) by toxins may also decrease the predation rate (β) and the growth rate of the predator (γ). The herbivore may even sequester plant toxins to defend itself against predators. Thus, the plant will only profit if it decreases the herbivore's growth rate proportionally stronger than the predation rate and the growth rate of the predator.

Another message of the equations is that plants should promote predators with high predation and growth rates. Often, these demands conflict, because the predation rate tends to increase with body size, whereas the intrinsic rate of increase tends to decrease with body size. Such trends are clear for predators of phytophagous thrips, such as mirids, anthocorids, predatory thrips, and predatory mites (Figs. 15.3(a) and (b); Sabelis & Van Rijn, 1997). At lower taxonomic levels (within-family, within-genus) the picture may be different. For example, within the Phytoseiidae – a family of plant-inhabiting predatory mites – positive high correlations exist between β and γ (Figs. 15.3(d) and (e); Janssen & Sabelis, 1992; Sabelis & Janssen, 1994). One may wonder whether the plant could selectively admit more effective predators over others, but it is not clear how, given that predators will go for the most profitable prey!

Predators alone decide whether it is profitable to stay on a plant or not. The pancake predation model assumes that predators stay until all prey are eaten. This scenario is not implausible, because it may be risky to disperse and search for new herbivore patches. In fact, it is frequently observed, e.g., in interactions among predatory mites and spider mites (Figs. 15.4(a) and (b); Sabelis & Van der Meer, 1986). Predators may of course leave before the exact moment of prey extinction. This would relieve predation pressure on herbivores, and predator-to-

prey ratios may become so low that the herbivore population rebounds. This could cause perpetuating cycles, comparable to the Lotka–Volterra model. The plant would then accumulate damage over the predator–prey cycles whereas it would be better off when predators exterminate the herbivores or maintain them at a very low level. Examples of these various outcomes are shown in Fig. 15.4 for interactions between phytoseiid predators and spider mites, phytoseiid mites and thrips, and anthocorid predators and thrips (Sabelis & Van Rijn, 1997). All these examples convincingly show that local herbivore populations are strongly suppressed by predators. Thus, the 'pancake predator' model seems to be a good caricature of the initial predator–prey population cycle, and in some special cases also for the end phase (Fig. 15.4(a)-(e), but in others the end phase differs (e.g., a stable, non-zero, but low level: Fig. 15.4(f)). Thus, for all cases where colonization of predator and prey go through one cycle before ending in extermination or reaching a very low equilibrium, this model helps one to understand the role of indirect and direct plant defenses.

Predator arrestment on herbivore patches is decisive in the success of indirect defense strategies of the plant. For example, if we assume a constant predator dispersal rate (i.e., independent of prey availability), then dispersal decreases the effective population growth rate of the predators. As shown in Fig. 15.5(a) small decreases of the growth rate have dramatic consequences for the duration of the interaction and for the number of herbivores attacking the plant. The overall damage to the plant increases nonlinearly with a reduction of the predator's population growth rate. Hence, it is important that several species of predators are strongly arrested in herbivore-colonized patches until the prey population is driven near extinction (Sabelis & Van der Meer, 1986; Sabelis & Van Rijn, 1997).

The herbivores can always make the last move in the game. They may not only develop resistance against the predators and overcome barriers raised by the plant, but ultimately, they may decide to leave the plant. If the plant gets rid of herbivores by attracting predators, and also by stimulating herbivore disper-

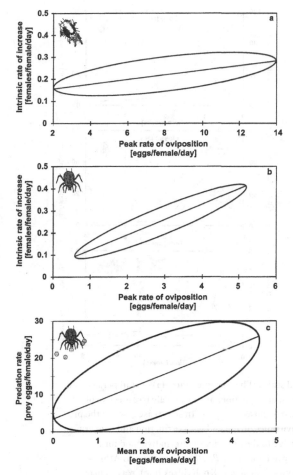

mites (Acari: Tetranychidae) plotted against the peak rate of oviposition (Sabelis, 1990); (b) intrinsic rate of population increase (females/female/day) of predatory mites (Acari: Phytoseiidae) plotted against the peak rate of oviposition (Sabelis & Janssen, 1994); (c) *per capita* predation rate of adult female, predatory mites on eggs/larvae of spider mites, plotted against the mean rate of oviposition (Janssen & Sabelis, 1992), (d) intrinsic rate of population increase (females/female/day) of predators of thrips (Acari: Phytoseiidae; Insecta: Aeolothripidae, Anthocoridae, Miridae, Nabidae, Coccinellidae), plotted against adult size (Sabelis & Van Rijn, 1997), (e) *per capita* predation rate of arthropod predators on young thrips larvae, plotted against adult size (Sabelis & Van Rijn, 1997). Note that (c) shows the predation rate of the adult females only, whereas (e) gives a weighted estimate, calculated as the sum of the products of predator-stage-specific predation rates and the share in the predator population assuming a stable age distribution.

Figure 15.3. Examples of ranges of the *per capita* population growth (prey, α; predator, γ) and predation (β) parameters in the pancake predation model for different groups of plant-inhabiting arthropods (herbivores, predators). The ranges are presented as ellipses around regression lines of the model parameters against ovipositional rate (as in (a) and (b) or against adult size (as in (d) and (e)). (a) Intrinsic rate of population increase (females/female/day) of spider

sal, then it gains disproportionately, as illustrated in Fig. 15.5(b).

In conclusion, plants can benefit in many ways by influencing the behavior and dynamics of predators and herbivores. Direct plant defenses do not merely slow down the herbivore growth rate, they may also affect predator impact, either positively (higher predator-to-herbivore ratio) or negatively (plant toxins protecting herbivore against predators). Indirect defenses also do not merely affect predator performance, they may also affect selection on herbivores, either positively (enemy avoidance) or negatively (resistance to predators). Hence, to understand the plant's allocation to direct and indirect defenses

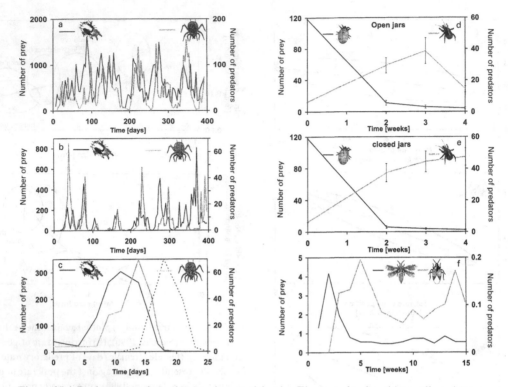

Figure 15.4. Predator–prey dynamics at various spatial scales. The examples show (a) overall persistence of predator–prey populations, but (b) and (c) show frequent extinctions at a local scale irrespective of whether the local populations reside in open ((c) and (d)) or closed (e) space. In some cases, where the prey have invulnerable stages or may hide in refuges, local dynamics do not end with extinction and converge to stable equilibria, but these stable levels can be very low due to the presence of alternative food for the predators (f). (a) Overall population fluctuations of the predatory mite *Phytoseiulus persimilis* Athias-Henriot (gray line) and the two-spotted spider mite *Tetranychus urticae* Koch (black line) in a circular system of six interconnected trays, each with 20 Lima-bean plants maintained at the two-leaf stage (by frequent removal of the apex and replacement of plants exhausted as a food source: Janssen *et al.*, 1997). (b) Predator–prey dynamics showing several events of prey (and next also predator) extinction on one of the six interconnected trays from the experiment described in (a) (Janssen *et al.*, 1997). (c) Local dynamics of the two-spotted spider mite *T. urticae* (black line) and the predatory mite *P. persimilis* (gray line) showing that first the prey population and then the predator population go extinct and that aerial dispersal of the predatory mites (broken line) takes place near the moment of prey extermination (B. Pels, unpublished data; see also Sabelis & Van der Meer, 1986). (d) and (e) dynamics of the bulb mite *Rhizoglyphus robini* Claparède and the soil predatory mite *Hypoaspis aculeifer* in either open (d) or closed (e) jars with scales of three lily bulbs (Lesna *et al.*, 1996), showing that the populations of bulb mites are driven close to extinction in closed as well as open jars, and that the population of predatory mites in the open jars starts to decline (due to dispersal) only after the bulb mite population is decimated, whereas the predator population in the closed jars does not decline. (f) Local dynamics of populations of the phytophagous thrips *Thrips palmi* Karny (black line) and the predatory bug *Orius* sp. (gray line) on eggplants (aubergines), showing that the thrips population first goes through a cycle and then persists at a very low level, whereas the predator population after an initial cycle persists for several months at a higher level than its prey (Kawai, 1995). This may well be an example where the predator population is promoted due to alternative food (i.e., pollen), thereby strongly suppressing the prey. The prey is not driven to extinction, but persists possibly due to the presence of invulnerable stages (i.e., eggs inserted in leaves and pupae in the soil: Sabelis & Van Rijn, 1997).

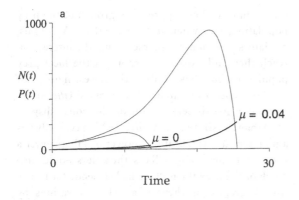

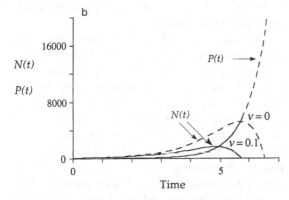

Figure 15.5. Sensitivity of local predator–prey dynamics according to the pancake model for changes in (a) the *per capita* population growth rate of the predator (γ) (or similarly, when predators disperse during the interaction period with rate μ, the effective predator growth rate $\gamma - \mu$). Parameter values are: $\mu = 0$ or 0.04; $\alpha = 0.3$; $\beta = 3$; $\gamma = 0.25$; $N_0 = 30$, $P_0 = 1$. (b) The *per capita* population growth of the herbivore (α) (or similarly, when prey disperse during the interaction period with rate n, the effective prey growth rate $\alpha - \nu$). Parameter values are: $\nu = 0$ (broken lines) or 0.1 (continuous lines), $\alpha = 1$, $\beta = 1$, $\gamma = 1.5$, $N_0 = 50$, $P_0 = 1$).

we should not only assess the costs, but also elucidate how these two types of defenses interact in their overall impact on herbivore damage: are they synergistic or antagonistic?

Tritrophic game theory and metapopulation structure

The evolution of direct and indirect plant defenses against arthropod herbivores is not a simple process. For one thing we have to take into account the defense and exploitation strategies of all three trophic levels. For another, time- and spatial scale considerations force us to consider tritrophic metapopulation models that add plant dynamical behavior to the repertoire of ditrophic models.

Moreover, metapopulation models are needed to understand evolution in structured populations. This is because evolutionary strategies play against each other at the patch level, their relative success determines metapopulation dynamics, and metapopulation processes in turn determine which strategies will compete again at the patch level. We refer to this chain of processes as population-dynamical feedback.

To gain insight into this complex problem carefully planned simplifications are required. Our first consideration is the evolution of traits at one trophic level in pairwise interactions with all others (predator–herbivore; herbivore–predator; plant–herbivore; herbivore–plant; plant–predator; predator–plant), assuming a steady state at the metapopulation level. Then, we tentatively discuss evolutionary outcomes in the full tritrophic system, again assuming a steady state metapopulation. Finally, we highlight some interesting evolutionary consequences of population dynamics of tritrophic systems.

The predator's dilemma: to milk or to kill?

Plants need to attract and arrest predators that are effective in suppressing the herbivore populations on them. The effectiveness of indirect plant defenses depends on the herbivore-exploitation strategies present in the predator population. These exploitation strategies can be varied. The pancake model for local predator–herbivore dynamics can be used to illustrate this. The strategy set is determined by combinations of the *per capita* growth rate of the predator (γ), the *per capita* predation rate (β), and, in spatially structured environments, the *per capita* dispersal rate of the predator. Of course, all predators will disperse when the herbivores are exterminated, but dispersal can also occur during the interaction. Both decreased predation and increased dispersal of the predator relieve the prey of predation pressure, thereby providing a larger future food source for the predator. Dispersal during the interaction, however,

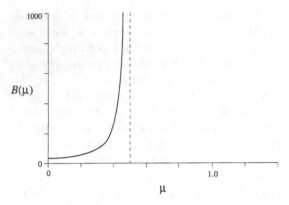

Figure 15.6. The relation between the overall production of predator dispersers (B) and the *per capita* rate of dispersal (m) during the predator–prey interaction period (from predator invasion to prey elimination: Van Baalen & Sabelis, 1995).

also increases the founding of new populations by the dispersers and in what follows we will focus on this dispersal trait alone.

Assume for simplicity that the *per capita* dispersal rate is a constant, given by parameter μ. The effective *per capita* population growth rate of the predators is reduced by μ, and the equation for the predator growth rate now reads:

$$\frac{dP}{dt} = (\gamma - \mu)P \qquad (15.5)$$

Since decreased values of γ have been shown to drastically alter local predator–prey dynamics, so will increased values of μ. As illustrated in Fig. 15.5(a), a small increase of μ from 0 to 0.04/day, greatly alters the area under the prey curve and thereby also the damage to the plant. Hence, a plant would benefit by retaining predators until all prey are eliminated, but success depends on the advantage to the predators (as well as on population-dynamical feedback). The success of a predator's exploitation strategy may be expressed as the number of dispersers produced during the interaction with prey, plus those dispersing after prey extermination. As shown in Fig. 15.6, the production of dispersers increases disproportionally with μ and reaches an asymptote when the *per capita* dispersal rate is so high that the preda-

tors cannot further suppress the growth of the prey population, i.e., when $\mu = \gamma - \alpha + \beta \, P_0/N_0$. Thus, predators that do not disperse during the interaction reach their full capacity to suppress the local prey population, but they produce the lowest number of dispersers per prey patch. This so-called *killer* strategy does worse (in terms of the production of dispersers) than the strategy of a *milker*, which typically has a non-zero dispersal rate. However, if killers enter a prey patch containing milkers, the killers would steal much of the prey the milkers had set aside for future use. Therefore, if there is a risk of invasions by killers, this will give rise to selection favoring killer-like exploitation strategies.

The outcome of the milker–killer dilemma is determined by a complex interplay of local competition between killers and milkers and global or meta-population dynamics. It depends on the probability of co-invasions in the same prey patches, on the resulting production of dispersers per prey patch and on metapopulation dynamics, as the latter in turn determines the probability of (co)-invasion. This population-dynamical feedback is complex; extensive book-keeping is required to continuously track the numbers of milkers and killers competing in local populations, dispersing into the global population and invading into local populations. Hence, simplification is necessary to gain insight. For example, Van Baalen & Sabelis (1995) assumed that all patches start with exactly the same number of predators and prey (the assumption of a metapopulation-wide equilibrium) and that the predators had time to reach their full production potential per prey patch (the assumption of sequential interaction rounds). In this setting they compared the reproductive success of one mutant predator clone with a *per capita* dispersal rate (μ_m) to the mean success of predators in the resident clone with another *per capita* rate of dispersal (μ_{res}). The question is then whether there exists a value of μ_{res} for which it does not pay any mutant to deviate (Maynard Smith, 1982). In particular, Van Baalen & Sabelis (1995) calculated the combination of parameters (P_0, N_0, α, β, γ) for which it does not pay to increase μ away from zero, i.e., the conditions favoring selection for killers. As illustrated in Fig. 15.7, killers are usually favored,

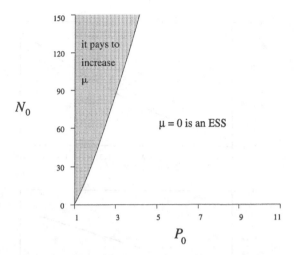

Figure 15.7. When does it pay to increase the *per capita* dispersal rate of the predator (μ) away from zero? A diagram showing that milker strategies are only possible when P_0 is low and N_0 sufficiently high (modified from Van Baalen & Sabelis, 1995).

except when the number of predator-foundresses is low and the number of prey-foundresses is high. In other words, the milkers are favored as long as they have a sufficiently large share in the local population to maintain control over the time to prey elimination (τ).

This analysis needs elaboration, taking into account (i) asynchrony in local dynamics, (ii) stochastic variation in predator and prey colonization rates (since these are probably low!), (iii) an upper bound to prey population size set by local resources, and (iv) metapopulation dynamics. Such extensions are likely to show that milkers achieving a longer interaction period are also exposed longer to subsequent predator invasions (and thus, face competition with killers sooner or later), that stochastic rather than uniform invasions will help to isolate milkers, thereby allowing them to gain full advantage of their exploitation strategy, and that food limitation on the prey will decrease opportunities for milkers, as they lose control over the interaction period (τ). As these factors have opposing effects, either killers or milkers could win the battle, or they may even coexist. Clearly, more theoretical work is needed. In

addition, experimental analysis of variation in predator exploitation strategies is important. Such an analysis carried out on the predatory mite *Phytoseiulus persimilis* Athias-Henriot, revealed that laboratory cultures harbor exclusively predators of the killer type (Sabelis & Van der Meer, 1986), whereas field-collected populations in the Mediterranean exhibit some variation in dispersal (B. Pels, unpublished data): in some populations from *Ricinus* plants on Sicily dispersal begins before the prey are eliminated, but it is still suppressed during most of the interaction period. Hence, evidence is equivocal that milkers exist in laboratory and field mite populations, but they may also be present in ants that tend colonies of honeydew-producing homopterans. Ants present on individual plants may be kin-related, as they probably originate from the same nest. Hence, the conditions for the evolution of milker-like exploitation strategies are met.

Obviously, the prevalence of killers is of great importance to the evolution of indirect plant defenses. By providing protection and food to predators and by signaling herbivore attack, plants increase the predator invasion rate into young herbivore colonies. Increased invasion, in turn, increases the probability of co-invasions of milkers and killers, which – all else being equal – ultimately favors the latter. So far, neither theoretical nor experimental analyses have addressed the possibility of more flexible strategies, such as: 'milk when exploiting the prey patch alone, and kill when other (e.g., non-kin) predators have entered the same patch'.

The herbivore's dilemma: to stay or to leave?

Like the predators, the herbivores are independent players in the tritrophic game. When they are discovered by milkers, they can still achieve reproductive success, especially when the milker has such a high dispersal rate that it cannot suppress the herbivore population. However, when killers enter the herbivore population, herbivores may invest in defense or leave the prey patch in search of enemy-free space. For simplicity we will consider the last type of response only.

Consider the pancake predation model again, but

now extended with a *per capita* dispersal rate (ν) of the herbivore:

$$\frac{dN}{dt} = (\alpha - \nu)N - \beta P$$

$$\frac{dP}{dt} = \gamma P \qquad (15.6)$$

As shown in Fig. 15.5(b), an increase in ν decreases the time to prey elimination, the area under the herbivore curve, and the number of dispersing predators. Is there an evolutionarily stable dispersal rate? To answer this we first define reproductive success as the *per capita* dispersal rate, ν, multiplied by the area (A) under the herbivore curve (which itself depends on ν), divided by the initial number of herbivores (N_0). This fitness measure is always maximized at intermediate values of ν, because A decreases rapidly with ν and obviously ν increases linearly with ν. Suppose all patches start with the same initial number of predators and herbivores (assumption of metapopulation-wide equilibrium) and that each patch is invaded by N_0 herbivore clones with ν_{res} and one mutant clone with ν_m. Further assume that the two clones are attacked in proportion to their relative abundance, but that the mutant is so rare that $N_{res} + N_m \approx N_{res}$. Thus, herbivore dynamics in the patch and the time to prey elimination (τ) depend entirely on the resident population. The mutant's presence does not influence the growth of the resident herbivore population. Now, is there a resident herbivore population with ν_{res} that cannot be invaded by a mutant with another value of ν_m? The results (Fig. 15.8) show that the evolutionarily stable (ES) dispersal rate ν^* increases with a decrease in time to prey elimination (τ) and, thus, with (i) decreasing *per capita* population growth rate of the herbivores (α), (ii) increasing *per capita* predation rate (β) and (iii) increasing *per capita* population growth rate of the predators (γ). Moreover, the larger the initial number of herbivores, the longer the time to prey elimination and the smaller the ES dispersal rate ν^*.

These results provide some important clues as to how the ES dispersal rate ν^* will change with the exploitation strategy of the predators. This is because milkers have an effectively lower *per capita*

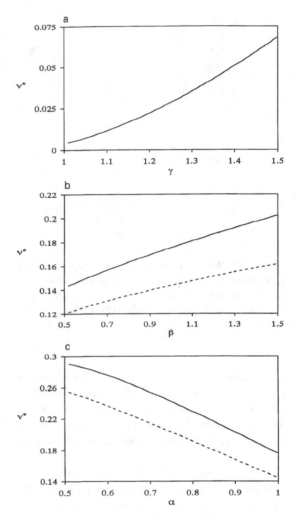

Figure 15.8. (a) The optimal dispersal rate (ν^*) as a function of predator growth rate (γ). A milker-like predator strategy (low γ) favors a low prey dispersal rate (predation tolerant prey). A killer-like predator strategy (high γ) favors a high prey dispersal rate (predator avoidance behavior). Other parameter values: $NI_0 = 49$; $N2_0 = 1$; $P_0 = 1$; $\alpha = 1$ and $\beta = 1$ (where NI_0 is the value of NI at time $= 0$, etc). (b). The relation between ESS dispersal rate and β. Continuous lines: $NI_0 = 99$; broken lines: $NI_0 = 199$. In both cases the mutant $N2_0 = 1$. Other parameter values: $P_0 = 1$, $\alpha = 1$, $\gamma = 1.5$. (c). The relation between ESS dispersal rate and α. Continuous lines: $NI_0 = 99$, broken lines: $NI_0 = 199$. In both cases the mutant $N2_0 = 1$. Other parameter values: $P_0 = 1$, $\beta = 1$, $\gamma = 1.5$.

rate of population growth due to non-zero dispersal $(\gamma - \mu)$ and the lower the predator's population growth rate, the lower the ES dispersal rate ν^* of the herbivore will be. Thus, a prevalence of milkers will select for lower dispersal rates of the herbivores and an increased tendency to stay in the herbivore aggregation, whereas a prevalence of killers will select for higher dispersal rates of the herbivores. Of course, these higher ES dispersal rates can not increase beyond the *per capita* population growth rate of the herbivores (α), but this is feasible in a large part of the parameter domain. Hence, the presence of herbivore aggregations in the presence of killer-like predators is not in conflict with our theory.

There is much to be learned about how the evolution of plant defense strategies interferes with the evolution of herbivore dispersal. Increased investment in direct plant defense will decrease the *per capita* rate of herbivore population growth (α), and as a by-product this will trigger selection for a higher ES dispersal rate (ν^*) of the herbivores. Thus, the effective *per capita* rate of herbivore population growth ($\alpha - \nu^*$) is decreased even more. This will pave the way for the evolution of feeding deterrents. The same applies to increased investments in indirect plant defenses. When plants promote the *per capita* predation rate (β) or the *per capita* rate of predator population growth (γ), the by-product will be that selection favors increased ES dispersal rates of the herbivores, thereby lowering the effective rate of herbivore population growth ($\alpha - \nu^*$). Thus, plants may also invest in releasing herbivore deterrents signaling a high risk of being eaten by predators, and the herbivores will be selected for vigilance in detecting the presence of predators.

The plant's dilemma: direct/indirect defense or no defense at all?

By investing in direct and indirect defenses, a plant gains protection against herbivory, but in doing so, it also benefits its neighbors. If neighbors are close kin, a plant also increases its inclusive fitness by investment in defense, but if not, it may promote the fitness of non-related individuals competing for the same space and nutrient sources. In either event, the neighbor gains associational protection (Atsatt & O'Dowd, 1976; Hay, 1986; Pfister & Hay, 1988). This leads to the plant's dilemma: should it defend itself, thereby benefiting its neighbors as well, or decrease its defensive efforts? The solution is simple: the defenses should protect the plant without benefiting palatable neighbors too much.

When herbivores do not discriminate between palatable or well-defended plants, there are three possibilities: (i) all plants are palatable, (ii) all plants are well-defended or (ii) there is a stable mixture of palatable and well-defended plants. Coexistence is possible when either of the two types increases when rare; a rare palatable plant can gain cheap protection by associating with a population of well-defended plants, whereas a rare, well-defended plant does better in a population of palatable plants as it benefits only a few of the palatable individuals and, hence, increases the average fitness very little. When either type increases its relative abundance, the costs and benefits ultimately balance, resulting in a stable polymorphism (Sabelis & De Jong, 1988; Augner *et al.*, 1991; Tuomi & Augner, 1993).

The conditions for polymorphisms are broad, but some of the assumptions are not generally valid. For example, herbivores are likely to distinguish between palatable and well-defended plants. If so, an individual plant may benefit from direct defenses as herbivores select the more palatable plants. It should be noted that this mechanism does not apply to indirect defenses. Predatory arthropods are usually more mobile than their prey, and they will readily move from the plant they are guarding to a neighboring plant that is under herbivore attack. Thus, even in the case of selective herbivores, palatable plants will gain from proximity to a plant defended by predatory arthropods. Clearly, this mechanism promotes polymorphisms (Sabelis & De Jong, 1988). However, when defensive plant strategies are continuous, not discrete (i.e., they cover the full range of possible investment levels), then there may be no polymorphism because ultimately all plants will exhibit the best average defensive response.

The latter case was analysed by Tuomi *et al.* (1994). They assumed that the cost of defense increases linearly with the probability of killing the

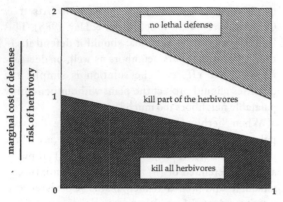

Figure 15.9. The effect of herbivore mobility (*x*-variable) on the parameter areas where the evolutionarily stable strategy (ESS) is to kill the herbivore (lower shaded region), to kill some of the herbivores (intermediate, open region) and not to invest in killing herbivores at all (upper shaded region). The areas depend on the *y*-variable expressed in terms of the ratio of marginal costs of plant defense and the initial risk of herbivory (redrawn from Tuomi *et al.*, 1994). The cost of herbivory equals unity if the herbivore survives, and 0.5 if it dies.

herbivore, the slope being the marginal cost of defense (i.e., rate of cost increases with the impact of defense on the herbivore). The ES lethality level depends on the risk of herbivory, the marginal cost of defense, and the mobility of herbivores between neighbor-plants, (Fig. 15.9). When herbivory risks are low and marginal costs of defense are high, it does not pay to kill the herbivore. But when risks are high and marginal costs are sufficiently low, it pays to kill the herbivore. For intermediate ratios of marginal defense costs and risk of herbivory, ES lethality depends on the inter-plant mobility of the herbivore (Tuomi *et al.*, 1994). Obviously, high mobility will cause neighboring plants to share the same herbivores and this will select for lower lethality, whereas low mobility will select for increased lethality. Since neighboring plants may communicate *via* damage-related signals (Bruin *et al.*, 1992, 1995; Shonle & Bergelson, 1995; Shulaev *et al.*, 1997), defense strategies should be conditional upon the neighbor's state, i.e., by whether it is actually under attack and by its

defensive response. This needs further theoretical and experimental work.

Whether defensive efforts develop in discrete jumps or more gradually determines whether there will be polymorphisms, but the most important message is that in both cases associational protection may lead to a lower average investment in defenses. This applies to direct defenses (Tuomi *et al.*, 1994), as well as indirect defenses (Sabelis & De Jong, 1988).

But coevolution may act as a boomerang . . .

The most important unknown of all is the interplay between metapopulation dynamics and evolution at all three trophic levels. This interplay can be illustrated with a thought experiment where plants evolve to invest more in direct and indirect defenses, and predators are initially killers. First, increased defense by the plant will decrease the size of the herbivore's metapopulation. Subsequently, the size of the predator's metapopulation will decrease, which in turn will decrease the rate of invasion of predators into herbivore patches. Consequently, the probability of co-invasion of predators with different prey-exploitation strategies will decrease, thereby providing a selective advantage to milker-like predators. This will render indirect plant defenses less effective. Hence, selection will favor lower investment in indirect defenses.

A similar thought experiment can be carried out for the case when a plant's investment in direct and indirect defenses also benefits competing neighbors (Sabelis & De Jong, 1988; Augner *et al.*, 1991). Here, increased plant defense will cause a boomerang effect, because neighboring plants profit and allocate the energy saved directly to increased individual fitness.

Boomerang coevolution arises *via* the impact of plant defenses on alternative allocation strategies of neighbouring plants and *via* selection for milker-like exploitation strategies. This may explain why plants generally channel so little of their energy into either direct or indirect defense against herbivores (Beattie, 1985; Simms & Rausher, 1987; Dicke & Sabelis, 1989). Low investment in plant defenses should not be taken as an argument for a low impact of defenses and predators. This will depend entirely on the quantitative details of the evolutionary trajectory of the tritrophic system and is largely an open question. A definitive explanation for the 'world is green' phenomenon (Hairston *et al.*, 1960) can not yet be given.

How dynamical complexities arise from including plant dynamics in predator–herbivore models

Let us now assume an evolutionary *status quo* and consider the dynamical consequences of plant defenses for plant-inhabiting arthropod predator–prey interactions. Models of these systems usually ignore the population-dynamical interactions with plants, based on the assumption that the generation times of arthropods are much shorter than those of plants. Changes in plant density will be trivial on the time-scale of the arthropods' lifespan. However, when the herbivores form clusters of colonies, and predators cannot easily move between prey clusters, what matters is the time-scale of the interaction between the herbivore colony and the plant, or between the predators and herbivores. These time-scales exceed arthropod generation times and approach the lifespan of leaves on perennial plants or the generation times of annual plants. In these cases the dynamics of leaf abundance of perennials or of annual plants should be incorporated into a tritrophic model (Sabelis *et al.*, 1991; Jansen & Sabelis, 1992, 1995).

In general, tritrophic systems can exhibit complex dynamics, such as low frequency cycles with bursts of high frequency cycles, chaos and quasi-periodic behavior (Abrams, 1993; Abrams & Roth, 1994a,b; Klebanoff & Hastings, 1994a,b; McCann & Yodzis, 1994a,b 1995; Jansen, 1995; Kuznetsov & Rinaldi, 1996; Rinaldi *et al.*, 1996; and see Chapter 9). This complexity arises when pairwise trophic interactions show oscillations and the frequencies of these oscillations differ due to the time-scale of the dynamic processes at each trophic level. Combining these pairwise interactions into a tritrophic model causes these periodic frequencies to mix, and this inter-connectedness of periodic frequencies produces complex dynamical behavior. Indeed, these complexities arise when top trophic level dynamics are slow

relative to the lower trophic levels. For predator–herbivore systems on plants we argue that the time-scales of the local predator–prey interactions are close to the time-scale of leaf biomass changes of perennial plants (e.g., trees) and annual plant dynamics. Complex dynamical behavior may not arise under natural conditions due to the stabilizing influence of particular structural changes, such as some types of refuges or predator interference (Eisenberg & Maszle, 1995). However, note that other structural changes, such as a predator dispersal phase (Jansen, 1995; Rinaldi *et al.*, 1996), seasonally driven traits of the organisms, or environmental stochasticity (Ives & Jansen, 1998), can destabilize the entire system. It is not known how these complex dynamics alter the evolution of defense and exploitation strategies of plants, herbivores, and predators.

Equilibrium tritrophic models show that the number of plants remaining free of herbivory increases with a decrease in (i) detectability of uninfested plants for herbivores, (ii) search rate of herbivores for uninfested plants, (iii) extinction rate of predator-inhabited prey patches, and (iv) with an increase in search rate of predators for herbivore-infested plant patches. Thus, when some plants promote the effectiveness of the enemies of their herbivores, all other plants will eventually profit. Individual plants can only gain from increased defenses when they partially monopolize the gains, thereby increasing their fitness relative to other plant genotypes.

Tritrophic models can have another simultaneously stable state, one in which the predators are absent and the plants and herbivores go through stable limit cycles (Jansen & Sabelis, 1992, 1995; McCann & Yodzis, 1994a,b; Jansen, 1995). The crucial factor for bistability is that the two-species limit cycle is resistant to predator invasion unless moderately large numbers are introduced. The loss of predators produced by the perturbation therefore is permanent. Why this type of bistability should be a common feature of tritrophic systems can be easily understood: when few predators are introduced during plant and herbivore oscillations, these predators have to survive the troughs in herbivore density. With relatively short peaks in prey density the losses suffered during the troughs cannot be compensated, and the predators will decrease over every cycle and eventually go extinct. When, however, sufficient predators are introduced they can effectively control the herbivore density and thus stabilize the herbivore–plant interaction, which results in coexistence. Pesticide applications represent an example of how disturbance can cause a permanent reversal from a state with plants, herbivores, and predators to one with repeated herbivore outbreaks. The disturbance caused by pesticides can therefore decrease predators, which eventually go extinct, leaving the pest to show sustained outbreaks (Fig. 15.10).

There is little empirical evidence that such bistable tritrophic systems exist, but the theoretical arguments for their existence are general. One possible example is the interaction between plants, spider mites, and predatory mites. When not controlled by natural enemies or pesticides, spider mites can defoliate the plants they infest. After defoliation the densities drop dramatically (Burnett, 1979). When the plants regenerate new leaves, such outbreaks are repeated. Small populations of natural enemy mites may not survive in such an environment (Burnett, 1979). Yet, it has been demonstrated experimentally that the three trophic levels can coexist (Van der Klashorst *et al.*, 1992; Janssen *et al.*, 1997). However, the dynamics under coexistence were not stable; a phenomenon that needs more detailed analysis.

Bistability in tritrophic systems is not of purely population-dynamical interest. It also has evolutionary consequences. Instantaneous perturbations – a brief spell of extremely high or low temperatures, intense showers of rain, or a treatment with knockdown pesticides – may bring the system from a tritrophic steady state to a ditrophic herbivore–plant cycle, and a substantial invasion of predators is needed for return to the tritrophic state. Any predator mutant that resists the conditions during the perturbation gains in relative fitness, but it will not persist when the system is perturbed to move to a ditrophic cycle. This mechanism may cause lags in the evolution of resistance due to perturbation events at the level of the predators.

A related problem, demographic stochasticity, arises when a ditrophic system (or tritrophic, for that matter) cycles. If densities cycle, local densities may become low enough for populations to go extinct. However, a discussion of this aspect falls outside the scope of this chapter.

Implications for biological pest control

Selection of candidate agents for biological pest control

In selecting predators for biological control, it is common practice to assess their rate of predation and their intrinsic rate of increase. For short-term control of local pest populations these traits are informative, because pest densities within the infestation are high, thereby allowing the control agents to reach their maximal predation rate and their maximum population growth rate. As illustrated by the pancake predation model, estimates of these rates can be used to calculate the minimum predator-to-prey ratios required for satisfactory pest control (Janssen & Sabelis, 1992). These calculations assume that predators are strongly arrested on prey-infested plants until the pest population is decimated or eradicated. The traits underlying arrestment are usually not explicitly measured, because it is not clear which behavioral responses to measure and at which spatial scale such observations should be performed. However, without arrestment predators will not realize their potential, and they may even be inadequate for pest control. The assessment of arrestment responses is one reason why small-scale experiments are necessary before embarking on large-scale biological control.

It is becoming increasingly clear that plants influence predator-to-prey ratios by providing protection, food, and alarm calls signaling herbivore attack. These alarm signals help predatory arthropods not only to find their prey, but also to stay in or near the infested area of the plants. The important point is that failure of a promising candidate for pest control in small-scale experiments may be due to inadequate responses of the predator and/or inadequate involvement of the plant. In the first case the problem may be solved by selecting new predator strains with adequate responses to plant-provided facilities or by conditioning the predators to fine-tune their response (e.g., *via* associative learning). In the

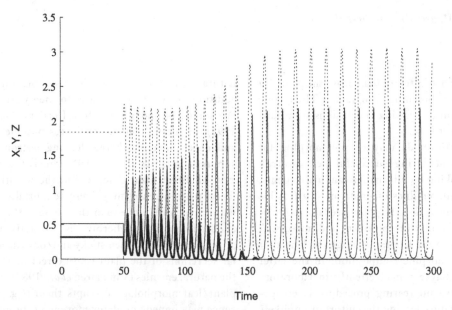

Figure 15.10. Bistability in a tritrophic patch-occupancy model (Jansen & Sabelis, 1992, 1995). Consider a simple patch-occupancy model for a pest that forms colonies (Sabelis *et al.*, 1991, Jansen & Sabelis, 1992, 1995). The variables represent densities of patches rather than of individuals. The variable X represents the density of vacant plant patches (i.e., leaves or clusters of leaves without prey or predators). Plant patches grow logistically up to a carrying capacity, c. A herbivore that discovers a vacant plant patch can start a colony and thus transform this patch into a herbivore patch. The density of herbivore patches is described by Y. New herbivore patches are formed at the rate $a_y YF(X)$, where $F(X)$ gives the fraction of space occupied when the density of vacant plant patches is X, and a_y is a proportionality constant. Herbivore patches disappear when the herbivores have depleted all plant material in a patch (with rate kY) or when they are transformed into a predator-invaded prey patch upon discovery by a predator (often referred to as a predator patch). Analogous to the discovery of empty patches by herbivores, predators discover herbivore patches with rate $a_z Z F(Y)$. Here, Z represents the density of colonies with herbivores and predators, and a_z is a proportionality constant. Patches with herbivores and predators disappear with rate mZ. The full system of equations reads:

$$\frac{dX}{dt} = rX\left(1 - \frac{X}{c}\right) - a_y YF(X)$$

$$\frac{dY}{dt} = a_y YF(X) - kY - a_z ZF(Y)$$

$$\frac{dZ}{dt} = a_z ZF(Y) - mZ$$

where the response of predators to detect prey patches is given by an Ivlev function:

$$F(x) = 1 - e^{-d_x x}$$

Where $d_x x$ is the effective size of a patch of type x. This tritrophic model describes an undisturbed system. Perturbations can be modeled as single, instantaneous reductions of herbivore and predator densities; such perturbations may be caused by sudden weather changes of short duration or by pesticide applications in agricultural systems, where the pesticides have only an instantaneous knock-down effect. Between perturbation events, the system and its population dynamics are described by the equations given above. The graph shows a simulation run that starts in the neighborhood of a stable three species equilibrium. At day 50 the herbivore and predator populations are both reduced by 60%. Eventually, the predators disappear while the plants and herbivores show sustained oscillations. Broken line, density of vacant plant patches (X); thin continuous line, density of herbivore patches (Y); thick continuous line, density of patches with herbivores and predators (Z). Parameters used for the simulation run: $r = 1$, $c = 4$, $d_x = 1$, $a_y = 2.3$, $d_y = 0.0075$, $a_z = 390$, $k = 1$, $m = 1.5$.

second case one may attempt to improve the plant *via* plant resistance breeding programs. Thus, one should be careful in drawing conclusions from failures to control the pest.

Predatory arthropods may be polymorphic for traits vital for achieving successful pest control. Rather than looking for alternative candidate agents, selection for suitable lines of predators is a feasible and a worthwhile option. Suppose predator populations consist of a mixture of killers and milkers. A sample brought to the laboratory might be biased or reflect the polymorphic composition. In the latter case selection may act inadvertently in laboratory cultures, because no rearing procedure is escape-proof, and predators leaving the culture are unlikely to return to it. Especially when provided with ample prey, laboratory cultures cause strong selection for killers because they do not leave the food source before exterminating most, or all, prey. This inadvertent selection probably occurs in all laboratory cultures and mass rearings and favors biological control aiming at a fast suppression of the pest. Whether selection for killers is indeed important for understanding success in biological control is an open question. Evidence might come from analysing genetic variability for exploitation strategies in natural populations. Other evidence might come from analysing selection processes in laboratory cultures and mass rearing units.

Plant breeding for pest resistance and biological pest control: do they always act synergistically?

The objective of plant-resistance breeding is to reduce the feeding rate and the population growth rate of pests. In most cases this involves only improvements in direct defenses of plants. Rarely has care been taken to investigate whether increased direct plant defenses interfere with the effectiveness of natural enemies (Boethel & Eikenbary, 1986), let alone if the plant traits affecting indirect plant defenses have been altered in the process of plant breeding. Increased direct defense may have positive and negative effects on the impact of natural enemies. On the positive side is that reduced population growth rates – all else being equal – enable the

natural enemies to suppress the pest more quickly. In addition, the vulnerable phases to enemy attack may last longer (Isenhour *et al.*, 1989, Loader & Damman, 1991), or the individual herbivores become weaker and more vulnerable, thereby leading to stronger pest suppression (Price *et al.*, 1980). This is the indirect gain of direct plant defense. On the negative side direct plant defenses may decrease mortality due to natural enemies more than they reduce the growth and feeding rate of the herbivores. A negative balance arises, for example, when sticky and/or toxic glandular secretions of the plant cause direct mortality of the natural enemies (Van Haren *et al.*, 1987), or when plant/leaf morphology disrupts them (e.g., pubescence may impede predator movement, or glabrous, very smooth leaves can reduce adhesion (Eigenbrode & Espelie, 1995)). The critical point, however, is not whether the indirect effects on natural enemies are negative *per se*, but rather whether the overall balance of defensive efforts is positive for the plant. If direct plant defenses alleviate predation pressure on the herbivores, one may still wonder how they compare with the direct and indirect gains of reducing the population growth of the herbivore. This can be illustrated using the pancake predation model. Consider joint proportional changes (p) in population growth rate ($p\alpha$) and feeding rate ($p\delta$) of the herbivores, as well as joint proportional changes (q) in predation rate ($q\beta$) and predator population growth ($q\gamma$). Taking the product of the total number of herbivore days (D) until herbivore extermination and the herbivore's consumption rate as a measure for plant damage, it can be shown that :

$$\frac{\delta D(\alpha,\beta,\gamma)}{p\,\delta D(p\alpha,q\beta,q\gamma)} = \frac{D(\alpha,\beta,\gamma)}{\frac{p}{q}D\left(\frac{p}{q}\alpha,\beta,\gamma\right)} \qquad (15.7)$$

Thus, when the proportional changes in all parameters ($p\delta$, $p\alpha$, $q\beta$, $q\gamma$) are equal ($p = q$), the right-hand ratio equals unity, which means that the amount of plant damage remains unaltered. This defines the critical borderline where positive and negative effects of proportional changes cancel out. When $p < q$, the plant suffers less damage, for $p > q$ it suffers more. In words: plants do not always profit from increased direct defense against herbivores. They can only profit when the reduction in feeding and growth rates of the herbivore is stronger than the concomitant reduction in predation and growth rate of the predator.

Direct and indirect defenses interact antagonistically when a decrease in p causes q to drop below unity. For synergism to occur a decrease in p should cause q to become larger than unity. This may occur when the plant's direct defenses make the herbivore more vulnerable to predation. For example, when herbivores get temporarily caught in the secretions of glandular trichomes, they are an easy prey for predators, that are less hindered by the secretions. Plants well-defended by glandular hairs also provide enemy-poor space. For a predator with a juvenile phase vulnerable to intraguild predators, it may be profitable to deposit its eggs on such plants despite the negative effects of glandular secretions. This may explain why some anthocorid predators readily deposit eggs on tomato (Coll, 1996; Ferguson & Schmidt, 1996) despite negative effects on foraging success due to dense masses of glandular hairs (Coll & Ridgway, 1995). There may be many more synergisms between direct and indirect defenses. In a review of published studies on interactions between plant resistance and biological control, Hare (1992) found that out of 16 cases, 6 were antagonistic, 8 neutral, and 2 synergistic. The underlying mechanisms are poorly known, however. No doubt, it is a challenge to plant breeders and biological control workers to identify these mechanisms and consider them when designing plant breeding programs.

It would be naive to think that there are only short-term effects of plant-resistance breeding on biological control. Long-term effects can also be manifested in two essentially different ways: *via* population-dynamical processes or *via* evolutionary responses. First, the defenses may decrease herbivore populations – and possibly also predator populations – to such an extent that the predators cannot maintain their population and go extinct. Second, as argued earlier, the lowered levels of herbivore populations may trigger selection favoring predators of the 'milker' type. These potential effects warrant long-term studies on changes in the effectiveness of biological control agents.

Pesticide-induced outbreaks in tritrophic systems

Bistability and evolution of milker–killer strategies may also be essential in understanding outbreaks of target and non-target pests following pesticide applications. Such pesticide-induced outbreaks have been reported for very different groups of phytophagous arthropods and for several types of pesticides (Ripper, 1956; Penman & Chapman, 1988; Barbosa & Schultz, 1987; Roush & Tabashnik, 1990), but the underlying mechanisms are only partially understood.

One explanation for pesticide-induced outbreaks is that, for reasons rooted in their evolutionary past, phytophagous pest species are less susceptible to pesticides than their natural enemies (Huffaker, 1971; Georghiou, 1972; Croft & Morse, 1979; Mullin *et al.*, 1982; Tabashnik & Johnson, 1990). Hence, pesticide treatment will have a more drastic effect on natural enemies than on pests. However, measurements of susceptibility to direct pesticide doses reveal that natural enemies are generally *not* more vulnerable to pesticides than their prey species (Croft & Brown, 1975; Morse & Croft, 1981; Anber & Oppenoorth, 1989; Hoy, 1990) and may even show a trend towards lower susceptibility (Croft & Brown, 1975).

A better explanation follows from consideration of the coupled population dynamics of predator and prey (May, 1985; May & Dobson, 1986). After pesticide application the densities of phytophagous arthropods are reduced. As soon as the harmful effects of the pesticide abate, conditions are favorable for herbivores: they suffer less predation and possibly experience reduced competition. Hence, the numbers of pests and potential secondary pests will increase at a rate close to their maximum growth rate and populations of predators can only increase after prey have attained sufficiently high densities (Tabashnik, 1986, 1990; Waage *et al.*, 1985).

This is a powerful hypothesis for pesticide-induced outbreaks, but it entails the view that the outbreaks are transient events (Ripper, 1956); after pesticide use is stopped, restoration of predator–prey balance is expected. However, this need not be the case. Persistent outbreaks can result from the simultaneous existence of two stable states;

one in which the predator controls the pest, and another in which the predator exists at very low levels, or goes extinct, and the pest is not controlled. After a small disturbance the system will return to its original state, but after larger disturbances the system may stabilize at either state, depending on the details of the disturbance. Bistability is a rather well-known phenomenon in predator–prey models in which either the prey growth is negative at low densities or the predators' functional response is sigmoid (Noy-Meir, 1975; Southwood & Comins, 1976; May, 1977). Bistability can also occur in models in which some of the stable states are limit cycles, under conditions less strict than those needed for bistable steady states. When an uncontrolled pest population exhibits cycles, perturbations can reduce predator populations beyond recovery and cause the herbivores to exhibit repeated outbreaks (Godfray & Chan, 1990).

Tritrophic food chains generally tend to be bistable. This is independent of the precise model formulation, as can be shown by analysis of the normal forms of tritrophic models (Klebanoff & Hastings, 1994a,b) and has been demonstrated in various tritrophic models (Jansen & Sabelis, 1992, 1995; Jansen, 1995; McCann & Yodzis, 1995; and see Chapter 9). In one stable state three species coexist while in the other the third trophic level is absent and the first and second trophic level exhibit sustained oscillations. Sufficiently large disturbances can bring about a change from one stable state to the other. Thus, pest outbreaks may arise by perturbations from pesticide applications, but pest outbreaks may also disappear when the outbreaks attract predators from the surrounding environment and the system subsequently moves to a stable state with all three trophic levels present.

Poor food conditions for the predators may also impose selection on how predators exploit local prey populations. Clearly, milker-like strategies will be favored as their local populations will persist for a longer time and are more productive than populations of killer-like predators. Hence, milkers create a stronger foothold population after pesticide applications. Whether such evolutionary changes in predator populations occur is entirely conjectural, but it is worthwhile to assess not only the changes in resistance after pesticide applications, but also the exploitation strategies of the survivors.

Pest management requires an understanding of the population dynamics and the evolutionary changes in populations of pests and their natural enemies. Repeated use of pesticides can diminish the food supply to such an extent that predators go extinct. This is not only due to mortality from the pesticide, but also to the impoverished food conditions for the surviving resistant predators. The same holds for pesticides that are less harmful to natural enemies, since these will also cause a reduction of prey densities. To evaluate the effects of pesticides it is not sufficient to measure the susceptibility of predators, because this ignores the population and evolutionary processes after spraying. Even when pesticides are non-persistent, it is of crucial importance to assess the fate of the predator populations in the period following the application.

Biological control in tritrophic systems: implications for predator introduction

Bistability is probably a crucial factor for the (re-) introduction of natural enemies. When the time scale of herbivore colony dynamics is similar to the plant's generation time (or the lifetime of leaves of a tree), the system will show herbivore–plant limit cycles in the absence of predators. When prey densities are low over sufficiently long time periods in the troughs of the limit cycles, introduction of too few predators can result in a population growth too small for practical purposes or even in their extinction. The introduction will only succeed if the predator finds enough prey to survive. Predators introduced in large quantities can stabilize the plant–herbivore interaction and establish themselves permanently.

Bistability might offer an explanation for the variable outcomes of field releases of some natural enemies. It may easily go unnoticed, as there usually are many potential reasons for the failures of introductions. This is illustrated by the following citation from a paper by Alam *et al.* (1971) on releases of parasitoids (*Lixophaga*, a tachinid fly) against the sugar cane borer on Barbados :

Extensive efforts have been made over the past ten years to explain which ecological factors were responsible for the repeated failures of *Lixophaga* and other Tachinids to become established in Barbados, and for its low abundance for several years after its establishment. Although several factors, such as climate, the number and timing of releases, diseases, host plants of adults etc. have been investigated, there has never been a satisfactory answer to this. Nor do we have an answer as to why *Lixophaga* is now apparently thriving. All we can say at this stage is that apparently the parasite has become genetically better adapted to the ecological conditions currently prevailing in Barbados and that the pessimistic view held previously that Barbados is somehow environmentally resistant to the permanent establishment of this fly no longer applies. The prolonged efforts made with *Lixophaga* in Barbados, and the success eventually achieved despite initial failures, demonstrate clearly the merits of persevering with a programme of biological control over a long period of time.

Clearly, it may be worthwhile to consider the role of alternative stable states in biological control. Ignorance of these effects may lead to misjudgements of the potential of some natural enemies or pesticide-resistant predators (Hoy, 1985, 1990) for control of arthropod herbivore pests.

References

Abrams, P. A. (1993) Effect of increased productivity on the abundances of trophic levels. *American Naturalist* **141**: 351–71.

Abrams, P. A. & Roth, J. D. (1994a) The effects of enrichment on three-species food chains with non-linear functional responses. *Ecology* **75**: 1118–30.

Abrams, P. A. & Roth, J. D. (1994b) The responses of unstable food chains to enrichment. *Evolutionary Ecology* **8**: 150–71.

Alam, M. M., Bennett, F. D. & Carl, K. P. (1971) Biological control of *Diatraea saccharalis* (F.) in Barbados by *Apanteles flavipes* Cam. and *Lixophaga diatraeae* T.T. *Entomophaga* **16**: 151–8.

Anber, H. A. I. & Oppenoorth, F. J. (1989) A mutant esterase degrading organophosphates in a resistant strain of the predacious mite *Amblyseius potentillae* (Garman). *Pesticide Biochemistry and Physiology* **33**: 283–97.

Atsatt, P. R. & O'Dowd, D. J. (1976) Plant defense guilds. *Science* **193**: 24–9.

Augner, M., Fagerström, T. & Tuomi, J. (1991) Competition, defence and games between plants. *Behavioural Ecology and Sociobiology* **29**: 231–4.

Barbosa, P. & Schultz, J. C. (1987) *Insect Outbreaks*. New York: Academic Press.

Beattie, A. J. (1985) *The Evolutionary Ecology of Ant–Plant Mutualisms*. Cambridge: Cambridge University Press.

Bentley, B. L. (1977) Extrafloral nectaries and protection by pugnacious bodyguards. *Annual Review of Ecology and Systematics* **8**: 407–27.

Berryman, A. A., Dennis, B., Raffa, K. F. & Stenseth, N. Ch. (1985) Evolution of optimal group attack, with particular reference to bark beetles (Coleoptera: Scolytidae). *Ecology* **66**: 898–903.

Boethel, D. J. & Eikenbary, R. D. (eds) (1986) *Interactions of Plant Resistance and Parasitoids and Predators of Insects*. Chichester: Ellis Horwood.

Bruin, J., Dicke, M. & Sabelis, M. W. (1992) Plants are better protected against spider mites after exposure to volatiles from infested conspecifics. *Experientia* **48**: 525–9.

Bruin, J., Sabelis, M. W. & Dicke, M. (1995) Do plants tap SOS-signals from their infested neighbours. *Trends in Ecology and Evolution* **10**: 167–70.

Burnett, T. (1979) An acarine predator–prey population infesting roses. *Researches on Population Ecology* **29**: 227–34.

Coll, M. (1996) Feeding and oviposition on plants by an omnivorous insect predator. *Oecologia* **105**: 234–40.

Coll, M. & Ridgway, R. L. (1995) Functional and numerical responses of *Orius insidiosus* (Heteroptera: Anthocoridae) to its prey in different vegetable crops. *Annals of the Entomological Society of America* **88**: 732–8.

Croft, B. A. & Brown, A. W. A. (1975) Responses of arthropod natural enemies to insecticides. *Annual Review of Entomology* **20**: 285–335.

Croft, B. A. & Morse, J. G. (1979) Recent advances on pesticide resistance in natural enemies. *Entomophaga* **24**: 3–11.

De Jong, M. C. M. & Sabelis, M. W. (1988) How bark beetles avoid interference with squatters: an ESS for colonization by *Ips typographus*. *Oikos* 51: 88–96.

De Jong, M. C. M. & Sabelis, M. W. (1989) How bark beetles avoid interference with squatters: a correction. *Oikos* 54: 128.

Dicke, M. & Sabelis, M. W. (1988) How plants obtain predatory mites as bodyguards. *Netherlands Journal of Zoology* 38: 148–65.

Dicke, M. & Sabelis, M. W. (1989) Does it pay plants to advertise for bodyguards? Towards a cost–benefit analysis of induced synomone production. In *Variation in Growth Rate and Productivity of Higher Plants*, ed. H. Lambers, M. L. Cambridge, H. Konings & T. L. Pons, pp. 341–58. The Hague: SPB Academic Publishing BV.

Dicke, M. & Sabelis, M. W. (1992) Costs and benefits of chemical information conveyance: proximate and ultimate factors. In *Insect Chemical Ecology, An Evolutionary Approach*, ed. B. Roitberg & M. Isman, pp. 122–55. London: Chapman and Hall.

Drukker, B., Scutareanu, P. & Sabelis, M. W. (1995) Do anthocorid predators respond to synomones from *Psylla*-infested pear trees under field conditions? *Entomologia Experimentalis et Applicata* 77: 193–203.

Eigenbrode, S. D. & Espelie, K. E. (1995) Effects of plant epicuticular lipids on insect herbivores. *Annual Review of Entomology* 40: 171–94.

Eisenberg, J. H. & Maszle, D. R. (1995) The structural stability of a three-species food chain model. *Journal of Theoretical Biology* 176: 501–10.

Ferguson, G. M. & Schmidt, J. M. (1996) Effect of selected cultivars on *Orius insidiosus*. *IOBC/WPRS Bulletin* 19: 39–42.

Georghiou, G. P. (1972) The evolution of resistance to pesticides. *Annual Review of Ecology and Systematics* 3: 133–68.

Godfray, H. C. J. & Chan, M. S. (1990) How insecticides trigger single-stage outbreaks in tropical pests. *Functional Ecology* 4: 329–37.

Hairston, N. G., Smith, F. E. & Slobodkin, L. B. (1960) Community structure, population control and competition. *American Naturalist* 94: 421–5.

Hare, J. D. (1992) Effects of plant variation on herbivore–natural enemy interactions. In *Plant Resistance to Herbivores and Pathogens. Ecology, Evolution and Genetics*, ed. R. S. Fritz & E. L. Simms, pp. 278–98. Chicago: University of Chicago Press.

Hassell, M. P. (1978) *The Dynamics of Arthropod Predator–prey Systems*. Princeton: Princeton University Press.

Hay, M. E. (1986) Associational defenses and the maintenance of species diversity: turning competitors into accomplices. *American Naturalist* 128: 617–41.

Hoy, M. A. (1985) Recent advances in genetics and genetic improvement of the Phytoseiidae. *Annual Review of Entomology* 30: 345–70.

Hoy, M. A. (1990) Pesticide resistance in arthropod natural enemies: variability and selection responses. In *Pesticide Resistance in Arthropods*, ed. R. T. Roush & B. E. Tabashnik, pp. 203–36. New York: Chapman and Hall.

Huffaker, C. B. (1971) The ecology of pesticide interference with insect populations. In *Agricultural Chemicals. Harmony, Discord for Food, People and the Environment*, ed. J. E. Swift, pp. 92–107. Berkeley: University of California, Division of Agricultural Sciences.

Isenhour, D. J., Wisman, B. R. & Layton, R. C. (1989) Enhanced predation by *Orius insidiosus* (Hemiptera: Anthocoridae) on larvae of *Heliothis zea* and *Spodoptera frugiperda* (Lepidoptera: Noctuidae) caused by prey feeding on resistant corn genotypes. *Environmental Entomology* 18: 418–22.

Ives, A. A. R. & Jansen, V. A. A. (1998). Complex dynamics in stochastic tritrophic models. *Ecology*. (In press).

Jansen, V. A. A. (1995) Effects of dispersal in a tritrophic metapopulation model. *Journal of Mathematical Biology* 34: 195–224.

Jansen, V. A. A. & Sabelis M. W. (1992) Prey dispersal and predator persistence. *Experimental and Applied Acarology* 14: 215–31.

Jansen, V. A. A. & Sabelis, M. W. (1995) Outbreaks of colony-forming pests in tri-trophic systems: consequences for pest control and the evolution of pesticide resistance. *Oikos* 74: 172–6.

Janssen, A. & Sabelis, M. W. (1992) Phytoseiid life-histories, local predator–prey dynamics and strategies for control of tetranychid mites. *Experimental and Applied Acarology* 14: 233–50.

Janssen, A., Van Gool, E., Lingeman, R., Jacas, J. & Van de Klashorst, G. (1997) Metapopulation dynamics of a persisting predator–prey system in the laboratory: time series analysis. *Experimental and Applied Acarology* 21: 415–30.

Janzen, D. H. (1966) Coevolution of mutualism between ants and acacias in Central America. *Evolution* **20**: 249–75.

Jolivet, P. (1996) *Ants and Plants: An Example of Coevolution.* Leiden: Backhuys Publishers.

Kawai, A. (1995) Control of *Thrips palmi* Karny (Thysanoptera: Thripidae) by *Orius* spp. (Heteroptera: Anthocoridae) on greenhouse eggplant. *Applied Entomology and Zoology* **30**: 1–7.

Klebanoff, A. & Hastings, A. (1994a) Chaos in three-species food chains. *Journal of Mathematical Biology* **32**: 427–51.

Klebanoff, A. & Hastings, A. (1994b) Chaos in one-predator-two-prey models: general results from bifurcation theory. *Mathematical Biosciences* **122**: 221–33.

Kuznetsov, Yu. A. & Rinaldi, S. (1996) Remarks on food chain dynamics. *Mathematical Biosciences* **134**: 1–33.

Lesna, I., Sabelis, M. W. & Conijn, C. G. M. (1996) Biological control of the bulb mite, *Rhizoglyphus robini*, by the predator mite, *Hypoaspis aculeifer*, on lilies: predator-prey interactions at various spatial scales. *Journal of Applied Ecology* **33**: 369–76.

Loader, C. & Damman, H. (1991) Nitrogen content of food plants and vulnerability of *Pieris rapae* to natural enemies. *Ecology* **72**: 1586–90.

May, R. M. (1977) Thresholds and breakpoints in ecosystems with a multiplicity of stable states. *Nature* **269**: 471–7.

May, R. M. (1985) Evolution of pesticide resistance. *Nature* **315**: 12–13.

May, R. M. & Dobson, A. P. (1986) Population dynamics and the rate of evolution of pesticide resistance. In *Pesticide Resistance: Strategies for Tactics and Management*, ed. National Research Council, pp. 170–93. Washington DC: National Academy Press.

Maynard Smith, J. (1982) *Evolution and the Theory of Games.* Cambridge: Cambridge University Press.

McCann, K. & Yodzis, P. (1994a) Biological conditions for chaos in a three-species food chain. *Ecology* **75**: 561–4.

McCann, K. & Yodzis, P. (1994b) Nonlinear dynamics and population disappearances. *American Naturalist* **144**: 873–9.

McCann, K. & Yodzis, P. (1995) Bifurcation structure of a three species food chain model. *Theoretical Population Biology* **48**: 93–125.

Morse, J. G. & Croft, B. A. (1981) Developed resistance to azinphosmethyl in a predator–prey mite system in greenhouse experiments. *Entomophaga* **26**: 191–202.

Mullin, C. A., Croft, B. A., Strickler, K., Matsumura, F. & Miller, J. R. (1982) Detoxification enzyme differences between a herbivorous and predatory mite. *Science* **217**: 1270–2.

Noy-Meir, I. (1975) Stability of grazing systems: an application of predator–prey graphs. *Journal of Ecology* **63**: 459–81.

Penman, D. R. & Chapman, R. B. (1988) Pesticide-induced outbreaks: pyrethroids and spider mites. *Experimental and Applied Acarology* **4**: 265–76.

Pfister, C. A. & Hay, M. E. (1988) Associational plant refuges: convergent patterns in marine and terrestrial communities result from different mechanisms. *Oecologia* **77**: 118–29.

Price, P. W., Bouton, C. E., Gross, P., McPheron, B. A., Thompson, J. N. & Weis, A. E. (1980) Interactions among three trophic levels: influence of plants on interactions between insect herbivores and natural enemies. *Annual Review of Ecology and Systematics* **11**: 41–65.

Rinaldi, S., Da Bo, S. & De Nittis, E. (1996) On the role of body size in a tritrophic metapopulation model. *Journal of Mathematical Biology* **35**: 158–76.

Ripper, B. E. (1956) Effect of pesticides on balance of arthropod populations. *Annual Review of Entomology* **1**: 403–38.

Roush, R. T. & Tabashnik, B. E. (eds) (1990) *Pesticide Resistance in Arthropods.* New York: Chapman and Hall.

Sabelis, M. W. (1990) Life history evolution in spider mites. In *The Acari: Reproduction, Development and Life-History Strategies*, ed. R. Schuster & P. W. Murphy, pp. 23–50. New York: Chapman and Hall.

Sabelis, M. W. (1992) Arthropod predators. In *Natural Enemies, The Population Biology of Predators, Parasites and Diseases*, ed. M. J. Crawley, pp. 225–64. Oxford: Blackwell Scientific.

Sabelis, M. W. & De Jong, M. C. M. (1988) Should all plants recruit bodyguards? Conditions for a polymorphic ESS of synomone production in plants. *Oikos* **53**: 247–52.

Sabelis, M. W. & Janssen, A. (1994) Evolution of life-history patterns in the Phytoseiidae. In *Mites. Ecological and Evolutionary Analyses of Life-History Patterns*, ed. M. A. Houck, pp. 70–98. New York: Chapman and Hall.

Sabelis, M. W. & Van der Meer, J. (1986) Local dynamics of the interaction between predatory mites and two-spotted spider mites. In *Dynamics of Physiologically Structured Populations*, ed. J. A. J. Metz & O. Diekmann, pp. 322–44. Berlin: Springer-Verlag.

Sabelis, M. W. & Van Rijn, P. C. J. (1997) Predation by insects and mites. In *Thrips as Crop Pests*, ed. T. Lewis, pp. 259–354. Wallingford: CAB International.

Sabelis, M. W., Diekmann, O. & Jansen, V. A. A. (1991) Metapopulation persistence despite local extinction: predator–prey patch models of the Lotka–Volterra type. *Biological Journal of the Linnean Society* 42: 267–83.

Shonle, I. & Bergelson, J. (1995) Interplant communication revisited. *Ecology* 76: 2660–3.

Shulaev, V., Silverman, P. & Raskin, I. (1997) Airborne signalling by methyl salicylate in plant pathogen resistance. *Nature* 385: 718–21.

Simms, E. L. & Rausher, M. D. (1987) Costs and benefits of plant resistance to herbivory. *American Naturalist* 130: 570–81.

Southwood, T. R. E. & Comins, H. N. (1976) A synoptic population model. *Journal of Animal Ecology* 45: 949–65.

Strong, D. R., Lawton, J. H. & Southwood, T. R. E. (1984) *Insects on Plants*. Cambridge: Harvard University Press.

Tabashnik, B. E. (1986) Evolution of pesticide resistance in predator–prey systems. *Bulletin of the Entomological Society of America* 32: 156–61.

Tabashnik, B. E. (1990) Modelling and evaluation of resistance management tactics. In *Pesticide Resistance in Arthropods*, ed. R. T. Roush & B. E. Tabashnik, pp. 203–36. New York: Chapman and Hall.

Tabashnik, B. E. & Johnson, M. W. (1990) Evolution of pesticide resistance in natural enemies. In *Principles and Application of Biological Control*, ed. T. Fisher *et al.* Berkeley: University of California Press.

Tuomi, J. & Augner, M. (1993) Synergistic selection of unpalatability in plants. *Evolution* 47: 668–72.

Tuomi, J., Augner, M. & Nilsson, J. (1994) A dilemma of plant defences: is it really worth killing the herbivore? *Journal of Theoretical Biology* 170: 427–30.

Turlings, T. C. J., Loughrin, J. H., McCall, P. J., Rose, U. S. R., Lewis, W. J. & Tumlinson, J. H. (1995) How caterpillar-damaged plants protect themselves by attracting parasitic wasps. *Proceedings of the National Academy of Sciences of the USA* 92: 4169–74.

Van Baalen, M. & Sabelis, M. W. (1995). The milker–killer dilemma in spatially structured predator–prey interactions. *Oikos* 74: 391–413.

Van der Klashorst, G., Readshaw, J. L., Sabelis, M. W. & Lingeman, R. (1992) A demonstration of asynchronous local cycles in an acarine predator–prey system. *Experimental and Applied Acarology* 14: 179–85.

Van Haren, R. J. F., Steenhuis, M. M., Sabelis, M. W. & de Ponti, O. B. M. (1987) Tomato stem trichomes and dispersal success of *Phytoseiulus persimilis* relative to its prey *Tetranychus urticae*. *Experimental and Applied Acarology* 3: 115–21.

Waage, J. K., Hassell, M. P. & Godfray, H. C. J. (1985) The dynamics of pest–parasitoid–insecticide interactions. *Journal of Applied Ecology* 22: 825–38.

Walter, D. E. (1996) Living on leaves: mites, tomenta, and leaf domatia. *Annual Review of Entomology* 41: 101–14.

16

A Darwinian view of host selection and its practical implications

ROBERT F. LUCK AND LEONARD NUNNEY

Introduction

The process by which parasitoids select their hosts can be viewed in two different, but complementary, ways. The first seeks to understand the proximate cause of events by which parasitoids select hosts (the 'how' questions). This involves behavioral and physiological studies that show how the parasitoid interacts with its environment to locate its hosts. The second way in which to view host selection is to understand the ultimate reasons why a parasitoid selects a host (the 'why' questions). This involves studies that show how different strategies influence the parasitoid's reproductive success and, therefore, the Darwinian fitness of the searching parasitoid. Thus, the researcher aims to understand why the parasitoid chooses a particular host type among those that are available.

Our aim in this chapter is to emphasize the 'why' of host selection by parasitoids and how this understanding can aid in identifying suitable natural enemies for biological control. It can also aid in creating conditions within a crop that maximize their success. We will also emphasize that a successful understanding of the ultimate causation of host selection requires a thorough knowledge of both 'how' parasitoids select their hosts and the natural history of the system under study. To illustrate this point, we focus on two parasitoid systems of practical importance, the *Trichogramma*–lepidopteran system and the *Aphytis*–red scale system.

The process of host selection

For parasitic Hymenoptera, like all parasitoids, offspring production requires the location and selection of hosts in which to oviposit. Historically, this has been viewed as a sequence of independent behavioral steps that progressively directs a female to the most suitable host (Salt, 1935, 1937; Doutt, 1964; Vinson & Iwantsch, 1980; Beckage *et al.*, 1993). Thus, host selection can be viewed as three behavioral steps: host-habitat location, host location and host acceptance. More recently, host selection has been recast as a process in which a natural enemy responds to a variety of information sources (chemical, visual, sound, and/or vibrational: Dicke & Sabelis, 1988; Vet & Dicke, 1992; Godfray, 1994; van Alphen & Jervis, 1996). The dominant information source utilized by most natural enemies in this process appears to be chemical. The infochemicals involved are called allelochemicals, i.e., chemicals that mediate interspecific interactions (Vet & Dicke, 1992; Godfray, 1994; van Alphen & Jervis, 1996). The attraction to the host-habitat and the location of a host within the habitat are mediated by (i) attractant stimuli, i.e., those stimuli that work over long distances to orientate the natural enemy to areas that either contain, or are associated with, the likely presence of a host, and (ii) arrestant stimuli, i.e., those that work over very short distances to keep a natural enemy in an area likely to harbor hosts (van Alphen & Jervis, 1996). A more detailed discussion of these ideas and the research supporting them can be found in reviews by Nordlund *et al.* (1981), Vet & Dicke (1992), Godfray (1994) and van Alphen & Jervis (1996).

The initial interest in the mechanism of host selection arose from two sources. The first was practical. Early practitioners found that a number of parasitoids could be reared on a variety of host species in the laboratory or insectary and yet many of these hosts were never utilized by the parasitoid in the field (Doutt, 1964). The second was theoretical. The most efficacious biological control agents were thought to be those whose population densities were tied to and depended on those of their hosts (Huffaker & Messenger, 1964; DeBach & Rosen, 1991). This view predicted that host specificity would be an important attribute of an efficacious agent, hence host specificity, and in particular specificity in host selection, became an important research focus in biological control (Doutt, 1964). Until recently this research focus has largely been one of understanding the proximate causes in host selection, that is, the mechanism of *how* the parasitoid locates a host (Charnov & Skinner, 1984, 1985; Vet & Dicke, 1992). The approach has been extremely insightful and of immense practical and theoretical value, but it has not told us *why* a parasitoid should narrow its host choice, what host it should select or how many eggs of what sex it should allocate to the host. Nonetheless, the mechanism of host selection is important from a Darwinian perspective; it is the variability within these mechanisms that provides the potential for adaptation. For example, those individuals that are more likely to encounter acceptable hosts because of detecting and learning the appropriate infochemicals will leave more offspring on average. These new attributes will be favored by natural selection if the proximal mechanism of host location is heritable from parent to offspring.

The Darwinian perspective

A second research focus, the *why* of host selection, has emerged largely during the last decade and a half (MacArthur & Pianka, 1966; Charnov, 1979, 1982; Charnov & Skinner, 1984, 1985; Godfray, 1994). To answer this question it is necessary to adopt a Darwinian perspective and to seek insight into the consequence of a female parent's choice of hosts in terms of her total fitness. Total fitness can be viewed as having three components: clutch size, a females lifetime production of clutches, and her influence on the fitness of her offspring. This last component is particularly important in parasitoids since offspring usually develop entirely within or on a single host. Thus, a female can influence the fitness of her offspring through her choice of a host, since the qualities of the host can directly affect offspring fitness.

In males, fitness is measured by mating success and in females, by realized fecundity. Of course, implicit in this Darwinian argument is the assumption that this behavior is inherited from parent to offspring.

An important feature of the Darwinian approach is that the mechanistically independent components of host selection may not be evolutionarily independent. They may conflict and thus involve trade-offs. For example, when clutch size is considered in isolation, the predicted course of evolution is clear – clutch size should increase indefinitely. It does not do so because as clutch size increases it begins to trade-off against some other factors. At the optimum a small increase in clutch size does not increase fitness because the resulting fitness gain is matched by the deterioration in other fitness components. Thus, Lack (1947) noted that increasing the number of eggs in a clutch beyond a certain point leads to a reduction in the number of offspring produced and he argued that the size of a single clutch would evolve to a level that maximized productivity. In reality, the optimal clutch size is often less than this ideal 'Lack clutch size', due to the influence of fitness trade-offs not considered in Lack's (1947) original model. There are a number of such effects (reviewed by Roff, 1992), but three are relevant here (and each will be discussed in more detail later):

1. Clutch size may trade-off with the quality of the offspring as well as their number (Klomp & Teerink, 1967). As noted above, the fitness of a female includes her influence on the fitness of her offspring. Charnov & Skinner (1984) stressed that maternal productivity must be scaled to account for quality differences among offspring.
2. A trade-off can exist between the time involved in laying the number of eggs allocated to a clutch and that available to find other hosts in which to lay additional clutches. Charnov & Skinner (1984, 1985) argued that parasitoids should maximize productivity per unit time and that the productivity of a single clutch must be scaled by the inter-clutch interval, i.e., the time between the end of laying one clutch and the end of laying the next. The inter-clutch interval includes both the time to produce a clutch and the search time that it

takes to find the host. They showed that their model predicted an optimum clutch size less than the Lack clutch size, because adding eggs to a clutch increases the laying time and hence the inter-clutch interval. The result is that productivity per unit time increases less than we would expect based on clutch size alone. Another way of viewing the same result, is that over a fixed time period increasing the clutch size decreases the search time available to find clutches.

3. A trade-off between clutch size and longevity (Charnov & Krebs, 1973). This is most generally applicable to large-bodied vertebrates, but has relevance in ectoparasitoids with large yolk-rich eggs, such as those of *Aphytis* species. For such species the allocation of a Lack clutch size to a host may shorten the life of the female, since under some circumstances (discussed below) eggs may be resorbed to supply some of her metabolic needs.

Reproductive biology of *Trichogramma* and *Aphytis*

In this chapter we discuss mostly *why* particular hosts are selected for oviposition and *why* a characteristic number of offspring are allocated to them. In doing so, however, we seek to ground this explanation in the practical, that is, in the use of natural enemies as biological control agents in pest management. We first turn our attention to *Trichogramma* species as an example. *Trichogramma* are polyphagous, gregarious, and mostly parasitoids of lepidopteran eggs (Clausen, 1940; Pinto & Stouthamer, 1994). They are of practical interest because a number of species are mass-produced and widely used as augmentative biological control agents (Li, 1994). It is our tenet that the quality of the *Trichogramma* produced for mass release, as well as that of other biological control agents, and their effectiveness as biological control agents depends on understanding *why* a characteristic number of offspring are allocated per host and *why* they select a particular host. After discussing *Trichogramma* species we next turn our attention to an aphelinid parasitoid, *Aphytis melinus* DeBach. This species is a

parasitoid of armored scale (Diaspididae), including California red scale, *Aonidiella aurantii* (Maskell), and several of its congeners (Rosen & DeBach, 1979). Red scale is a pest of citrus in arid and semi-arid regions of the world. In southern California, this exotic pest is suppressed by an introduced and permanently established biological control agent. The agent displaced a previously introduced and established congener, *Aphytis lingnanensis* Compere. We will illustrate how this displacement occurred through the perspective of host selection and *why* the two congeners choose the hosts they do. It will also provide insight into how to evaluate the status of biological control in a citrus grove from a practical perspective. The host selection behavior of *A. melinus* also suggests a testable hypothesis to explain why this parasitoid is only intermittently effective in suppressing red scale on citrus in central California, whereas it is consistently effective in suppressing the scale insect in southern California citrus.

These two parasitoid genera also offer an interesting contrast in reproductive biologies. While both are idiobionts, i.e., they arrest their hosts' development following oviposition (Strand, 1986; Rosen & DeBach, 1979), *Trichogramma* species are endoparasitoids and largely pro-ovigenic, i.e., they lay their eggs inside the host, and they emerge with most of their egg complement. In contrast *Aphytis* are synovigenic ectoparasitoids. Synovigenic parasitoids emerge with a fraction of their potential fecundity (about 10% in the case of *Aphytis*) and produce additional eggs from nutrients gained from subsequent host-feeding.

Trichogramma produce 80% of their eggs within 2 days of emergence (Hohmann et al., 1988; Bai et al., 1992; C. L. Hohmann, unpublished results), and they produce a few additional eggs if they expend the initial egg complement early in their lifespan (C. L. Hohmann, unpublished results). Their initial egg complement depends on their size; the larger the wasp, the more eggs it contains (Waage & Ng, 1984; Hohmann et al., 1988; Bai et al., 1992). A female's lifespan as an adult depends on a carbohydrate source on which to feed. In the laboratory, a female *Trichogramma* lives for about 2 days in the absence of honey, with or without hosts, whereas it lives for

7 days in the presence of honey, with or without hosts (Smith *et al.*, 1986; Hohmann *et al.*, 1988; Bai *et al.*, 1992; Leatemia *et al.*, 1995; McDougall & Mills, 1997). Although *Trichogramma* are known to host-feed, that is, they imbibe fluids from the host egg, the value of host-feeding in *Trichogramma* is not understood (Heimpel & Collier, 1996). In particular, it does not appear to increase a female's lifespan in the laboratory (Bai *et al.*, 1992; Heimpel & Collier, 1996; C. L. Hohmann, unpublished results). Thus, *Trichogramma* species are largely pro-ovigenic, emerging with most of their complement of yolk-poor (alecithal) eggs, which absorb material from the host egg prior to larval eclosion (Klomp & Teerink, 1967; Strand, 1986).

In contrast, *Aphytis* species are synovigenic ectoparasitoids (of armored scale insects). Typically, synovigenic ectoparasitoids paralyze their hosts and the larvae feed through the host's epidermis but usually in a protected situation. In the case of *Aphytis* the protection is provided by the armored scale's waxy cover that overlies the scale body (Rosen & DeBach, 1979; Foldi, 1990). Female *Aphytis* lay one, occasionally two, large, yolk-rich (lecithal) eggs after they have paralyzed and arrested the red scale's development (Rosen & DeBach, 1979). Female *A. melinus* continue to produce eggs during their lifetime by periodically feeding on the body fluids of immature hosts and using the body fluid as nutrients to mature additional eggs (Collier, 1995a; G. E. Heimpel, D. Kattari & J. A. Rosenheim, unpublished data cited in Heimpel & Collier, 1996; Heimpel *et al.*, 1996; see also Jervis & Kidd, 1986; and Chapter 8). They usually mature their first batch of eggs within 24 hours of emergence from resources they carry over from the larval stage (Heimpel & Collier, 1996). Their initial egg complement depends on wasp size, the larger the wasp, the larger the initial egg complement (Opp & Luck, 1986; Collier, 1995a). About 1.3 eggs are produced from a host-feeding after 12–18 hours if the female has not recently oviposited, or about 2.7 eggs are produced if the female has recently oviposited (Collier 1995a). Host-feeding when combined with a carbohydrate source, also prolongs life beyond that promoted by feeding on honey alone; but it does not prolong life in the absence of honey

(Collier, 1995a; Heimpel & Collier, 1996). Thus, host-feeding appears to supply nutrients for both metabolic maintenance and egg production (Collier, 1995a). This is also suggested by the absence of eggs at death when *Aphytis* females are held with honey but do not have access to hosts for oviposition or host-feeding; the female resorbs about 1 egg per day (Collier, 1995a). However, in the absence of honey, egg resorption apparently occurs too slowly to supply the metabolic needs of the wasp (Collier, 1995a). Thus, in summary, *Aphytis* species are synovigenic, emerging with only a small portion of their potential lifetime complement of yolk-rich (lecithal) eggs. They rely on host-feeding to provide the resources to develop additional eggs. Host-feeding also supplies resources for the wasp's metabolic needs, which increases its lifespan, but the benefits of host-feeding are only realized if the wasp also feeds on a carbohydrate source.

Trichogramma and *Aphytis* provide an interesting contrast in both reproductive biology and host selection. First, an average *Trichogramma* has 75–80% of its egg load 24–36 hours after emergence and it may mature additional eggs if the initial egg complement is expended (Hohmann *et al.*, 1988; Bai *et al.*, 1992; C. L. Hohmann, unpublished results), whereas *Aphytis melinus* has only 10–15% of its lifetime egg complement 24 hours after emergence and additional eggs are produced as a consequence of host-feeding (Opp & Luck, 1986; Collier, 1995a). Second, the hosts used by *Trichogramma* (primarily lepidopteran eggs) are generally suitable for parasitoid development during the first 75% of the hosts' embryological development (Pak, 1986; Strand, 1986; Honda, 1995), although other factors such as a thick chorion or a sticky coating may also influence host acceptance and suitability (see Honda, 1995, and below). In contrast, *Aphytis* host (armored scale insects) grow through a window of availability, improving in quality as they enter the window and get larger, but then suddenly becoming unavailable as they mature further.

Bearing these differences in mind, we now consider each of these species in turn to examine the question of why each selects the types of hosts it does.

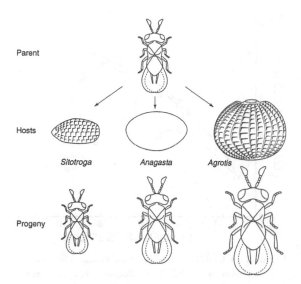

Parent

Hosts

Sitotroga   *Anagasta*   *Agrotis*

Progeny

Figure 16.1. The effect of host size on offspring size in *Trichogramma evanescens* Westwood. The relative size is shown of solitary female offspring arising from one of three host species that differ in size. (Redrawn from Salt, 1940).

Trichogramma species

The *why*, or Darwinian perspective of host selection was spawned by the classic work of Klomp & Teerink (1962, 1967) using *Trichogramma embryophagum* Hartig and its subsequent reformulation and generalization by Charnov & Skinner (1984, 1985). However, host selection research using *Trichogramma* species began earlier with the work of Salt (1934, 1935, 1937, 1940). In one of these studies, Salt (1940) used three host species, each of which differed in size. The size of a solitary *Trichogramma* that emerged depended on the size of its host (Fig. 16.1). But *Trichogramma* is facultatively gregarious (Clausen, 1940) and the number of offspring allocated to a host is a consequence of host size. Klomp & Teerink (1962) demonstrated this by transferring a *Trichogramma* female to a larger or smaller host after it had conducted its initial assessment of a host, but before it had oviposited. In their experiments, a female *Trichogramma* assessed a host by repeatedly walking over it and drumming it with its antennae. The transferred wasp laid the number of eggs in the new host that was characteristic of the host from which it was transferred.

Under natural conditions, a female wasp, after assessing and accepting the host, engages its ovipositor on the host's surface, drills through the chorion, and oviposits one or more eggs (= clutch size) (Salt, 1935; Schmidt & Smith, 1985). *Trichogramma* allocates a characteristic clutch size (= normal clutch) to a given host species based on the size of the host and, in general, the normal clutch size is related to the size of the host egg. *Trichogramma* measures the host's size during its initial transit across the host (Schmidt & Smith, 1985, 1987). If its transit is interrupted by placing a barrier in its path, the wasp lays fewer eggs in the host than if she is allowed to complete her initial transit across the host.

So why should a *Trichogramma* measure the size of its host? The hypothesis that suggests itself is that, for a given size of host egg, there exists an optimal clutch size of wasp eggs that maximizes maternal fitness by maximizing the number of reproducing offspring produced from that host, i.e., a wasp is expected to produce the 'Lack clutch size' discussed earlier (see Godfray, 1994 for a more detailed discussion of this idea). Klomp & Teerink (1967) set out to test this hypothesis. As a measure of the reproductive prospects of offspring from a host they counted the total number of eggs present in all of the emerging progeny from a given host. The wasp species they used in these experiments only produced female offspring (parthenogenetically) so they did not have to worry about the production of male offspring in a host as a confounding factor. To determine the effects on wasp progeny of belonging to a smaller than normal clutch, they interrupted an ovipositing female before she had completed her egg-laying. To determine the effects on wasp progeny of belonging to a larger than normal clutch, they increased the ratio of wasps to hosts so that the wasps would lay more eggs than normal in a host. With these experiments they documented that (i) the size of the emerging wasps was inversely related to the clutch size allocated to a host (Fig. 16.2(a)) and (ii) mortality increased substantially with increasing clutch size (Fig. 16.2(b)). Since the reproductive potential of the emerging offspring strongly depended on their size (Fig. 16.2(c)), the Darwinian fitness of a female measured as the summed reproduction of the offspring

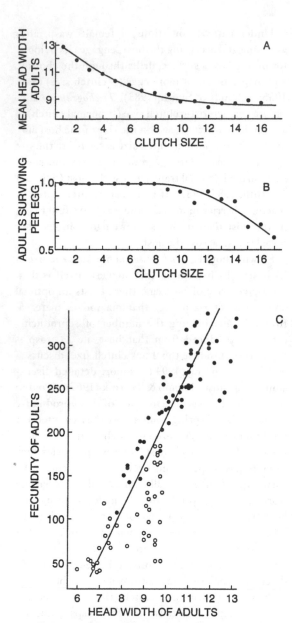

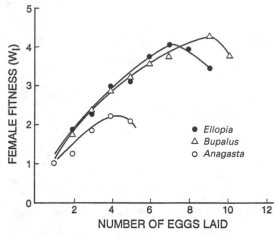

Figure 16.3. Clutch size and maternal fitness. The relationship between the clutch size allocated to each of three host species that differ in size and the fitness of female *Trichogramma embryophagum* Htg. arising from that clutch is shown. Fitness is measured as the sum of the fecundities of all the females arising from a single clutch. (Redrawn from Charnov & Skinner, 1994). ●, *Ellopia*; △, *Bupalus*; ○, *Anagasta*.

Figure 16.2. The relationships linking clutch size to maternal fitness. (a) Shows the mean head width (wasp size) of *Trichogramma embryophagum* Htg. adults emerging from clutches of different sizes; (b) shows the survival of *T. embryophagum* during immature development in clutches of different sizes; and (c) shows the fecundity of *T. embryophagum* and adult female size (Redrawn from Klomp & Teerink, 1967).

from a single clutch was strongly dependent upon clutch size. Klomp and Teerink (1967) used these relationships to predict the clutch size that maximized the parental female's fitness and found that this value was notably higher than the clutch size normally observed.

Charnov & Skinner (1984, 1985) further developed Klomp & Teerink's (1967) idea of incorporating the reproductive success of offspring in the evaluation of parental fitness. The contribution of a single clutch to a female's fitness (w_i) = (clutch size) × (proportion emerging offspring) × (offspring size) × (fecundity per offspring size). From Klomp & Teerink's (1967) data, Charnov & Skinner determined the curve relating clutch size and maternal fitness for three of the host species used by Klomp & Teerink (Fig. 16.3). The Lack clutch size is the clutch size defining the maximum of these curves. For the different host species, the curves predicted a clutch of 4 for *Anagasta*, 7 for *Ellopia* and 9 for *Bupalus*, which are broadly larger than those observed (1–2 for *Anagasta*, 5–8 for *Ellopia*, and 5–8 for *Bupalus*). Waage & Ng (1984) conducted an experiment similar to that of Klomp & Teerink (1967) using a different *Trichogramma* species that was bisexual. Even when they constrained their estimates of clutch size to include sufficient males to mate with the females, they also predicted clutch sizes that were greater than those they observed. This has generally been the case for predicted clutch sizes in *Trichogramma* and other gregarious parasitoid species (Godfray, 1994 and references therein). As noted earlier, this general pattern suggests that one or more important fitness trade-offs are not included in the model defining the Lack clutch size, even after it has been corrected for changes in offspring fecundity.

Charnov & Skinner (1984) suggested three hypotheses that might explain the discrepancy between the predicted and observed clutch sizes: a trade-off between clutch size and subsequent maternal survival; an underestimate of the trade-off between clutch size with offspring fitness; and a trade off between clutch size and the inter-clutch interval. First, the clutch size laid by the female may affect her subsequent survival; the larger the clutch, the less likely she will survive to lay additional eggs. The initial formulation by Klomp & Teerink and Charnov & Skinner assumed that a parental female's survival was independent of the clutch size she laid. In wasps this trade-off between current and future reproduction appears especially appropriate for a wasp with a reproductive biology similar to that of *A. melinus*, i.e., synovigenic species with large yolk-rich eggs. These wasps depend on host-feeding to produce additional eggs and to support their metabolic needs (Jervis & Kidd, 1986; see also Chapter 8; Collier, 1995a; Heimpel & Collier, 1996). In the absence of host-feeding the wasp resorbs mature eggs and uses these resources instead. Thus, a choice to lay an additional egg in a clutch, or in a poor quality host, is a choice to reduce the potential metabolic resources available *via* egg resorption in the event that subsequent host-feeding opportunities are unavailable. Thus the problem is one of trading-off future and current reproduction.

The same issue can arise even in a pro-ovigenic or non-host-feeding species when host quality varies, measured in terms of the size of the wasps that are produced. Committing an additional egg to a clutch prevents its allocation to a host of higher quality later in life. The decision to lay or postpone oviposition is mediated by such factors as egg load, nutritional reserves, life expectancy (age), and mortality risks (e.g., predation) (Mangel, 1989a,b). As egg load increases, a parasitoid is predicted to be less selective in its choice of hosts. The reduced selectivity is presumed to stem from the rather large reproductive potential that would be lost if the wasp should die early in life without laying or only laying a few eggs. As the egg load declines from oviposition and eggs become limiting, the parasitoid is predicted to become more selective. Her death at this stage is not such a reproductive catastrophe as would be the case if she were to die early in life. Recent results with several *Aphytis* species indicate this to be the case (Rosenheim & Rosen, 1991; Collier, 1995a,b; Heimpel & Rosenheim, 1995; Heimpel & Collier, 1996; Heimpel *et al.*, 1996). Moreover, any factor that shortens life expectancy is expected to reduce a parasitoid's host selectivity (see Mangel, 1989a,b; Minkenberg *et al.*, 1992; Roitberg *et al.*, 1993; Godfray, 1994; Collier, 1995b; and Heimpel & Collier, 1996, for further discussion of these ideas and those related to feeding *versus* ovipositing on a host). Consequently, the need to balance the trade-offs between future and current reproduction is one potential explanation for the discrepancy between the predicted and observed clutch sizes.

The second hypothesis proposed by Charnov & Skinner (1984) to explain the quantitative difference between the predicted and observed clutch size was that the measure of offspring fitness is incomplete. They had measured offspring fitness in terms of it's fecundity; however, it is possible that the size of a wasp influences fitness in other ways. For example, size could affect dispersal and/or searching behavior (Waage & Ng, 1984). The result would be that smaller wasps would have a much lower realized fecundity under natural conditions. The experiments of Kazmer & Luck (1995) support this hypothesis. These workers tested whether a wasp's size was related to its reproductive success under field conditions. They measured reproductive

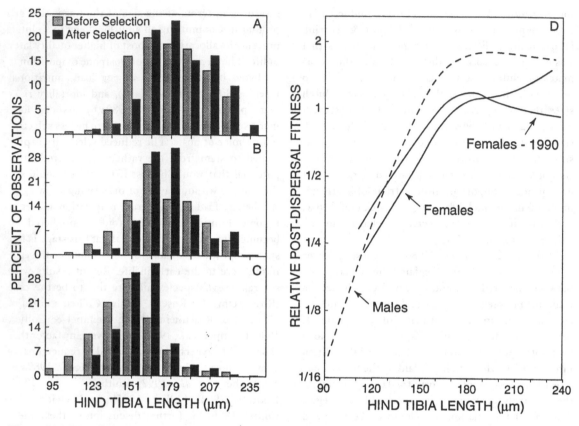

Figure 16.4. The effect of wasp size on their ability to find hosts and mates in the field. (a)–(c) show the frequency distribution of *Trichogramma pretiosum* Riley sizes emerging from field collected eggs, i.e., 'before selection' for dispersal ability (■) and sizes of adults dispersing to egg cards placed in the field after wasp emergence, i.e., 'after selection' (■). Hind tibia lengths were used as an index of wasp size: (a) data for females in 1988, (b) for females in 1990, and (c) for males in 1990. (d) shows the data expressed as the relative success of wasps in locating hosts or mates as a function of wasp size. (Redrawn from Kazmer & Luck, 1995).

success as the ability of a female *Trichogramma* to find hosts under field conditions. The hosts in this case were cabbage looper eggs, *Trichoplusia ni* (Hübner), glued to cards that were placed on tomato plants in a field. The searching females were those that emerged naturally in the field. To estimate the distribution of wasp sizes emerging from field-parasitized hosts, Kazmer & Luck sampled host eggs, *Heliocoverpa zea* (Boddie), *T. ni*, and *Manduca sexta* (L.), from the tomato field on one date and reared the *Trichogramma* from each parasitized host in the laboratory. The wasps that emerged were mounted on a microscope slide and their hind tibia lengths (HTL)

were measured as an index of their size. When this sampled generation of wasps was due to emerge in the field, the cards with *T. ni* eggs were set out in the same field from which the wasps had been collected. The female wasps that arrived at the cards were collected and measured. The size distribution of female wasps emerging from the hosts parasitized in the field was significantly different to that of female wasps arriving at the eggs. The size distribution of wasps collected at the egg cards had a greater percentage of larger wasps than those that emerged (Fig. 16.4(a)), suggesting that the smaller wasps were less able to find hosts in the field than their larger counterparts. Using these size distributions, the relative fitness of small *versus* large females were compared (Kazmer & Luck, 1995). It increased with female size until a threshold value of 0.18 mm HTL after which it leveled off (Fig. 16.4(d)). This experiment was repeated a second year with similar results (Figs 16.4(a) and (b), see also Fig. 16.4(d)). Two other studies on different species of wasps in different systems have been conducted to compare the relative host-finding success of large *versus* small wasps in the field (Visser, 1994; West *et al.*, 1996). Both studies showed that the larger female wasps were more successful.

In addition to larger females finding more host patches, they also appear to search a larger area within patches. Large female *Trichogramma platneri* Nagarkatti search 30% more area per unit time than their smaller counterparts (Honda, 1995). *Trichogramma* mostly forage for hosts by walking on plant substrates (R. F. Luck, personal observation) thus, all else being equal, a larger area searched per unit time results in an increased likelihood of encountering hosts. This further emphasizes how important it is for a female to produce large offspring and that the ability of *Trichogramma* to measure host size is very important.

The third hypothesis proposed by Charnov & Skinner (1984) to explain the discrepancy between the predicted and observed clutch sizes depends on how long it takes to find a host (search time) relative to the time it takes to lay a clutch of eggs. As the time (or energy) a wasp spends searching for hosts increases, the optimum clutch size also increases (Fig. 16.5). At the limit, when search time is long, the optimum approaches the Lack clutch size (the clutch size defining the peak of the curve in Fig. 16.5).

Charnov & Skinner (1984), employing the marginal value theorem, recognized that the optimum clutch size maximizes a female's gain in fitness over the time involved in parasitizing hosts (the inter-clutch interval). This equals the time a female spends searching for hosts, t_s, (the left-hand side of the x-axis in Fig. 16.5) and the time it takes her to assess and oviposit in a host, t, (the right-hand side of the x-axis in Fig. 16.5). In their model, two fitness trade-offs interact. First, there is the well-established trade-off

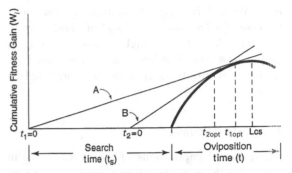

Figure 16.5. A graphical solution of Charnov & Skinner's (1984) marginal value theorem that defines the optimal clutch maximizing a female's gain in fitness over the time involved in parasitizing a host (the inter-clutch interval). The inter-clutch interval equals the time a female spends searching for hosts, t_s, and the time it takes her to assess and oviposit in a host, t. The longer a wasp spends searching for a host (i.e., when t_1 rather than t_2 defines the start of the search), the larger the clutch she should allocate to that host. The Lack clutch size (Lcs) is defined by the peak of the curve in the right-hand side of the graph. (See the text for further explanation) (Redrawn from Charnov & Skinner, 1984).

between clutch size and offspring number and quality. The fitness gain of a female laying additional eggs in a host is one of diminishing returns, since as the clutch size is increased, the average quality of the offspring decreases. This is represented by the dome-shaped curve in the right-hand side of Fig. 16.5. Second, there is the trade-off between clutch size and the inter-clutch interval, because as the clutch size increases the oviposition time also increases. As a result, the optimum solution depends upon the search time. Line A in Fig. 16.5 represents a longer search time than line B (the distance from the vertical line in the middle of the x-axis to the point where each line intersects the abscissa in the left-hand side of the graph). When these lines are extended into the right-hand side of the graph so that they just touch (are tangent to) the dome-shaped curve, the point at which they touch indicates the optimal clutch size that should be laid if time (energy expended) is the only consideration. If a wasp contacts hosts frequently, then it should allocate relatively smaller clutches to a host than if it has difficulty finding hosts. Waage & Ng's (1984) results suggest that this is the case. If a *Trichogramma* was given a host every 30 minutes, it laid a larger clutch per host than if it was given a host every 5 or less minutes. Thus, observed results qualitatively agree with those predicted by theory.

Practical implications

The practical implications of these findings lie in predicting the ideal phenotype to produce as an augmentative biological control agent. *Trichogramma* are mass-produced on the angoumois grain moth, *Sitotroga cerealella* (Oliver) or the Mediterranean flour moth, *Anagasta* (= *Ephestia*) *kuehnienella* (Zeller). Both are small hosts that, when parasitized by *Trichogramma*, issue a single offspring (Salt, 1940; Fig. 16.1). *Anagasta* has a larger egg than *Sitotroga* (Salt, 1940) and thus produces a larger wasp (Fig. 16.1), but both types are in the lower quartile of the size ranges of wasps emerging from field-collected hosts. Female wasps emerging from *Anagasta* average 140 μm HTL, whereas those emerging from *Sitotroga* average 120 μm HTL (Bai *et al.*, 1992; see also Bigler *et al.*, 1987). If these wasps' sizes are compared to the relative fitnesses of those in the field study of Kazmer & Luck (1995), they have between a quarter to a half of the fitness of the fittest wasp (≥180 μm HTL; see Fig. 16.4(d)). Thus they are much less likely to find hosts in the field than larger wasps, especially if they are released from a single point. This assumes, of course, that the mobility of the mass-produced *Trichogramma* is the same as that of *Trichogramma* emerging in the field. Based on observations of mass-produced *Trichogramma* (van Bergeijk *et al.*, 1989; Bigler, 1994) mobility may be reduced and host acceptance altered, exacerbating the problem even more.

Honda (1995) investigated how readily *T. platneri* utilized the eggs of two periodic avocado pests in southern California, the avocado leaf folder, *Amorbia cuneana* Walsingham, and the omnivorous looper, *Sabulodes aegrotata* (Guenée). His study illustrates two important points: first, the effect of wasp quality can vary depending on specific attributes of a host (e.g., chorion thickness), and second, the degree to which an inefficiently exploited host is utilized can decrease in the presence of an efficiently exploited alternate host.

Oatman *et al.* (1983) had found that *T. platneri* was a frequent component of the natural enemy complex on these species in the avocado groves, although parasitism of looper eggs by *T. platneri* was more variable than that of leaf folder eggs. Oatman & Platner (1985) developed an augmentative release program against both moth species using *T. platneri* mass-produced on *Sitotroga* eggs. The leaf folder lays an egg-mass of about 30 scale-like, overlapping eggs, whereas the looper lays somewhat flattened spherical eggs individually but in a loose cluster of about 18 eggs (Oatman *et al.*, 1983; Honda, 1995). In comparing the fitness benefits accruing to *T. platneri* females in exploiting these eggs, only variation in exploitation efficiency is relevant since offspring arising from looper eggs were of similar fecundity and size as those arising from the leaf folder eggs (Honda, 1995). Thus, Honda (1995) recorded and compared the time it took a large *T. platneri* female derived from a standard *Trichoplusia ni* egg to oviposit in a looper *versus* a leaf folder egg. A large *T. platneri* female averaged 54.4(±4.6) s (mean(±SE)) to drum, drill and lay 5.12(±2.10) eggs in a looper egg; but, only 29% of these first attempts were successful. A total of 50% of the wasps were successful in their second attempt, adding an additional 21.0 (±1.56) s to the time that was incurred during the first drilling attempt. The remaining 21% of the wasps were successful on their third attempt but they incurred an additional cost of 24.6(±5.19) s. A weighted average of these times indicated that a large *T. platneri* (HTL = 180(±2) μm) can expect to lay 1 egg for every 14.6(±2.3) s it invests in oviposition attempts on a looper egg. In contrast, on leaf folder eggs, these wasps were always successful in a single attempt and averaged 6.9(±2.2) s to drum, drill and oviposit 1.5(±1.0) eggs per leaf folder egg, i.e., 1 egg per 4.6(±1.2) s. The difference is even more pronounced for smaller wasps (HTL = 150(±1)μm) attempting to oviposit on looper eggs. A small wasp ovipositing in a leaf folder egg-mass can expect to lay 1 egg for every 5.3(±1.4) s invested in an oviposition attempt, whereas it can expect to lay 1 egg for every

20.4($\pm$ 3.1) s invested in an oviposition attempt in a looper egg. Of these smaller wasps, 17% were unable to penetrate the chorion of a looper egg even after three attempts; they gave up and left the area containing the eggs.

When ovipositing in a leaf folder egg-mass, *T. platneri* parasitized on average 76% of the leaf folder egg-mass, expending 78% of its initial egg complement of about 40–55 mature eggs (the egg-load 24 hours after emergence). On the egg clusters of the looper, large female *T. platneri* only parasitized on average 35% of the eggs. Laying an average of 5.12 eggs per host, these females expended an average 61% of their egg load. Smaller *Trichogramma*, with an initial complement of about 20–35 eggs, were less able to penetrate a looper egg, expending 33% of their initial complement compared to the 86% of eggs expended by similar wasps parasitizing leaf folder eggs. Using residual egg load as an index of host quality from *T. platneri*'s perspective, the wasp is more reluctant to use looper eggs than leaf folder eggs. This is to be expected given the low efficiency with which the wasp can exploit looper eggs. The hard chorion of the looper egg makes the 5.1 offspring allocated to the looper egg more expensive per egg per unit time than the 1.6 offspring allocated to a leaf folder egg and this ignores the additional time spent by *T. platneri* moving between the looper eggs in a cluster.

The implications of Honda's evaluation is that indigenous *T. platneri* are unlikely to be an effective biological control agent of omnivorous looper, especially if alternative hosts, such as avocado leaf folder egg-masses, are present. Moreover, if the smaller, commercially-produced *Trichogramma* (HTL = 120 μm from *Sitotroga* (Bigler *et al.*, 1987; Hohmann *et al.*, 1988; Bai *et al.*, 1992) are released against the looper eggs there is even less prospect for success. At best, only a few of these tiny wasps will be able to parasitize looper eggs.

Aphytis species

Two species of *Aphytis* were introduced into California to control California red scale, *Aonidiella aurantii* (Maskell), one of the major pests of citrus in California. One species, *A. lingnanensis*, was imported from Guangdong (Canton) Province, Peoples' Republic of China and established in southern California in 1947. The second species, *A. melinus*, was imported from northwestern India and southwestern Pakistan and established in southern California in 1956–7 (Rosen & DeBach, 1979). Unlike *Trichogramma*, which uses eggs that have a relatively constant size for a given species, *Aphytis* uses several life history stages of red scale. The resulting variation in host size has important implications, and to understand host selection in *Aphytis* it is necessary to understand the life history of the California red scale.

The female California red scale passes through three instars; the male passes through five (Fig. 16.6). The first instar in both sexes is mobile until it finds a suitable site on which to settle ($\leq$ 24 hours). Once settled, a female remains sedentary at that site for the rest of its life. The male has a second mobile stage, as an adult ($\leq$ 24 hours), during which it seeks out females with which to mate. The sexes cannot be distinguished during the first instar and a half, after which they differentiate morphologically. The male molts at the end of the second instar to a prepupa and from a prepupa to a pupa before it emerges as an adult coincident with the occurrence of adult virgin females with whom it mates. Only the second instar male and the second and adult virgin female scales are potential hosts for *A. melinus*. After insemination, a virgin female transforms into a gravid female, which is invulnerable to *Aphytis*. It develops a hardened scale cover which *Aphytis* cannot penetrate. Even under the best of circumstances, the virgin females, the stage on which most *Aphytis* females are produced, comprise only 17% of the scale insect's lifespan (Yu & Luck, 1988).

Host size and sex allocation

In parasitic Hymenoptera, offspring sex ratios are known to be determined at oviposition *via* haplodiploidy (arrhenotoky). Fertilized eggs are diploid and female; unfertilized eggs are haploid and male (Flanders, 1965). Most female parasitoids are thought to mate once and store sperm in a spermatheca (Allen

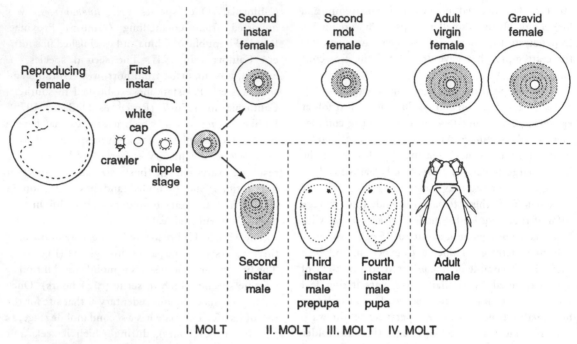

Figure 16.6. The ontogeny of California red scale, *Aonidiella aurantii* (Maskell). Second instar male and female and virgin adult female scales are the stages used as hosts by *Aphytis melinus* DeBach (and *A. lingnanensis* Compere).

et al., 1994). Thus, female parasitoids have the potential to manipulate the sex ratio of their offspring by controlling the access of sperm to the egg at oviposition (Flanders, 1965; Suzuki *et al.*, 1984). This ability is sometimes compromised, however, when females are infected with selfish genetic elements (see Chapter 14).

In the case of *Aphytis* the available hosts (red scale) vary in size, both because of developmental stage and nutritional condition. This variation in host quality generates variation in fitness among emerging wasps, since in general larger hosts provide more resources for developing parasitoids, which are consequently of higher quality. As a result host-size variation is predicted to influence the sex ratio of offspring allocated to a host (Charnov, 1979, 1982; Charnov *et al.*, 1981; Green, 1982). Charnov *et al.* (1981, see also Charnov, 1982; King, 1987; Godfray, 1994) modeled this effect and their results suggest that parental females do best when they allocate daughters to large hosts and sons to small hosts, on the assumption that the fitness

difference between being a large or a small wasp is less for a male than a female, i.e., that reducing a wasp's body size lowers its fitness, but it reduces the fitness of a daughter more than it does that of a son. Additional assumptions made by the model include the following: (i) the parasitoid uses a continuous range of host sizes, (ii) a parasitoid oviposits a single egg on the host, and (iii) large wasps arise from large hosts and small wasps arise from small hosts (which is generally the case whenever the host is paralyzed when the parasitoid oviposits on it, or the host is in a non-growing stage).

To demonstrate that wasps allocate eggs according to the quality-dependent sex allocation hypothesis it is necessary to show that females manipulate the primary sex ratio of offspring in response to host size (since secondary sex ratio differences could arise from resource-dependent sex differences in mortality, as discussed below) and that the fitness consequence of being small differs between males and females. To test the first assumption, R. F.

Luck, D. S. Yu & W. W. Murdoch (unpublished results) determined the sex of *Aphytis melinus* offspring allocated to California red scale of different sizes in the field. *Aphytis* paralyzes its host prior to an oviposition thereby arresting further development of the host. Thus, the food available to the developing offspring is that contained in the host at the time the scale insect is parasitized (Rosen & DeBach, 1979). R. F. Luck *et al.* (unpublished results) sampled scale insects from field populations on citrus trees near Fillmore, California. Parasitized scale insects were measured (length × width of the scale body) and the young *A. melinus* larva was then transferred from the scale on which it had been laid to a larger oleander scale, *Aspidiotus nerii* Bouché, using the method described by Luck & Podoler (1985). Oleander scale is a factitious host used to mass-produce *A. melinus* in a commercial insectary (Rosen & DeBach, 1979).

The transfers were made to exclude the possibility that female eggs of *A. melinus* allocated to small hosts exhausted their food before they reached the critical size for pupation. Males of *A. melinus* are smaller and, thus, require less food than females to reach pupation weight. This sex difference in resource need could result in an excess of males emerging from small hosts due to the higher mortality of females and not because the parental females were allocating only males to small hosts. The transfers were designed to determine whether eggs of both sexes were equally likely to be laid on small hosts (scale insects).

Figure 16.7 shows that they were not. Male wasps were allocated to small hosts and male and female offspring to large hosts. Of the female offspring allocated to the scale insect in the field, 93% were allocated to hosts larger than about 0.39 mm^2. Over 90% of the offspring allocated to the small hosts (<0.39 mm^2) were male and about 30% of the offspring allocated to the large hosts were also male. The transfers showed that the parental female 'decided' to lay sons on small hosts and almost all daughters on large hosts. This pattern was the same as that observed in the laboratory for *A. melinus* and for a congener, *A. lingnanensis* (Luck & Podoler, 1985). *Aphytis lingnanensis*, however, does not allocate daughters to California red scale until the host reaches a (body) size of 0.55 mm^2. The size of both male and female *A. melinus* depends on the size of the scale from which they emerge: the larger the scale the larger is the wasp (Opp & Luck, 1986; Yu, 1986). Lifetime fecundity of the *A. melinus* females increases with its size (Fig. 16.8).

The next step is to demonstrate that females lose more fitness than males by being small. This has not been done experimentally, in large part because of the difficulty of establishing the fitnesses of males of varying size. It is generally assumed to be the case because large females are more fecund than small females and it is assumed that there is little sexual selection among males, that could, if present, create very large fitness differences among males of different sizes.

Thus, it seems probable that *A. melinus* satisfies the assumptions required by Charnov *et al.*'s (1981) model. It uses a range of host sizes, oviposits a single egg on most hosts, paralyzes the host before ovipositing, large wasps arise from large hosts and small wasps arise from small hosts and females probably lose proportionately more fitness than males by being small.

A minor complication is that large hosts sometimes received two eggs (Fig. 16.7). Since both *A. melinus* and *A. lingnanensis* are facultatively gregarious, their clutch size increases with increasing host size (Luck *et al.*, 1982). About 15% of the hosts in the field under some circumstances receive two eggs.

The Charnov *et al.* (1981) model predicts that the size at which a wasp shifts from laying male to laying female eggs is relative to the overall size distribution of hosts. In the *Aphytis*–California red scale system the size distribution of hosts changes with the season (Yu & Luck, 1988), with citrus variety (Hare *et al.*, 1990) and with location (plant part) within the tree (Luck & Podoler, 1985; Hare *et al.*, 1990). The scale insect is smallest if it grows on a scaffolding branch or the tree trunk; it is of intermediate size if it grows on the leaf; and it is largest if it grows on a fruit (Luck and Podoler 1985, Hare *et al.* 1990). Whether the wasps alter their size threshold for laying male *versus* female eggs with these patterns has not been tested.

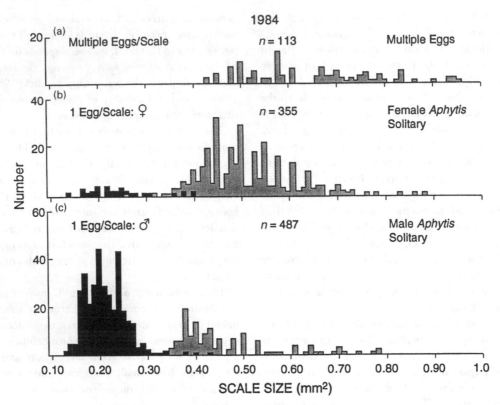

Figure 16.7. The size distribution of California red scale, *Aonidiella aurantii* (Maskell), parasitized by *Aphytis melinus* DeBach. The scale were from field samples taken at a Ventura County, California, citrus grove. The wasps laid male (c) and female (b) eggs on second (■) and third (■) instar scales. (a) depicts the size distribution of second and third instar scales on which multiple eggs were laid of either mixed or single sexes.

Practical implications

California red scale, *Aonidiella aurantii* (Maskell) is one of the major pests of citrus in California. It is exotic to the State and presumably arrived on citrus seedlings imported from China *via* Australia in the latter half of the nineteenth century (Quayle 1938). The scale insect's pest status prompted an extensive biological control campaign that sought out its natural enemies leading to the introduction of *A. lingnanensis* and *A. melinus*. Although *A. lingnanensis* was generally established throughout southern California prior to the introduction of *A. melinus*, it only provided economic suppression of the armored scale in citrus groves along the coast and in the inland

coastal valleys. Successful suppression of the armored scale in southern California finally came about with the introduction of *A. melinus*. DeBach and his colleagues followed the course of establishment and spread of these congeners by sampling and rearing the parasitoids associated with red scale in different locations within southern California (DeBach & Sundby, 1963). They documented the establishment and spread of *A. lingnanensis* and the subsequent establishment and spread of *A. melinus*. It was during this sampling program that DeBach & Sundby noticed the displacement of *A. lingnanensis* by *A. melinus*.

DeBach & Sundby (1963) considered the two species of *Aphytis* to be ecological homologs, that is,

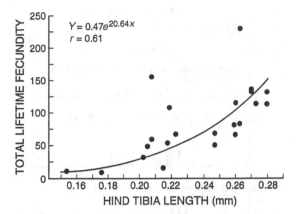

Figure 16.8. The relationship between lifetime fecundity and the size of a female *Aphytis melinus* DeBach. The hind tibia length (HTL) of a female was used as an index of wasp size.

they used the same scale stages in the same fashion to produce offspring. In tracking the displacement of *A. lingnanensis* by *A. melinus*, DeBach & Sundby noticed that it occurred in the presence of an abundance of virgin third instar scales, the putative host stage for the two parasitoid species. They interpreted this observation as indicating that the densities of the two *Aphytis* species were not limited by the densities of their host and concluded that *A. melinus* displaced *A. lingnanensis* because it was a better searcher (DeBach & Sundby, 1963). This conclusion conflicted with the prevailing theory of how biological control was achieved. Natural enemy populations were assumed to drive a host population to densities that, in turn, limited the densities of the natural enemy population. If these two congeners were not limited by their host's density, then how was biological control achieved?

To test whether the two species of *Aphytis* were ecological homologs, Luck and colleagues conducted a series of host selection experiments (Luck *et al.*, 1982; Luck & Podoler, 1985). They showed that the two congeners used different scale sizes for the production of daughters. *Aphytis melinus* produces daughters on smaller scales than *A. lingnanensis* (0.39 mm^2 *versus* 0.55 mm^2, length × width of the scale body, respectively; Fig. 16.9). Transfers of first

instar wasp larvae from the smaller hosts selected by the parental wasp to larger hosts (*Aspidiotus nerii*) showed that the parental female had 'chosen' to allocate only male offspring to the smaller scale. The parental females allocated most of their daughters to hosts larger than 0.39 mm^2 in the case of *A. melinus* or 0.55 mm^2 in the case of *A. lingnanensis* (Fig. 16.9). The scale grew into the size range used by *A. melinus* to produce daughters before it reached the size required by *A. lingnanensis* for daughters. If the scale was parasitized at this smaller size by *A. melinus*, it ceased growing; thus, it never entered the size range used by *A. lingnanensis* for the production of daughters. Luck & Podoler (1985) concluded that *A. melinus* displaced *A. lingnanensis* because it out-competed its congener for the daughter-producing resource and this resource was in short supply. This hypothesis, however, remains to be field tested. Murdoch *et al.* (1996, see also Chapter 2) have modeled this hypothesis with a stage-structured model and found that *A. melinus* rapidly displaced *A. lingnanensis* based on the difference in the required daughter-producing resource.

This information is of practical value because it defines the resource needed by *A. melinus* to produce daughters. It is the daughters that perpetuate the *Aphytis* population and the resources that are required to produce them are frequently scarce. These resources are defined as the size range of female scales that lie between 0.39 mm^2, the lower threshold for daughter production, and the size at which the scale transforms to a gravid female (at insemination). The size at transformation, however, depends on the conditions under which the scale insect grows. If the scale insect is stressed (high temperatures or growing on woody tissue), it is small when it transforms. If it grows on fruit or during the spring or fall it is large when it transforms (Fig. 16.10). Thus, of the virgin third instar scales that DeBach & Sundby (1963) presumed to be hosts, some were too small for either *A. melinus* or *A. lingnanensis* to use for the production of daughters, e.g., those that inhabited the wood; hence, they were not a contested resource.

An important aspect of this resource window (the size range of scales used by *A. melinus* to produce

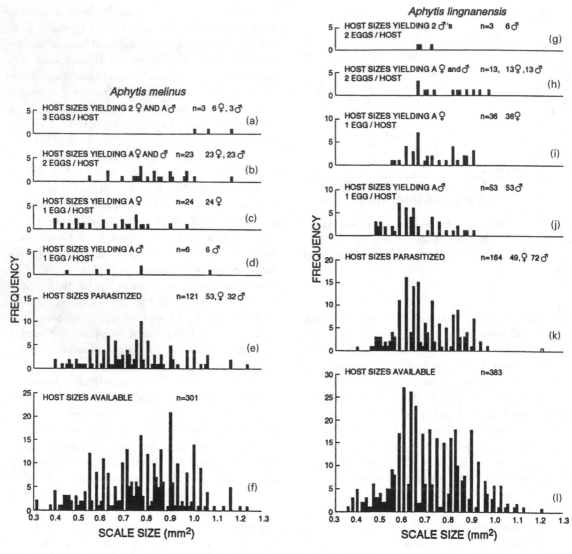

Figure 16.9. The size distributions of third instar California red scale, *Aonidiella aurantii* (Maskell), and of those parasitized by *Aphytis melinus* DeBach ((a)–(f)) and *A. lingnanensis* Compere ((g)–(l)). The scale were parasitized with (a) and (g) 2 male eggs; (b) and (h) 1 male and 1 female egg; (c) and (i) 1 female egg; or (d) and (j) 1 male egg, which together define (e) and (k) all parasitized scale chosen from (f) and (l) the scale size distribution offered.

daughters) is its value in terms of the quality of the daughters that arise from this resource. If the scale insect is stressed and transformation to a gravid third instar female occurs when it is relatively small, the daughter produced from this resource is small. Moreover, this smaller scale insect is only suitable for

the production of female wasps for a very short period, and is therefore less at risk from parasitism (Fig. 16.10). Laboratory experiments indicate that ovipositing *A. melinus* will not exploit these smaller scale as readily as the larger unstressed scales (R. F. Luck *et al.* unpublished results). Unstressed scales

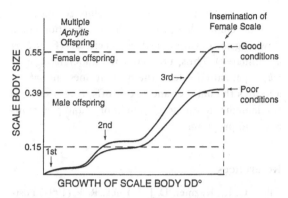

Figure 16.10. A comparison of the body size of California red scale, *Aonidiella aurantii* (Maskell), grown under good conditions (moderate temperatures or on citrus fruits) and poor conditions (hot temperatures or on branches or the trunk of citrus trees). DD, degree days

are a larger, and hence better, resource before they transform to gravid females. As the preferred host, they are more at risk from parasitism.

An example of the difference in quality of the available resource is illustrated by the size distribution of virgin third instar females at two locations in California: Ventura County and Tulare County. Ventura County abuts the southern California coast and is characterized by a mild climate with occasional hot days but cool evenings. Biological control of the scale in this region, as well as throughout southern California's inland coastal valleys, is consistently effective in the absence of disruptive effects such as dust, ants, and pesticides (DeBach & Sundby, 1963; DeBach & Huffaker, 1971). Tulare county is located in California's San Joaquin Valley and it is characterized by greater temperature fluctuations, higher summer-time temperatures, and lower humidities than are southern California's inland coastal valleys. Biological control has been inconsistent in this region (Haney *et al.*, 1992). Figure 16.11 shows the size distribution of adult virgin scales on three substrates (fruits, leaves, and branches). These scale grew during the hottest part of the summer at both locations. More of the scale insects are greater than 0.39 mm^2 in Ventura than in Tulare County on all three substrates. This suggests that in Tulare County usable hosts are relatively scarce and, of those available, a larger percentage are of lower

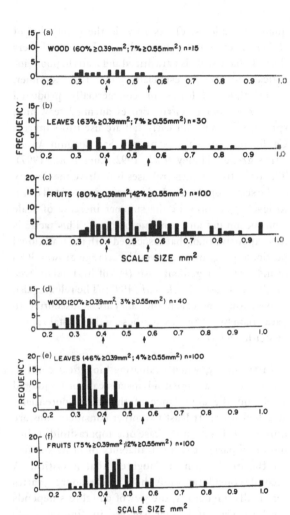

Figure 16.11. A comparison of the size distribution in September 1979 of virgin female California red scale, *Aonidiella aurantii* (Maskell), on wood, leaves, and fruits from Valencia citrus trees in Limoncoa, Ventura County, a California coastal area ((a), (b), and (c)), and in October 1979 in Woodlake Tulare County, an interior central California location ((d), (e), and (f)). The left hand arrow beneath each abscissa indicates the smallest scale size on which *Aphytis melinus* DeBach lays female eggs. The right hand arrow beneath each abscissa indicates the smallest scale size on which *A. lingnanensis* Compere lays female eggs.

quality (smaller size). As a result, the population of *Aphytis* is expected to be small, which in turn lowers the risk that a scale is parasitized and leads to poor biological control, especially in the summer. Augmentative releases of commercially produced *Aphytis melinus* in citrus groves during late winter, spring, summer, and early fall are used to suppress California red scale in mature San Joaquin Valley citrus groves (Haney *et al.*, 1992; Luck *et al.*, 1997). The objective of these releases is to drive the density of the scale population during the spring and fall down to levels that prevent the summer increase of scale from becoming an economic problem. This tactic is less expensive than that associated with the traditional pesticide program and saves the average grower 30% of his pest management costs (6% of his total on-tree production costs: Luck *et al.*, 1997). The releases also indicate that the scale in the valley are as susceptible to parasitism by *Aphytis* as are those in the coastal valleys of southern California

In summary, when evaluating biological control, hosts used by a parasitoid species cannot be judged solely on the abundance of the stages potentially attacked by it. At least two other considerations are important. First, in the case of ectoparasitoids (idiobionts in part), it is the availability of hosts suitable for the production of daughters that is critical. A second consideration is how this resource varies seasonally. Generally, the size of daughters depends on host size and their success in finding hosts increases with their size, at least to a threshold value (Visser, 1994; Kazmer & Luck, 1995; West *et al.*, 1996). The pattern we observed with the red scale and *Aphytis* is not unique. It is often typical of idiobionts (Clausen, 1939; Charnov *et al.*, 1981; Charnov, 1982; King, 1987; Godfray, 1994). Moreover, understanding this interaction has allowed the development of an ecologically based pest management program for California red scale (Haney *et al.*, 1992; Luck *et al.*, 1997). The program is based on assessing the availability of the daughter-producing resource and its quality for *Aphytis* in the citrus groves. The degree to which this resource is exploited by *Aphytis* is then monitored, which allows the manager to determine the likely prospects for an economically

damaging scale population to develop. A guide has been developed that provides this information and a flow diagram outlines the decision criteria (Forster *et al.*, undated). Thus, a fundamental understanding of *why* a parasitoid selects the hosts it does and *why* it allocates the number and sex of offspring it does, is fundamental to the development of sound pest management programs.

References

Allen, G. R., Kazmer, D. J. & Luck, R. F. (1994) Post-copulatory male behaviour, sperm precedence and multiple mating in a solitary parasitoid wasp. *Animal Behaviour* 48: 635–44.

Bai, B., Luck, R. F., Forster, L., Stephens, B. & Janssen, J. A. M. (1992) The effect of host size on quality attributes of the egg parasitoid, *Trichogramma pretiosum. Entomologica Experimentalis et Applicata* 64: 37–48.

Beckage, N. E., Thompson, S. N. & Federici, B. A. (eds) (1993) *Parasites and Pathogens of Insects. Volume 1: Parasites.* San Diego: Academic Press.

Bigler, F. (1994) Quality control in *Trichogramma* production. In *Biological Control with Egg Parasitoids*, ed. E. Wajberg & S. A. Hassen, pp. 93–111. Wallingford: CAB International.

Bigler, F., Meyer, A. & Bosshart, S. (1987) Quality assessment in *Trichogramma maidis* Pintureau et Voegele reared from eggs of the factitious hosts, *Ephestia kuehniella* Zell. and *Sitotroga cerealella* (Oliver). *Journal of Applied Entomology* 101: 64–85.

Charnov, E. L. (1979) The genetical evolution of patterns of sexuality: Darwinian fitness. *American Naturalist* 113: 465–80.

Charnov, E. L. (1982) *The Theory of Sex Allocation.* Princeton: Princeton University Press.

Charnov, E. L. & Krebs, J. R. (1973) On clutch size and fitness. *Ibis* 116: 217–19.

Charnov, E. L. & Skinner, S. W. (1984) Evolution of host selection and clutch size in parasitoid wasps. *Florida Entomologist* 67: 5–21.

Charnov, E. L. & Skinner, S. W. (1985) Complementary approaches to the understanding of parasitoid ovipositions. *Environmental Entomology* 14: 383–91.

Charnov, E. L., Los-den Hartogh, E. L., Jones, W. T. & van den Assem, H. (1981) Sex ratio evolution in a variable environment. *Nature* 289: 27–33.

Clausen, C. P. (1939) The effect of host size upon the sex ratio of hymenopterous parasites and its relation to methods of rearing and colonization. *Journal of the New York Entomological Society* **47**: 504–16.

Clausen, C. P. (1940) *Entomophagous Insects*. New York: Hafner.

Collier, T. M. (1995a) Host feeding, egg maturation, resorption, and longevity in the parasitoid *Aphytis melinus* (Hymenoptera: Aphelinidae). *Annals of the Entomological Society of America* **88**: 206–14.

Collier, T. M. (1995b) Adding physiological realism to dynamic state variable models of parasitoid host feeding. *Evolutionary Ecology* **9**: 217–35.

DeBach, P. & Huffaker, C. B. (1971) Experimental techniques for evaluation of the effectiveness of natural enemies. In *Biological Control*, ed. C. B. Huffaker, pp. 113–40. New York: Plenum Press.

DeBach, P. & Rosen, D. (1991) *Biological Control by Natural Enemies*, 2nd edn. Cambridge: Cambridge University Press.

DeBach, P. & Sundby, R. A. (1963) Competitive displacement between ecological homologues. *Hilgardia* **34**: 105–66.

Dicke, M. & Sabelis M. W. (1988) Infochemical terminology: based on a cost benefit analysis rather than origin of compounds? *Functional Ecology* **2**: 131–9.

Doutt, R. L. (1964) Biological characteristics of entomophagous adults. In *Biological Control of Insect Pests and Weeds*, ed. P. DeBach, pp. 145–67. London: Chapman and Hall.

Flanders, S. E. (1965) On the sexuality and sex ratios of hymenopterous populations. *American Naturalist* **99**: 489–94.

Foldi, I. (1990) The scale cover. In *Armored Scale Insects, their Biology, Natural Enemies, and Control*, ed. D. Rosen, pp. 43–54. Elsevier, Amsterdam.

Forster, L. D., Luck, R. F. & Grafton-Cardwell, E. E. (Undated) *Life Stages of California Red Scale and its Parasitoids*. Publication no. 21529, 12pp. University of California, Division of Agricultural and Natural Resources, Berkeley.

Green, R. F. (1982) Optimal foraging and sex ratio in parasitic Hymenoptera. *Journal of Theoretical Biology* **95**: 43–8.

Godfray, H. C. J. (1994) *Parasitoids: Behaviour and Evolutionary Ecology*. Princeton: Princeton University Press.

Haney, P. B., Morse, J. G., Luck, R. F., Griffths, H., Grafton-Cardwell, E. E. & O'Connell, N. V. (1992) *Reducing Insecticide Use and Energy Costs in Citrus Pest Management*. Integrated Pest Management Publication no. 15 University of California, Berkeley: Statewide Integrated Pest Management Project, Division of Agriculture and Natural Resources.

Hare, J. D., Yu, D. S. & Luck, R. F. (1990) Variation in life history parameters of California red scale on different citrus cultivars. *Ecology* **71**: 1451–60.

Heimpel, G. E. & Collier, T. R. (1996) The evolution of host-feeding behaviour in insect parasitoids. *Biological Reviews* **71**: 373–400.

Heimpel, G. E. & Rosenheim, J. A. (1995) Dynamic host feeding by the parasitoid *Aphytis melinus*: the balance between current and future reproduction. *Journal of Animal Ecology* **64**: 153–67.

Heimpel, G. E., Rosenheim, J. A. & Mangel, M. (1996) Egg limitation, host quality and dynamic behavior by a parasitoid in the field. *Ecology* **77**: 2410–20.

Hohmann, C. L., Luck, R. F. & Oatman, E. R. (1988) The effects of honey and of *Sitotroga cerealella* and *Trichoplusia ni* eggs on the fecundity and longevity of the egg parasitoid, *Trichogramma platneri*. *Journal of Economic Entomology* **81**: 1307–12.

Honda, J. (1995) The Biological Control Potential of *Trichogramma platneri* Nagarkatti on the Omnivorous Looper, *Sabulodes aegrotata* (Guenée) and the Western Avocado Leafroller, *Amorbia cuneana* Walsingham. PhD thesis, University of California, Riverside.

Huffaker, C. B. & Messenger, P. S. (1964) The concept and significance of natural control. In *Biological Control of Insect Pests and Weeds*, ed. P. DeBach, pp. 74–117. London: Chapman and Hall.

Jervis, M. A. & Kidd, N. A. C. (1986) Host feeding strategies in hymenopteran parasitoids. *Biological Reviews* **61**: 395–434.

Kazmer, D. J. & Luck, R. F. (1995) Field tests of the size-fitness hypothesis in the egg parasitoid, *Trichogramma pretiosum*. *Ecology* **76**: 412–25.

King, B. H. (1987) Offspring sex ratios in parasitoid wasps. *Quarterly Review of Biology* **62**: 367–96

Klomp, H. & Teerink, B. J. (1962) Host selection and number of eggs per oviposition in the egg-parasite, *Trichogramma embryophagum* Htg. *Nature* **195**: 1020–1.

Klomp, H. & Teerink, B. J. (1967) The significance of oviposition rates in the egg parasite, *Trichogramma embryophagum* Htg. *Archives Nederlandaises de Zoologie* 17: 350–75.

Lack, D. (1947) The significance of clutch size. *Ibis* 89: 309–52.

Leatemia, J. A., Laing, J. E. & Corrigan, J. E. (1995) Effects of adult nutrition on longevity, fecundity, and offspring sex ratios of *Trichogramma minutum* Riley (Hymenoptera: Trichogrammatidae). *Canadian Entomologist* 127: 245–54.

Li, L.-Y. (1994) Worldwide use of *Trichogramma* for biological control on different crops: a survey. In *Biological Control with Egg Parasitoids*, ed. E. Wajberg & S. A. Hassen, pp. 37–53. Wallingford: CAB International.

Luck, R. F. & Podoler, H. (1985) Competitive exclusion of *Aphytis lingnanensis* by *A. melinus*: potential role of host size. *Ecology* 66: 904–13.

Luck, R. F., Podoler, H. & Kfir, R. (1982) Host selection and egg allocation behavior by *Aphytis melinus* and *A. lingnanensis*: comparison of two facultatively gregarious parasitoids. *Ecological Entomology* 7: 397–408.

Luck, R. F., Forester, L. D. & Morse, J. G. (1997) An ecologically based IPM program for citrus in California's San Joaquin Valley using augmentative biological control. In *Proceedings of the International Society of Citriculture: 8th International Citris Congress, 12–17 May 1996*, vol. 1, pp. 499–503. Sun City, South Africa: International Society of Citriculture.

MacArthur, R. H. & Pianka E. R. (1966) On the optimal use of a patchy environment. *American Naturalist* 100: 603–9.

Mangel, M. (1989a) An evolutionary interpretation of the 'motivation to oviposit'. *Journal of Evolutionary Biology* 2: 157–72.

Mangel, M. (1989b) Evolution of host selection in parasitoids: does the state of the parasitoid matter? *American Naturalist* 133: 688–705.

McDougall, S. J. & Mills, N. J. (1997) The influence of hosts, temperature and food sources on the longevity of *Trichogramma platneri*. *Entomologia Experimentalis et Applicata* 83: 195–203.

Minkenberg, O. P. J. M., Tatar, M. & Rosenheim, J. A. (1992) Egg load as a major source of variability in insect foraging and oviposition behavior. *Oikos* 65: 134–42.

Murdoch, W. W., Briggs, C. J. & Nisbet, R. M. (1996) Competitive displacement and biological control in parasitoids: a model. *American Naturalist* 148: 807–26.

Nordlund, D. A., Jones, R. L. & Lewis, W. J. (eds) (1981) *Semiochemicals: Their Role in Pest Control.* New York: J. Wiley and Sons.

Oatman, E. R. & Platner, G. R. (1985) Biological control of two avocado pests. *California Agriculture* 39: 21–3.

Oatman, E. R., McMurtry, J. A., Waggonner, M., Platner, G. A. & Johnson, H. G. (1983) Parasitization of *Amorbia cuneana* (Lepidoptera: Tortricidae) and *Sabulodes aegrotata* (Lepidoptera: Geometridae) on avocado in southern California. *Journal of Economic Entomology* 76: 52–3.

Opp, S. B. & Luck, R. F. (1986) Effects of host size on selected fitness components of *Aphytis melinus* DeBach and *A. lingnanensis* Compere (Hymenoptera: Aphelinidae). *Annals of the Entomological Society of America* 79: 700–4.

Pak. G. A. (1986) Behavioral variations among strains of *Trichogramma* spp.: a review of the literature on host-age selection. *Journal of Applied Entomology* 101: 55–64.

Pinto, J. D. & Stouthamer, R. (1994) Systematics of the Trichogrammatidae with emphasis on *Trichogramma*. In *Biological Control with Egg Parasitoids*, ed. E. Wajberg & S. A. Hassen, pp. 1–36. Wallingford: CAB International.

Quayle, H. J. (1938) *Insects of Citrus and other Subtropical Fruits.* Ithaca: Comstock Publishing Company.

Roitberg, B. D., Sircom, J., Roitberg, C. A., van Alphen, J. J. M. & Mangel, M. (1993) Life expectancy and reproduction. *Nature* 364:108.

Roff, D. A. (1992) *The Evolution of Life Histories.* New York: Chapman and Hall.

Rosen, D. & DeBach, P. (1979) *Species of* Aphytis *of the World (Hymenoptera: Aphelinidae).* Boston: W. Junk.

Rosenheim, J. A. & Rosen, D. (1991) Foraging and oviposition decisions in the parasitoid *Aphytis lingnanensis*: distinguishing the influence of egg load and experience. *Journal of Animal Ecology* 60: 873–93.

Salt, G. (1934) Experimental studies in insect parasitism. II. Superparasitism. *Proceedings of the Royal Society of London, Series B* 114: 455–76.

Salt, G. (1935) Experimental studies in insect parasitism. III. Host selection. *Proceedings of the Royal Society of London, Series B* 117: 413–35.

Salt, G. (1937) The sense used by *Trichogramma* to distinguish between parasitized and unparasitized hosts. *Proceedings of the Royal Society of London, Series B* 122: 57–75.

Salt, G. (1940). Experimental studies in insect parasitism. VII. The effects of different hosts on the parasite *Trichogramma evanescens* Westw. (Hym. Chalcidoidea). *Proceedings of the Royal Entomological Society of London, Series A* 15: 81–124.

Schmidt, J. M. & Smith, J. J. B. (1985) Host volume measurement by the parasitoid wasp *Trichogramma minutum*: the roles of curvature and surface area. *Entomologia Experimentalis et Applicata* 39: 287–94.

Schmidt, J. M. & Smith, J. J. B. (1987) The measurement of exposed host volume by the parasitoid wasp *Trichogramma minutum* and the effects of wasp size. *Canadian Journal of Zoology* 65: 2837–45.

Smith, S. M., Hubbes, M. & Carrow, J. R. (1986) Factors affecting inundative releases of *Trichogramma minutum* Ril. against the spruce budworm. *Journal of Applied Entomology* 101: 29–39.

Strand, M. R. (1986) The physiological interaction of parasitoids with their hosts and their influence on reproductive strategies. In *Insect Parasitoids*, ed. J. K. Waage & D. J. Greathead, pp. 97–136. San Diego: Academic Press.

Suzuki, Y., Tsuji, H. & Sasakawa, M. (1984) Sex allocation and effects of superparasitism on secondary sex ratios in the gregarious parasitoid, *Trichogramma chilonis*. *Animal Behaviour* 32: 478–84.

van Alphen, J. J. M. & Jervis, M. A. (1996) Foraging Behaviour. In *Insect Natural Enemies*, ed. M. A. Jervis & N. A. C. Kidd, pp. 1–62. London: Chapman and Hall.

van Bergeijk, K. E., Bigler, F., Kaashoek, N. K. & Pak, G. A. (1989) Changes in host acceptance and host suitability as an effect of rearing *Trichogramma maidis* on a factitious host. *Entomologia Experimentalis et Applicata* 52: 229–38.

Vet, L. E. M. & Dicke, M. (1992) Ecology of infochemicals used by natural enemies in a tritrophic context. *Annual Review of Entomology* 37: 141–72.

Vinson, S. B. & Iwantsch, G. F. (1980) Host suitability for insect parasitoids. *Annual Review of Entomology* 25: 397–419.

Visser, M. E. (1994) The importance of being large – the relationship between size and fitness in females of the parasitoid *Aphaereta minuta* (Hymenoptera, Braconidae). *Journal of Animal Ecology* 63: 963–78.

Waage, J. K. & Ng, S. M. (1984) The reproductive strategy of a parasitic wasp. I. Optimal progeny and sex allocation in *Trichogramma evanescens*. *Journal of Animal Ecology* 53: 401–15.

West, S. A., Flanagan, K. E. & Godfray, H. C. J. (1996) The relationship between parasitoid size and fitness in the field, a study of *Achrysocaroides zwoelferi* (Hymenoptera: Eulophidae). *Journal of Animal Ecology* 65: 631–9.

Yu, D. S. (1986) The Interactions Between California Red Scale, *Aonidiella aurantii* (Maskell), and its Parasitoids in Citrus Groves of Inland Southern California. PhD thesis, University of California, Riverside.

Yu, D. S. & Luck, R. F. (1988) Temperature dependent size and development of California red scale (Homoptera: Diaspididae) and its effect on host availability for the ectoparasitoid, *Aphytis melinus*. *Environmental Entomologist* 17: 154–61.

PART V

Microbes and pathogens

In the past, biological control mostly depended on the use of arthropods as control agents, and the long history of applied ecological theory reflects this emphasis. The use of pathogens has been much more restricted, with myxoma virus for rabbit control representing one of the few exceptions. Consequently, the theoretical development of host–pathogen interactions has only occurred within the past 20 years. This imbalance is likely to shift in the next century for at least two reasons, (i) pathogens are typically highly specific, which should reduce their impact on non-target organisms, and (ii) pathogen genomes are easier to engineer than those of predators or parasitoids, so the attributes that facilitate control can more easily be manipulated. Further, the scope of biological control is being expanded, and pathogens also are beginning to be targeted. This section begins with an overview of the development of host–pathogen theory by Godfray & Briggs (Chapter 17). Despite differences in the biologies of pathogens and other types of natural enemies, many of the processes that are being explored in predator–prey and host–parasitoid theory also characterize pathogens. For example, Godfray & Briggs discuss the importance of host refuges, host density dependence, spatial heterogeneity, multiple-species interactions, and equilibrium *versus* non-equilibrium dynamics. These have been addressed in other sections of this book, but here the focus is on biological control using pathogens as agents.

Given that most insect biological control projects involve parasitoids, it is likely that many pathogens will be introduced into systems where they must coexist with parasitoids. In Chapter 18, Begon *et al.* discuss theoretical developments in host–pathogen–parasitoid interactions. Their discussion considers both mathematical theory and biological theory, the latter based on experimental work on a stored-product pest, a parasitoid, and a granulosis virus. An important message arising from both approaches is that coexistence between the natural enemies cannot be taken for granted, because the potential for interference competition is high, which can reduce overall control of the pest. But the outcome of the interaction depends critically on the specific biological attributes of the enemies involved. Thus, multispecies biological control systems need to be well understood before their dynamics can be predicted with any degree of certainty.

Up to this point, this book has focused entirely on the control of arthropod pests, largely reflecting the historical emphasis of theory on insect population dynamics. Onstad & Kornkven (Chapter 19) begin with a discussion of the use of pathogens for the biological control of weeds. Their model incorporates spatial structure, and as amply demonstrated for insects, space is also crucial for the persistence of pathogens and their success against weed populations. Onstad & Kornkven then further develop the theme introduced by Begon *et al.*, to a biologically rich, spatially explicit model of European corn borer, a parasitoid, and a microsporidium. Their goal is to understand the efficacy of the parasitoid in competition with the ubiquitous pathogen. Their work provides another example of how general constructs can be incorporated into detailed models to understand the functioning of specific biological control systems.

Not all pest species are amenable to classical biological control. One rapidly developing alternative is to use pathogens as biotic insecticides. However, pathogens cannot be treated simply as chemicals; they are living organisms, and as with more traditional techniques their success depends on understanding the underlying dynamics between the agent and the host. Locusts have historically been extraordinarily difficult to control. In Chapter 20, Thomas *et al.* describe a theoretical approach to evaluate the potential of a fungal biopesticide to control them. This chapter provides another example of how theory is currently being used as an important research tool to solve real field problems.

In Chapter 21, Johnson shifts the discussion from pathogens as agents to pathogens as targets. The biological control of weed pathogens is a relatively young field, and hence, the development of relevant theory is still in its early stages. Johnson extends standard dose–disease response models used to evaluate disease antagonists by incorporating a refuge that protects the disease from attack. We have already seen the effects of refuges in the context of host–parasitoid interactions. Thus, the book closes with an example of how conceptual constructs common to a diverse range of systems are making biological control, slowly but surely, less of an art and more of a science.

17

The dynamics of insect–pathogen interactions

H. CHARLES J. GODFRAY AND CHERYL J. BRIGGS

Introduction

While the potential for insect pathogens to help control pests has long been appreciated, until recently biological control programs have tended to concentrate on predators and parasitoids. Perhaps for this reason, a theory of how predators and parasitoids might regulate the numbers of their hosts was developed as long ago as the 1930s, while what are now considered the classical models of insect–pathogen interactions are barely 15 years old. The last 20 years or so have seen an enormous boom in interest in biological control using insect pathogens, both classical biological control but more especially augmentative biological control and the use of pathogens as bioinsecticides. Part of the reason for the increase in interest in pathogens is the molecular biological revolution that has enabled their natural history to be studied for the first time other than at rather gross anatomical detail. This same revolution has opened the possibility of genetic manipulation of insect pathogens, the relatively simple genomes of many making this a far more realistic prospect than the manipulation of parasitoids or predators.

Biological control is applied population dynamics, and it has always been appreciated that an understanding of the dynamic interactions between pests and their natural enemies will help the efficient implementation of control programs. That being said, there is less agreement about how to set about this exercise, with some pessimists suggesting that the dynamics of natural populations are too contingent to be subject to general theory. We are guardedly more optimistic and our aims in this chapter are to describe current ideas about the population dynamics of a major group of insect pathogens, those whose

hosts become infected by eating food contaminated with spores, viruses or other free-living pathogen stages. We base our discussion on simple strategic models developed by population ecologists, but consider how these may be developed and adapted to answer the concrete questions facing biological control practitioners

Many of the ideas we discuss in this chapter are most easily expressed using the shorthand of mathematics. This has both advantages and disadvantages. Advantages in the ease in which complicated ideas can be expressed and in the reduction of ambiguity; but disadvantages in that it reduces the audience of the work. One of the chief goals of this chapter is to present current thinking on modeling insect–pathogen interactions in a largely non-mathematical format so that both its strength and weakness are clear to readers unwilling to plough through the formal literature.

There are a number of different approaches available for modeling insect–pathogen interactions. We shall largely concentrate on the 'population biology' approach, which we define as the construction of models based on the classical models of theoretical ecology – the Nicholson–Bailey model, Lotka–Volterra model and, in the case of insect pathogens, the Anderson–May model. By based we mean that the model, even if it is very complicated and system specific, can be reduced to a classical model, at least as a limiting case. The ability to extract simpler models as limits is significant as it means that the analytical insights gained from their full analysis can be used to interpret the dynamic behavior of the more complicated model, which typically must be solved numerically. Berryman (Chapter 1), Briggs *et al.* (Chapter 2), and Hochberg & Holt (Chapter 4) also belong to this school of population biology. The second main technique we call the 'systems' approach and has it roots in the development of systems ecology in the 1960s. The problem is studied by writing simulation models of the system under study that typically are large and 'holistic', incorporating many details of pathogen, insect, host plant and other biological players. A good example of this approach applied to insect–pathogen studies is that of Onstad and colleagues (Onstad & Maddox, 1989; Onstad *et al.*, 1990) who have worked on the interaction between the European corn borer (*Ostrinia nubilalis*) and its microsporidian pathogen, *Nosema pyrausta*. Their largest model included 160000 state and rate equations and a simulation over 25 years required 500 min of CPU time on a Cray supercomputer. The equations in a systems model, like a population biology model, are designed to represent real biological interactions. Onstad & Kornkven (Chapter 19) and van Lenteren & van Roermund (Chapter 7) provide further examples and discussion of the systems approach, while Barlow (Chapter 3) reviews a wide range of different approaches. The final approach, the 'statistical' approach, simply tries to predict pest and pathogen population levels given biotic and abiotic data. The statistical model, typically a regression model or a more sophisticated relative, is optimized using one data set and then tested against other data sets or used to make predictions. This method is at the core of many within-season pest management systems and is invaluable for short-term predictions in well monitored systems.

There are many different circumstances in which a biological control worker might need an understanding of the dynamics of an insect–pathogen system. Four possible scenarios are:

1. A pathogen is used as a biological insecticide with the goal of killing insects on a temporary crop. Secondary infections by the pathogen do not occur or are unimportant.
2. A pathogen is used as in (1), but unlike a conventional insecticide it is hoped that secondary infections will occur and help limit pest density.
3. A pathogen is used to control the density of an outbreak pest in a semi-permanent agricultural or forestry system such as orchard, plantation or forest.
4. A pathogen is introduced as a classical biological control agent with the aim of providing long-term control of a pest, typically in a semi-permanent agroecosystem.

It is interesting that these four scenarios are probably numbered in order of decreasing importance in pest management, although in increasing order of the amount of attention given to them. Scenario 4 has been particularly well studied because it can be analysed using the well-developed methodologies of theoretical ecology, and because it is essentially the same question as how might pathogens regulate their hosts in natural ecosystems. Because of the fulcral role played by issues of pathogen regulation in the development of the subject, we begin with scenario 4 and move towards non-equilibrium problems later in the chapter. Many of the general issues of the role of population dynamic theory in biological control are also discussed by Hochberg & Holt (Chapter 4).

Conceptual frameworks

The Anderson–May model

Can a pathogen exert long-term control of its host, and more specifically can it regulate the abundance of a pest below its economic threshold? This question was first posed in a quantitative framework by Anderson & May (1980, 1981) and we begin by a discussion of their work, which forms the basis of much recent study that is considered next. Anderson & May (1981) described a suite of models that could be applied to invertebrate–pathogen interactions. The most important, both in terms of the number of systems to which it can be applied and the interest it generated, was Model G, the model concerned with pathogens possessing a free-living infectious stage in the environment. The model predicted that under certain circumstances the host and pathogen may show long-term cycles with a period of many years. This offered a potential explanation for long-term cycles in forest insects, a point explored in a *Science* paper (Anderson & May, 1980).

The Anderson–May model is expressed as a system of differential equations that describe the rate of change of numbers of infected and uninfected hosts, and of the free-living infectious stage (the latter might be fungal spores, baculovirus occlusion bodies etc). The simplest models make two important assumptions. First, that the environment is constant – there is no seasonality or fluctuations in abiotic conditions. Second, that the host population is not structured by age, genotype or spatial location. All hosts are identical except that they may or may not be infected. Now clearly these are highly unrealistic assumptions, but the value of a simple model such as this is that it says what is possible in a simple abstracted system when virtually everything has been removed except the actual interaction of interest. Before the model can be applied to particular systems, much more realism must be added. But as we shall see, it is impressive how accurately a very simple model such as this can predict the behavior of much more complicated, detail-rich models.

Let us now consider the three equations that constitute the Anderson–May model. The first is:

$$\text{Rate of change of healthy hosts} = \begin{cases} \text{births} - \text{deaths} - \text{infections} \\ + \text{recovery} \end{cases} \quad (17.1)$$

The number of healthy hosts increase due to births: the basic model assumes that all individuals, both infected and uninfected produce offspring at a fixed rate per time. The number of healthy hosts declines due to deaths, which are assumed to occur with a constant probability per unit time. There is no density-dependent increase in mortality due to, for example, resource competition. Healthy hosts also decrease in density due to infection. Anderson & May modeled infection as a 'binary collision process'. Suppose that hosts moved at random through the environment encountering infectious disease particles that themselves are randomly distributed in the environment. The rate at which infection occurs, per individual, is then proportional to the density of the pathogen in the environment (this is also called 'linear transmission'). The constant of proportionality (the slope of the line relating risk of infection to pathogen density) is called the transmission coefficient. The mode of infection is important in determining the dynamics and obviously this simple form is an idealization. Yet many much more complicated forms of transmission can be approximated by linear transmission, which is clearly a sensible place to start. (For recent evidence of non-linear transmission and other complexities in insects see Dwyer, 1991; D'Amico *et al.*, 1996; and Knell *et al.*, 1996 for an entry into the theoretical literature see Liu *et al.*, 1986 and Hochberg, 1991.) Note that as we discussed above, the model framework assumes that all hosts are equally susceptible to infection, regardless of age, sex, genotype or location. Finally, the number of healthy hosts can increase due to recovery. For nearly all the pathogens of interest to biological control, recovery does not occur and thus we shall ignore it.

Now consider the second equation:

$$\text{Rate of change of infected hosts} = \begin{cases} \text{infections} - \text{deaths} \\ (\text{background} + \text{disease}) \\ - \text{recovery} \end{cases} \quad (17.2)$$

The numbers of infected hosts increase through infection (the same term as in eqn (17.1)) and decrease through mortality and (were we not to ignore it) recovery. Mortality is divided into two components, the background mortality experienced by healthy and infected hosts alike, and an extra source of mortality brought about by the disease. Note that all infected hosts are equally susceptible to disease mortality, irrespective of age, genotype etc, and also irrespective of the time since infection. Thus there is no fixed incubation time and the risk of succumbing from the disease is constant no matter how long you have been infected.

The final equation represents the free-living infectious stage:

$$\text{Rate of change of infectious stage} = \begin{pmatrix} \text{Number} \\ \text{of infected} \\ \text{hosts} \end{pmatrix} \begin{pmatrix} \text{pathogen} \\ \text{discharge} \\ \text{rate} \end{pmatrix} - \begin{matrix} \text{pathogen} \\ \text{decay} \end{matrix} \quad (17.3)$$

The model assumes that infected hosts release infectious pathogen stages into the environment at a fixed rate per unit time (an alternative would have been to assume that pathogens are only released on the death of the host). These infectious stages remain in the environment, their numbers decaying away at a certain fixed rate. (The original model also assumed the free-living pathogen stage decreased in numbers due to consumption by the host; for most realistic parameterizations of the model this source of mortality will be small and safely can be ignored).

Having constructed the model, we can begin to ask questions of it. First, when will a disease be able to establish itself in the population? For this to occur, we need to count the number of secondary infections caused by an initial infection, a quantity known universally among epidemiologists as R_0.

$$R_0 = \left(\begin{array}{c} \text{pathogen} \\ \text{discharge rate} \end{array} \right) \left(\begin{array}{c} \text{lifespan of} \\ \text{infected host} \end{array} \right)$$

$$\left(\begin{array}{c} \text{lifespan of} \\ \text{infectious stage} \end{array} \right) \left(\begin{array}{c} \text{transmission} \\ \text{coefficient} \end{array} \right) \left(\begin{array}{c} \text{susceptible} \\ \text{host density} \end{array} \right)$$

$$\text{(17.4)}$$

Given that an infection occurs, the number of infectious stages released into the environment is simply the pathogen discharge rate multiplied by the expected lifespan of the infected host. The latter will be influenced both by the background mortality and by the extra disease-induced mortality. A highly virulent pathogen will kill the host quickly and few pathogens will be released. The probability that any of these pathogens in the environment will cause further infections is a product of three quantities. First, the lifespan of the infectious particle in the environment (the longer they survive the more likely infection is to occur). Second, the transmission coefficient that determines the efficiency of transmission. This will be influenced by numerous factors including the mobility of the host and the chance that infection occurs after ingestion of (or other contact with) the pathogen stage. Third, the (susceptible) host density. If hosts are abundant, secondary infections are more likely.

Asking when $R_0 > 1$ allows us to predict whether the infection will spread. Basically anything that increases the magnitude of any of the five bracketed terms in eqn (17.4) will make spread more likely. Now for any particular disease, the parameters in the first four bracketed terms will be fixed aspects of the life history of the pathogen and the host. But the last

term, host density, is an ecological factor that may be highly variable and determine whether or not $R_0 > 1$ is satisfied. We can define a certain host density, the host threshold, at which $R_0 = 1$. Below the threshold, the host is too rare to support the disease, which dies out. Above the threshold the disease is able to spread. R_0 and the host threshold are crucial tools in all models of insect epidemiology and we shall make extensive use of them here.

Let us now assume that the disease is spreading through the population. Note that in the basic model there is no density dependence so that in the absence of the disease the population will increase at a rate determined by the size of the (births−deaths) term in eqn (17.1). For the disease to prevent this happening, and exert control, the net death rate of infected hosts must exceed their net birth rate. If this condition is not met, the disease may either fail to establish, or reduce the geometric rate of increase of the host population.

Having established that the disease will invade and prevent the population exploding, we can ask what are the equilibrium levels of healthy and infected hosts, and of the free-living pathogen stage. Again, the concept of R_0 is valuable. At equilibrium, every infection must result in exactly one further infection: if this were otherwise the numbers of infected hosts would grow or decline. Thus we can write down the equilibrium by setting $R_0 = 1$ and rearranging eqn (17.4):

$$\begin{array}{c} \text{Equilibrium} \\ \text{susceptible} \\ \text{host density} \end{array} = 1 \left/ \left[\left(\begin{array}{c} \text{pathogen} \\ \text{discharge rate} \end{array} \right) \left(\begin{array}{c} \text{lifespan of} \\ \text{infected host} \end{array} \right) \right. \right.$$

$$\left. \left. \left(\begin{array}{c} \text{lifespan of} \\ \text{infectious stage} \end{array} \right) \left(\begin{array}{c} \text{transmission} \\ \text{coefficient} \end{array} \right) \right] \right. \text{(17.5)}$$

Exactly the same aspects of the pathogen and host life history that increase the chance of pathogen spread also lead to a lower (susceptible) host equilibrium. At equilibrium, the ratio of infected to susceptible hosts is (births−deaths)/(disease-induced mortality):1. When disease-induced mortality is high, the total population density is very near the host threshold.

The existence of a feasible equilibrium does not

guarantee that the equilibrium is stable. Minute deviations from the equilibrium may be magnified by the system's dynamics leading to diverging oscillations (and hence, in a biological context, to extinction of one or both partners of the interaction); to persistent population cycles, or to even more exotic dynamics such as quasiperiodicity or chaos. A battery of mathematical techniques can be applied to models such as this to investigate their local and global stability.

Anderson & May (1981) found that their model predicted a stable equilibrium as long as the half-life of the infectious stage in the environment was not too long, and as long as the host rate of increase was not too high. However, when these conditions were not met, the equilibrium was unstable and the model predicted long-term cycles with a period of typically 3–10 years. What happens is that the pathogen reduces the density of susceptible hosts some way below the host threshold. The host is now too rare to support the pathogen, although the latter can persist by virtue of its slow decay rate in the environment. In time, the host population begins to rise again and then rapidly surpasses the host threshold allowing the pathogen again to increase in density. But because pathogen densities are now so low, there is a time lag before the host population can be controlled and the resulting epizootic is severe and involves a large reduction in host density.

Forest insects display the type of long-term cyclic fluctuations in density predicted by the Anderson–May model and are also subject to the type of pathogen that promotes this dynamic behavior (in particular nuclear polyhedrosis virus (NPV) baculoviruses). Typically, these pathogens kill the host before they breed while the Anderson–May model assumes infected hosts reproduce. But relaxing this assumption increases the region of parameter space in which population cycles occur. However, there are alternative explanations of forest insect cycles (see Berryman, 1996 for a recent review), and not all developments of the Anderson–May model predict this type of cycle. We shall return to this issue after we have discussed some other models of insect–pathogen dynamics.

Discrete-generation models

The Anderson–May model is written as differential equations in continuous time. In this section we discuss a different but related framework for studying insect–pathogen dynamics, that of discrete-time models based on the Nicholson–Bailey equations. These models have been extensively investigated as a means of studying host–parasitoid dynamics, but can also be used to study host–pathogen dynamics (see Chapters 2, 4, and 13).

Consider the following pair of equations:

$$\begin{pmatrix} \text{Hosts at} \\ \text{time } t+1 \end{pmatrix} = \begin{pmatrix} \text{Hosts at} \\ \text{time } t \end{pmatrix} (\text{fecundity}) \begin{pmatrix} \text{Fraction} \\ \text{uninfected} \end{pmatrix}$$

(17.6a)

$$\begin{pmatrix} \text{pathogens at} \\ \text{time } t+1 \end{pmatrix} = \begin{pmatrix} \text{Hosts at} \\ \text{time } t \end{pmatrix} \left[1 - \begin{pmatrix} \text{Fraction} \\ \text{uninfected} \end{pmatrix} \right]$$

$$\begin{pmatrix} \text{Pathogen yield} \\ \text{per host} \end{pmatrix}$$

(17.6b)

The number of hosts in the next generation (time $t+1$) is simply those that escape the disease multiplied by *per capita* net fecundity (not births per unit time as in the Anderson–May model but total fecundity adjusted for mortality between seasons). The number of pathogens in the next generation is simply the number of infected hosts multiplied by the total number of pathogens released over the course of the infection (again adjusted for mortality between seasons). How are we to model the fraction infected? The simplest assumption is to assume that within the generation pathogen numbers remain the same and the transmission process is the binary collision (linear) form assumed by Anderson & May. In that case the mean risk of infection is simply the product: (pathogen density)×(transmission coefficient)× (length of time exposed to the pathogen) and the fraction that escape parasitism is the zero term of the Poisson distribution with this mean.

This model has some advantages over the Anderson–May model but also some substantial disadvantages. The advantages are that it is more straightforward to apply to species such as forest insects with discrete generations, and which are only susceptible to infection during part of their life cycle. The major disadvantage is that it assumes constant pathogen densities during the season – in biological terms that the density of infectious particles remains constant during the summer until they are all destroyed in the autumn to be replaced by the pathogens produced by the infected insects of the current year. Below we shall modify the model to remove these absurdities (although note that these assumptions exactly fit many univoltine parasitoids) but in so doing we will destroy the model's analytical tractability. It is thus worth pausing to look at the dynamic behavior of eqns (17.6a,b), which will be an important limiting case of our more realistic model (for much fuller analysis in a parasitoid context see Hassell & Pacala, 1990, and references therein).

With the assumption of linear transmission, the model (eqns (17.6a,b)) is invariably unstable. The pathogens severely reduce host densities and then themselves crash in density. From a very low base the hosts increase in density and, initially unchecked, reach very high densities before the pathogen becomes abundant enough to control them. Essentially, we are getting the same boom and bust dynamics as caused the cycles in the Anderson–May model; but here instead of converging on a persistent cycle, divergent oscillations occur. Discrete-generation consumer–resource models have an inherent propensity to be unstable compared with their continuous-time analogs because of the presence of time lags that prevent the consumer, the pathogen in this case, from responding quickly to changes in host density (May, 1974).

A large number of variants of eqns (17.6a,b) have been analysed in the host–parasitoid literature that do allow coexistence of host and parasitoid. The largest class incorporate a form of non-linear parasitoid attack (i.e., transmission) that results in the risk of parasitism (infection) increasing at a less than linear rate with parasitoid (pathogen) density. There are many biological submodels that can lead to non-linear parasitoid attack: for example spatial heterogeneity in parasitoid attack, the presence of a host refuge from the parasitoid, heterogeneity in the host's ability to defend itself against parasitism, behavioral interference among parasitoids at high density (see also Chapters and 4). Most workers in the field believe that some form of non-linear parasitoid attack is responsible for persistent host–parasitoid interactions.

As we said above, it is possible to remove the most objectionable features of eqns (17.6a,b) (from a pathogen point of view) but with the loss of mathematical tractability (Briggs & Godfray, 1996a). We now allow the pathogen to have variable dynamics within the season. During the season, the densities of infectious particles in the environment decay at a constant rate, but are added to as hosts are infected and release pathogens. We also remove the restriction that the pathogen can survive a maximum of 1 year in the environment. Unlike the simple model, hosts now suffer variable risks of infection within the season. If the pathogen completes its life cycle relatively fast, it can go through two or more generations within the season. Exactly this type of within-season dynamics has been studied and modeled by Dwyer & Elkinton (1993) with gypsy moth (*Lymantria dispar*) and its NPV.

So, does the addition of more realistic pathogen biology stabilize the unstable parasitoid model, and are long-term population cycles of the Anderson–May type observed? The answers are somewhat and no. The ability of the pathogen to cause secondary infections within the season contributes towards the stability of the system but not enough to give a persistent interaction in the face of a linear transmission term. To obtain a persistent interaction in these simple models, some non-linearity in transmission must be included, although not as much as in the case of a host–parasitoid model with synchronized generations (Fig. 17.1). However, the modified models still do not show the long-term population cycles displayed by the Anderson–May model.

Age structure

So far we have discussed two approaches to the study of insect–pathogen dynamics, the Anderson–May model, which is constructed in continuous time, frequently predicts a persistent interaction, and can show long-term limit cycles; and discrete-time

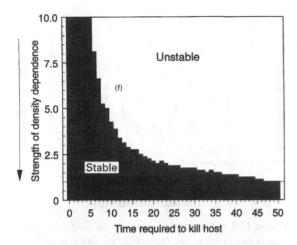

Figure 17.1. Stability analysis of the discrete-generation model. When the pathogen only has a single generation a year, the interaction is only stable when quite substantial density dependence (in the form of non-linear transmission) is included in the model. As the pathogen is able to have several rounds of infection per season (because it takes a relatively short time to kill the host), the interaction requires less density dependence for stability.

models, which are much less stable, and which at least in their simplest form do not show long-term cycles.

We can reconcile the two models by adding a little more biology to the Anderson–May model. In this model, all hosts are identical (except that they may be infected or uninfected) and give birth at a constant rate to new susceptible individuals. Now suppose that there is a time-lag in host development: hosts give birth at a constant rate but their offspring only recruit to the susceptible class after a fixed period of time. Further, assume that infected individuals do not start releasing pathogen particles into the environment immediately but that there is a second fixed delay, this one representing incubation. In effect, what we are doing is adding age-structure to the model. Technically, there are a variety of ways that this can be done but the account here is based on our work which uses the lumped-age class formalism of Gurney *et al.* (1983; see also Chapter 2) to produce a model phrased as a system of delay-differential equations (Briggs & Godfray, 1995). The critical assumption of this approach is that the life history of

the host and pathogen can be divided into periods of fixed length within which the demographic parameters are constant. This assumption is often met by insects whose life cycle falls naturally into major stages (egg, larva, pupa, adult) or instars.

Age-structure can be included in a simpler manner if one relaxes the assumption that there is a minimum time period in each developmental stage, and a minimum incubation period (e.g., Anderson & May, 1981; Vezina & Peterman, 1985; Hochberg & Waage, 1991). The resulting model is expressed as a system of ordinary differential equations without time lags. However, care must be taken with this approach as real insects do have minimum developmental times, and these can be important in determining the predicted dynamics. For example, Vezina & Peterman (1985) concluded that an Anderson–May type model could not explain the type of population cycles displayed by tussock moth (*Orgyia pseudotsugata*), possibly driven by an insect virus. To make the model more realistic, they included an incubation period, but without a minimum period, and this would have had the effect of dampening any cycles.

The introduction of age-structure and the concomitant time lags can have both stabilizing and destabilizing effects. However, unless there is a long-lived invulnerable stage (see below), age-structured models tend to be less stable than the corresponding non-age-structured models. Thus, introducing age-structure into the Anderson–May model typically leads to the prediction of diverging population oscillations. Of course, the interactions can be made stable by the inclusion of some additional density dependence such as non-linear transmission. But even so the basic model tends to behave like the discrete-time models discussed above and does not show persistent limit cycles. The presence of time lags also explains why discrete-time models tend to be less stable than Anderson–May counterparts. Indeed, age-structured models in continuous time can be constructed that have almost identical stability properties to the discrete-time models. How important is it to include the time lags? The answer depends very much on the biology of the system. Species that are subject to pathogen attack throughout most of their larval and adult development period will have dynamics most closely approximated by the Anderson–May model, those that are subject to attack during a restricted part of their life cycle will have dynamics better described by an age-structured model.

Models with age-structure in continuous time have some properties absent in the simpler models discussed above. First, consider an insect that is attacked by a pathogen during an early developmental stage but which if it avoids infection goes on to become an adult. Suppose also that the adult is relatively long-lived in comparison to its juvenile stages. Many homopterans and beetles have this type of life history, while lepidopterans and dipterans tend to be short-lived as adults. The presence of a long-lived, immune life history stage has a marked stabilizing influence on the dynamics (Murdoch *et al.*, 1987; Briggs & Godfray, 1995). This occurs because the immune stage acts as a buffer preventing destabilizing boom and bust cycles: even when the pathogen is very abundant, hosts in the long-lived stage are protected from attack and prevent a drastic collapse in host densities.

A second prediction of the age-structured model concerns hosts with exactly the opposite natural history: a short adult lifespan compared with the juvenile period. As was first realised for host–parasitoid interactions, persistent cycles can occur with a very short period, approximately equal to the length of the host developmental period and usually called generation cycles (Godfray & Hassell, 1987, 1989; Briggs & Godfray, 1995). For this to occur in a pathogen context, the incubation time of the disease must be approximately one-half the host developmental period. The cycles are caused by an interference phenomenon resulting from the interaction of two unequal time lags (Fig. 17.2). In fact, the model predicts cycles with a period of $1/2, 1/3 \ldots 1/n$ the host developmental lag for decreasing incubation periods although with the possible exception of the $1/2$ cycles, the dynamic behavior is too fragile to be observed in nature. The prediction of generation cycles by host–parasitoid and host–pathogen models is more than a mathematical curiosity as exactly this type of dynamic behavior is observed in a range of tropical plantation pest insects whose biology at least superficially matches their assumptions. Unfortunately, detailed field studies to confirm the mechanism

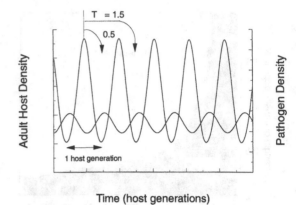

Figure 17.2. Schematic description of the mechanism producing generation cycles. Generation cycles can be produced when the time required to kill an infected host (T) is around 0.5 or 1.5 times the host developmental period. Most infections occur at the peak of host density and if the incubation period is approximately equal to 0.5 or 1.5 times the developmental period, there is an interference phenomenon resulting in persistent cycles.

and distinguish between different natural enemies are yet to be performed.

Exploring the biology

In the previous three sections we have described a series of template models that can be used as a framework for understanding insect–pathogen dynamics. Most of the biology of the system has been stripped away in order to concentrate on the bare bones of the insect–pathogen interaction. We thus have normative cases with which we can compare the predictions of more complicated models. In this section we review a variety of studies that have considered the influence on dynamics of particular aspects of the insect–pathogen interaction. The majority of studies have used the simple Anderson–May model as the starting point and it is important to consider whether their conclusions also apply to the other templates.

Pathogen refuges

The pathogens we are considering in this chapter have a free-living stage in the environment that causes infection when eaten by a host. The life expectancy of the free-living stage may vary considerably depending on where it is in the environment. In some places, exposed to ultra-violet (UV) light and other potential mortality sources, the density of pathogens is likely to decay rapidly, while in refuges such as the soil it may survive for long periods of time. Baculoviruses, for example, have been isolated from soils a decade after infection.

Hochberg (1989, see also Chapter 4) used an Anderson–May framework to explore the dynamics of a pathogen that existed in the environment in a refuge, where it decayed slowly but could not infect hosts (for example, the soil) and in more exposed sites where transmission was possible (for example, on leaves). Pathogens are transported between the two sites at a constant rate by abiotic or biotic agencies (washed off by rain, moved by birds or other animals, etc). The presence of a refuge has an important stabilizing effect, something that is also found when a refuge is included in discrete-time or age-structured models (Hochberg *et al.*, 1990; Briggs & Godfray, 1995, 1996a). Like long-lived, immune life history stages, refuges act as a buffer that prevent diverging boom and bust population dynamics. After a drop in host densities, a steep crash in pathogen densities is prevented by translocation of the free-living stage from the refuge in the environment. Pathogen particles are thus available to limit the next boom in host densities. Hochberg (1989) argued that the presence of environmental reservoirs was characteristic of those agricultural systems where pathogens could exert long-term control on host density. A discrete-time model with age-structure designed to mimic a forest insect–pathogen interaction predicted limit cycles or chaotic behavior when a long-lived reservoir was included (Briggs & Godfray, 1996a).

Vertical transmission

A purely vertically transmitted pathogen cannot spread through a population unless it posseses some drive mechanism (Anderson & May, 1981). Drive mechanisms are a means by which the pathogen can infect more than the single daughter that survives in an equilibrium population. To see why, consider a population at equilibrium where every female produces over its lifespan exactly one daughter to replace itself. A purely vertically transmitted pathogen with perfect transmission will infect that daughter and hence have an R_0 of unity, not enough to allow spread. In fact, most vertically transmitted pathogens cause reduced fecundity or have imperfect transmission and hence will have $R_0 < 1$.

The types of drive mechanism that have been reported in insects include (i) segregation distortion; (ii) sex ratio distortion; (iii) feminization; and (iv) unidirectionality incompatibility (sperm only accepted from infected females). There has been much interest in this subject in the last 10 years because of the discovery of genetic parasites such as *P* elements and other transposable elements that spread within the host genome, and of a variety of more traditional pathogens that are transmitted cytoplasmically and have $R_0 > 1$ due to sex ratio distortion or unidirectional incompatibility. The intracellular bacterium, *Wolbachia*, for example, spreads through unidirectional incompatibility and may be found in 10–20% of all insects (see Chapter 14). Interest in this type of parasite has also been raised by the possibility that they might be used as drive elements to introduce artificially-engineered transgenes into populations. Space does not allow a discussion of the fascinating dynamics of these elements; an entrance into the literature can be found in Turelli (1994).

A number of normally horizontally transmitted pathogens also show a restricted amount of vertical transmission and this can have quite marked effects on population dynamics (Anderson & May, 1981; Anderson, 1982; Vezina & Peterman, 1985). In the context of the original Anderson–May model, Anderson (1982) found the inclusion of vertical transmission tended to promote the region of parameter space in which long-term cycles occurred. We included vertical transmission in a discrete-time forest insect–pathogen model (Briggs & Godfray, 1996a). We assumed a certain fraction of hosts entered an exposed but not infectious class and that these individuals became adults that transmitted the pathogen vertically to a certain fraction of their offspring. Unlike the basic discrete-time model, this version did predict long-term cycles, at least for some parameter values. However, the chief reason that cycles are found is not the presence of vertical transmission but that of the exposed but not infectious class that acts as a refuge for the host population at times of high pathogen density.

Pathogens can persist if they go through an obligate vertically transmitted stage (perhaps over winter), as long as they also have horizontal transmission at other times (for example during the summer) (Régnière, 1984; Briggs & Godfray, 1996b). The horizontal transmission takes the place of the drive mechanisms described in the last paragraph in allowing the pathogen to achieve $R_0 > 1$. Models of this type can also predict long-term population cycles, especially when host reproductive rates are moderate.

Evidence is accumulating that some insect pathogens, previously assumed to be invariably fatal, can cause sublethal infections (e.g., Boots & Begon, 1994; Rothman & Myers, 1994, 1996). Even if females infected in this way do not transmit the pathogen vertically to the next generation, such infection can influence population dynamics if it results in reduced fecundity or increased mortality (Anderson & May, 1981; Ginzburg & Taneyhill, 1994).

Host density dependence

So far the only density dependence in the models that we have considered has come about through the interaction of the host and the pathogen, and even then the models frequently predict diverging oscillations in host and pathogen density. In the real world, diverging oscillations do not occur because some other form of density dependence comes into play at high host densities, typically competition among hosts for food. What is the consequence of combining host–pathogen dynamics and host density dependence, and in particular are long-term cycles more or less likely?

This problem has been addressed by Vezina & Peterman (1985), Bowers *et al.* (1993) and Dwyer (1994) who all worked with modifications of the basic Anderson–May model. Dwyer concluded that in the

presence of host density dependence, long-term population cycles were more likely to occur and would tend to be of a longer period. Vezina & Peterman and Bowers *et al.* found long-term cycles to be predicted over a narrower range of parameter values, and the latter found the cycles to be of a period that depended quite critically on precise parameter values. The reason for the discrepancy in the two predictions is that Vezina & Peterman and Bowers *et al.* added density dependence to the original Anderson–May model while Dwyer both added density dependence *and* assumed infected hosts did not reproduce (White *et al.*, 1996b). How the addition of density dependence affects the likelihood of population cycles in different modeling frameworks, more appropriate for forest insects, is not clear.

A striking example of the interplay of host density dependence and pathogen dynamics is provided by the work of Begon and colleagues (see Chapter 18) on a model laboratory system consisting of the moth *Plodia interpunctella* and its granulosis virus. In the absence of the virus, the host shows strong population cycles with a period of approximately one host generation. Such cycles are found in a number of model stored product systems of which *Tribolium* is probably the most famous (e.g., Descharnais & Liu, 1987). The cause of the cycles is intraspecific larval competition (Gurney *et al.*, 1983) although the exact mechanism is not yet completely clear. When the virus is added, a persistent interaction occurs in which the cycles continue to be found, although their period is slightly longer. We (with M. Begon's group) are currently trying to explain these dynamics using the type of age-structured population models described above.

Space

There is growing interest in the incorporation of a spatial dimension in host–pathogen models. The simplest way to include spatial heterogeneity is in a phenomenological manner through a modification of the transmission term. Thus recall that in our discussion of the simplest discrete-time model, the fraction of hosts escaping disease was the zero term of a Poisson model. Underlying this choice of statistical distribution is the assumption of random mixing of hosts and pathogens in the environment. If mixing was non-random, another statistical distribution could be used, the negative binomial being a popular choice (May, 1978). The assumption that pathogens are distributed in the environment in a clumped rather than a random manner results in a non-linear transmission term which can be a powerful stabilizing influence.

A more realistic way of including space is to consider a set of spatially discrete, weakly connected populations. By weakly connected we mean that they exchange migrants at a rate sufficient to prevent isolation, but too low to cause complete synchronization. This type of population structure is called a metapopulation and has been widely studied by theoretical ecologists (see Chapter 10 and Chapter 12), although not specifically in the context of insect–pathogen interactions. One broad feature of this type of model is that regional persistence can occur in the face of local instability as long as populations fluctuate out of synchrony. A local population in which host or pathogens are extremely rare or extinct is rescued by immigration from populations in a different phase of the dynamic cycle.

Finally, space can be included in the most explicit way by working with models that describe the changes of host and pathogen density at actual spatial coordinates. Until recently, most work in this area has used diffusion models and we discuss an application of this to insect–pathogen interactions in a moment. However, in the last 10 years a suite of new techniques have become available for studying spatial processes including lattice models, cellular automata models, interacting particle models, as well as more sophisticated forms of diffusion models (Hastings, 1996). For host–parasitoid interactions, lattice and cellular automata models predict a curious menagerie of phenomena including spirally propagating waves, and disordered spatial structures that appear chaotic (Hassell *et al.*, 1994). An extension of this work to an insect–pathogen system (using a discrete-time model but allowing the natural enemy to persist potentially for more than one generation) predicts the same basic patterns, although with some interesting variations (White *et al.*, 1996a). However, a different

insect–pathogen lattice model constructed by Wood & Thomas (1996, see also Chapter 20) gives quite different results, although reinforcing the observation that the incorporation of a spatial dimension can have a radical effect on a model's prediction. Wood & Thomas' model is based on detailed observations of an orthopteran–fungal interaction (see also below) and unlike the other models discussed here assumes no dispersal by the pathogen. They conclude that generalizations based on simple spatial models must be treated with caution. They also show that very virulent pathogens that could not persist in a homogeneous population (because their $R_0 < 1$) may persist in a spatially distributed interaction by virtue of the complex spatial interactions that are generated. The problem with the new approaches is the almost impossibility of testing their predictions with the limited spatial data we currently possess. Hopefully, the presence of more sophisticated (and more realistic models) may stimulate the collection of such data.

The best spatial data we currently possess refers to the rate of spread of certain insect epizootics. To attempt to predict the rate of spread, Dwyer (1994) took the Anderson–May model and added host density dependence and a diffusion term (he also assumed infected hosts did not reproduce). The equation that describes the numbers of susceptible hosts is now of the form:

$$\text{Rate of change of healthy hosts at } x,y = \begin{array}{l} \text{births} - \text{deaths (a function} \\ \text{of } N) - \text{infections} \pm \\ \text{diffusions } (x, y) \end{array} \qquad (17.7)$$

The equation describes changes in density at each spatial location, x, y. Numbers increase due to births and fall due to deaths (now a function of local host (N) density) and infection. There is now a new term, diffusion that may be either positive or negative: individuals are assumed to move randomly so that there will be a net flow of individuals from areas of high density to low density just as after a drop of ink is placed in a glass of water, the ink particles slowly diffuse from an initial area of high density. The rate at which mixing occurs is governed by the diffusion parameter.

The major prediction of most reaction–diffusion

models, including Dwyer's, is that a disease epizootic will spread from an initial focus and that after initial transients have died away, the speed and shape of the wave front (or traveling wave) will be a constant. Dwyer calculated the speed of advance of the disease front as a function of the parameters in the model and compared the prediction with data on several insect diseases. A strict test of the theory was not possible as estimates of all parameter values were not available, but Dwyer was able to show that the speed of advance of the disease was consistent with the model. One system for which quite good data were available was the nuclear polyhedrosis virus (NPV) of European spuce sawfly, *Gilpinia hercyniae* (Entwistle *et al.*, 1983). It had previously been thought that the spread of the disease required dispersal by some external agency such as birds. However, Dwyer's model suggested that simple diffusion by sawfly movement was sufficient to account for the spread. The model also suggested that in the long-term, the spatial interaction between the host and the pathogen might take the form of persistent waves moving through space (patterns similar to those predicted by the lattice models described above). Again good enough data are not currently available to test this prediction.

In an experimental study, Dwyer (1992) found that a diffusion model of the type just described was successful in predicting the local spread of the NPV of Douglas-fir tussock moth (*Orgyia pseudotsugata*). In another system, Dwyer & Elkinton (1995) experimentally studied the rate of spread of the NPV of gypsy moth after deliberate release of the virus in an uninfected population. They reasoned that the main dispersal route would be through ballooning first instar larvae (ballooning occurs when larvae secrete long silk threads that are blown by the wind). Independent estimation of dispersal rates through ballooning could explain the initial rate of spread of the infection, but not its sustained rate. Incorporation of shorter-scale larval movement through crawling still failed to account for the rate of spread. They conclude that another, as yet unidentified, factor causes the faster than expected spread. One candidate is mechanical vectoring by parasitoid flies. Long distance transfer of pathogens by agencies such as birds and humans has been implicated in the spread of a number of other insect diseases (Entwistle *et al.*, 1983; Fuxa & Richter, 1994) although, as discussed above, it is rare for the null hypothesis of simple diffusion to be eliminated.

More than two species

The abstraction of one insect species and a single type of pathogen is useful as a conceptual starting point and also as a model of some agriculture systems. However, in more natural systems and in many applied systems the insect–pathogen interaction is embedded within a much larger community of interacting species. In this section we briefly review some studies of three and more species interactions.

Suppose that two hosts are attacked by one species of pathogen. Even though the two hosts may never directly encounter each other in the field, their population dynamics are still linked through the shared disease. The general issue of the dynamic interactions of two species attacked by a common natural enemy was first analysed by Holt (1977). He found that the effect of one species on the other had much in common with resource competition: if the numbers of one species increase, this allows the natural enemy density to go up, and the second species suffers. The growth rates of each prey population are negatively affected by an increase in density of the other, as in competitive interactions. Because of these similarities, Holt (1977) called reciprocal negative effects mediated by natural enemies *apparent competition*. One of the most convincing examples of apparent competition in the field involves two woodlice (Isopoda) that share a common virus (Grosholz, 1992).

The first explicit model of indirect interactions mediated *via* pathogens was that of Holt & Pickering (1985, see also Begon *et al.*, 1992), which considered a directly transmitted disease. Coexistence of the disease and the two hosts was possible if the force of interspecific infection was less than the force of intraspecific infection. Because interspecific effects are weaker than intraspecific, each species is at a relative advantage when rare. Bowers & Begon (1991) studied the dynamics of two insect species that did not interact directly but which were both susceptible to a disease that was contracted through contact with a free-living independent stage (i.e., the common type of insect disease considered in this chapter). Unlike the cases considered by Holt & Pickering, the force of the infection depends solely on the density of the free-living infectious stage, not on whether a susceptible individual encounters a conspecific or allospecific. In consequence, coexistence is impossible and the host that survives is the one that at equilibrium supports the highest density of the infectious free-living stage. To see why this occurs, consider the inferior species at its equilibrium in the absence of the other species. The rate of increase of the host is zero when the density of the infectious pathogen stage in the environment is at equilibrium. Now add the second host, if this species tends to lead to an increase in the infectious stage above that present at the first host's equilibrium, the latter's rate of increase will be depressed below zero and hence it will go extinct. As Bowers & Begon point out, there is a marked similarity between the outcome of apparent competition among hosts and of traditional competition for resources. Competition between species mediated by resource depletion leads to the exclusion of all species except the one that drives the resource to the lowest equilibrium level (Tilman's (1982) R^* rule). Similar results are also found for hosts that share a common parasitoid (Holt & Lawton, 1993). Holt & Lawton (1993) point out that what appear to be monophagous prey—predator or specific host—pathogen interactions may arise through population dynamic interactions (*dynamic monophagy*) and thus not reflect underlying physiological specialization. An important corollary for biological control workers is that the establishment of a pathogen as a biological control agent could lead to the extinction of other species that it also infects (Bowers & Begon, 1991). The more complicated outcomes that result when host density dependence is also included have been discussed by Begon & Bowers (1994).

In contrast to two-host—one-pathogen dynamics, the dynamic interactions between two pathogens attacking a single host have not been studied for insect diseases with a free-living stage. The (quite complicated) literature on this subject for directly transmitted pathogens has been reviewed by Dobson

(1985) and Begon & Bowers (1995). The literature on two parasitoids attacking a common host is also relevant here (e.g., Godfray, 1994). Extrapolating from this literature, we can predict that for coexistence to occur, both species need to have an advantage when rare. Thus it will be impossible for two pathogens that exploit their host in the same way to coexist. However, if the two pathogens tend to attack different sectors of the host population, coexistence may be possible. By sectors we mean hosts in different habitats, emerging at different times, or of different genotypes. In order to understand fully the interaction between two pathogens, the crucial parameters will be the covariance in the risk of infection and the outcome of multiple infection of the host by the two pathogens.

Finally, we note briefly the results of a model that combined a host, pathogen and parasitoid (Hochberg *et al.*, 1990). The model is interesting both because of its structure and because of the dynamics it predicts. It is phrased in discrete time with the number of hosts surviving to the adult stage those that escape both parasitism and the disease. Each generation, an epizootic occurs from which a certain fraction of host survive. The results are complicated, with both natural enemy coexistence and exclusion occurring for different parameter values. The system can also display very complicated dynamics including chaos. One of the main messages of this work is that an understanding of host—pathogen—parasitoid dynamics will require the detailed study of specific systems (see Chapter 19). Recently, Begon *et al.* (1996) have published the first results of the dynamics of a laboratory model system including a host (*Plodia interpunctella*, a small phycitid stored product moth), a pathogen (a granulosis virus) and a parasitoid wasp. The addition of the wasp leads to very different dynamics to that displayed by the host and pathogen alone, although the reasons for this are not yet understood. The interactions between hosts, pathogens and parasitoids are treated in much more detail by Begon *et al.* in Chapter 18.

Non-equilibrium dynamics

Until now, we have almost exclusively concentrated on models whose main purpose has been to study the persistent behavior of the population, answering such questions as when does a pathogen regulate its host, and is the host population regulated at a stable equilibrium or does it display limit cycles. We now turn to models that have different goals, the study of host–pathogen interactions in situations where long-term persistent pathogen regulation is either absent or irrelevant.

Pathogens as insecticides

One use of pathogens in pest control is as biological insecticides. Without doubt, the most significant pathogen from a commercial point of view is *Bt*, *Bacillus thuringiensis*. Common isolates of the bacterium produce a toxin that is poisonous to a variety of Lepidoptera, and strains have also been discovered that attack other insects such as Coleoptera and Diptera. The bacterium is available in several formulations as an insecticide suitable for spraying on different crops. The toxin is a protein and the gene that codes for it has been introduced into plants to produce resistant varieties. However, after spraying, there is very little secondary infection and the effect of *Bt* on the pest can be studied just as if it were a traditional chemical insecticide.

Another pathogen that is being developed as a sprayable insecticide is the fungus *Metarhizium flavoviride*, which attacks locusts and other large Orthoptera. The fungal spores are not very stable in the environment and until recently it was not thought to be very useful against locusts. However, it has been discovered that using an oil-based formulation can greatly increase its commerical value. Thomas *et al.* (1995) have used parameterized population dynamic models to compare the efficacy of biological pesticides and non-persistent chemical pesticides (persistent chemicals are now banned). The models are structurally similar to the two-stage models described above with an age-structured within-season component, and a discrete-time between-season component. The fungus kills the orthopteran and further infections are caused by healthy hosts encountering the infectious cadaver. The infectedness of the dead insect depends in a complicated manner on the time since infection, but a statistical description of this process based on experiments was incorporated in the model. In comparisons of biopesticides and non-persistent chemical pesticides with similar initial kill rates, the former is markedly superior because of the extra deaths that result from secondary infections. Moreover, the advantage of biopesticides is greatest at high host population densities, exactly when control is most needed.

Outbreak species

Some insect diseases are only recorded infrequently, typically after their hosts have experienced a population outbreak which they often bring to an end. There is of course a rather mundane explanation for this observation – perhaps the insect is only studied by entomologists after an outbreak or perhaps the host and the disease are too rare during the endemic phase to be noted. But is it possible for an insect disease that can only increase in density during an outbreak of its host to persist in the environment?

Suppose that for most of the time the host is regulated by unspecified factors at a density someway below the host threshold (recall this is the density at which the disease can replace itself and begin to spread). During this time, the disease is present in the environment, perhaps in a protected refuge, at densities that are decaying at a fixed rate. At infrequent intervals, host regulation fails and the host increases in density exponentially. When it exceeds the host threshold, the pathogen is able to increase in abundance and eventually first stems the increase in host density and then reverses it. Once host densities have dropped below the host threshold, the original unspecified factors regulating the host at its endemic level switch back in. The pathogen can now no longer increase in abundance, although its density has been substantially increased by the free-living stages released during the host outbreak.

We have constructed a series of models exploring diseases with this natural history (Godfray & Briggs, 1995). In the simplest model, we assume all outbreaks are of constant size resulting in the release of the same number of free-living stages. Outbreaks occur randomly with a constant probability per unit time and the pathogen is assumed to go extinct if its densities drop below a certain threshold. In more complicated models, the dynamics within an outbreak are represented by Lotka–Volterra or Anderson–May dynamics. In the simplest cases, the mean and variance of the expected time to pathogen extinction can be calculated analytically, and these results are good approximations to the outcome of simulating the more complicated cases. Because outbreaks occur randomly, eventually a time will come when the gap between two outbreaks is so long that the pathogen goes extinct, but this may not happen for a very long time. The versions that include explicit within-outbreak dynamics make two predictions. After an outbreak there may be a refractory phase when even if the extrinsic regulation breaks down, the density of infectious stages in the environment is sufficiently great to prevent a second outbreak. Also, the size of an outbreak (maximum host density) is positively correlated with the length of time since the last outbreak. There is evidence from the larch budmoth (*Zeiraphera diniana*) that supports this last prediction, although we suspect that other models explaining *Zeiraphera* outbreaks may make similar predictions. Other aspects of the dynamics of insect outbreaks are discussed by Hastings in Chapter 12.

Discussion

Insect–pathogen dynamics is an exciting field with the molecular revolution in insect pathology both providing new tools for answering old questions, and new dynamic problems that require solution. For reasons of space, in this chapter we have concentrated on the dynamic models and the predictions they make and have said rather little about the biology of the pathogens involved or of the experimental work that provides parameter estimates. Our concentration on the more abstract aspects of the subject should not be interpreted as a belief that theory alone is a panacea for understanding insect–pathogen dynamics. We see theory as making two contributions. The first is to provide general conceptual models at a high level of abstraction that allow the quantitative investigation of different dynamic processes, and the second is to help understand the dynamics of particular insect–pathogen

interactions. The two are linked in that a major role of the more simple models of the Anderson–May school is to help to interpret the output of the more specific models. While we would not argue that we have all the simple, conceptual models that we need, we do feel that the most exciting challenges for population dynamicists lie in understanding the dynamics of actual insect–pathogen interactions, both in the field and the laboratory.

We believe that the best way to understand specific insect–pathogen interactions is by building models of intermediate complexity within what we called at the beginning of the chapter the population-biology approach. Models of intermediate complexity are more complicated than the classical models such as the Anderson–May, but simpler than the models of the systems approach. Their distinguishing feature is that the classical models can be obtained as limiting cases, something that greatly assists in the investigation of their dynamics.

We illustrate our argument in a shamelessly *parti pris* manner by describing our recent attempts to model the interaction between the microsporidian, *Nosema pyrausta*, and its host, the European corn borer, *Ostrinia nubilalis* (Briggs & Godfray, 1996b). The interaction is complicated and does not fall happily into any of the simple models discussed above. The system is highly seasonal with the host having more or less two discrete generations a year. The pathogen is transmitted both vertically and horizontally, the latter occurring when the larvae consume infectious spores. The spores cannot survive the winter and hence vertical transmission is essential for year-to-year survival. The structure of the interactions is illustrated in Fig. 17.3.

We extended the discrete generation model with intra-generational dynamics (see above) so that it described the *Ostrinia–Nosema* interaction. The model required about 15 equations (delay-differential for within-season, difference for between-season) and 24 parameters for its specification. Analytical results were impossible. But of the 24 parameters, estimates of 21 were available in the literature. The model provided a good qualitative prediction of within-generation dynamics and an excellent quantitative fit when tuned by one of the unknown

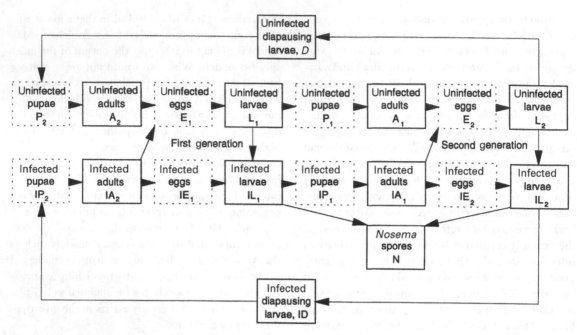

Figure 17.3. Diagram of the European corn borer–*Nosema pyrausta* model. Explicit equations are needed only for the stages enclosed in solid boxes.

parameters (the others having little influence; see Fig. 17.4). It thus provided testable predictions of the magnitude of the important but unknown parameter, as well as predictions about the importance of the other parameters for which estimates were available. The most likely long-term dynamics are persistent population cycles. Despite the complexity of the model, its predictions were remarkably in accord with those of the Anderson–May model, reinforcing the value of simple conceptual models.

One reason we chose to develop a model of intermediate complexity for this system is that two previous models were available for comparison; one much simpler and one much more complicated. The simpler model (Cavalieri & Koçak, 1994) was a discretized version of the Anderson–May model that predicted chaotic population. However, we had problems in understanding the derivation of this model (Briggs & Godfray, 1996b). The more complicated model is the work of Onstad and colleagues that we mentioned at the beginning of the chapter (Onstad & Maddox, 1989; Onstad *et al.*, 1990). Their model suf-

fered from a lack of empirical data in exactly the same areas that ours did. Because of the enormous complexity of their model, very little exploration of parameter space was possible, and it was not possible to understand why the model predicted the long-term dynamics it did (stable equilibrium or 7 year cycle). However, unlike our model, their's was able to make predictions about the consequence of spatial heterogeneity of the corn borer.

We finish by mentioning a few topics that are important but which we have hardly touched upon. Insects and their pathogens act as important selective agents on each other and evolutionary processes may operate on population dynamic time-scales (Anderson & May, 1980, 1982; May & Anderson, 1983; and see Chapter 13). Topics that might benefit from studies employing a joint evolutionary and population dynamic approach include the evolution of virulence and latency (Antia *et al.*, 1994; Myers & Rothman, 1995) and the evolution of pathogen resistance (Boots & Begon, 1993; Bowers *et al.*, 1993). We have also said very little about the dynamic

(a) corn borer larval density

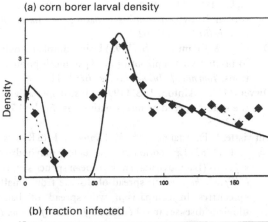

(b) fraction infected

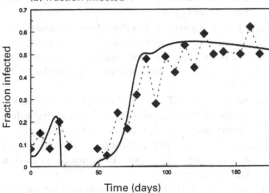

Time (days)

Figure 17.4. Comparison of the within-season dynamics predicted by the model (continuous line) with data from Andreadis (1987, broken line): (a) the total density of corn borer larvae (uninfected and infected) in the first and second generation; (b) the fraction of larvae infected with *Nosema pyrausta*.

consequences of the genetic manipulation of insect diseases. The incorporation of *Bt* genes into plants and of insect toxins into baculoviruses are both high profile examples of modern agrotechnology that have led to much public debate on ethics and safety. Many of the safety issues boil down to questions of population dynamics and population genetics (Crook & Winstanley, 1995; Godfray, 1995; Huber, 1995; Whitten, 1995; Godfray *et al.*, 1996). Although one option is to abstain from such debates on the grounds that any genetic manipulation is unconscionable; the fact that they are likely to go ahead anyway makes it important that we understand their dynamic consequences.

References

Anderson, R. M. (1982) Theoretical basis for the use of pathogens as biological control agents of pest species. *Parasitology* **84**: 3–33.

Anderson, R. M. & May, R. M. (1980) Infectious diseases and population cycles of forest insects. *Science* **210**: 658–61.

Anderson, R. M. & May, R. M. (1981) The population dynamics of microparasites and their invertebrate hosts. *Philosophical Transactions of the Royal Society of London, Series B* **291**: 451–524.

Anderson, R. M. & May, R. M. (1982) Coevolution of hosts and parasites. *Parasitology* **85**: 411–26.

Andreadis, T. G. (1987) Horizontal transmission of *Nosema pyrausta* (Microsporida: Nosematidae) in the European cornborer, *Ostrinia nubilalis* (Lepidoptera: Pyralidae). *Environmental Entomology* **16**: 1124–9.

Antia, R., Levin, B. R. & May, R. M. (1994) Within-host population dynamics and the evolution and maintenance of microparasite virulence. *American Naturalist* **144**: 457–72.

Begon, M. & Bowers, R. G. (1994) Host–host–pathogen models and microbial pest control: the effect of host self regulation. *Journal of Theoretical Biology* **169**: 275–87.

Begon, M. & Bowers, R. G. (1995) Beyond host–pathogen dynamics. In *Ecology of Infectious Diseases in Natural Populations*, ed. B. T. Grenfell & A. P. Dobson, pp. 478–509. Cambridge: Cambridge University Press.

Begon, M., Bowers, R. G., Kadianakis, N. & Hodgkinson, D. E. (1992) Disease and community structure: the importance of host self-regulation in a host–host–pathogen model. *American Naturalist* **139**: 1131–50.

Begon, M., Sait, S. M. & Thompson, D. J. (1996) Predator–prey cycles with period shifts between two- and three-species systems. *Nature* **381**: 311–15.

Berryman, A. A. (1996) What causes population cycles of forest Lepidoptera? *Trends in Ecology and Evolution* **11**: 28–32.

Boots, M. & Begon, M. (1993) Trade-offs with resistance to a granulosis virus in the Indian meal moth, examined by a laboratory evolution experiment. *Functional Ecology* **7**: 528–34.

Boots, M. & Begon, M. (1994) Resource limitation and the lethal and sublethal effects of a viral pathogen in the Indian meal moth, *Plodia interpunctella*. *Ecological Entomology* **19**: 319–26.

Bowers, R. G. & Begon, M. (1991) A host–host–pathogen model with free living infective stages applicable to microbial pest control. *Journal of Theoretical Biology* **148**: 305–29.

Bowers, R. G., Begon, M. & Hodgkinson, D. E. (1993) Host–pathogen populaton cycles in forest insects? Lessons from simple models reconsidered. *Oikos* **67**: 529–38.

Briggs, C. J. & Godfray, H. C. J. (1995) The dynamics of insect-pathogen interactions in stage-structured populations. *American Naturalist* **145**: 855–87.

Briggs, C. J. & Godfray, H. C. J. (1996a) Models of intermediate complexity in insect–pathogen interactions: population dynamics of the European corn borer, *Ostrinia nubilalis*, and its microsporidian pathogen, *Nosema pyrausta*. *Parasitology* **111**: S78–S89.

Briggs, C. J. & Godfray, H. C. J. (1996b) The dynamics of insect–pathogen interactions in seasonal environments. *Theoretical Population Biology* **50**: 149–77.

Cavalieri, L. F. & Koçak, H. (1994) Chaos in biological control systems. *Journal of Theoretical Biology* **169**: 179–87.

Crook, N. E. & Winstanley, D. (1995) Benefits and risks of using genetically engineered baculoviruses as insecticides. In *Biological Control: Benefits and Risks*, ed. H. M. Hokkanen & J. M. Lynch, pp. 223–30. Cambridge: Cambridge University Press.

D'Amico, V., Elkinton, J. S., Dwyer, G., Burand, J. P. & Buonaccorsi, J. P. (1996) Virus transmission in gypsy moths is not a simple mass action process. *Ecology* **77**: 201–6.

Desharnais, R. A. & Liu, L. (1987) Stable demographic limit cycles in laboratory populations of *Tribolium castaneum*. *Journal of Animal Ecology* **56**: 885–906.

Dobson, A. P. (1985) The population dynamics of competition between parasites. *Parasitology* **91**: 317–47.

Dwyer, G. (1991) The roles of density, stage, and patchiness in the transmission of an insect virus. *Ecology* **72**: 559–74.

Dwyer, G. (1992) On the spatial spread of insect pathogens: theory and experiment. *Ecology* **73**: 479–84.

Dwyer, G. (1994) Density dependence and spatial structure in the dynamics of insect pathogens. *American Naturalist* **143**: 533–62.

Dwyer, G. & Elkinton, J. S. (1993) Using simple models to predict virus epizootics in gypsy moth populations. *Journal of Animal Ecology* **62**: 1–11.

Dwyer, G. & Elkinton, J. S. (1995) Host dispersal and the spatial spread of insect pathogens. *Ecology* **76**: 1262–75.

Entwistle, P. F., Adams, P. H. W., Evans, H. F. & Rivers, C. F. (1983) Epizootiology of a nuclear polyhedrosis virus (Baculoviridae) in European spruce sawfly (*Glipinia hercyniae*): spread of disease from small epicentres in comparison wth spread of baculovirus diseases in other hosts. *Journal of Animal Ecology* **20**: 473–87.

Fuxa, J. R. & Richter, A. R. (1994) Distance and rate of spread of *Anticarsia gemmatalis* (Lepidoptera: Noctuidae) nuclear polyhedrosis virus released into soybean. *Environmental Entomology* **23**: 1308–16.

Ginzburg, L. R. & Taneyhill, D. E. (1994) Population cycles of forest Lepidoptera – a maternal effect hypothesis. *Journal of Animal Ecology* **63**: 79–92.

Godfray, H. C. J. (1994) *Parasitoids, Behavioral and Evolutionary Ecology*. Princeton: Princeton University Press.

Godfray, H. C. J. (1995) Field experiments with genetically manipulated insect viruses: ecological issues. *Trends in Ecology and Evolution* **10**: 465–9.

Godfray, H. C. J. & Briggs, C. J. (1995) The population dynamics of pathogens that control insect outbreaks. *Journal of Theoretical Biology* **176**: 125–36.

Godfray, H. C. J. & Hassell, M. P. (1987) Natural enemies can cause discrete generations in tropical insects. *Nature* **327**: 144–7.

Godfray, H. C. J. & Hassell, M. P. (1989) Discrete and continuous insect populations in tropical environments. *Journal of Animal Ecology* **58**: 153–74.

Godfray, H. C. J., O'Reilly, D. R. & Briggs, C. J. (1996) A model of nucleopolyhedrovirus (NPV) population genetics applied to co-occlusion and the spread of the few polyhedra (FP) phenotype. *Proceedings of the Royal Society of London, Series B* **264**: 315–22.

Grosholz, E. D. (1992) Interactions of intraspecific, interspecific and apparent competition with host–pathogen population dynamics. *Ecology* **73**: 507–14.

Gurney, W. S. C., Nisbet, R. M. & Lawton, J. H. (1983) The systematic formulation of tractable single-species population models incorporating age structure. *Journal of Animal Ecology* **52**: 479–96.

Hassell, M. P. & Pacala, S. W. (1990) Heterogeneity and the dynamics of host–parasitoid interactions. *Philosophical Transactions of the Royal Society of London (Biology)* **330**: 203–20.

Hassell, M. P., Comins, H. N. & May, R. M. (1994) Species coexistence and self-organizing spatial dynamics. *Nature* **370**: 290–2.

Hastings, A. (1996) Models of spatial spread: is the theory complete? *Ecology* **77**: 1675–9.

Hochberg, M. E. (1989) The potential role of pathogens in biological control. *Nature* **337**: 262–5.

Hochberg, M. E. (1991) Non-linear transmission rates and the dynamics of infectious disease. *Journal of Theoretical Biology* **153**: 301–21.

Hochberg, M. E. & Waage, J. K. (1991) A model for the biological control of *Oryctes rhinoceros* (Coleoptera: Scarabaeidae) by means of pathogens. *Journal of Applied Ecology* **28**: 514–31.

Hochberg, M. E., Hassell, M. P. & May, R. M. (1990) The dynamics of host–parasitoid–pathogen interactions. *American Naturalist* **135**: 74–94.

Holt, R. D. (1977) Predation, apparent competition and the structure of prey communities. *Theoretical Population Biology* **12**: 197–229.

Holt, R. D. & Lawton, J. H. (1993) Apparent competition and enemy-free space in insect host–parasitoid communities. *American Naturalist* **142**: 623–45.

Holt, R. D. & Pickering, J. (1985) Infectious disease and species coexistence, a model in Lotka–Volterra form. *American Naturalist* **124**: 337–406.

Huber, J. (1995) Opportunities with baculoviruses. In *Biological Control: Benefits and Risks*, ed. H. M. Hokkanen & J. M. Lynch, pp. 201–6. Cambridge: Cambridge University Press.

Knell, R. J., Begon, M. & Thompson, D. J. (1996) Transmission dynamics of *Bacillus thuringiensis* infecting *Plodia interpunctella*: a test of the mass action assumption with an insect pathogen. *Proceedings of the Royal Society of London, Series B* **263**: 75–81.

Liu, W., Levin, S. & Iwasa, Y. (1986) Influence of non-linear incidence rates upon the behavior of SIRS epidemiological models. *Journal of Mathematical Biology* **23**: 187–204.

May, R. M. (1974) *Stability and Complexity in Model Ecosystems*. Princeton: Princeton University Press.

May, R. M. (1978) Host–parasitoid systems in patchy environments: a phenomenological model. *Journal of Animal Ecology* **47**: 833–43.

May, R. M. & Anderson, R. M. (1983) Epidemiology and genetics in the coevolution of parasites and hosts. *Proceedings of the Royal Society of London, Series B* **219**: 281–313.

Murdoch, W. W., Nisbet, R. M., Blythe, S. P., Gurney, W. S. C. & Reeve, J. D. (1987) An invulnerable age class and stability in delay-differential parasitoid–host models. *American Naturalist* **129**: 263–82.

Myers, J. H. & Rothman, L. E. (1995) Virulence and transmission of infectious diseases in humans and insects: evolutionary and demographic patterns. *Trends in Ecology and Evolution* **10**: 194–8.

Onstad, D. W. & Maddox, J. V. (1989) Modeling the effects of the microsporidium, *Nosema pyrausta*, on the population dynamics of the insect, *Ostrinia nubilalis*. *Journal of Invertebrate Pathology* **53**: 410–21.

Onstad, D. W., Maddox, D. W., Cox, D. J. & Kornkven, E. A. (1990) Spatial and temporal dynamics of animals and the host-density threshold in epizootiology. *Journal of Invertebrate Pathology* **55**: 76–84.

Régnière, J. (1984) Vertical transmission of disease and population dynamics of insects with discrete generations: a model. *Journal of Theoretical Biology* **107**: 287–301.

Rothman, L. D. & Myers, J. H. (1994) Nuclear polyhedrosis virus treatment effect on reproductive potential of western tent caterpillar (Lepidoptera: Lasiocampidae). *Environmental Entomology* **23**: 864–9.

Rothman, L. D. & Myers, J. H. (1996) Debilitating effects of viral diseases on host Lepidoptera. *Journal of Invertebrate Pathology* **67**: 1–10.

Thomas, M. B., Wood, S. N. & Lomer, C. J. (1995) Biological control of locusts and grasshoppers using a fungal pathogen: the importance of secondary cycling. *Proceedings of the Royal Society of London, Series B* **259**: 265–70.

Tilman, D. (1982) *Resource Competition and Community Structure*. Princeton: Princeton University Press.

Turelli, M. (1994) Evolution of incompatibility-inducing microbes and their hosts. *Evolution* **48**: 1500–13.

Vezina, A. & Peterman, R. M. (1985) Tests of the role of a nuclear polyhedrosis virus in the population dynamics of its host, Douglas-fir tussock moth, *Orgyia pseudotsugata* (Lepidoptera: Lymantriidae). *Oecologia* **67**: 260–6.

White, A., Begon, M. & Bowers, R. G. (1996a) Host–pathogen systems in a spatially patchy environment. *Proceedings of the Royal Society of London, Series B* 263: 325–32.

White, A., Bowers, R. G. & Begon, M. (1996b) Host–pathogen cycles in self-regulated forest insect systems: resolving conflicting predictions. *American Naturalist* 148: 220–5.

Whitten, M. J. (1995) An international perspective for the release of genetically engineered organisms for biological control. In *Biological Control: Benefits and Risks*, ed. H. M. Hokkanen & J. M. Lynch, pp. 253–60. Cambridge: Cambridge University Press.

Wood, S. N. & Thomas, M. B. (1996) Space, time and persistence of virulent pathogens. *Proceedings of the Royal Society of London, Series B* 263: 673–80.

18

Host–pathogen–parasitoid systems

MICHAEL BEGON, STEVEN M. SAIT AND DAVID J. THOMPSON

Introduction

The words 'theoretical' and 'mathematical' are not synonymous. Hence, in this chapter, theoretical approaches to the role of host–pathogen–parasitoid interactions in biological control are taken to include not only mathematical models, but also 'biological models'. Mathematical models in ecology (and elsewhere) are simplified reconstructions of the natural world, carried out by the modeler with twin aims pulling in opposite directions. On the one hand, the model should capture enough of the essence of the natural system to make the output from the model enlightening (interpreted cautiously). On the other hand, though, the model must be simple enough both to be tractable (i.e., to allow interpretable conclusions to be drawn from the assumptions built into the model) and to make apparent the connection between those assumptions and conclusions.

Biological models in ecology, in the sense used in this chapter, are also simplified reconstructions of the natural world, carried out by an experimenter, which also have twin aims pulling in opposite directions. The first, again, is to capture enough of the essence of the natural system to make the output from the model enlightening (interpreted cautiously). The second in this case, though, is to be simple enough both to generate repeatable and statistically interpretable results and to connect variations in those results with variations in experimental treatment. Thus, biological, just like mathematical models, tell us, at least, what 'might be' or what 'can be' – in short, what is theoretically possible. Indeed, good models go further and tell us what 'would be – if'. In this way, they offer the opportunity of reaching out from the model to the natural world.

The importance of parasitoids and pathogens as determinants of their host's population dynamics has long been recognized (e.g., Hassell, 1978; Anderson & May, 1981), as well as their potential for pest control (e.g., May & Hassell, 1988; Payne, 1988; Hochberg, 1989). However, the tradition of studying these two-species interactions independently is clearly a simplification of the natural world and the potentially complex associations that can exist between a shared host/prey and a variety of natural enemies. Indeed, inter-kingdom competition, such as that between parasitoids and pathogens, may be extremely common (Hochberg & Lawton, 1990).

Practical biological control of insect pests requires the application of population and community theory, demanding understanding of food web ecology (e.g., Pimm *et al.*, 1991) and how species interact in multi-trophic systems (e.g., Gange & Brown, 1997 and chapters therein). However, in order to determine the population dynamic consequences of using two (or more) natural enemies to control a pest species, and in order to predict the potential for reducing the abundance of that pest, it is essential to understand how they interact at the individual level. Theoretical studies of multispecies systems (see below) illustrate that a rich variety of dynamic patterns are possible. Host suppression is only one of the outcomes. Any particular outcome will be entirely dependent on the biology of the interacting species, and the nature of the competition between the two natural enemies.

This chapter, therefore, is set out by moving from mathematical population models, to biological population models and then to interactions between pathogens and parasitoids at the level of the individual host. In this way our discussion is gradually broadened. The generality of the conclusions from the mathematical models are constrained by the limitations of the assumptions incorporated. The generality of the conclusions from the biological population model are similarly constrained, because of the particularities of the only system of which we are aware that has been studied in sufficient detail. Studies at the individual level then begin with the same model system as that used to study population dynamics, before opening out to a more widespread review, in an attempt to assess how general the conclusions from the models might be.

Mathematical models

A general host–pathogen–parasitoid model – Hochberg et al. (1990)

Only one mathematical model, that of Hochberg *et al.* (1990), has examined in a general manner some of the possible population dynamic consequences of host–pathogen–parasitoid interactions. A number of other models, however, have looked at similar interactions (host–pathogen–pathogen, host–parasite–parasite, host–parasitoid–parasitoid) and Hochberg *et al.*'s results will therefore be examined in the light of these.

Note at the outset that their model assumes discrete host generations (as in a seasonal environment) so that host–parasitoid dynamics can be described by a series of recurrence relations (as in standard Nicholson–Bailey type models). The pathogen is then assumed to cause epidemics within a season, with pathogen particles accumulated at the end of the season (but not infected hosts) passing from one generation to the next. For many host–parasitoid–pathogen interactions, however, for instance the (biological) model populations considered in detail in this chapter, environments are non-seasonal and generations intrinsically continuous. Hence, Hochberg *et al.*'s results may need to be applied with some caution.

Their model equations are:

$$N_{t+1} = FN(1 - I)f(P_t) \tag{18.1}$$

$$P_{t+1} = cN_t\{(1 - \phi I)[1 - f(P_t)]\} \tag{18.2}$$

$$W_{t+1} = g\left[Q_t - \frac{\Omega}{\beta}\right]\ln(1 - I) \tag{18.3}$$

$$Q_{t+1} = \frac{hW_{t+1}}{g} \tag{18.4}$$

Here, N and P are the host and parasitoid population densities. In the absence of disease, F is the host's finite rate of increase, c is the average number of female parasitoid progeny per host attacked (assumed in practice to be one) and $f(P_t)$ is the fraction of hosts that escape attack by the parasitoid, given by:

$$f(P_t) = \left(1 + \frac{aP_t}{k}\right)^{-k} \tag{18.5}$$

where a is a measure of the *per capita* searching efficiency of the parasitoid and k is the 'clumping parameter' of attacks (with an assumed negative binomial distribution).

Each generation, t, of hosts is assumed to suffer a separate epidemic in which a varying fraction $(1 - I)$ of the hosts escape infection. The within-season dynamics giving rise to the expression $(1 - I)$ are based on standard epidemiological assumptions; details are omitted here but may of course be found in Hochberg *et al.* (1990). The parameter ϕ, appearing in eqn (18.2) but not eqn (18.1), can vary between 0 (complete dominance of parasitoid) and 1 (complete dominance of pathogen). Thus, it expresses the idea of interference between pathogen and parasitoid in hosts that harbor both.

The density of free-living infective stages of the pathogen, capable on consumption of giving rise to infection, is given by W. There is also another population of pathogens, density Q, in a reservoir (for example, the soil in the case of leaf-dwelling hosts) where they are protected (a death rate of zero) but unable to infect hosts. The term in square brackets in eqn (18.3) represents the density of pathogens in the reservoir at the end of the infectious season in generation t. Hence, eqns (18.3) and (18.4) express the idea that, of these, by the beginning of the next season, a fraction, g, are translocated from the reservoir to the transmissable population (for example, back to the leaf surfaces) and a fraction, h, remain (neither being translocated nor dying).

Finally, then, the square-bracketed term in eqn (18.3) combines the pathogens present at the start of the season, Q_t, with those added during the season (note that $\ln(1 - I)$ is always negative). During each season, pathogens are transferred from the free-living pool to the reservoir at a rate Ω. The parameter β is the transmission coefficient, determining the *per capita* rate at which hosts acquire infection on contact with infectious particles, but also the rate of loss of particles as a result of host consumption. Hence, the number added to the reservoir population increases with the fraction infected, I, and the rate of transfer, Ω, but decreases with the rate of pathogen consumption, β.

Analysis of the model

Hochberg *et al.* (1990) show, through an analysis of invasion, that this model system can display three possible patterns of behavior:

1. The parasitoid and pathogen may coexist, with either constant, cyclic or chaotic dynamics;
2. One natural enemy (i.e., 'predator') may predictably exclude the other; or
3. The pathogen may exclude the parasitoid or *vice versa*, in a manner contingent on initial densities.

The conditions determining which of these outcomes applies concern four main factors:

1. The relative extrinsic (outside-the-host) potentials of the parasitoid and pathogen (parasitoid searching efficiency, pathogen transmission, etc);
2. Their relative intrinsic (inside-the-host) potentials (competition within hosts, timing of attacks, specifically ϕ);
3. The aggregation of parasitoid attacks; and
4. The host's finite rate of increase.

The most fundamental conclusion, perhaps, is that three-species coexistence requires an approximate balance between the extrinsic superiority of one of the 'predators' (the pathogen and parasitoid) in exploiting healthy hosts and the intrinsic superiority of the other in exploiting hosts attacked by both predators. This echoes the results from host–parasite–parasite models (Dobson, 1985) and host–pathogen–pathogen models, especially that of Hochberg & Holt (1990), where coexistence occurs

between a virulent, efficient exploiter of infected hosts, and a less virulent, efficient exploiter of healthy hosts, more capable than the virulent pathogen of depressing host abundance.

In all of these cases, if one of the predators clearly displays both an extrinsic and an intrinsic superiority, then it will exclude the other predator. Elimination of one or other predator, contingent on initial densities, however, tends to occur, in the host–parasitoid–pathogen model at least, when one predator has an extreme superiority in the exploitation of healthy hosts and the other an equally extreme superiority in exploiting hosts that both have attacked.

The aggregation of parasitoid attacks promotes three-species coexistence further, by concentrating parasitoid competitive effects more against other parasitoids than against pathogens (a point that also emerges, for parasites, from Dobson's (1985) model). It is safe to assume, therefore, that heterogeneities in pathogen transmission (some hosts susceptible to repeated infection), had they been incorporated into the model, would have had the same effect.

Finally, coexistence appears to become more likely as the host's finite rate of increase itself increases, presumably because low rates of increase imply a relatively limited supply of hosts, potentially intense competition between parasitoid and pathogen, and hence a particularly fine balance required for coexistence.

The dynamics of coexistence may be either constant or oscillatory (regular or apparently chaotic) – a range also displayed, broadly, by both the host–pathogen and the host–parasitoid dynamics. It is noteworthy, though, that parameter values that give rise to one pattern in a two-species context can give rise to quite a different pattern once all three species are involved. For example, a tendency towards an aggregated search pattern amongst the parasitoids (low k) is stabilizing (constant dynamics) in the host–parasitoid interaction but can destabilize the pattern of host–pathogen–parasitoid dynamics. In short, dynamic patterns can emerge in the three-species system that are 'unexpected', given knowledge only of the component two-species systems.

Messages for biological control

From the point of view of biological control, the key questions concern the circumstances in which a second natural enemy (i) undermines the control exerted by the first, or (ii) enhances that control by not only coexisting with the first predator, but also adding to its controling effect. In the model, undermining of control could be represented either by a higher host density in the presence of both predators than with at least one of the predators alone, or by the exclusion by one predator of another predator that was a more effective biocontrol agent, or perhaps by the replacement of one pattern of dynamics by another, less stable, more 'outbreak' pattern. Enhancement would be represented by a host density that was lower in the presence of both (coexisting) predators than it would be with either predator alone.

Coexistence (a necessary precondition for control-enhancement, but clearly not sufficient) is favored by high host rates of increase (typical of most pests) and by heterogeneities in the attack-patterns of the predators (often suggested as a desirable quality of biological control agents even in a single predator context (Beddington *et al.*, 1978)). However, these same characteristics tend to push the dynamics towards instability – typically, multigeneration cycles.

Coexistence is also favored by a complementarity between parasitoid and pathogen in terms of their extrinsic and intrinsic qualities. In other words, a second natural enemy being considered for release as an additional biocontrol agent should not only not be clearly inferior to an already established agent – it should also be better than the established agent, either intrinsically or extrinsically.

Once coexistence is established, then, enhancement is certainly possible (though by no means certain) in the model. However, while two predators together never give rise to a higher host density than either predator alone, host density in the presence of both predators is often only intermediate between the two one-predator densities. In other words, introduction of a second natural enemy, when both persist, can often lead to an increase in pest density. This type of undermining of control therefore

appears to be a genuine danger. When, though, do two coexisting predators enhance control, and when does one undermine the control of the other?

In the first place, enhanced control is more likely to occur when the extrinsic qualities of both predators are high (search, transmission, and survival rates). This is perhaps not surprising, but it does emphasize that there is no simple additivity between agents. On the contrary, a second but mediocre agent is apparently more likely to give rise to an unwanted increase in pest density than it is to make an additional, albeit modest, contribution to pest depression. More particularly, it is more likely that both predators will contribute to the depression in host abundance (enhancement of control) when the pattern of attack of one of them is aggregated while the other has a moderate dominance within dually-infected hosts.

However, it is possible in the model for a second agent to undermine control drastically, by excluding an established, more effective agent. As Briggs (1993) points out, in the context of her host–parasitoid–parasitoid model, this would not be possible if the interaction were framed in terms of conventional exploitative competition between the predators for a simple, single resource (the prey) – the predator that could coexist with the prey at the lowest prey abundance would always win. However, the explicit stage-structure in Briggs' model allows an egg parasitoid (by getting in early) to exclude a larval parasitoid that would have given rise to lower adult host densities. Similarly, in Hochberg *et al.*'s (1990) host–parasitoid–pathogen model, stage-structuring is implicitly included, in that an interference advantage (high or low ϕ) may represent the ability of one of the predators to regularly pre-empt prey by attacking them earlier in their life than the other predator. Thus, we may deduce, for example, that a parasitoid with a marked interference advantage may exclude a pathogen that would otherwise have been capable of effecting a greater reduction in host density. Indeed, Hochberg *et al.* show this to be possible for very low ϕ (parasitoid strongly dominant within hosts), and we can expect a similar outcome, but with the roles reversed, for very large ϕ values.

In summary, then, the model suggests that caution

is appropriate whenever there are plans to supplement an established parasitoid with a pathogen in biological control, or *vice versa*, since there are many circumstances in which it will do harm rather than good (a view echoed by Briggs (1993) for two parasitoids). This, though, will not always be the case: biological control using a combination of a pathogen and a parasitoid is most likely to be a sound strategy:

1. When the attack patterns of both are moderately clumped; too much or too little aggregation may tend to prevent coexistence or promote instability;
2. When both natural enemies have high rates of search, transmission, and survival, so that both are effective exploiters of healthy hosts; and
3. When there is some degree of overlap in the timing of attacks of the predators, and/or competition within the host is not invariably one-sided (see also, therefore, Chapter 19, which considers infection of parasitoids by host pathogens).

Specific models

A number of modeling approaches have been used to assess the impact of two or more natural enemies in very specific systems. Berryman *et al.* (1990), using a simple logistic model, examined the dynamic consequences of a baculovirus invasion of an established host–parasitoid interaction, based on data collected for the blackheaded budworm, *Acleris variana*. A virus epizootic destabilized the host–parasitoid equilibrium interaction and caused continuous outbreak cycles in the host. The period of these cycles was 10 years, as observed in the field. Similarly, in a multitrophic interaction between the Douglas-fir tussock moth, *Orgyia pseudotsugata*, its host plant, a parasitoid and a baculovirus, the moth host was found to exhibit wide amplitude, cyclical outbreak dynamics, as found in the field (Mason & Torgersen, 1987), when the parasitoid enhanced the transmission of the virus. In the absence of the virus the system always exhibited a stable equilibrium. These two models, therefore, illustrate clearly one of the general points to emerge from Hochberg *et al.*'s model – that 'unexpected' dynamics may emerge in three-species systems.

Finally, in an agricultural system, Gutierrez *et al.* (1990) and Gutierrez (1992) developed a series of complex multitrophic models based around a system consisting of the blue alfalfa aphid *Acyrthosiphon kondoi*, the pea aphid *A. pisum* and several natural enemies including a parasitoid specific to *A. pisum*, *Aphidius smithi*, and a fungal pathogen *Pandora neo-aphidis*, also specific to the pea aphid. They suggest that the pathogen allows the blue alfalfa aphid to coexist at high levels with the pea aphid, and that the concurrent rarity of the parasitoid is due both to the severe depression of the pea aphid population through interspecific competition for resources and to pathogen-induced mortality. This demonstration of the complex interplay between natural enemies and a shared host does not deal explicitly with biological control, all species being present in the system at the outset, but it does provide a reminder that in many cases the extension of theory to even three species may still not be sufficient.

Biological models

Background

Our biological model for understanding three-species population dynamics is a laboratory system involving a stored product moth, its granulosis virus (GV) and a parasitoid. It is a system in which, as we shall see, the dynamics of the host alone and of two separate two-species systems are well understood (Sait *et al.*, 1994b; Begon *et al.*, 1995, 1996, 1997). We are conscious that the work described is a laboratory system not a field or an orchard. However, we would argue that the real world for a stored product pest of grain is probably a warehouse maintained at constant temperature with a single species food supply and that our population cages mirror such a food storage facility rather closely. We are also conscious that the messages we wish to convey for population dynamics and for biological control are not based on reviews of many three-species systems but rather they reflect the clear messages that emerge from one simple system.

We begin by describing the population dynamics of the host, its GV and the parasitoid, maintained in various combinations in population cages. For full details of the experimental system and of the species' biology see Harvey *et al.* (1994), Sait *et al.* (1994b,c, 1995, 1997), Begon *et al.* (1995). Briefly, the host is a pyralid moth, *Plodia interpunctella*, a cosmopolitan pest of stored products. It has five larval instars and, including an adult reproductive life of around 4 days, its generation length under the conditions of the present work is around 41 days. The parasitoid is a solitary, effectively parthenogenic ichneumonid wasp, *Venturia canescens*, that lays eggs singly in the later instars of its host: there is a transition from invulnerability to vulnerability over instars II and III, but fitness costs to the parasitoid are considerably less if it attacks the later stages (Sait *et al.*, 1995). The parasitoid synchronizes its development with that of the host, taking between 21 and 25 days depending on the instar parasitized, but will only complete its own development once the host has reached the final instar, regardless of the stage that was initially attacked (Harvey *et al.*, 1994). The pathogen, *Plodia interpunctella*-GV (hereafter PiGV), infects larval hosts (only) when consumed by them with their food. The lethal dose increases with larval instar, making instar V effectively invulnerable to infection (Sait *et al.*, 1994c). Infection itself proceeds from the midgut to most of the rest of the body, giving rise, typically 7–14 days later, to a cadaver almost full of infective viral particles, which may itself be eaten or may contaminate nearby food. Infective particles are not shed prior to host death.

Host, host–pathogen and host–parasitoid dynamics

Throughout, for brevity, dynamic patterns are illustrated for just one (typical) population, but descriptions refer to the full range of replicates (for further details see Begon *et al.*, 1996). The dynamics of the host alone are illustrated in Fig. 18.1. After an initial period of population growth, populations (three replicates) fluctuated with marked regularity around an average density of roughly 120 moths, with typical maxima and minima of around 230 and 60. The period of these fluctuations was 6 or 7 weeks, approximately one host generation. The strength of this

periodicity for each replicate population was measured by the value of the autocorrelation function (ACF) at its peak, and of the length of the period measured by the lag at which the peak ACF occurred, and, where data were sufficient, by the peak period in a spectral analysis. In each case the fluctuations in the ACF were strongly damped with increasing period suggesting an endogenous cause for the fluctuations provided by the populations themselves, rather than an exogenous cause provided by the environment in which they were maintained.

Host dynamics in host–pathogen systems (three replicates) differed very little from those of the host alone (Fig. 18.2), a point to which we shall return. In particular, the regularity and period of the fluctuations were effectively unaltered, but average host density (around 150) was slightly though not significantly higher ($t = 1.58$, $p = 0.19$). The amplitudes of the fluctuations, however, were somewhat increased, with typical maxima and minima of around 320 and 40, and coefficients of variation of abundance significantly higher ($t = 2.90$, $p < 0.05$). Pathogen dynamics cannot be portrayed on a scale directly comparable with that for the host. Instead, the number of infected larvae in a subsection of the food are shown. Their dynamics showed less clear evidence of regular periodicity, though some signs of cycles comparable with those of the host were apparent in two of the three replicates (including the one depicted) once an overall increasing trend in pathogen abundance had been taken into account. The numbers of infected larvae, combined with a typical host fecundity of around 200 eggs (Sait *et al.*, 1994a) suggest a prevalence of infection of around 1% and certainly less than 5%.

Host–parasitoid dynamics (three replicates, see Fig. 18.3) displayed regular fluctuations with periods again similar to those in the host-alone and host–pathogen systems. Moreover, they were exhibited clearly not only by the host but also by the parasitoid. The parasitoid cycles were typically slightly behind those of the host. The parasitoid also significantly reduced host abundance (to a mean value of around 10) and substantially increased the amplitudes of the fluctuations (on a log scale), with typical maxima and minima of around 100 and 0.

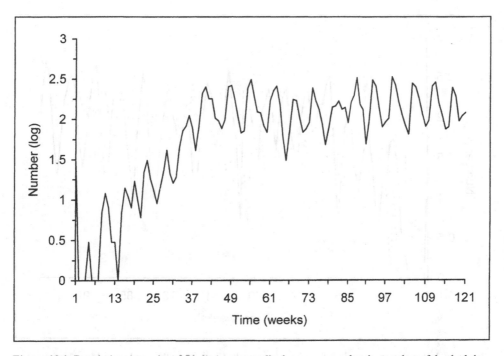

Figure 18.1. Population dynamics of *Plodia interpunctella* alone, measured as the number of dead adults per weekly census. The mean population density ($\pm$ SE) of this representative culture, after an initial, settling-down period of 37 weeks, was 147.5($\pm$7.6). The cycle period was 6 and 6.1 weeks (from the peak autocorrelation function (ACF) and spectral analysis, respectively) and the value of the peak ACF (a measure of the strength of the periodicity) was 0.61 ($p < 0.05$).

Thus the host–pathogen and host–parasitoid systems both exhibited patterns of endogenous host generation-length cycles. A pattern of one-host-generation cycles, as opposed to the classical multi-generation cycles, has been predicted by theory under the conditions pertaining in the *Plodia–Venturia* system (Godfray & Hassell, 1989; Gordon *et al.*, 1991; Begon *et al.*, 1995). The reduction of an order of magnitude in host density suggests strongly that these generation cycles were indeed host–parasitoid cycles. Host generation cycles have also been predicted for host–pathogen interactions when the host has an explicit stage structure, as here (Briggs & Godfray, 1995). Previous empirical support for both these predictions has been scarce (Godfray & Hassell, 1989; Gordon *et al.*, 1991; Reeve *et al.*, 1994; Sait *et al.*, 1994b).

On the other hand, generation cycles have also been predicted in single species populations in which larval competition acts immediately (rather than with a delay), and especially where competition is markedly asymmetric (Gurney & Nisbet, 1985), as it is in *P. interpunctella* (Sait *et al.*, 1994b), and, as we have seen, generation cycles were demonstrable here when the host was maintained alone. Clearly, therefore, it would be wrong to rush too readily into claiming that the present data support models predicting host–parasitoid or host–pathogen generation cycles. No proper explanation can fail to ignore the cycles generated by the host itself. More generally, this illustrates that caution may be necessary in ascribing causes to dynamical patterns in data sets involving two species when little or nothing is known about the dynamical behavior of those species alone. This is a theme to which we return later in the context of three-species dynamics.

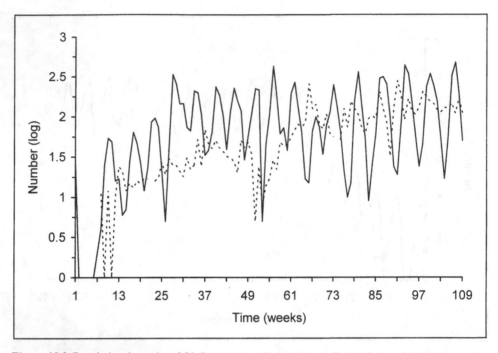

Figure 18.2. Population dynamics of *Plodia interpunctella* (continuous line) and granulosis virus–infected larvae (broken line), measured as the number of dead adult moths and infected larvae per weekly census. The mean host density ($\pm$ SE) of this population, after a settling-down period of 26 weeks, was 134.7 ($\pm$ 11.6). The cycle period was 6 and 6.4 weeks (ACF and spectral analysis, respectively), and the peak ACF was 0.52 ($p < 0.05$). Other than a steady increase in the level of infection , there was no obvious pattern in the dynamics of the pathogen.

Host–pathogen–parasitoid dynamics – a lack of persistence

Host–pathogen–parasitoid populations were constructed either by adding parasitoids to an existing host–pathogen interaction (Fig. 18.4) or by adding pathogen (infected hosts) to an existing host–parasitoid interaction (Fig. 18.5). The first notable contrast with previous patterns is the lack of persistence. Whereas all the previously described single- and two-species populations persisted either until they were abandoned or were used for other purposes, of 14 host–pathogen–parasitoid populations, 12 failed to persist and the remaining two showed patterns that engendered no confidence in their long-term persistence. Moreover, in each extinction, the host disappeared first, followed by

its enemies, rather than the population collapsing from a three- to a two-species, predator–prey system.

Three of the populations were constructed as a host–pathogen system maintained for almost 800 days to which the parasitoid was then added. In these, the number of infected larvae found each week had risen to around 100. The mean time ($\pm$ SE) to extinction of the host once all three species had been combined was 82 ($\pm$ 12) days.

In the other 11 three-species populations, the third species was added earlier – as soon as the two-species system had become established. By this time, there were typically around 35 infected larvae each week in the host–pathogen populations. One host–parasitoid + pathogen population was still extant after around 370 days as a three-species

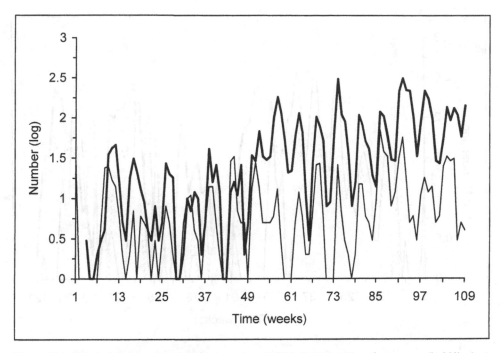

Figure 18.3. Population dynamics of *Plodia interpunctella* (thin line) and *Venturia canescens* (bold line), measured as the number of dead adult moths and parasitoids per weekly census. Following a settling-down period of 7 weeks, the mean host density ($\pm$ SE) of this population was 9.6 ($\pm$ 1.2); the mean parasitoid density was 55.4 ($\pm$6.5). The cycle periods were 6 and 6.3 weeks for the host; 6 and 6.5 weeks for the parasitoid (ACF and spectral analysis, respectively). The peak ACF values were 0.53 and 0.56 for the host and parasitoid, respectively ($p < 0.05$).

system, and one host–pathogen + parasitoid population was unavoidably terminated after around 250 days. If the times to extinction of these are conservatively estimated as 370 and 250 days, respectively, then the mean times to extinction ($\pm$ SE) are host–pathogen + parasitoid: 286($\pm$67) days, and host–parasitoid + pathogen: 259($\pm$36) days. Thus, amongst these 11, there appears to be not only a qualitative consistency irrespective of the manner in which they were constructed (all went extinct or were apparently heading towards extinction) but also an overall quantitative consistency (average time to extinction was approximately the same in the two groups). On the other hand, extinction was clearly less rapid in these host–pathogen + parasitoid populations, on average, than when the initial prevalence of infection was higher.

The three-species population dynamics here are therefore far less stable than those of their component two-species systems. This is a striking result given the consistency of behavior of the host-alone, host–pathogen and host–parasitoid populations. It emphasizes that the population dynamics of three-species systems may be difficult to predict even when the dynamics of their component subsystems are known in detail. In the present case, this is true both at the level of dynamical patterns (the loss of generation cycles) and also at the level of population-existence: the host was consistently driven to extinction in the three-species system, even though the host–parasitoid system would apparently cycle indefinitely, and the host–pathogen system would do so too, with no apparent effect of the pathogen on host abundance.

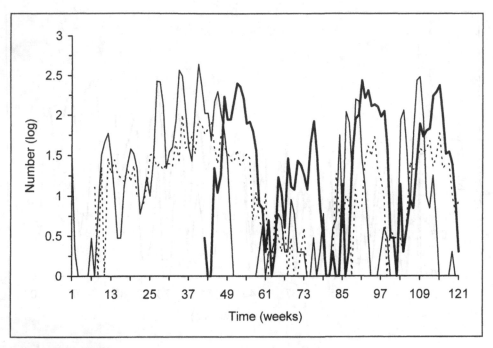

Figure 18.4. Population dynamics of *Plodia interpunctella* (thin line), granulosis virus-infected larvae (broken line) and *Venturia canescens* (bold line), measured as a weekly census of dead adults and infected larvae, following the introduction of the parasitoid (after 42 weeks) to an established host–pathogen system. The cycle periods (host, parasitoid, pathogen) were 20, 20, and 64 weeks, respectively. Similarly, the peak ACF values were 0.24, 0.53, and 0.15 ($p < 0.05$). The data series for the three species system was too short for spectral analysis.

Multigeneration cycles

The dynamics of the various three-species systems were consistent, then, in their lack of persistence and loss of generation cycles. On the other hand, it is notable that their detailed patterns differed according to how they were constructed. The host–parasitoid + pathogen populations displayed no consistent pattern. Most striking of all, however, the host–pathogen + parasitoid populations showed clear evidence of multigeneration cycles. This is seen most clearly in Fig. 18.4, the longest run of data, where after the introduction of the parasitoid, both host and parasitoid ran through four cycles of abundance, with significant peak ACF values at 20 weeks, prior to the extinction of the host. Three further replicates passed through two population cycles prior to host extinction or termination. These had

peak periods, respectively, for host and parasitoid, at 26 and 25 weeks, both significant, 25 and 25 weeks, the former not quite significant, and 21 and 17 weeks, both significant. On the other hand, the dynamics of virus-infected larvae in these populations showed no comparable cycles or regularities. In addition, a fifth replicate and also the three populations to which the parasitoid was added later at higher disease prevalence, all went extinct after a single rise and fall in parasitoid and host abundance, lasting a similar length of time to the cycles in the previous replicates (17–25 weeks in the case of the parasitoid).

Overall, therefore, a clear and remarkable picture emerges from the eight host–pathogen + parasitoid populations: cycles of abundance in both host and parasitoid, roughly 3–4 host generations in length, and of such amplitude that host extinction is likely following any one of them. The 'classical' expecta-

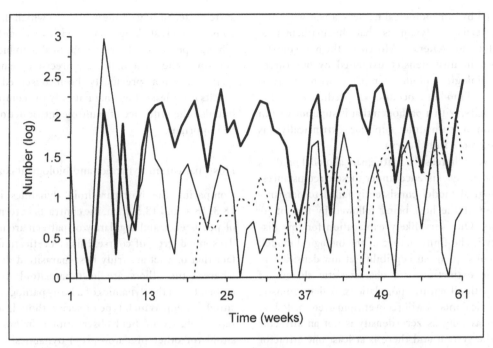

Figure 18.5. Population dynamics of *Plodia interpunctella* (thin line), granulosis virus-infected larvae (broken line) and *Venturia canescens* (bold line), measured as a weekly census of dead adults and infected larvae, following the introduction of granulosis virus-infected larvae (after 18 weeks) to an established host–parasitoid system. The data series for the three species system was too short for spectral analysis.

tion when a parasitoid and host (or any other predator and prey) interact alone is for 'Lotka–Volterra' cycles, several generations in length. Here, on the contrary, the parasitoid and host only exhibit such cycles in the presence of a third species, a pathogen, with which they interact strongly. In the absence of this third species they exhibit even clearer cycles -but these are roughly one (not several) host generations in length.

The mathematical and biological models

The laboratory microcosm described here is unique in that it provides population data for the single and two-species component systems as well as for the three-species interaction, thereby placing the complex system dynamics in their true context. By adopting this 'modular' approach in a biological model, akin to that used by Gutierrez *et al.* (1990)

and Gutierrez (1992) in their mathematical models, it is possible to determine the contribution of each natural enemy individually to pest regulation, as well as to assess the consequences of adding each enemy, and the significance of the sequence in which it is added, to biological control. Clearly, this is a luxury not afforded to field biologists and pest controllers but, as with theory, lessons can be learned from experimental models.

The ecological scenarios represented by Hochberg *et al.*'s (1990) mathematical model and the present biological model are different in important respects – most notably the seasonal dynamics envisaged by the former and the continuous breeding exhibited by the latter. Nonetheless, there are at least some points on which the results are in agreement. First, the message from the mathematics, that three-species systems may display dynamics 'unexpected' on the basis of what is known of their components, is supported by our biological model – albeit with specific patterns of dynamics that the mathematical model does not generate. Moreover, the more drastic reduction in host density exhibited by our three-species biological model compared with the two-species 'submodels' occurs, as predicted by the mathematics, with a pathogen that is superior within hosts and a parasitoid superior extrinsically (as explained more fully below).

However, whereas the mathematical model predicts three-species coexistence under these circumstances, the biological model consistently failed to persist – extinction of the host being followed by that of its predators. One very likely explanation for this discrepancy is the finite nature of the biological model, and its consequent susceptibility, at low densities, to stochastic extinction – a characteristic shared, of course, with all natural populations. A deterministic mathematical model will recover from even the lowest densities as long as zero-density is not an attractor within the system and there is at least one attractor with a positive host density. In this respect, mathematical models may be misleading. Real populations, as our biological model reminds us, are strongly prone to extinction if they are forced down to low densities.

Furthermore, contrary to the predictions of the mathematical model, the biological model shows that two predators may have a marked effect on host density between them, even when one of them alone (in this case the pathogen) has a negligible effect. This provides a reminder, if any was needed, that the predictions of a model are contingent on the validity of that model's assumptions, and that exact connection between assumption and prediction may not always be apparent. In the present case, the most likely explanation (M. Hochberg, pers. comm.) is a lack of any real niche differentiation in the attack distributions of the two natural enemies in the Hochberg *et al.* (1990) model, since with two parasitoids, a fairly benign parasitoid can 'tip the balance' in terms of host density if the attack distributions of the two are sufficiently different (Hochberg, 1996).

Finally, the biological model raises possibilities that are not predicted by existing mathematical models. In particular, the contrast in the dynamic patterns displayed by the host–pathogen–parasitoid

systems depending on the manner in which they were constructed is striking (compare Figs 18.4 and 18.5). Though perhaps not directly applicable to biological control in the sense that these precise dynamic patterns have not previously been observed, these results do at least show that not only the composition but also the sequence of multiple introductions may be important.

Host–parasitoid dynamics and biological control

Comparing the host-parasitoid dynamics in Figs 18.3, 18.4, and 18.5, then, the contrasts in terms both of persistence and regularity of pattern are marked. To some degree, of course, this is a reflection of the fact that one set are truly host–parasitoid dynamics whereas the others are host–parasitoid dynamics embedded in the dynamics of a host–pathogen–parasitoid system. Which type of series, though, is more representative of field observations on host–parasitoid (or other predator–prey) dynamics? Or – to pose the question another way – how often will the host species in such interactions also be affected by pathogens, which nonetheless remain undetected simply because the hosts are not subjected to postmortem examination?

We contend that such undetected infections may be common, especially when the prevalence of overtly infected hosts is low, and when host-alone dynamics are only weakly affected by the presence or absence of infection – both of which we have shown to be the case here. Indeed, natural pathogens are notoriously difficult to locate and identify in the field, and it is rare to observe disease epizootics, the clearest demonstration of their persistence in the host population and the environment. These data suggest, therefore, that interactions may commonly be classified as 'two-species' when they are actually more complex, and that whereas such simplification may often be justified on grounds of pragmatism or because interactions with further species are only weak, it may also often provide a barrier to the proper understanding of the dynamics of the populations concerned.

Parasitoids are the most frequently used biological control agents. The pests that they are intended to control may or may not carry low level infections of pathogens. Such low levels of pathogens, if they behaved in a similar manner to our laboratory model system, would generate population dynamics that were almost indistinguishable from single species dynamics and, given the vagaries of the field may, indeed, be indistinguishable. In instances of successful biological control, it is entirely possible that success may be brought about because there is a three-species interaction – not the two-species interaction that the pest controller imagines is happening. The alternative view is just as intriguing. If exotic insects become pests in new localities, they may arrive without both their parasitoids and their pathogens. Then what would appear to be an unsuccessful attempt at biological control by a parasitoid that would normally appear to exercise control in the country of origin, would in fact be the typical reaction of host and parasitoid *alone* and the failure to achieve biological control would be attributable to the absence of the pathogen.

Ecologists know that the species with which they are primarily concerned exist in a web of interactions with other species. Most population ecologists, most of the time, isolate one or two species from this web. They do this not only as a practical necessity, but also because they indulge in an act of faith, which goes along the line that the isolation retains enough reality to forge an acceptable understanding of the species' dynamics. The present data allow us, perhaps for the first time, to put such acts of faith to the test – at least in the context of one particular system. Faith seems to have been misplaced. The message for population ecologists and for biological control is not a new one – know your organisms, know your system.

Host–pathogen–parasitoid interactions at the individual level

Insect pathogens, comprising bacteria, viruses, fungi, nematodes, and protozoa, have been studied experimentally for many years in relation to their potential for pest control (for reviews, see Laird *et al.*, 1990), and more recently in relation to their interactions with predators and parasitoids (Brooks, 1993, Rosenheim *et al.*, 1995). Pathogens and parasitoids are in direct competition for a limited resource, the host, but this interaction is more than simply one of conventional competition (between the predators) and predation (of the host). Since the act of infection or parasitism is also the only means by which these natural enemies reproduce, the within-host features have particularly profound population dynamic consequences. Moreover, given the intimate physiological association between a host and its pathogens and endoparasitoids in particular (e.g., Lawrence, 1990; Tanada & Kaya, 1993), it is not surprising that the outcome of within-host competition between the natural enemies is complex. Neither is it surprising that these outcomes influence to a great extent their value as integrated pest control agents.

Most pathogens rarely infect or have a direct effect on the competing endoparasitoid, and the detrimental effects arise more typically because an infected host either dies before the parasitoid can complete its development, or is nutritionally or physiologically unsuitable for parasitoid development. Naturally, there are exceptions. Pathogens with broad insect host ranges, like heterorhabditid and steinernematid nematodes (Akhurst, 1990), and fungi, like *Beauveria bassiana* and *Metarhizium anisopliae* (Goettel *et al.*, 1990), are capable of infecting both the host and the parasitoid. Microsporidia also infect the competing parasitoid, but the disease is often chronic, with adverse consequences more in terms of increased development time, malformations or reduced fecundity than direct mortality (Brooks, 1973; Huger, 1984; Cossentine & Lewis, 1987). There are no records of any direct baculovirus infection of a developing parasitoid in coinfected hosts. Kaya & Tanada (1972, 1973) did find that certain strains of nuclear polyhedrosis virus and granulosis virus produced a 'factor' that was toxic to the parasitoid larva. However, common to all pathogen–parasitoid interactions, the outcome of this interference competition is often settled by which parasite invades the host first, and the time interval between coinfection, giving rise to priority effects.

A model laboratory system

The system to be described in detail here is, as previously explained, the same as that of the biological

Table 18.1. *When infection precedes parasitism. The stage of development, and number, reached by parasitoids in infected larvae that died of the disease in the fifth instar, following parasitism of hosts with increasing levels of infection. Overt disease symptoms appear after approximately 4 days. The percentage emergence of parasitoid adults from diseased hosts compares with approximately 90% emergence from uninfected hosts.*

Infection period (days)	Number parasitized	Stage reached by developing parasitoids at host death (adult emergence given as a percentage of original number parasitized)							
		Egg	1st	2nd	3rd	4th	5th	Pupa	Adult (%)
1	7	0	0	0	0	0	0	7	85.7
2	12	1	0	0	0	0	1	10	33.3
3	31	2	2	1	2	0	5	19	19.4
4	40	3	3	5	1	1	10	17	20.0
5	25	0	0	0	1	1	18	5	8.0
6	25	0	0	3	1	1	15	5	4.0
7	25	6	3	2	2	5	7	0	0
8	23	7	7	5	0	1	3	0	0
9	24	7	5	7	2	1	1	1	0
10	15	10	4	1	0	0	0	0	0
11	15	14	1	0	0	0	0	0	0
16	15	13	2	0	0	0	0	0	0

model of population dynamics. In this particular three-species interaction, the respective host exploitation strategies of each natural enemy lead to a clear cut distinction in one aspect of the competitive interaction; if the host exhibits overt disease symptoms before the final instar (thereby preventing further development), regardless of whether *V. canescens* or the baculovirus enters the host first, the parasitoid larva is invariably doomed. However, if disease symptoms appear in the final instar, successful emergence of the parasitoid is determined by the time interval between coinfection, i.e., as the interval between infection and parasitism increases (and hence the level of infection increases) parasitoid survivorship decreases dramatically (Table 18.1). Thus, despite the possibility that parasitoids can emerge from infected hosts when the disease attacks the host first, the 'window of opportunity' for parasitoid survival is small in this particular system and, like many such pathogen–parasitoid interactions, within-host competition is markedly asymmetric in favor of the pathogen when infection occurs first (e.g., bacteria (Niwa *et al.*, 1987; McDonald *et al.*,

1990), baculoviruses (Irabagon & Brooks, 1974; Hochberg, 1991), fungi (Powell *et al.*, 1986; Goh *et al.*, 1989), nematodes (Kaya & Hotchkin, 1981; Kaya, 1984)).

If *V. canescens* attacks the host first, the outcome is somewhat more balanced (as long as the host reaches the final instar) and a number of outcomes are possible (see Fig. 18.6 for a summary). First, parasitized larvae are more resistant to subsequent baculovirus infection, this reduction becoming more enhanced as the period between parasitism and infection increases, and parasitoids are able to develop successfully without adverse affects. However, there are three outcomes of interference competition when hosts do exhibit disease symptoms (Fig. 18.6). (i) The pathogen wins outright, inhibiting any parasitoid development. (ii) There is no clear victor since the parasitoid larva consumes the entire virus-infected host, thereby reducing substantially pathogen production, but is itself unable to complete its own development and dies. (iii) The parasitoid wins outright, and an adult wasp emerges, but at some considerable cost. These

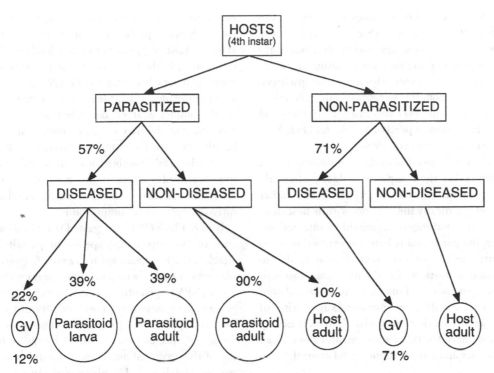

Figure 18.6. Flow diagram documenting the fates of individual *Plodia interpunctella* (in percentage terms) when either parasitized by *Venturia canescens* and susequently infected with granulosis virus (GV) in the fourth instar, or when only infected with virus. The figures of 12% and 71% at the bottom of the diagram refer to percentages of the original cohorts succumbing to granulosis virus infection alone when parasitized or not parasitized, respectively. The full figures for parasitoid adults, emerging from diseased and non-diseased hosts, respectively, were: weights 0.74 ($\pm$0.07) mg, 1.13 ($\pm$0.03) mg (means $\pm$ SE), $t = 7.5$, $p < 0.01$; development times 26.2 ($\pm$1.3) days, 24.7 ($\pm$2.1) days (means $\pm$ SE), $t = 2.4$, $p < 0.01$.

wasps take longer to develop and are significantly smaller than their counterparts that emerge from uninfected hosts.

Qualitatively similar results are obtained from other host–pathogen–parasitoid interactions, the crucial factor again being the time interval between coinfection (e.g., bacteria (Temerak, 1980; McDonald *et al.*, 1990; Nealis & van Frankenhuyzen, 1990), baculoviruses (Santiago-Alvarez & Caballero, 1990; Murray *et al.*, 1995; Nakai & Kunimi, 1997), fungi (King & Bell, 1978; Folegatti *et al.*, 1990; Fransen & van Lenteren, 1994)).

Given the extreme costs to the parasitoid of coinfection it is surprising that, as with many host–pathogen–parasitoid interactions, parasitoids readily attack infected hosts, which constitutes a seemingly maladaptive waste of eggs and time (Sait *et al.*, 1996). Some parasitoids do exhibit some degree of infected host avoidance, though this is by no means perfect (e.g., bacteria (Temarak, 1980; Weseloh & Andreadis, 1982), baculoviruses (Beegle & Oatman, 1975; Caballero *et al.*, 1991), fungi (Brobyn *et al.*, 1988; Fransen & van Lenteren, 1993)). The key factor again appears to be the timing of attack since discrimination is observed more readily in heavily infected hosts, presumably because they are more unlike the healthy hosts that parasitoids are adapted to locating and attacking. Perhaps more significant in terms of the population dynamics of the three-species interaction, parasitoids that attack infected hosts subsequently transmit the pathogen to other susceptibles (e.g., bacteria (Bell *et al.*, 1974), baculoviruses

(Young & Yearian, 1990), fungi (Fransen & van Lenteren, 1993), protozoa (Siegel *et al.*, 1986)). These hosts may become new foci of infection, magnifying enormously the number of pathogens available in the environment. Moreover, the pathogen benefits further from this parasitoid-enhanced transmission since it will lower the host density threshold necessary for pathogen persistence (Anderson & May, 1981; Onstad & Carruthers, 1990).

Over the whole host life cycle it would appear that the pathogen has the competitive edge, particularly since it is more effective at attacking the earlier stages, and is typically the superior within-host competitor in pathogen–parasitoid interactions. However, the parasitoid is better at exploiting uninfected prey in the environment since it is able to search for new hosts, whilst the pathogen must wait for chance encounters (analogous to a sit-and-wait predator). This division between superiority of intrinsic and extrinsic competitive abilities is exactly what the models described above predict should lead to coexistence and enhanced biological control.

Field studies

Much of what is known about the consequences of pathogen–parasitoid interference competition is the result of laboratory studies. Unfortunately there are very few field studies that have explored either the population dynamic or the biological control consequences of the regulation of the host species by one and by several natural enemies. Furthermore, of the examples that do exist, observations are rarely for more than a single generation or season. Many problems are associated with these investigations, not least the many other biotic and abiotic factors that may influence host population dynamics (e.g., Myers, 1988), clouding the role that any one natural enemy may play in pest regulation. Nonetheless, whilst much can only be inferred from the data that have been collected, the overall picture to emerge from field studies is one of the parasitoid species in the interaction faring badly.

A well studied field system is the interaction between the European corn borer, *Ostrinia nubilalis*, the microsporidium *Nosema pyrausta*, and the braconid parasitoid *Macrocentrus grandii*. The parasitoid shows a preference for intermediate larval stages, whilst *N. pyrausta* can attack all larval stages of the host. In the field, *M. grandii* parasitism of the host was inversely correlated with *N. pyrausta* infection (Andreadis, 1982; Siegel *et al.*, 1986). Thus, as in the model system described above, there is considerable overlap of co-exploited stages (our baculovirus–ichneumonid interaction consisting of an 'early–late' combination: this microsporidium–braconid interaction consisting of an 'early–intermediate' combination), resulting in intense interference competition.

Bird & Elgee (1957) suggested that the disappearance of two introduced species of parasitoid that attacked the European spruce sawfly, *Gilpinia hercyniae*, was caused by a concurrent nuclear polyhedrosis virus (NPV) epizootic, the mechanism presumably being suppression of the host's density to a level too low to support the parasitoids. This example of 'invading' species being rebuffed by the resident is one of the predicted outcomes of a number of multispecies models (e.g., Hochberg *et al.*, 1990). In other studies, the use of baculoviruses as bioinsecticides has also resulted in reduced rates of parasitism (e.g., Vail *et al.*, 1972; Hamm & Hare, 1982), and fungal epizootics are often negatively correlated with parasitism rates (Los & Allen, 1983; Goh *et al.*, 1989).

Other pathogen–parasitoid studies yield conflicting results. In field trials using *Bacillus thuringiensis*, reduced parasitism has been recorded (Hamel, 1977), as well as no effect on parasitism (e.g., Andreadis *et al.*, 1983; Niwa *et al.*, 1987), and elevated levels (Andreadis *et al.*, 1983; Weseloh *et al.*, 1983). These different outcomes suggest that for each potential biological control scenario, system-specific details are likely to be important, and may imply that generalizations regarding the suitability or otherwise of a particular natural enemy are not possible.

Towards an integrated biological control strategy

Whilst the within-host competitive interactions between pathogens and parasitoids often appear to be detrimental to both natural enemies in the coinfected host, it may be possible to select life history attributes that will reduce these undesirable effects, enhancing the coexistence of the natural enemies and consequently promoting the success of the biological control strategy. The groups of pathogens with potential for biological control are constrained by specific requirements that ensure infection of the host. Typically they possess long-lived external stages, which are static in the environment and must wait for encounters with susceptibles for horizontal transmission to take place. Also, they need to gain access to the internal environment of the host for reproduction which, with the exception of fungi, requires their being ingested. Parasitoid species, on the other hand, display a wide range of host exploitation strategies, such as endoparasitic or ectoparasitic lifestyles, solitary or gregarious development, and as such are possibly the most amenable to selection for specific pest control needs that can be designed around a particular kind of microparasite. Conversely, application of a particular pathogen may be determined by which predator/parasitoid is deemed to be the major regulatory factor of the insect pest, though as was found in the three-species system decribed above, a seemingly unimportant partner in a multispecies interaction may have a profound influence on the system dynamics. This reiterates the need for a sound and detailed knowledge of the pest and its web of interactions.

Parasitoids can be broadly classified according to whether they prevent further growth of their host once it is parasitized, representing a fixed resource (idiobiont), or whether they allow the host to continue development until it reaches an optimal size for parasitoid development (koinobiont) (see Askew & Shaw, 1986, for a full explanation of these terms). In terms of probability of infection, hosts attacked by koinobiont parasitoids, such as *V. canescens*, are vulnerable to pathogen infection at any time following parasitism and so may suffer future loss to the parasitoid population.

The host stage attacked is possibly the most important characteristic to consider. In the three-species system described above the virus attacked early larval stages, whilst the parasitoid attacked later instars. There was, however, considerable overlap between the prey stages coexploited (larvae), leading to extensive and complex interactions between the endoparasites. This illustrates the general principle that it is crucial to understand the age-dependent nature of the host–pathogen–parasitoid interaction (e.g., Briggs, 1993; Briggs & Godfray, 1995; see also Chapters 2, 17, and 19). As the model system and many other empirical studies demonstrate, such detrimental competitive interactions are likely to occur when natural enemies have similar host exploitation strategies. These characteristics could perhaps be used to eliminate potentially incompatible species for pest control. Coexistence, on the other hand, is promoted by resource partitioning (e.g., Yu *et al.*, 1990; Briggs, 1993; Hochberg, 1996), which in this case means preference for different host stages. For example, host eggs and pupal stages are invulnerable to (horizontal) pathogen infection so parasitoids that attack these stages are likely to avoid damaging confrontations in coinfected hosts. Early larval stages that have escaped parasitoid attack as eggs may nonetheless be highly vulnerable to subsequent pathogen infection.

Age-dependence in parasitism or infection could be exploited further by introducing a second natural enemy, as far as it is possible, at a time that circumvents detrimental interactions. For example, a parasitoid could be introduced to a pest–pathogen system when it coincides with the decline in the primary phase of infection so that (i) the parasitoid can attack the residual host population, and (ii) the parasitoid can assist in the transmission of the pathogen, thereby enhancing secondary cycling and facilitating pest control for a longer period and possibly over a greater area. Indeed, where reduced rates of parasitism have followed the application of pathogens as bioinsecticides (see above), it is believed that careful timing of pathogen spraying may avoid this, though no details are given and it is likely to be system-specific. This reiterates the need for detailed knowledge of the pest and its interacting species.

Linked to age-structure, the dynamics of the pest and its interacting species may be exploited. The inherent cyclical nature of the single and two-species systems in our model system suggests there may be an opportunity to add a second natural enemy when it is either most likely to exert a strong effect on the pest or, conversely, when it is least likely to have a detrimental effect on the resident natural enemy. The pathogen displayed rather consistent levels of prevalence in the host population, as in other systems that may become saturated with time (e.g., Fleming *et al.*, 1986), but diseases of forest insects do appear to exhibit cycles that correspond with those of their hosts (though the link between disease and insect cycles remains equivocal (e.g., Myers, 1988; Berryman, 1996)).

Hassell & May (1986), examining the invasion capabilities of interacting parasitoids, predicted a stable three-species coexistence if an invading specialist acts on the host's life cycle before the resident generalist. The same principles apply to host–pathogen–parasitoid interactions where the pathogen often constitutes a specialist natural enemy, whilst the parasitoid will often be a generalist. A generalist parasitoid will also be buffered against low host densities because it can switch to alternative prey, which will promote long-term classical biological control. However, for the successful invasion and coexistence of an established host–natural enemy interaction by a second natural enemy, it is commonly assumed in the models that the 'invader' is rare compared with the 'resident' (e.g., Hochberg *et al.*, 1990). In contrast, the inundative release of natural enemies in a pest control context will typically involve the release of large numbers, particularly in the case of pathogens. This type of scenario would benefit from a theoretical analysis to determine the population dynamic consequences of such a strategy.

The decision to release a second natural enemy should only occur if the regulation of the insect host by a resident natural enemy is not sufficient economically (otherwise the insect clearly does not constitute a pest) since it runs the risk of disrupting the regulation already in place. Parasitoids, for example, often depress and maintain low host population levels (e.g., Waage & Hassell, 1982) so it is important to introduce a second natural enemy, such as a pathogen,

only if these risks are small. Whether the implementation of a pathogen and a parasitoid is predicted to be successful at biological control depends largely on the ultimate aim. If long-term classical biological control is desired then clearly the potential lack of coexistence and persistence is detrimental to this aim. Similarly, the wildly fluctuating, non-equilibrium population dynamics of the interacting species away from equilibrium levels, of the pest in particular, are undesirable characteristics of this form of pest control. However, if rapid suppression of host abundance is required and the continuous between-season re-application or re-release of the pathogen and parasitoid is economically viable, analagous to conventional pesticide use, then clearly a number of natural enemy combinations will possess these attributes.

Given that species exist within a web of interactions with other species, some of which consist of predators, parasitoids, and pathogens, any successful biological control strategy that has seemingly involved the introduction of a single natural enemy has clearly invoked a suite of multispecies and multitrophic population and community processes. Any approach to biological control, combining both theoretical and empirical elements, must recognize and seek to understand this complexity in order to ensure the desired control of a pest, and to avoid the many traps that lie in wait for the unwary and the uninformed.

References

Akhurst, R. J. (1990) Safety to non-target invertebrates of nematodes of economically important pests. In *Safety of Microbial Insecticides*, ed. M. Laird, L. A. Lacey & E. W. Davidson, pp. 232–40. Boca Raton, FL: CRC Press.

Anderson, R. M. & May, R. M. (1981) The population dynamics of microparasites and their invertebrate hosts. *Philosophical Transactions of the Royal Society of London, Series B* **291**: 451–524.

Andreadis, T. G. (1982) Impact of *Nosema pyrausta* on field populations of *Macrocentrus grandii*, an introduced parasitoid of the European corn borer, *Ostrinia nubilalis*. *Journal of Invertebrate Pathology* **39**: 298–302.

Andreadis, T. G., Dubois, N. R., Moore, R. E., Anderson, J. F. & Lewis, F. B. (1983) Single applications of high concentrations of *Bacillus thuringiensis* for control of gypsy moth (Lepidoptera: Lymantriidae) populations and their impact on parasitism and disease. *Journal of Economic Entomology* **76**: 1417–22.

Askew, R. R. & Shaw, M. R. (1986) Parasitoid communities: their size, structure and development. In *Insect Parasitoids*, ed. J. K. Waage & D. J. Greathead, pp. 225–64. London: Academic Press.

Beddington, J. R., Free, A. C. & Lawton, J. H. (1978) Characteristics of successful natural enemies in models of biological control of insect pests. *Nature* **273**: 513–19.

Beegle, C. C. & Oatman, E. R. (1975) Effect of a nuclear polyhedrosis virus on the relationship between *Trichoplusia ni* (Lepidoptera: Noctuidae) and the parasite, *Hyposoter exiguae* (Hymenoptera: Ichneumonidae). *Journal of Invertebrate Pathology* **25**: 59–71.

Begon, M., Sait, S. M. & Thompson, D. J. (1995) Persistence of a parasitoid–host system: refuges and generation cycles? *Proceedings of the Royal Society of London, Series B* **260**: 131–7.

Begon, M., Sait, S. M. & Thompson, D. J. (1996) Predator–prey cycles: period shifts between two- and three-species systems. *Nature* **381**: 311–15.

Begon, M., Sait, S. M. & Thompson, D. J. (1997) Two's company, three's a crowd: host–pathogen–parasitoid dynamics. In *Multitrophic Interactions in Terrestrial Systems*, ed. A. C. Gange & V. K. Brown, pp 307–32. Oxford: Blackwell Scientific.

Bell, J. V., King, E. G. & Hamalle, R. J. (1974) Interactions between bollworms, a braconid, and the bacterium *Serratia marcescens*. *Annals of the Entomological Society of America* **67**: 712–14.

Berryman, A. A. (1996) What causes population cycles of forest Lepidoptera? *Trends in Ecology and Evolution* **11**: 28–32.

Berryman, A. A., Millstein, J. A. & Mason, R. R. (1990) Modelling Douglas-fir tussock moth population dynamics: the case for simple theoretical models. In *Population Dynamics of Forest Insects*, ed. A. D. Watt, S. R. Leather, M. D. Hunter & N. A. C. Kidd, pp. 369–80. Andover: Intercept.

Bird, F. T. & Elgee, D. E. (1957) A virus disease and introduced parasites as factors controlling the European spruce sawfly, *Diprion hercyniae* (Htg.), in central New Brunswick. *Canadian Entomologist* **89**: 371–8.

Briggs, C. J. (1993) Competition among parasitoid species on a stage-structured host and its effect on host suppression. *American Naturalist* **141**: 372–97.

Briggs, C. J. & Godfray, H. C. J. (1995) The dynamics of insect–pathogen interactions in stage-structured populations. *American Naturalist* **145**: 855–87.

Brobyn, P. J., Clark, S. J. & Wilding, N. (1988) The effect of fungus infection of *Metopolophium dirhodum* (Hom.: Aphidae) on the oviposition behaviour of the aphid parasitoid *Aphidus rhopalosiphi* (Hym.: Aphidiidae). *Entomophaga* **33**: 333–8.

Brooks, W. M. (1973) Protozoa: host–parasite–pathogen interrelationships. *Miscellaneous Publications of the Entomological Society of America* **9**: 105–11.

Brooks, W. M. (1993) Host–parasitoid–pathogen interactions. In *Parasites and Pathogens of Insects. Volume 2: Pathogens*, ed. N. E. Beckage, S. N. Thompson & B. A. Federici, pp. 231–72. New York: Academic Press.

Caballero, P., Vargas-Osuna, E. & Santiago-Alvarez, C. (1991) Parasitization of granulosis-virus infected and noninfected *Agrotis segetum* larvae and the virus transmission by three hymenopteran parasitoids. *Entomologia Experimentalis et Applicata* **58**: 55–60.

Cossentine, J. E. & Lewis, L. C. (1987) Development of *Macrocentrus grandii* Goidanich within microsporidian-infected *Ostrinia nubilalis* (Hübner) host larvae. *Canadian Journal of Zoology* **65**: 2532–5.

Dobson, A. P. (1985) The population dynamics of competition between parasites. *Parasitology* **91**: 317–47.

Fleming, S. B., Kalmakoff, J., Archibold, R. D. & Crawford, K. M. (1986) Density-dependent virus mortality in populations of *Wiseana* (Lepidoptera: Hepialidae). *Journal of Invertebrate Pathology* **48**: 193–8.

Folegatti, M. E. G., Alves, S. B. & Botelho, P. S. M. (1990) Pathogenicity of the fungus *Metarhizium anisopliae* (Metsch.) Sorok for pupae and adults of *Apanteles flavipes* (Cam.). *Pesquisa Agropecuaria Brasileira* **25**: 247–51.

Fransen, J. J. & van Lenteren, J. C. (1993) Host selection and survival of the parasitoid *Encarsia formosa* on greenhouse whitefly, *Trialeurodes vaporariorum*, in the presence of hosts infected with the fungus *Ashersonia aleyrodis*. *Entomologia Experimentalis et Applicata* **69**: 239–49.

Fransen, J. J. & van Lenteren, J. C. (1994) Survival of the parasitoid *Encarsia formosa* after treatment of parasitized greenhouse whitefly larvae with fungal spores of *Ashersonia aleyrodis*. *Entomologia Experimentalis et Applicata* **71**: 235–43.

Gange, A. C. & Brown, V. K. (eds) (1997) *Multitrophic Interactions in Terrestrial Systems*. Oxford: Blackwell Scientific.

Godfray, H. C. J. & Hassell, M. P. (1989) Discrete and continuous insect populations in tropical environments. *Journal of Animal Ecology* **58**: 153–74.

Goettel, M. S., Poprawski, T. J., Vandenberg, J. D., Li, Z. & Roberts, D. W. (1990) Safety to non-target invertebrates of fungal biocontrol agents. In *Safety of Microbial Insecticides*, ed. M. Laird, L. A. Lacey & E. W. Davidson, pp. 209–31. Boca Raton, FL: CRC Press.

Goh, K. S., Berberet, R. C., Young, L. J. & Conway, K. E. (1989) Mortality of the parasite *Bathyplectes curculionis* (Hymenoptera: Ichneumonidae) during epizootics of *Erynia phytonomi* (Zygomycetes: Entomophthorales) in alfalfa weevil. *Environmental Entomology* **18**: 1131–5.

Gordon, D. M., Nisbet, R. M., De Roos, A., Gurney, W. S. C. & Stewart, R. R. (1991) Discrete generations in host–parasitoid models with contrasting life cycles. *Journal of Animal Ecology* **60**: 295–308.

Gurney, W. S. C. & Nisbet, R. M. (1985) Fluctuation periodicity, generation separation, and the expression of larval competition. *Theoretical Population Biology* **28**: 150–80.

Gutierrez, A. P. (1992) Physiological basis of ratio-dependent predator–prey theory: the metabolic pool model as a paradigm. *Ecology* **73**: 1552–63.

Gutierrez, A. P., Hagen, K. S. & Ellis, C. K. (1990) Evaluating the impact of natural enemies: a multitrophic perspective. In *Critical Issues in Biological Control*, ed. M. Mackauer, L. E. Ehler & J. Roland, pp. 81–107. Andover: Intercept.

Hamel, D. R. (1977) The effects of *Bacillus thuringiensis* on parasitoids of the western spruce budworm, *Choristoneura occidentalis* (Lepidoptera: Tortricidae), and the spruce coneworm, *Dioryctria reniculelloides* (Lepidoptera: Pyralidae), in Montana. *Canadian Entomologist* **109**: 1409–15.

Hamm, J. J. & Hare, W. W. (1982) Application of entomopathogens in irrigation water for control of fall armyworms and corn earworms (Lepidoptera: Noctuidae) on corn. *Journal of Economic Entomology* **75**: 1074–8.

Harvey, J. A., Harvey, I. F. & Thompson, D. J. (1994) Flexible larval growth allows the use of a range of host sizes by a parasitoid wasp. *Ecology* **75**: 1420–8.

Hassell, M. P. (1978) *The Dynamics of Arthropod Predator–Prey Systems*. Princeton: Princeton University Press.

Hassell, M. P. & May, R. M. (1986) Generalist and specialist natural enemies in insect predator–prey interactions. *Journal of Animal Ecology* **55**: 923–40.

Hochberg, M. E. (1989) The potential role of pathogens in biological control. *Nature* **337**: 262–5.

Hochberg, M. E. (1991) Intra-host interactions between a braconid endoparasitoid, *Apanteles glomeratus*, and a baculovirus for larvae of *Pieris brassicae*. *Journal of Animal Ecology* **60**: 51–63.

Hochberg, M. E. (1996) Consequences for host population levels of increasing natural enemy species richness in classical biological control. *American Naturalist* **147**: 307–18.

Hochberg, M. E. & Holt, R. D. (1990) The coexistence of competing parasites. I. The role of cross-species infection. *American Naturalist* **136**: 517–41.

Hochberg, M. E. & Lawton, J. H. (1990) Competition between kingdoms. *Trends in Ecology and Evolution* **5**: 367–71.

Hochberg, M. E., Hassell, M. P. & May, R. M. (1990) The dynamics of host–pathogen–parasitoid interactions. *American Naturalist* **135**: 74–94.

Huger, A. M. (1984) Susceptibility of the egg parasitoid *Trichogramma evanescens* to the microsporidium *Nosema pyrausta* and its impact on fecundity. *Journal of Invertebrate Pathology* **44**: 228–9.

Irabagon, T. A. & Brooks, W. M. (1974) Interaction of *Campoletis sonorensis* and a nuclear polyhedrosis virus in larvae of *Heliothis virescens*. *Journal of Economic Entomology* **67**: 229–31.

Kaya, H. K. (1984) Effect of the entomogenous nematode *Neoaplectana carpocapsae* on the tachinid parasite *Compsilura concinnata* (Diptera: Tachinidae). *Journal of Nematology* **16**: 9–13.

Kaya, H. K. & Hotchkin, F. G. (1981) The nematode *Neoaplectana carpocapsae* Weiser and its effect on selected ichneumonid and braconid parasites. *Environmental Entomology* **10**: 474–8.

Kaya, H. K. & Tanada, Y. (1972) Response of *Apanteles militaris* to a toxin produced in a granulosis-virus-infected host. *Journal of Invertebrate Pathology* **19**: 1–17.

Kaya, H. K. & Tanada, Y. (1973) Hemolymph factor in armyworm larvae infected with a nuclear poly-hedrosis virus toxic to *Apanteles militaris. Journal of Invertebrate Pathology* **21**: 211–14.

King, E. G. & Bell, J. V. (1978) Interactions between a braconid, *Microplitis croceipes*, and a fungus, *Nomuraea rileyi*, in laboratory-reared bollworm larvae. *Journal of Invertebrate Pathology* **31**: 337–40.

Laird, M., Lacey, L. A. & Davidson, E. W. (1990) *Safety of Microbial Insecticides*. Boca Raton, FL: CRC Press.

Lawrence, P. O. (1990) The biochemical and physiolog-ical effects of insect hosts on the development and ecology of their insect parasites: an overview. *Archives of Insect Biochemistry and Physiology* **13**: 217–28.

Los, L. M. & Allen, W. A. (1983) Incidence of *Zoophthora phytonomi* (Zygomycetes: Entomophthorales) in *Hypera postica* (Coleoptera: Curculionidae) larvae in Virginia. *Environmental Entomology* **12**: 1318–21.

McDonald, R. C., Kok, L. T. & Yousten, A. A. (1990) Response of fourth instar *Pieris rapae* parasitized by the braconid *Cotesia rubecula* to *Bacillus thuringiensis* subsp. *kurstaki* endotoxin. *Journal of Invertebrate Pathology* **56**: 422–3.

Mason, R. R. & Torgersen, T. R. (1987) Dynamics of a nonoutbreak population of the Douglas-fir tussock moth (Lepidoptera: Lymantriidae) in Southern Oregon. *Environmental Entomology* **16**: 1217–27.

May, R. M. & Hassell, M. P. (1988) Population dynam-ics and biological control. *Philosophical Transactions of the Royal Society of London, Series B* **318**: 129–69.

Murray, D. A. H., Monsour, C. J., Teakle, R. E., Rynne, K. P. & Bean, J. A. (1995) Interactions between nuclear polyhedrosis virus and three larval para-sitoids of *Helicoverpa armigera* (Hübner) (Lepidoptera: Noctuidae). *Journal of the Australian Entomological Society* **34**: 319–22.

Myers, J. H. (1988) Can a general hypothesis explain population cycles in forest Lepidoptera? *Advances in Ecological Research* **18**: 179–232.

Nakai, M. & Kunimi, Y. (1997) Granulosis virus infec-tion of the smaller tea tortrix (Lepidoptera: Tortricidae): effect on the development of the endoparasitoid, *Ascogaster reticulatus* (Hymenop-tera: Braconidae). *Biological Control* **8**: 74–80.

Nealis, V. & van Frankenhuyzen, K. (1990) Interactions between *Bacillus thuringiensis* Berliner and *Apanteles fumiferanae* Vier. (Hymenoptera: Braconidae), a parasitoid of the spruce budworm, *Choristoneura fumiferana* (Clem.) (Lepidoptera: Tortricidae). *Canadian Entomologist* **122**: 585–94.

Niwa, C. G., Stelzer, M. J. & Beckwith, R. C. (1987) Effects of *Bacillus thuringiensis* on parasites of western spruce budworm (Lepidoptera: Tortricidae). *Journal of Economic Entomology* **80**: 750–3.

Onstad, D. W. & Carruthers, R. I. (1990) Epizootiological models of insect diseases. *Annual Review of Entomology* **35**: 399–419.

Payne, C. C. (1988) Pathogens for the control of insects: where next? *Philosophical Transactions of the Royal Society of London, Series B* **318**: 225–48.

Pimm, S. L., Lawton, J. H. & Cohen, J. E. (1991) Food web patterns and their consequences. *Nature* **350**: 669–74.

Powell, W., Wilding, N., Brobyn, P. J. & Clark, S. L. (1986) Interference between parasitoids (Hym.: Aphidiidae) and fungi (Entomophthorales) attacking cereal aphids. *Entomophaga* **31**: 293–302.

Reeve, J. D., Cronin, J. T. & Strong, D. R. (1994) Parasitism and generation cycles in a salt marsh planthopper. *Journal of Animal Ecology* **63**: 912–20.

Rosenheim, J. A., Kaya, H. K., Lester, L. E., Marois, J. J. & Jaffee, B. A. (1995) Intraguild predation among biological control agents: theory and evi-dence. *Biological Control* **5**: 303–35.

Sait, S. M., Begon, M. & Thompson, D. J. (1994a) The effects of a sublethal baculovirus infection in the Indian meal moth, *Plodia interpunctella. Journal of Animal Ecology* **63**: 541–50.

Sait, S. M., Begon, M. & Thompson, D. J. (1994b) Long-term population dynamics of the Indian meal moth, *Plodia interpunctella*, and its granulosis virus. *Journal of Animal Ecology* **63**: 861–70.

Sait, S. M., Begon, M. & Thompson, D.J. (1994c) The influence of larval age on the response of *Plodia interpunctella* to a granulosis virus. *Journal of Invertebrate Pathology* **63**: 107–10.

Sait, S. M., Andreev, R. A., Begon, M., Thompson, D. J., Harvey, J. A. & Swain, R. D. (1995) *Venturia canescens* parasitizing *Plodia interpunctella*: host vulnerability – a matter of degree. *Ecological Entomology* **20**: 199–201.

Sait, S. M., Begon, M., Thompson, D. J. & Harvey, J. A. (1996) Parasitism of baculovirus-infected *Plodia interpunctella* by *Venturia canescens* and subsequent virus transmission. *Functional Ecology* 10: 586–91.

Sait, S. M., Begon, M., Thompson, D. J., Harvey, J. A. & Hails, R. S. (1997) Factors affecting host selection in an insect host–parasitoid interaction. *Ecological Entomology* 22: 225–30.

Santiago-Alvarez, C. & Caballero, P. (1990) Susceptibility of parasitized *Agrotis segetum* larvae to a granulosis virus. *Journal of Invertebrate Pathology* 56: 128–31.

Siegel, J. P., Maddox, J. V. & Ruesink, W. G. (1986) Impact of *Nosema pyrausta* on a braconid, *Macrocentrus grandii*, in central Illinois. *Journal of Invertebrate Pathology* 47: 271–6.

Tanada, Y. & Kaya, H. K. (1993) *Insect Pathology*. New York: Academic Press.

Temerak, S. A. (1980) Detrimental effects of rearing a braconid parasitoid on the pink borer larvae inoculated by different concentrations of the bacterium, *Bacillus thuringiensis* Berliner. *Zeitschrift für Angewandte Entomologie* 89: 315–19.

Vail, P. V., Soo Hoo, C. F., Seay, R. S., Killinen, R. G. & Wolf, W. W. (1972) Microbial control of lepidopterous pests of fall lettuce in Arizona and effects of chemicals and microbial pesticides on parasitoids. *Environmental Entomology* 1: 780–5.

Waage, J. K. & Hassell, M. P. (1982) Parasitoids as biological control agents – a fundamental approach. *Parasitology* 84: 241–68.

Weseloh, R. M. & Andreadis, T. G. (1982) Possible mechanism for synergism between *Bacillus thuringiensis* and the gypsy moth (Lepidoptera: Lymantriidae) parasitoid, *Apanteles melanoscelus* (Hymeoptera: Braconidae). *Annals of the Entomological Society of America* 75: 435–8.

Weseloh, R. M., Andreadis, T. G., Moore, R. E. B., Anderson, J. F., Dubois, N. R. & Lewis, F. B. (1983) Field confirmation of a mechanism causing synergism between *Bacillus thuringiensis* and the gypsy moth parasitoid, *Apanteles melanoscelus*. *Journal of Invertebrate Pathology* 41: 99–103.

Young, S. Y. & Yearian, W. C. (1990) Transmission of nuclear polyhedrosis virus by the parasitoid *Microplitis croceipes* (Hymenoptera: Braconidae) to *Heliothis virescens* (Lepidoptera: Noctuidae) on soybean. *Environmental Entomology* 19: 251–6.

Yu, D. S., Luck, R. F. & Murdoch, W. W. (1990) Competition, resource partitioning and coexistence of an endoparasitoid *Encarsia perniciosi* and an ectoparasitoid *Aphytis melinus* of the California red scale. *Ecological Entomology* 15: 469–80.

19

Persistence of natural enemies of weeds and insect pests in heterogeneous environments

DAVID W. ONSTAD AND EDWARD A. KORNKVEN

Introduction

Classical microbial control involves the introduction of a pathogen into an ecosystem with the objective of long-term persistence of, and pest control by, the pathogen. In some cases, classical control uses several coexisting biological control agents to maintain a pest population at low levels. In managed as well as unmanaged ecosystems, populations of weeds and insects are distributed heterogeneously over space and change phenologically over time. Furthermore, the movement of pathogens and insects contributes to and is influenced by these heterogeneous environments. Mathematical models and concepts derived from model analyses should account for this variability in space and time (see Chapters 10 and 12).

Our goal in this chapter is to derive hypotheses about persistence, endemicity, and long-term coexistence of pathogens and parasitoids (Walde & Nachman (Chapter 10) emphasize the persistence of predators). After a review of weed microbial control literature, we discuss a hypothetical model of a weed species and its disease. Then we describe a case-specific model of European corn borer control by a pathogen and a parasitoid. We finish the chapter with a few remarks concerning the value of models and the need for temporal and spatial scales in hypotheses derived from models.

Microbial control of weeds

There are two dominant strategies for the utilization of pathogens to reduce weed infestations in a variety of habitats (TeBeest et al., 1992). The classical approach emphasizes the importation and release of host-specific pathogens to suppress a targeted weed by reducing reproduction, competitive ability, or increasing mortality (Politis et al., 1984; Callaway et al., 1985; Bruckart & Dowler, 1986, Paul & Ayres, 1986, 1987; Trujillo et al., 1988; Bruckart & Hasan, 1991). This strategy is often characterized by the reduction of host densities, but not elimination of the host, and the inability to predict the probable long-term levels of host density after release of the pathogen. The second strategy utilizes endemic diseases and initiates epidemics in weed populations by application of inoculum as mycoherbicides. Charudattan (1991) pointed out that although more than 160 fungal pathogens have been studied as potential mycoherbicides only three have been successful in the field. In an attempt to improve this success rate, Yang & TeBeest (1993) advocated investigation of epidemiological and ecological processes to account for the roles of both primary and secondary infection after the application of mycoherbicides.

The goal of weed management is usually not eradication but regulation of the weed density at a low level (TeBeest et al., 1992). To achieve this goal, microbial control requires a theoretical foundation that focuses on sparse plant populations and finite host densities. For example, knowledge about host density effects, such as thresholds, can be useful in planning inoculative or inundative releases of pathogens for weed biocontrol. To control the weeds before they reach homogeneous monoculture levels, or at least maintain them once their densities are reduced, we need models that do not define the host population as a non-limiting, uniformly distributed resource for pathogen growth and spread even though this has been the tradition in plant epidemiology (Van der Plank, 1963, 1975; Jeger & van den Bosch, 1994). Because diseases in sparse populations such as those in natural systems and urban systems are rarely investigated, new hypotheses are needed to help explain the interactions between pathogen and host populations in low density situations (Alexander, 1989; Harper, 1990).

Burdon (1993) and Burdon *et al.* (1989) reviewed studies that focus on host density and spatial distributions of pathogen and host in natural communities. Supkoff *et al.* (1988) studied the effect of *Puccinia chondrillina* on the population densities of *Chondrilla juncea*, a weed of rangeland in California. Their data indicate that the pathogen may cause declines in host density after densities rise above a certain level. The data also suggest that prevalence of the pathogen declines at low host densities. Thus, the long-term persistence of the pathogen and host regulation is likely to depend upon the relationship between host density and the ability of the pathogen to grow and spread. Jennersten *et al.* (1983) found that *Ustilago violacea* could regulate its host, *Viscaria vulgaris*, and a host-density threshold may exist that prevents the pathogen from existing in low-density patches. Burdon *et al.* (1995) concluded that incidence of the rust *Triphragmium ulmariae* was strongly correlated with host population size in metapopulations of *Filipendula ulmaria*. Jarosz & Levy (1988) discovered that host density had a major impact on the epidemiology of a powdery mildew in natural plant populations, and Chin & Wolfe (1984) concluded that the density of susceptible seedlings was the major determinant of the spread of powdery mildew in barley. Empirical studies such as these are needed as well as work with mathematical models to derive general hypotheses about classical microbial control of weeds.

Models

Although several plant pathologists have modeled the temporal and spatial dynamics of real or hypothetical pathogens, they have not addressed the question of long-term persistence and other issues of population dynamics theory (Kiyosawa, 1976; Kampmeijer & Zadoks, 1977; Mundt & Leonard, 1986; Mundt *et al.*, 1986). Most often the emphasis was on the study of dispersal gradients, focus spread, and disease (not pathogen) progress within a single

year (Minogue, 1989; van den Bosch *et al.*, 1990; Zawolek & Zadoks, 1992).

Some authors have used traditional animal disease models to study the epidemiology of plant diseases. The core of these models is the mass-action transmission term or infection rate, which is a simple linear function of the product of susceptible (S) hosts multiplied by infectious (I) hosts (e.g., bSI). Yang & TeBeest (1992) use this approach to examine how disease reduction in host reproduction or survival influences yearly fluctuations of uniformly distributed annual weed populations. They demonstrated how the pathogen virulence can be used to select candidates for either the classical or mycoherbicide strategy for microbial control of weeds. May (1990) presents a variety of modeling ideas based on traditional zoological models and relates some of them to the work of Van der Plank (1963). May (1990) also discusses epidemics in heterogeneously distributed populations but does not analyze models concerning spatial dynamics or sparse plant populations.

Thrall & Jarosz (1994) tested several two-equation models representing the temporal dynamics of a natural population of plants and a vectored disease. Their exponential disease transmission models, in which the probability of infection takes the form $1 - e^{-bI}$ instead of the linear form bI, matched the experimental field data better than the linear models.

The model presented here can be used to study persistence of weed pathogens. It simulates the kinds of systems and contains most of the same implicit and explicit assumptions of the models studied by Van der Plank (1963, 1975) and Zadoks (1971). However, neither these two scientists nor Jeger & van den Bosch (1994) analyzed models with spatial dynamics, spatial heterogeneity, and limiting host densities with the goal of understanding persistence. This model can be easily extended to study other questions of interest to weed management experts.

Our model consists of four differential equations representing each plant site in which S represents susceptible host tissue (leaflets/plant site), N symbolizes total host tissue (leaflets/plant site), and L, I, and D are the densities of leaflets with latent, infectious, and removed lesions, respectively (Onstad & Kornkven, 1992). Host tissue in the latent stage is not yet infectious. For simplicity, we assume that a lesion covers a single small leaflet by the end of the latent period.

$$\frac{dL_{tj}}{dt} = 0.2RI_{tj}\left(\frac{S_{tj}}{N_{tj}}\right) + \sum_{k=1}^{8} 0.1RI_{tk}\left(\frac{S_{tj}}{N_{tj}}\right) - \frac{L_{tj}}{p} \quad \text{(W19.1)}$$

$$\frac{dI_t}{dt} = \frac{L_t}{p} - \frac{I_t}{i} \quad \text{(W19.2)}$$

$$\frac{dD_t}{dt} = \frac{I_t}{i} \quad \text{(W19.3)}$$

$$\frac{dS_{tj}}{dt} = b - 0.2RI_{tj}\left(\frac{S_{tj}}{N_{tj}}\right) - \sum_{k=1}^{8} 0.1RI_{tk}\left(\frac{S_{tj}}{N_{tj}}\right) \quad \text{(W19.4)}$$

$$0.2RI_{tj}\left(\frac{S_{tj}}{N_{tj}}\right) + \sum_{k=1}^{8} 0.1RI_{tk}\left(\frac{S_{tj}}{N_{tj}}\right) \leq S_i \quad \text{(W19.5)}$$

The average latent period is p, the average infectious period is i, and the host growth rate is b. Both periods are expressed in days. R is the potential reproductive rate in terms of new inoculum per infectious lesion per day. If this inoculum lands on susceptible tissue, then it will produce infected leaflets or latent lesions. The subscripts j and k represent plant sites, and t is time.

Equation (W19.1) describes the increase in latent lesions due to reproduction and the decrease due to maturation. On average, a new lesion requires p days to completely cover its leaflet. In eqn (W19.2) infectious leaflets are increased by maturation from the latent stage and decreased by the rate of removal or development out of the stage. Equation (W19.4) describes host growth and infection. For host growth rate $b > 0$, new susceptible leaflets are produced each day; for $b = 0$, N is constant. The second term in eqn (W19.4) is the effective rate of infection or successful reproduction by the pathogen. Because $N = S + L + I + D$, S/N is equivalent to $1 - (L + I + D)/N$. The variable L is part of the infected tissue that reduces reproduction because when two or more propagules land on a latent leaflet, the leaflet will still be completely covered p days after germination of the first propagule.

To model the spatial dynamics of the pathogen in the heterogeneous and often sparse populations of weeds, we assigned the four state variables calculated by eqns (W19.1)–(W19.5) to each site, j, in a 64×128

grid of cells. Some or all of the sites may be occupied by a host plant. Thus, the distribution of hosts may or may not be uniform. Inoculum is dispersed among plants by assigning one-fifth of RI_j calculated at site j to site j and one-tenth to each of the eight adjacent sites ($k = 1,8$) in the square of nine sites. Propagules landing on non-host sites or on non-susceptible tissue are lost. Inoculum landing on host plants survives and germinates with probability S_k/N_k for plant k. To mimic the spread of disease in a much larger region, hosts on the edge of the grid are assumed to be adjacent to the sites on the opposite edge. No inoculum is dispersed out of the region. We modeled the absolute density of hosts in the region by changing the total density, or number of occupied sites. Details concerning the simulations of eqns (W19.1)–(W19.5) and their analysis are given by Onstad & Kornkven (1992).

Hypotheses and results concerning persistence of weed pathogens

Before studying the results and deriving hypotheses from the model, appropriate temporal and spatial scales must be chosen (Onstad, 1992b). According to Burdon & Chilvers (1982), the appropriate temporal scale for this type of study has a basic unit related to the reproductive cycles of the host and pathogen. Zadoks & Schein (1979) considered the infection cycle to be the basic temporal unit for plant epidemiology. For a pathogen with overlapping generations, the median age for reproduction can be used as a measure of generation time. Thus, generation time is an appropriate temporal unit for analyzing persistence of plant pathogens.

The choice of spatial scale for the analysis determines whether the pathogen was persistent in the model. The smallest spatial unit for the analysis of pathogen population dynamics should be based on the dispersal function (the average area over which a propagule disperses from the mother source). Thus, in our study, the minimum spatial unit is the unit of nine plant sites defined by eqn (W19.1) as the dispersal area. In the simulations, the pathogen was not always endemic (persistent) in all units. For example, with $iR = 2.3$, $p = 10$, $I = 10$, and 3000 growing weeds in a region of 882 spatial units of nine plant sites, an average of 799 spatial units had persistent pathogen occupancy; whereas 2774 host plants had living pathogen in the region after 1000 days. When the pathogen disappeared from a single unit, it persisted in the region by remaining endemic in other units. Spatial dynamics among units may keep the pathogen endemic at the regional scale as in most metapopulation models (Hess, 1996).

When Van der Plank (1975) and Zadoks & Schein (1979) analyzed the population dynamics of disease with a simpler model, they concluded that endemic disease has $iR = 1$ on average. Although Van der Plank's theorem, $iR = 1$, assumes that hosts are non-limiting, his models of disease progress do not make this assumption. Thus, iR in the theorem is actual reproduction under conditions that are rarely realistic, whereas iR in his models is potential reproduction (actual reproduction = potential reproduction × the probability of a propagule landing on susceptible tissue). Onstad & Kornkven (1992) concluded that the value of iR by itself is not sufficient to predict whether a pathogen will persist or not. In some cases, when iR is a constant or has an average value of 1, the pathogen did not persist in the simulations.

How long will a pathogen persist in a region larger than the area of pathogen propagule dispersal? As total potential reproduction, iR, decreases, a pathogen is less likely to persist for a given number of pathogen generations (persistence time will be less). With constant iR, a pathogen is more likely to persist over a given number of days when it has a longer generation time defined by the latent and infectious periods, but the period of persistence is fairly constant in terms of pathogen generations. With iR constant, the longer generation times slow down any decline in the pathogen population (if R, but not iR, is constant, total reproduction would be less for shorter infectious periods, and the ratio of p to i would be important).

For those implementing classical microbial control of weeds, the model results indicate that care should be taken in predicting effectiveness from laboratory/greenhouse studies in which potential pathogen reproduction is measured. Actual reproduction and

spread in heterogeneous and sparse weed populations may be much lower. Potential reproduction, iR, may need to be much greater than 1 for effective weed control.

The model demonstrated that conditions of the weed population will also influence persistence of the pathogen. One hypothesis is that variability of iR due to environmental or genetic heterogeneity of weed plants will always increase persistence relative to a homogeneous weed population with the average iR. A second conclusion is that lower weed densities decrease the number of generations a pathogen will persist. A third hypothesis is that continuous growth of susceptible weed tissue increases the probability of persistence for the pathogen.

Pathogen persistence is, of course, not the only issue in weed management. However, as Yang & TeBeest (1993) emphasized, the persistence of even mycoherbicidal pathogens can be important to effective control of many weeds.

Biological control of insects

Weed and insect control share many of the same issues. Host density, spatial heterogeneity, and spatial dynamics are just as important in insect communities as they are in weed communities. Also the persistence and coexistence of biological control agents is still of major concern. Fuxa & Tanada (1987) and Ignoffo (1988) review microbial control of insect pests. Viruses, fungi, and microsporidia have been utilized for classical microbial control in which persistence of the pathogen is important.

Many models have been created to study real and hypothetical systems of insects and their pathogens (Onstad & Carruthers, 1990). Godfray & Briggs (Chapter 17) provide an analysis of recent models of insect pathogens, ranging from the simplest to those with intermediate complexity. Thomas *et al.* (Chapter 20) emphasize the modeling of microbial insecticides.

Several recent models of insect pathogens have included spatial dynamics. Dwyer (1994) modeled host movement with a one-dimensional diffusion function to study the short- and long-term dynamics of disease wave fronts in space. Dwyer's approach can be used to study how fast a pathogen will spread from a point of release for microbial control. Wood & Thomas (1996) used an empirically based model simulated on a two-dimensional hexagonal grid to show how spatial heterogeneity and spatial dynamics of an insect host can lead to persistence of virulent pathogens that would otherwise be extirpated from a homogeneous region. Onstad *et al.* (1990) used a two-dimensional rectangular grid of fields and plant sites within each field to model the long-term dynamics and regulation of the European corn borer by a microsporidian disease. They used the model to show that a host-density threshold is sensitive to initial prevalence of disease and spatial dynamics of the host. The results concerning pathogen persistence are discussed below.

With regard to the modeling of parasitoids, several papers complement the specific approach and model described in this chapter. Begon *et al.* (Chapter 18) review their own work as well as the recent work of others on the interactions of parasitoids and pathogens. Hochberg *et al.* (1990) modeled pathogen–parasitoid competition in which the pathogen does not infect the parasitoid in homogeneous space. Briggs *et al.* (Chapter 2) review the extensive literature concerning the modeling of parasitoids. Only a few host–parasitoid models contain explicit spatial structure and spatial dynamics (Allen, 1975; Nachman, 1981; Casas, 1990).

Biological control of European corn borer

In the study presented below, our goal was to examine the persistence and coexistence of a pathogen and parasitoid in a heterogeneous environment. As Wood & Thomas (1996) emphasize, even habitats that appear homogeneous, such as a maize monoculture, can be heterogeneous because of aggregated distributions of host larvae, non-uniform distributions of disease, and movement of adults. Another factor is that chronic diseases often require a more complex model to handle the numerous disease types that should be modeled as separate cohorts. We believe the structure exemplified in Fig. 19.1 is useful for modeling such diseases. It has been used to model *Nosema whitei* infecting *Tribolium confusum* (Onstad

& Maddox, 1990) and *Nosema pyrausta* infecting the European corn borer (Onstad & Maddox, 1989). The procedure is useful in studying chronic diseases in age-structured host populations when the age of the host at the time of initial infection determines survival, behavior, and fecundity of surviving infected adults. Begon *et al.* (Chapter 18) discuss a laboratory study that demonstrates the value of modeling the timing of infection, the timing of parasitism, and the tracking of cohorts following these events.

Although the model may seem complex and the number of parameters may seem large compared to models presented by others (see Chapters 17 and 18), the parameters are relatively easy to measure and are often available for well-studied pests. By learning about parameters included in the model, readers may discover how their own data may make a valuable contribution to a model. Simpler models for the European corn borer and *Nosema pyrausta* have been created by Briggs & Godfray (1995) and Cavalieri & Koçak (1995).

The microsporidium, *Nosema pyrausta*, is arguably the most important classical biological control agent affecting the European corn borer, *Ostrinia nubilalis*, in the USA (Onstad & Maddox, 1989; Onstad *et al.*, 1990). *Macrocentrus grandii* is a polyembryonic braconid wasp that attacks the European corn borer. These three species were observed together in natural habitats in Europe many years ago (Parker, 1931), and now they occur in maize monocultures in the USA (Andreadis, 1980; Siegel *et al.*, 1986; Cossentine & Lewis, 1987). The pathogen may reduce parasitism by infecting the parasitoid and altering its mortality, behavior, and fecundity, or it may reduce herbivore density below the level needed to maintain the parasitoid in the field or region.

For 2 years in the fall, Mason *et al.* (1994) sampled larval parasitoids of the European corn borer across the eastern USA and found *Macrocentrus grandii* in eight states with a mean percent parasitism of 2% and a range of 0–5%. Siegel (1985) observed *M. grandii* in Illinois in two generations of *O. nubilalis* in each year of a 4 year period and reported the following mean percentages of parasitism: (7%, 2%), (0%, 0%), (18%, 5%), and (24%, 3%). Because *M. grandii* is specific to the European corn borer and because

N. pyrausta is so common and widespread, a spatial refuge away from the disease may not be possible. The foraging behavior of this parasitoid (Onstad *et al.*, 1991) and its interaction with *N. pyrausta* may cause it to be relatively scarce and insignificant as a biological control agent in the USA. We used our model to help understand why *M. grandii* is rare in communities of *O. nubilalis* found in the USA (Mason *et al.*, 1994).

We studied the relationship between the parasitoid and the disease using the following model. The model not only explicitly accounts for spatial heterogeneity and movement but also explicitly accounts for the timing of infection based on the age of the host. Onstad (1988) published the model for European corn borer population dynamics without natural enemies. Later, the basic model was expanded to include the spatial and temporal dynamics of the European corn borer under the influence of the microsporidium *N. pyrausta* (Onstad & Maddox, 1989; Onstad *et al.*, 1990).

All state variables are continuous variables except for the female parasitoids in the oviposition period. These parasitoids are modeled as individuals or integers. This record keeping is necessary to separate the healthy, infected, and vectoring parasitoids as they forage in the maize crop and to preserve heterogeneity.

The expanded model has the following general form for the temporal dynamics on each plant. The state variables are the densities of healthy herbivores, N, infected herbivores, I, parasitized herbivores, P, parasitized infected herbivores, PI. The parasitoid population is represented by C and CI for the healthy and infected immature stages after host death, respectively and by A, AV, and AI for the female parasitoids that are non-infected, temporary mechanical vectors, and infected, respectively. S is the amount of spore contamination in each plant. The subscripts i and j represent the age classes of the herbivore and parasitoid, respectively.

$$\frac{dN_{i=1}}{dt} = a(N_{i=\text{adult}}; I_{i=\text{adult}}) - f(N_{i=1}) - b(N_{i=1})$$
(I19.1)

$$\frac{dI_{i=1}}{dt} = a(I_{i=\text{adult}}) - f(I_{i=1}) - b(I_{i=1})$$
(I19.2)

$$\frac{dN_i}{dt} = f(N_{i-1}) - b(N_i) - f(N_i) - h(N_i,S)k(N_i,A,AI,AV)$$
(I19.3)

$$\frac{dI_i}{dt} = f(I_{i-1}) + h(N_i,S) - b(I_i) - f(I_i) - k(I_i,A,AI,AV)$$
(I19.4)

$$\frac{dP_i}{dt} = f(P_{i-1}) + k(N_i,A) - b(P_i) - f(P_i) - h(P_i,S)$$
(I19.5)

$$\frac{dPI_i}{dt} = f(P_{i-1}) + h(P_i,S) + k(I_i,N_i,A,AI,AV) - b(P_i) - f(PI_i)$$
(I19.6)

$$\frac{dC_j}{dt} = mf(P_{i=\text{last}}) - q(C_j) - w(C_j)$$
(I19.7)

$$\frac{dCI_j}{dt} = mf(PI_{i=\text{last}}) - q(CI_j) - w(CI_j)$$
(I19.8)

$$\frac{dA}{dt} = w(y,C_{j=\text{last}}) - q(A)$$
(I19.9)

$$\frac{dAV}{dt} = r(I,PI)A, \quad \text{with } AV = 0 \text{ at start of each day}$$
(I19.10)

$$\frac{dAI}{dt} = f(y,CI_{j=\text{last}}) - q(AI)$$
(I19.11)

$$\frac{dS}{dt} = g(I,PI), \quad \text{with } S = 0 \text{ at start of each year}$$
(I19.12)

The functions a, f, and b are the birth, development, and death rates for the herbivore, respectively. The infection rate and parasitism rate are functions h and k. Parameters m and y are the number of parasitoids emerging from each host and the proportion of females, respectively. Functions q and w are the rates of mortality and maturation, respectively for the parasitoid. The function r is the probability of a healthy parasitoid becoming a temporary mechanical vector of the pathogen. Function g is the contamination rate for the microsporidian spores. Each state variable may have a unique time-dependent function for one or more of these processes (see below).

The temporal dynamics on each plant are depicted by Figs 19.1 and 19.2. Figure 19.1 represents the dynamics simulated with the original model. The top row represents the healthy corn borers. Symbols E, P, M, and F are eggs, pupae, males, and females, respectively; roman numerals represent larval stadia. The

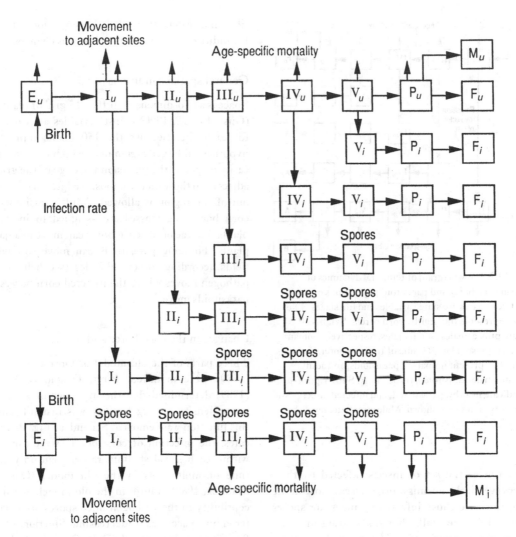

Figure 19.1. Flow diagram of temporal dynamics of European corn borer population on each plant. The squares are state variables and life stages. The top row represents the healthy corn borers. Symbols E, P, M, and F are eggs, pupae, males, and females, respectively; roman numerals represent larval stadia. Subscripts *u* and *i* indicate healthy and infected insects. The bottom row is the set of insects infected by their mothers, whereas those in the middle rows are those infected by ingesting spores distributed in the stalk. Only certain stages of infected larvae produce spores in their frass. These stages are denoted with the word 'spores' above the squares. Not all arrows indicating mortality are shown. The adult male stage can be omitted from the model.

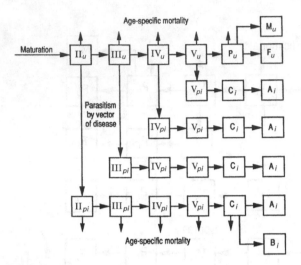

Figure 19.2. Flow diagram of temporal dynamics of European corn borer and parasitoid populations on each plant. The squares are state variables and life stages. The top row represents the healthy corn borers. Symbols P, M, and F are pupae, males, and females, respectively; roman numerals represent larval stadia of the corn borer. Symbols C, A and B represent immature parasitoids, and adult female and male parasitoids, respectively. Subscripts u, p and i indicate healthy, parasitized, and infected insects. Not all mortality arrows are shown. Male insects are not important.

bottom row is the set of insects infected by their mothers (vertical transmission), whereas those in the middle rows are those infected by ingesting spores distributed in the stalk (horizontal transmission). Only certain stages of infected larvae produce spores in their frass. These stages are denoted with the word 'spores' above the squares in Fig. 19.1. Similar parasitized stages also produce spores.

Figure 19.2 represents the cases in which a healthy larva can be attacked by a vector and simultaneously become parasitized and infected. This model structure is also used for parasitism of previously infected larvae and infection of previously parasitized larvae. This approach permits the model to keep track of the disease and parasitism status of cohorts according to the timing of infection or parasitism. Thus, the model has disease, parasitism, and age structure for the insect populations to permit more precise defini-

tion and calculation of behavior, mortality, and fecundity of cohorts after their status changes.

General spatial conditions

Space is a two-dimensional 10×15 grid of plant sites (Onstad *et al.*, 1990). Most variables are calculated for each site. Because the 150 plants represent a hypothetical larger region over which the moths can easily disperse, the plants on the edge of the grid are adjacent to those on the opposite edge. No movement out of the region is allowed. Adult parasitoids and corn borer egg masses are assigned to individual plants. Larvae of the corn borer can move to adjacent plants. Foraging parasitoids can move throughout fields according to several rules (see below). The pathogen moves where the infected corn borers and parasitoids move.

Changes in the corn borer and disease submodels

Three parts of the old model of Onstad & Maddox (1989) were modified. First, Onstad & Maddox (1989) determined that the influence of male corn borer density on egg fertility was insignificant, so this function was removed. Second, as Fig. 19.1 indicates, the infection of corn borer larvae can occur within each larval stadium rather than only at the time of molting as in the old model (Onstad & Maddox, 1989). Third, in the old model, larval susceptibility to the pathogen once spores are encountered in space was a decreasing function of age. Recent unpublished work (D. W. Onstad & L. Sotter) indicates that infectivity of the microsporidium is not age-dependent and that susceptibility is close to 100%.

June flight and foraging by overwintering generation

The distribution of female parasitoid foraging activity over time has the same shape as the distribution used for the corn borer oviposition in early summer (Onstad, 1988) but it is shifted 100 degree-days (base 10 °C) later to make the curve more synchronous with the larval stages.

Maturation

At the end of the fifth stadium, non-parasitized larvae pupate and parasitized larvae die. Parasitoid larvae can overwinter in the fifth instar larvae in diapause. The maturation of the parasitoid before host death is not modeled explicitly. The following three phases of parasitoid development are each modeled using a distributed delay (order seven) and a one-twentieth day time-step.

The internal phase of the parasitoid after host death in the model is an additional two times the length of the corn borer's fifth larval stadium. Data from Dittrick & Chiang (1982) demonstrate that the internal phase is not very sensitive to host instar attacked but it is influenced by temperature. We attempt to approximate this relationship between temperature and development of the parasitoid by using the temperature-dependent function of the fifth stadium for the corn borer (Onstad 1988; rate = $-0.097 + 0.00901 \times$ temperature). This approach seems to approximate the observations better for cooler than warmer temperatures. We chose not to segregate the immature parasitoids from the corn borer state variables during the internal phase before host death because of the need to continue calculating pathogen transmission by older parasitized larvae and possible future interests in modeling host larval behavior and damage to plants.

The emerging parasitoid larvae complete development outside the host as pupae in cocoon masses. Dittrick & Chiang (1982) and Parker (1931) collected data on external phase maturation that were used to calibrate the following temperature-dependent equation: rate of development $= -0.055 + 0.00535 \times T$, where the rate is a proportion per day between 0 and 1, and T is temperature in degrees Celsius at each time-step. This equation had an $r^2 = 0.98$ ($n = 8$) and both parameters were significantly different from zero ($t = -6.85; t = 15.3; p < 0.01$).

Development rate for the female adults before oviposition is simply 0.25/day based on a statement by Parker (1931). Parker (1931) also observed average lifespans of 15.4 days for males and 16.7 days for female parasitoids. We use 16 days for the oviposition period in the model.

Mortality

If the parasitoid is not killed by factors that kill the parasitized host, then the model calculates survival rates at each time-step from stage-specific rates for the three developmental phases described above. The survival rate for the internal phase parasitoid larvae (after host death) is 1.0 for mean daily temperatures above 20 °C and 0.33 for lower mean temperatures based on the data of Dittrick & Chiang (1982). For the external phase of immature parasitoids, data from Parker (1931) and Dittrick & Chiang (1982) indicate that survival per stage is 0.90 between mean daily temperatures of 18–25 °C and 0.50 for temperatures beyond these limits. Field observations by Parker (1931) support his laboratory data.

All of the adult females are assumed to survive unless they are infected. The greater the infection level, the lower the survival rate for the females in the pre-oviposition period. To approximate level of infection, we assumed that disease would be greatest in the parasitoids emerging from corn borers that had been infected longest. Thus, the model's structure relative to timing of initial infection (Figs 19.1 and 19.2) allows this type of information to be used. The survival rates for parasitoids emerging from corn borers infected as eggs to fifth instar larvae are 0.01, 0.05, 0.10, 0.50, 0.90, and 1.0, respectively, in the model. Andreadis (1980) reported that all infected females emerging from field-collected hosts died before the oviposition period. Siegel *et al.* (1986), using similar methods, observed 16% relative survival for parasitoids to adulthood after emergence from infected corn borers. Cossentine & Lewis (1987) recorded a 39% survival rate for eclosing parasitoids that had emerged from infected host larvae treated with *N. pyrausta* as second to third instars. Orr *et al.* (1994) observed 26% total survival rate for immature and adult parasitoids emerging from infected host larvae treated with *N. pyrausta* as first instars. Compared to the laboratory observations, the model's survival rates are lower for infected parasitoids; although stresses in the field are likely to be greater than in the laboratories. All mortality due to infection is included in this stage instead of incorporating it in all three phases separately.

Infection of parasitoids

In two field studies, Andreadis (1980) and Siegel *et al.* (1986) observed almost the same disease prevalence (percent infected) in the parasitoid and corn borer populations. In the model, all immature parasitoids emerging from infected corn borers are infected (Andreadis, 1980). In addition, all eggs oviposited by infected females are infected. Thus, vertical transmission occurs. Siegel *et al.* (1986) reported 100% vertical transmission, whereas Cossentine & Lewis (1987) observed no vertical transmission.

The disease harms the parasitoid population in several ways. As noted above, survival of infected parasitoids is lower after emergence from the host. Survival in infected hosts is also lower. Furthermore, we assume that rate of parasitism is lower for infected females (see below).

Attack rate

The following function calculates the proportion, Y, of each larval class, i, parasitized on a given plant during each time-step (one-twentieth day).

$$Y_i = 1 - e^{-z_i p_i A - 0.5 z_i p_i AI(1-\text{vert})} \qquad (I19.13)$$

where p is proportion of all larvae on the plant not already parasitized in class i; A and AI are the number of healthy and infected foraging parasitoids on the plant; and z is the larval-age-dependent attack rate per parasitoid (proportion attacked per parasitoid per single time-step). Most evidence suggests that second through fourth instars are preferred by the foraging parasitoids or are more easily attacked (Wishart, 1946; Dittrick & Chiang, 1982). Therefore, $z = 0$ when $i = 1$ (first instar). For the second through fourth instars, $z = 0.10$. Dittrick & Chiang (1982) reported that parasitoids had no difficulty parasitizing fifth instars, but Wishart (1946) stated that they are not easily attacked. Therefore, $z = 0.04$ when $i = 5$. Because Siegel *et al.* (1986) observed a much shorter lifespan for infected parasitoid adults but did not measure fecundity directly, the rates in the second term of the exponent for diseased parasitoids are half of the normal rates (e.g., 0.05 and 0.02). The

parameter vert, is the proportion of parasitoid offspring that are infected vertically from the mother. The standard value of vert is 1.0. With $z = 0.1$ one healthy parasitoid produces a maximum Y of 0.1. With two healthy parasitoids, Y reaches 0.18. The equation for the proportion of larvae that are attacked and become infected is:

$$YYi = 1 - e^{-z_i p_i AV - 0.5 z_i p_i AI\text{vert}} \qquad (I19.14)$$

where AV is the number of vectoring healthy parasitoids on the plant.

Host density does not influence attack rate on a plant, but it can influence departure rate from the plant (see below). The attack rate is neither a function of temperature nor age of female. Note that all parasitized host larvae die in the fifth stadium either in the current year or after overwintering in diapause.

Polyembryony and sex ratio

Parker (1931) observed an average of 21 parasitoid larvae emerging from each host larva. Wishart (1946) and Dittrick & Chiang (1982) observed several different emergence numbers depending on the environmental conditions, but the average seems to be independent of the number of eggs oviposited in each host. The value used in the model is 20 parasitoid larvae per host larva, an approximate average of these studies. Siegel *et al.* (1986) reported a small decrease in the number emerging from infected corn borers.

In the field, Wishart (1946) observed an overall sex ratio of 54% males in cocoon masses with only 8% of masses yielding both sexes. Because the laboratory colonies of Wishart (1946) and Parker (1931) had higher percentages of males than that observed by Wishart in the field, we used 60% males and 40% females emerging from corn borer larvae in the model. Orr *et al.* (1994) observed no significant differences in sex ratios for parasitoids emerging from infected and non-infected corn borers.

Parasitoid foraging over space

Ovipositing parasitoids are dispersed in four different ways in the model. After emergence and mating, female parasitoids are randomly distributed across the region. A uniform random number generator selects a plant (and field) to receive each individual parasitoid at the beginning of the foraging period. Multiple occupancy is not allowed unless the number of parasitoids in the region is high (see below). The fractional remainder of the female parasitoid density generated each time-step is held over to the next time-step until one extra parasitoid is obtained. The female parasitoid submodel is a model of individuals, not densities.

The probability of leaving a plant at each time-step equals e^{-lX} where X is the host larval density on a plant and l is a non-negative multiplier. When $l = 1.0$, the probability ranges from 1.0 for zero larvae to 0.36 and 0.02 for one and four larvae, respectively. If this probability is greater than a proportion generated from a uniform distribution, then the parasitoid leaves. For the parasitoids that leave their plants, three options can be used to move them from plant to plant at each time-step, but no forager can leave a field of 150 plants after beginning the oviposition period. If the number of ovipositing parasitoids in the region is lower than 25% of the number of plants, then the model uses one of two options. The parasitoid can simply be moved to one of the four adjacent plants selected at random. In this standard case multiple occupancy is allowed. Or it can be assumed that a current occupant will prevent occupation by others; if the selected adjacent plant is occupied, then the next plant in that same row or column (same direction of movement) is selected to receive the parasitoid.

If more than 37 female parasitoids (150/4) are active at the current time-step, then dispersal is completely random and all parasitoid females leave their current plants. A uniform random number generator selects a plant to receive one of the parasitoids. Multiple parasitoids may occupy a plant. This procedure was chosen to reduce the extensive computations required to move many individual parasitoids, especially when multiple occupations are not allowed.

Vectoring of pathogen

Spores can be dispersed by passive vectoring when spores from an infected host larva are carried by the ovipositor of a foraging parasitoid. Mechanical vectoring of the pathogen by the parasitoid complicates the model. We cannot use the average proportion of healthy parasitoids in the field that are mechanical vectors in the calculation of the functional response equation for each plant because this would violate our concern for spatial heterogeneity. Every parasitoid would be infectious. The solution is to calculate on each plant, the probability that a given parasitoid becomes a mechanical vector. This probability equals the probability of attacking a corn borer larva on a plant multiplied by the probability that the attacked larva is carrying spores. Then the resulting probability is compared to a randomly generated value to determine whether the probability is high enough to make the parasitoid a vector. The vectoring ability lasts 1 day with 100% effectiveness on each subsequent attack (D. W. Onstad, D. B. Orr, J. J. Obrycki & W. J. Lewis, unpublished results).

Superparasitism

In this model, parasitized larvae can be attacked again by female parasitoids, but we assume that those oviposited in the host first will out compete and kill younger immature parasitoids before emergence. The assumption of superparasitism influences two parts of the model. First, parasitized larvae can be attacked by infected parasitoids or by healthy vectors causing infection in the host and its parasitoids. Second, the density of all larval hosts is accounted for in the function that determines the probability of a healthy parasitoid becoming a vector. Note also that the total host density on a plant is used to determine the probability of a parasitoid's departure from the plant.

Computations

The stochastic model was simulated with average daily temperatures for central Illinois with the same pattern used each year. The model was programmed in FORTRAN and computed on the Cray Y/MP at the National Center for Supercomputing Applications in Illinois. A uniform random number generator that was part of the Cray FORTRAN 77 compiler was used for a variety of the calculations concerning spatial dynamics and vectoring. Euler integration was used to update state variables at each one-twentieth day time-step.

Results of simulations without natural enemies

As Onstad (1988) and Onstad & Maddox (1989) determined, the European corn borer population increases to an asymptotic density when no natural enemies exist. The asymptote is 2400 fifth instar larvae in diapause in the fall of each year (16 larvae/plant) or over 2100 egg masses per 150 plant field at the start of each season. All references to corn borer larvae and egg masses refer to these two points in time.

Results without parasitism

When the pathogen and herbivore persist over 25 years, damped fluctuations in densities result (Onstad & Maddox, 1989). The host-density threshold for pathogen persistence over 25 years in a similar heterogeneous region was below 0.18 susceptible egg masses per plant at the start of the 25 years for an initial prevalence at or below 40% (Onstad *et al.*, 1990). Thus, for an average density of susceptible egg masses greater than 0.18 per plant during the first spring, the pathogen did not become extinct. For a homogeneous version, the situation was more complicated. With initial prevalence less than 10%, the host-density threshold was between 0.3 and 0.5 egg masses per plant, but the threshold increased above 3.0 egg masses per plant with initial prevalence equal to 40% (Onstad *et al.*, 1990). Contrary to the conclusion of Mollison (1986), heterogeneous mixing of the population did not increase the threshold for persistence.

When the pathogen enters the system with initial conditions of 150 healthy and 2 infected egg masses, the long term densities of healthy and infected larvae fluctuate between 25–45 and 10–25, respectively, in a 150 plant field (Fig. 19.3). The corresponding egg mass densities are 30–40 healthy masses and 1–2 infected masses.

Results without disease

If the pathogen does not exist but the hypothetical larval parasitoid enters the system with initial conditions of 150 healthy corn borer egg masses and 6 female parasitoids, three complete cycles of the two insect species occur during the 25 year simulations (Fig. 19.4). The larval densities fluctuate between 75 and 250 per field for the healthy corn borers and between 5 and 120 for the parasitized borers. The parasitoid, as currently modeled, causes much greater fluctuations than the pathogen. In four out of five replications, cycling of the two populations occurred, but in a fifth replication the parasitoid went locally extinct after two cycles. For fall larvae in diapause, levels of percent parasitism vary from 3% to over 40%.

With the same initial conditions except for 40 (not 150) non-infected egg masses, the parasitoid persisted for 25 years in all five replications. The number of healthy parasitoids emerging to forage in early summer fluctuated between 3 and 13 per field of 150 plants over the 25 years of each replication.

Two more scenarios support a more general study by Hochberg & Lawton (1990). First, the parasitoid may sometimes be endangering itself by parasitizing too many corn borers. We raised the attack rate of the parasitoid for second–fourth instars to $z = 0.2$ for healthy parasitoids. All five replications resulted in extirpation of the parasitoid after only one cycle. The second scenario involved changing the parasitoid foraging rules to disallow multiple occupancy by foragers on a single plant. Extirpation of the parasitoid always occurred, but it happened either after one cycle (once), two cycles (twice), or three cycles (twice). These results may indicate that when corn borer densities drop very low, multiple occupancy is needed to permit parasitoid reproduction to be as high as possible on plants with hosts.

Results with parasitism and disease

When the pathogen and the parasitoid enter the system together in the standard version of the model, the hypothetical parasitoid always goes extinct and

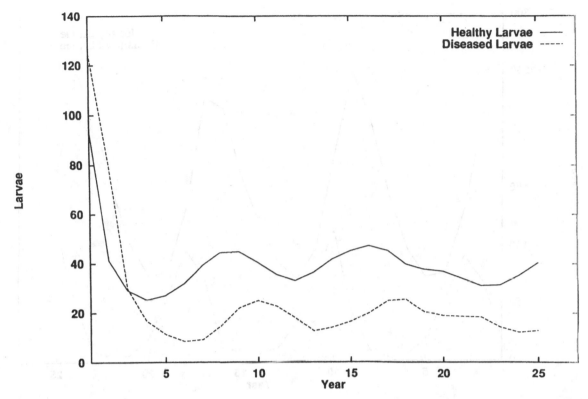

Figure 19.3. Number of healthy (——) and infected (- - -) corn borer larvae in diapause at the end of the year in a region of 150 plants. No parasitism.

the pathogen persists with dynamics eventually similar to those described above for the scenario not involving the parasitoid. This result is the same whether the initial density of healthy corn borer egg masses is 150 or 40 per field of 150 plants; although, extinction time for the parasitoid with 150 masses was 4–5 years, whereas the extinction time for three simulations with 40 initial masses ranged from 10–13 years (Table 19.1).

We studied two aspects of foraging behavior. We raised the attack rate, z, to 0.2 and 0.1 for healthy and infected parasitoids, respectively, attacking second to fourth instar hosts. With initial conditions of 40 healthy egg masses, 2 infected egg masses and 6 healthy parasitoids per 150 plant field, the parasitoid was extirpated in the second year and the pathogen still persisted for 25 years. Then we changed parameter l from 1.0 to 0.01 in the function calculating the

probability of a parasitoid leaving a plant, which decreased the probability of staying on a plant from almost 1.0 to about 0.1 with 10 larvae per plant. Again, the extinction time was 2 years for the parasitoid and the pathogen persisted.

Does the disease in the parasitoid population cause the parasitoid to go extinct or does the disease in the corn borer population indirectly harm the parasitoid? To answer this question we first eliminated infection in the modeled parasitoid. The initial conditions were always 40 healthy egg masses, 2 infected egg masses, and 6 healthy parasitoids. In four out of five replications the parasitoid coexisted for 25 years with the other two species (Table 19.1). In the other replication, the parasitoid was extirpated in the second year. Thus, the model suggests that the disease directly decreases persistence of the parasitoid.

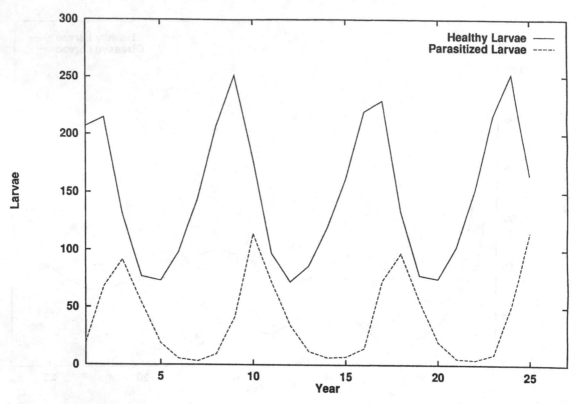

Figure 19.4. Number of healthy (—) and parasitized (- - -) corn borer larvae in diapause at the end of the year in a region of 150 plants. No disease.

Role of vertical transmission in parasitoid

Siegel *et al.* (1986) reported 100% vertical transmission, but Cossentine & Lewis (1987) observed no vertical transmission. We investigated this process by changing the value of vert in eqns (I13)–(I14) from 0.0 to 0.5 and 0.1. In all cases the initial conditions are 40 healthy egg masses, 2 diseased egg masses and 6 healthy parasitoids. With 50% vertical transmission (a compromise between the two studies) the parasitoid coexisted with the pathogen for 25 years in one of the five replications (Table 19.1). The other replications ended in extirpation after 6, 7, 9, and 15 years. With 10% vertical transmission, the parasitoid persisted for 25 years in three replications and went extinct after 7 and 12 years in the other two simulations (Table 19.1). The lower the proportion of females transmitting the disease to

their offspring or the lower the probability of all infected females passing on the disease to any of their eggs, the greater the likelihood of the modeled parasitoid persisting in the European corn borer community.

With 50% vertical transmission we lowered infectivity of the pathogen for the European corn borer and raised the fecundity of the herbivore in two separate simulations. When infectivity, v, was lowered to 0.90 from 1.0, the parasitoid coexisted with the pathogen in two replications (Table 19.1). This is one more replication compared to the scenario in which infectivity is 1.0. In the other three replications, extirpation occurred after 13, 14, and 22 years. (With 100% vertical transmission, the parasitoid goes extinct in 4–6 years when infectivity is 0.90.) Raising maximum fecundity from 27.6 egg masses per female to 40 masses had little effect on the

Table 19.1. *Simulations of pathogen–parasitoid interactions. Pathogen always persists*

Infection?	Vertical transmission	Extinction time	Persists 25 years
Yes	100%	3–13 years	No
No	—	Year 2, 1 rep	In 4 reps
Yes	10%	Years 7–12, 2 reps	In 3 reps
Yes	50%	Years 6–15, 4 reps	In 1 rep
Yes	50%	Years 13–22, 3 reps	In 2 reps[†]

Notes:
[†] Also 10% lower susceptibility of corn borer to disease.
rep(s), replicate(s).

results. The parasitoid persisted for 25 years in only one replication and was extirpated after 9 (twice), 10 and 18 years in the other replications.

Summary of results

Several hypotheses concerning *Macrocentrus grandii* persistence and coexistence can be drawn from our analysis. First, in the absence of *N. pyrausta*, *M. grandii* will persist for at least 50 generations in a region of corn over which a typical corn borer moth can disperse. Because the pathogen is so ubiquitous, this scenario is purely hypothetical. Second, infection in *M. grandii* increases the likelihood of regional extirpation of the parasitoid during a 25 year period in corn monocultures. In the unlikely event that a resistant strain of this parasitoid could be found or developed, parasitoid persistence might become more likely. Third, the lower the proportion of parasitoid females infecting their offspring (vertical transmission), the greater will be the probability of persistence by *M. grandii* and coexistence for at least 50 generations in a region of corn over which a typical corn borer moth can disperse. The vertical transmission observed in three empirical studies may have been different because of variability in microsporidian strain or because of temperature differences between Iowa, Illinois and Connecticut. Again, learning more about population and strain differences may be important for utilizing parasitoids to control the European corn borer. Nevertheless, the microsporidium and another

pathogen, *Beauveria bassiana* (Feng *et al.*, 1988) will likely continue to be the most important regulators of the European corn borer in corn monocultures in the USA.

Guidelines for using both parasitoids and pathogens to control insects

Interference between parasitoids and pathogens is not uncommon. Powell *et al.* (1986) observed interference between fungi and parasitoids attacking cereal aphids. This case and the more general studies of Begon *et al.* (Chapter 18) and Hochberg *et al.* (1990) demonstrate the importance of interference even when the parasitoid is not infected by the pathogen. Four additional points should be remembered by biocontrol specialists beyond the concern about natural enemy competition. First, determine whether a pest species is commonly infected with a disease that can affect a candidate parasitoid before spending too much time and money introducing the parasitoid. Second, check parasitoids in quarantine facilities for infection by chronic diseases caused by microsporidia and other protozoa. Third, a disease may not be as harmful to some parasitoid strains as others. Fourth, an effective microbial control agent may lower pest density below the level needed by a parasitoid species to persist and be effective. The success of microbial control is not threatened by this last situation but time and money may be wasted on parasitoids if they cannot coexist with the pathogens.

Concluding remarks

Onstad (1992b) emphasized the point that models are not hypotheses just as field experiments are not hypotheses. Modeling allows hypotheses to be created or tested. For example, from the analysis of his models, Van der Plank (1963, 1975) postulated that total potential reproduction per pathogen, iR, must be greater than 1.0 for an epidemic to occur. Onstad (1992a) tested this hypothesis under modeled conditions of spatial heterogeneity and limiting host densities. He found that the threshold value of iR, which determines whether the density of infected (latent and infectious) leaflets will increase over a pathogen's generation, increased as the initial density of susceptible hosts decreased. In other words, the threshold value of iR is not constant as Van der Plank suggested. Onstad (1992a) developed probabilistic formulas to predict the change in the number of infected hosts (not leaflets) over a pathogen generation. These formulas are based on the proportion of a region occupied by hosts, the initial proportion of hosts that are infected, and the number of sites within the pathogen propagule dispersal neighborhood.

Theorists should always define terms and express hypotheses so that biological control practitioners can implement the hypotheses and recommendations over appropriate spatial and temporal scales (Onstad, 1992b). Every hypothesis needs measurable but general temporal and spatial scales over which the phenomenon of interest can be studied and implemented. We have attempted to accomplish this with our own hypotheses described in this chapter. Model computation or analysis may require smaller units of time and space to ensure proper calculation of functions and stable results, but these computational units are not satisfactory for the conceptual units. Onstad (1992b) provides details for identifying appropriate scales. Two other perspectives on identifying spatial scales are presented by Fahrig (1992) and De Roos *et al.* (1991).

References

Alexander, H. M. (1989) Spatial heterogeneity and disease in natural populations. In *Spatial Components of Plant Disease Epidemics*, ed. M. J. Jeger, pp. 144–64. Englewood Cliffs: Prentice Hall.

Allen, J. C. (1975) Mathematical models of species interactions in time and space. *American Naturalist* 109: 319–42.

Andreadis, T. G. (1980) *Nosema pyrausta* infection in *Macrocentrus grandii*, a braconid parasite of the European corn borer, *Ostrinia nubilalis*. *Journal of Invertebrate Pathololology* 35: 229–33.

Briggs, C. J. & Godfray, C. J. (1995) Models of intermediate complexity in insect-pathogen interactions – population dynamics of the microsporidian pathogen, *Nosema pyrausta*, of the European corn borer, *Ostrinia nubilalis*. *Parasitology* 111 (Supplement): 871–89.

Bruckart, W. L. & Dowler, W. M. (1986) Evaluation of exotic rust fungi in the United States for classical biological control of weeds. *Weed Science* 34 (Supplement 1): 11–14.

Bruckart, W. L. & Hasan, S. (1991) Options with plant pathogens intended for classical control of range and pasture weeds. In *Microbial Control of Weeds*, ed. D. O. TeBeest, pp. 69–79. New York: Chapman and Hall.

Burdon, J. J. (1993) The structure of pathogen populations in natural plant communities. *Annual Review of Phytopathology* 31: 305–23.

Burdon, J. J. & Chilvers, G. A. (1982) Host density as a factor in plant disease ecology. *Annual Review of Phytopathology* 20: 143–66.

Burdon, J. J., Jarosz, A. M. & Kirby, G. C. (1989) Pattern and patchiness in plant-pathogen interactions – causes and consequences. *Annual Review of Ecolology and Systematics* 20: 119–36.

Burdon, J. J., Ericson, L. & Müller, W. J. (1995) Temporal and spatial changes in a metapopulation of the rust pathogen *Triphragmium ulmariae* and its host, *Filipendula ulmaria*. *Journal of Ecology* 83: 979–89.

Callaway, M. B., Phatak, S. C. & Wells, H. D. (1985) Studies on alternate hosts of the rust *Puccinia canaliculata*, a potential biological control agent for nutsedges. *Plant Disease* 69: 924–6.

Casas, J. (1990) Multidimensional host distribution and nonrandom parasitism: a case study and a stochastic model. *Ecology* 71: 1893–1903.

Cavalieri, L. F. & Koçak, H. (1995) Chaos – a potential problem in the biological control of insect pests. *Mathematical Bioscience* 127: 1–17.

Charudattan, R. (1991) The mycoherbicide approach with plant pathogens. In *Microbial Control of Weeds*, ed. D. O. TeBeest, pp. 24–57. New York: Chapman and Hall.

Chin, K. M. & Wolfe, M. S. (1984) The spread of *Erysiphe graminis* f. sp. *hordei* in mixtures of barley varieties. *Plant Pathology* 33: 89–100.

Cossentine, J. E. & Lewis, L. C. (1987) Development of *Macrocentrus grandii* Goidanich within microsporidian-infected *Ostrinia nubilalis* (Hübner) host larvae. *Canadian Journal of Zoology* 65: 2532–5.

De Roos, A. M., McCauley, E. & Wilson, W. G. (1991) Mobility versus density-limited predator–prey dynamics on different spatial scales. *Proceedings of the Royal Society of London, Series B* 246: 117–22.

Dittrick, L. E. & Chiang, H. C. (1982) Developmental characteristics of *Macrocentrus grandii* as influenced by temperature and instar of its host, the European corn borer. *Journal of Insect Physiology* 28: 47–52.

Dwyer, G. (1994) Density dependence and spatial structure in the dynamics of insect pathogens. *American Naturalist* 143: 533–62.

Fahrig, L. (1992) Relative importance of spatial and temporal scales in a patchy environment. *Theoretical Population Biology* 41: 300–14.

Feng, Z., Carruthers, R. I., Larkin, T. S. & Roberts, D. W. (1988) A phenology model and field evaluation of *Beauveria bassiana* (Bals.) Vuillemin (Deuteromycotina: Hyphomycetes) mycosis of the European corn borer, *Ostrinia nubilalis* (Hbn.) (Lepidoptera: Pyralidae). *Canadian Entomologist* 120: 133–44.

Fuxa, J. R. & Tanada, Y. (1987) *Epizootiology of Insect Diseases*. New York: John Wiley & Sons.

Harper, J. L. (1990) Pests, pathogens and plant communities: an introduction. In *Pests, Pathogens and Plant Communities*, ed. J. J. Burdon & S. R. Leather, pp. 3–14. Oxford: Blackwell Scientific.

Hess, G. (1996) Disease in metapopulation models: implications for conservation. *Ecology* 77: 1617–32.

Hochberg, M. E. & Lawton, J. H. (1990) Spatial heterogeneities in parasitism and population dynamics. *Oikos* 59: 9–14.

Hochberg, M. E., Hassell, M. P. & May, R. M. (1990) The dynamics of host–parasitoid–pathogen interactions. *American Naturalist* 135: 74–94.

Ignoffo, C. (1988) *Entomogenous Protozoa and Fungi*. CRC Handbook of Natural Pesticides, Part A of Volume 5, *Microbial Insecticides*. Boca Raton, FL: CRC Press.

Jarosz, A. M. & Levy, M. (1988) Effects of habitat and population structure on powdery mildew epidemics in experimental *Phlox* populations. *Phytopathology* 78: 358–62.

Jeger, M. J. & van den Bosch, F. (1994) Threshold criteria for model plant disease epidemics. II. Persistence and endemicity. *Phytopathology* 84: 28–30.

Jennersten, O., Nilsson, S. G. & Wastljung, U. (1983) Local plant populations as ecological islands: the infection of *Viscaria vulgaris* by the fungus *Ustilago violacea*. *Oikos* 41: 391–5.

Kampmeijer, P. & Zadoks, J. C. (1977) *EPIMUL, a Simulator of Foci and Epidemics in Mixtures of Resistant and Susceptible Plants, Mosaics and Multilines*. Wageningen: Pudoc.

Kiyosawa, S. (1976) A comparison by simulation of disease dispersal in pure and mixed stands of susceptible and resistant plants. *Japanese Journal of Breeding* 26: 137–45.

Mason, C. E., Romig, R. F., Wendell, L. E. & Wood, L. A. (1994) Distribution and abundance of larval parasitoids of the European corn borer (Lepidoptera: Pyralidae) in the east central United States. *Environmental Entomology* 23: 521–31.

May, R. M. (1990) Population biology and population genetics of plant–pathogen associations. In *Pests, Pathogens and Plant Communities*, ed. J. J. Burdon & S. R. Leather, pp. 309–25. Oxford: Blackwell Scientific.

Minogue, K. P. (1989) Diffusion and spatial probability models for disease spread. In *Spatial Components of Plant Disease Epidemics*, ed. M. J. Jeger, pp. 127–43. Englewood Cliffs: Prentice Hall.

Mollison, D. (1986) Modelling biological invasions: chance, explanation, prediction. *Philosophical Transactions of the Royal Society of London, Series B* 314: 675–93.

Mundt, C. C. & Leonard, K. J. (1986) Analysis of factors affecting disease increase and spread in mixtures of immune and susceptible plants in computer-simulated epidemics. *Phytopathology* 76: 832–40.

Mundt, C. C., Leonard, K. J., Thal, W. M. & Fulton, J. H. (1986) Computerized simulation of crown rust epidemics in mixtures of immune and susceptible oat plants with different genotype unit areas and spatial distributions of initial disease. *Phytopathology* 76: 590–8.

Nachman, G. (1981) A simulation model of spatial heterogeneity and non-random search in an insect host–parasitoid system. *Journal of Animal Ecology* **50**: 27–47.

Onstad, D. W. (1988) Simulation model of the population dynamics of *Ostrinia nubilalis* (Lepidoptera: Pyralidae) in maize. *Environmental Entomology* **17**: 969–76.

Onstad, D. W. (1992a) Evaluation of epidemiological thresholds and asymptotes with variable plant densities. *Phytopathology* **82**: 1028–32.

Onstad, D. W. (1992b) Temporal and spatial scales in epidemiological concepts. *Journal of Theoretical Biology* **158**: 495–515.

Onstad, D. W. & Carruthers, R. I. (1990) Epizootiological models of insect diseases. *Annual Review of Entomology* **35**: 399–419.

Onstad, D. W. & Kornkven, E. A. (1992) Persistence and endemicity of pathogens in plant populations over time and space. *Phytopathology* **82**: 561–6.

Onstad, D. W. & Maddox, J. V. (1989) Modeling the effects of the microsporidium, *Nosema pyrausta*, on the population dynamics of the insect, *Ostrinia nubilalis*. *Journal of Invertebrate Pathology* **53**: 410–21.

Onstad, D. W. & Maddox, J. V. (1990) Simulation model of *Tribolium confusum* and its pathogen, *Nosema whitei*. *Ecological Modeling* **51**: 143–60.

Onstad, D. W., Maddox, J. V., Cox, D. J. & Kornkven, E. A. (1990) Spatial and temporal dynamics of animals and the host-density threshold in epizootiology. *Journal of Invertebrate Pathology* **55**: 76–84.

Onstad, D. W., Siegel, J. P. & Maddox, J. V. (1991) Distribution of parasitism by *Macrocentrus grandii* (Hymenoptera: Braconidae) in maize fields infested by *Ostrinia nubilalis* (Lepidoptera: Pyralidae). *Environmental Entomology* **20**: 156–9.

Orr, D. B., Lewis, L. C. & Obrycki, J. J. (1994) Behavior and survival in corn plants of *Ostrinia nubilalis* (Lepidoptera: Pyralidae) larvae when infected with *Nosema pyrausta* (Microspora: Nosematidae) and parasitized by *Macrocentrus grandii* (Hymenoptera: Braconidae). *Environmental Entomology* **23**: 1020–4.

Parker, H. L. (1931) *Macrocentrus gifuensis* Ashmead, a polyembryonic braconid parasite in the European corn borer. *USDA Technical Bulletin* **230**, 1–62.

Paul, N. D. & Ayres, P. G. (1986) The impact of a pathogen (*Puccinia lagenophorae*) on populations of groundsel (*Senecio vulgaris*) overwintering in the field. *Journal of Ecology* **74**: 1069–84.

Paul, N. D. & Ayres, P. G. (1987) Survival, growth and reproduction of groundsel (*Senecio vulgaris*) infected by rust (*Puccinia lagenophorae*) in the field during summer. *Journal of Ecology* **75**: 61–71.

Politis, D. J., Watson, A. K. & Bruckart, W. L. (1984) Susceptibility of musk thistle and related composites to *Puccinia carduorum*. *Phytopathology* **74**: 687–91.

Powell, W., Wilding, N., Brobyn, P. J. & Clark, S. J. (1986) Interference between parasitoids (Hym.: Aphidiidae) and fungi (Entomophthorales) attacking cereal aphids. *Entomophaga* **31**: 293–302.

Siegel, J. P. (1985) The Epizootiology of *Nosema pyrausta* (Paillot) in the European Corn Borer, *Ostrinia nubilalis*, in Central Illinois. PhD Thesis: University of Illinois, Urbana-Champaign.

Siegel, J. P., Maddox, J. V. & Ruesink, W. G. (1986) Impact of *Nosema pyrausta* on a braconid, *Macrocentrus grandii*, in central Illinois. *Journal of Invertebrate Pathology* **47**: 271–6.

Supkoff, D. M., Joley, D. B. & Marois, J. J. (1988) Effect of introduced biological control organisms on the density of *Chondrilla juncea* in California. *Journal of Applied Ecology* **25**: 1089–95.

TeBeest, D. O., Yang, X. B. & Cisar, C. R. (1992) The status of biological control of weeds with fungal pathogens. *Annual Review of Phytopathology* **30**: 637–57.

Thrall, P. H. & Jarosz, A. M. (1994) Host–pathogen dynamics in experimental populations of *Silene alba* and *Ustilago violacea*. II. Experimental tests of theoretical models. *Journal of Ecology* **82**: 561–70.

Trujillo, E. E., Aragaki, M. & Shoemaker, R. A. (1988) Infection, disease development and axenic culture of *Entyloma compositarum*, the cause of hamakua pamakani blight in Hawaii. *Plant Disease* **72**: 355–7.

van den Bosch, F., Verhaar, M. A., Buiel, A. A. M., Hoogkamer, W. & Zadoks, J. C. (1990) Focus expansion in plant disease. IV. Expansion rates in mixtures of resistant and susceptible hosts. *Phytopathology* **80**: 598–602.

Van der Plank, J. E. (1963) *Plant Diseases: Epidemics and Control*. New York: Academic Press.

Van der Plank, J. E. (1975) *Principles of Plant Infection*. New York: Academic Press.

Wishart, G. (1946) Laboratory rearing of *Macrocentrus gifuensis* Ashm., a parasite of the European corn borer. *Canadian Entomologist* **78**: 78–82.

Wood, S. N. & Thomas, M. B. (1996) Space, time, and persistence of virulent pathogens. *Proceedings of the Royal Society of London Series B* **263**: 673–80.

Yang, X. B. & TeBeest, D. O. (1992) The stability of host-pathogen interactions of plant disease in relation to biological weed control. *Biological Control* **2**: 266–71.

Yang, X. B. & TeBeest, D. O. (1993) Epidemiological mechanisms of mycoherbicide effectiveness. *Phytopathology* **83**: 891–3.

Zadoks, J. C. (1971) Systems analysis and the dynamics of epidemics. *Phytopathology* **61**: 600–10.

Zadoks, J. C. & Schein, R. D. (1979) *Epidemiology and Plant Disease Management*. New York: Oxford University Press.

Zawolek, M. W. & Zadoks, J. C. (1992) Studies in focus development: an optimum for the dual dispersal of plant pathogens. *Phytopathology* **82**: 1288–97.

20

Application of insect–pathogen models to biological control

MATTHEW B. THOMAS, SIMON N. WOOD AND VERONICA SOLORZANO

Introduction

The development of insect pathogens as control agents requires a comprehensive understanding of pest–pathogen dynamics if their full potential is to be realized. Aspects of the ecology of both pest and pathogen, however, can be difficult to measure. As a consequence, identification of the key factors that influence the dynamics of the pest–pathogen interaction can be difficult. In this context, mathematical population models can provide useful tools to focus limited research capability on the collection of the most relevant information. Furthermore, adequately validated population models permit the investigation of control options without the need for full field implementation. As such they can prove useful in the development and evaluation of optimum control strategies within an Integrated Pest Management (IPM) framework.

In this chapter we describe the development and application of population dynamic models to investigate the potential of the fungal entomopathogen *Metarhizium flavoviride* Gams and Rozsypal (Deuteromycotina: Hyphomycetes) for the biological control of locusts and grasshoppers. The factors considered in model development are discussed and results of some parameterization experiments are presented. We show how the models are used to assess population fluctuations and reductions arising from spray applications of a biopesticide based on the pathogen and demonstrate how ecological approaches can provide a link between laboratory and field studies, thereby aiding the interpretation of field trial results. This latter point is important since direct experimental quantification is frequently hampered by factors such as slow mortality and multiple infection routes (see Lomer *et al.*, 1993;

Thomas *et al.*, 1996a, 1997). We then describe how models are used to investigate the potential for exploiting the biological properties of the pathogen and what effect this has on the impact of a single pathogen application. We conclude by placing our findings in the general context of future development and use of biopesticides.

Background to the system

Locusts and grasshoppers enjoy a notoriety shared by few other agricultural pests or diseases. During this century alone, eight major plagues of the desert locust, *Schistocerca gregaria* (Forskål), varying in length between 1 and 22 years, have threatened agricultural production across an invasion area of more than 20% of the world's land surface (Steedman, 1990). A suite of other locust and grasshopper species and species assemblages cause much more regular and, through their cumulative effects, significant damage throughout Africa, the middle East, Australia, parts of Asia and North and South America (Office of Technical Assessment, 1990; Steedman, 1990; Lomer & Prior, 1992; Jenkins, 1994).

In recent years, the challenge of controlling these ubiquitous pests has largely been confronted with synthetic chemical insecticides. Fenitrothion, a short-persistent organophosphate with a half-life of approximately 24 h (Sekizawa *et al.*, 1992), is one of the most widely used chemical insecticides for locust and grasshopper control (Steedman, 1990; Joffe, 1995). However, compared with persistent organochlorides such as dieldrin, the effectiveness of such non-persistent chemicals is limited, often requiring repeated applications within a season or large-scale blanket sprays to achieve more than temporary relief (Brader, 1988). Furthermore, the extensive use of even these non-persistent chemicals has led to substantial concerns over health and environmental impacts of this conventional approach (to humans, animals, birds, invertebrates, and soil microorganisms), particularly in fragile ecosystems such as the Sahel (Office of Technical Assessment, 1990; Joffe, 1995).

In response to these problems, there is now considerable interest in biological control, particularly microbial control, as an environmentally benign alternative to chemical spraying (for recent reviews, see Prior & Greathead, 1989; Streett & McGuire, 1990; Bidochka & Khachatourians, 1991; Lomer & Prior, 1992; Prior & Streett, 1997). To this end, a number of research programs around the world are developing biological pesticides, based on entomopathogenic hyphomycete fungi, for locust and grasshopper control (e.g., Cunningham, 1992; Johnson *et al.*, 1992; Lobo-Lima *et al.*, 1992; Prior *et al.*, 1992; Baker *et al.*, 1994; Zimmermann *et al.*, 1994). One of these programs is LUBILOSA (LUtte BIologique contre les LOcustes et SAuteriaux), which has been developing an oil-based biopesticide containing the entomopathogenic fungus, *M. flavoviride*. This pathogen acts through direct contact and a number of successful laboratory, semi-field and field tests have been conducted against a range of hosts under different ecological conditions in Africa (e.g., Bateman *et al.*, 1993; Lomer *et al.*, 1993; Douro-Kpindou *et al.*, 1995; Kooyman *et al.*, 1997). In this chapter we describe how host–pathogen models are being developed and applied in this program to help understand the consequences of spray applications and improve control.

Patterns of infection following pathogen application

In thinking about the dynamics of a biopesticide application we can identify three distinct routes of infection:

1. Direct contact with the initial spray application. The percentage infection or mortality resulting from this route is governed largely by application

techniques and physical environmental factors such as temperature, wind speed, and vegetation structure.

2. Infection *via* the spray residue. Once again application, formulation, and environmental factors, through their effects on the persistence and spatial distribution of the pathogen propagules, are important. However, in addition to these, biotic factors such as pathogenicity and natural survival of the pathogen are also important.

3. Horizontal transmission of the pathogen from individuals infected directly or by residue or by individuals themselves infected horizontally. The dynamics of this phase are governed by the factors that regulate natural host–pathogen interactions since infections result from natural pathogen delivery mechanisms.

Interestingly, although these different routes of infection might be expected, as will be discussed later, we find that this distinction is rarely considered in the use and evaluation of microbial pesticides. The drive to mimic conventional chemical pesticide performance and the use of standard chemical pesticide evaluation protocols and methodologies, means that the possibility and potential value of horizontal transmission in particular, is often overlooked. Our suggestion here is that if the consequences of an artificial application of pathogen on within- and between-season host–pathogen dynamics are to be investigated, the relative contribution of each of these separate routes of infection needs to be examined, taking into account both the biotic and abiotic aspects of the system. Since it is rather complicated (perhaps impossible) to do this through experimentation alone, examination of various aspects of the host–pathogen interaction is aided greatly by the use of population dynamic models.

In the following section we begin with the initial spray application, highlighting the effects of the spray residue. To illustrate our approach, we present as a case study a field trial conducted on *Hieroglyphus daganensis* Krauss, one of the most abundant grasshopper species in field sites in north Benin and southern Niger and a pest of rice and other crops throughout west Africa (Steedman, 1990). Full details of this study are presented in Thomas *et al.* (1997).

Evaluating the effects of residual pick-up of spores following spraying

Study system

Like most Sahelian grasshoppers, *H. daganensis* is univoltine. Nymphs emerge during the wet season (June–September) from eggs laid the previous year and pass through 5 or 6 instars to adulthood in 4–6 weeks. Reproduction and oviposition take place over 4–8 weeks and then the adults die. Eggs remain in diapause in the soil throughout the dry season (8 or 9 months) until the onset of the rains (Steedman, 1990). During this period the eggs are not susceptible to the pathogen. Therefore, any interaction between grasshopper and pathogen is restricted to the wet season.

The mycopesticide was applied to three 8 ha plots of natural grassland savanna vegetation comprising large flat areas of dense grass interspersed with ponds fringed with reeds, cultivated areas and occasional shrubs, at field sites near Malanville in north Benin. Areas such as these, adjacent to farmers' fields, usually support the majority of the grasshopper populations early in the season (Amatobi *et al.*, 1988). Because of difficulties in finding relatively uniform, accessible areas that were suitable for spraying, the replicated plots were positioned at three different villages along the Niger river in the Malanville area: Bodjecali, Madecali, and Birni Lafia. These sites were very similar with respect to vegetation and grasshopper densities. Each site contained an 8 ha control plot, separated from the treated plot by at least 200 m, which received no spray treatment.

Measurement of spray residue

Spores of *M. flavoviride* (isolate IMI 330189) were applied using hand-held Micro-Ulva sprayers in a 70:30 mixture of kerosene: peanut oil at 2 *l*/ha giving 5×10^{12} spores/ha. The persistence and infectivity of the spray residue was monitored using a field bio-

assay technique. Five field cages ($0.5 \, m^2 \times 0.4 \, m$ high with metal 1 mm mosquito mesh sides, a removable top and an open bottom) were placed within the central areas of each of the sprayed plots 2 h after treatment (i.e., while the spray residue was still very fresh but after any small droplets should have settled). Any naturally occurring grasshoppers and any predators were removed from the cages by hand. Twenty *H. daganensis* nymphs (predominantly L4, although with some L3 and L5) collected from the unsprayed control plots at each site were then introduced into each cage. After 72 h the grasshoppers were removed from the cages and were incubated in the laboratory for 21 days. During this period any mortality and any *Metarhizium* infections were recorded; only insects that showed positive mycosis were used to estimate levels of infection. All dead insects were removed from the incubation containers daily so that the only source of inoculum was from the initial spray application. It was intended to repeat this procedure seven times at each site so as to monitor the spray residue for a period of up to 21 days. However, heavy rains caused extensive flooding in the area and it was only possible to collect a full data set (i.e., introductions of untreated insects into cages at 0, 3, 6, 9, 12, 15, 18, and 21 days after treatment) at Madecali, which was the last site to flood. Data were collected until day 15 at Bodjecali and until day 12 at Birni Lafia. On each introduction date, the cages were moved to a previously undisturbed position to minimize the effects of the cages on treatment decay. Sweep net samples of 50 grasshoppers were also collected from sprayed and unsprayed areas on each date and were incubated in the laboratory to assess percentage infection of the resident field populations.

A population dynamic model for the impact of spray residue

The within-season component of this model begins after egg hatch, with a population of susceptible grasshopper hosts of density H/m^2. A spray event results in a proportion of these becoming infected. Spraying is assumed to take place on day 30 of the season, when the majority of hosts are at the 2–4 instar stage: this is the stage at which chemical pesticides would normally be applied. The proportion ultimately infected depends on spray efficiency and residual infection, and is described by a composite function made up of the direct contact proportion, S, and the probability of infection from the spray residue over the remaining 90 days of the season, r.

For the interval between seasons, we assume that the host population grows at a finite rate of increase, F (where those individuals available to breed are the fraction of susceptible hosts that survived the spray application). Density-independent mortality of the grasshoppers is built into the system *via* this finite rate of increase. Density-dependent mortality is ignored: a conservative assumption that will increase the difficulty of achieving control in the model system. At this stage, it is further assumed that the pathogen is lost from the system at the end of the season and that no horizontal or vertical infections occur. Thus, let $H_i(t)$ be the population of healthy grasshoppers in season i, t days after spraying and the number that escape direct spray contact, $H_i(0) = (1-S) \times$ (healthy population in season i before spraying).

We assume that the instantaneous risk of infection per healthy host per day as a result of contact with the spray residue is described by the negative exponential $r = P^{-at}$, where P is a measure of the initial infectivity of the residue and a is a measure of decay rate (see below and see Jenkins & Thomas, 1996). So, the rate of change of healthy hosts due to infections from the spray residue alone is given by:

$$\frac{dH}{dt} = -P^{-at}H \tag{20.1}$$

By solving this equation (dividing both sides by H and integrating with respect to time) we obtain the equation for the healthy hosts surviving at the end of the season:

$$H_i(d) = H_i(0)e\left[\frac{P}{a}(e^{-ad}-1)\right] \tag{20.2}$$

where d is the duration of the season after spraying.

From eqn (20.2), the between-season change in host population after reproduction is therefore given by:

$$H_{i+1}(0) = H_i(0)\left[\frac{P}{a}(e^{-ad}-1)\right]F(1-S) \tag{20.3}$$

Infectivity of spray residues

The results of the experiment to measure infectivity of the spray residue are presented in Fig. 20.1. This figure reveals that the levels and patterns of infection over the first 12 days were similar at each site. Initial values for the mean proportion of grasshoppers infected were 0.48 at Madecali and Bodjecali and 0.39 at Birni Lafia. By day 12, the proportions infected had fallen to 0.20, 0.19, and 0.08, respectively. For Madecali and Bodjecali, where samples were taken beyond day 12, the levels of infection were then observed to increase. Since, under field conditions, it takes about 12–15 days for an infected grasshopper to die and start producing spores (Lomer *et al.*, 1997), it is likely that these increases in infectivity were due to horizontal transmission from grasshoppers infected by the initial spray application. Therefore, to quantify the effects of the spray residue alone, infection data from the first 12 days only were used (i.e., the same observation period for each site), and a single estimate of r for the population dynamic model was obtained by fitting an exponential regression to the mean infection data for the three replicate sites. A single curve was fitted to all sites, since we were unable to find a significant site effect over the first 12 days (ANOVA $F_{2,8} = 3.78$; $p = 0.08$). By correcting these data to give risk of infection per day, the expression for the instantaneous risk of infection was calculated to be $r = 0.156^{-0.102t}$; $R^2 = 0.85$; half-life of residue $= 6.8$ days. Very low levels of natural infection were observed after 21 days incubation for the sweep net insect samples collected from the control plots. Mean percentage infections for the three control sites were 2% on day 0, 0.7% on day 3, 1.3% on day 6, 0% on day 9 and 0.7% on day 12. Disease levels in the field populations in the sprayed plots were much higher (36%) for insects sampled immediately after spraying (which is effectively the level of infection from direct contact with the spray) rising to a peak of 62% on day 3. Data for all the days are presented in Fig. 20.2.

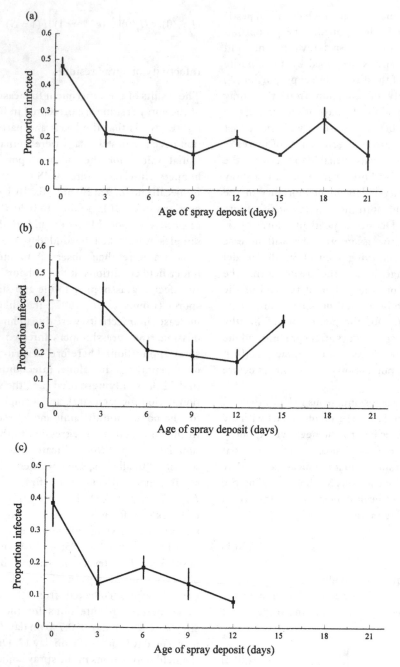

Figure 20.1. Mean infectivity of spray residues ($\pm$ SE) measured at three field sites in north Benin following application of oil formulations of *Metarhizium flavoviride*. Infectivity was measured by monitoring the proportion of grasshoppers infected after being placed in field cages for 3 day periods in the sprayed areas (see the text for further details). (a) Madecali, (b) Bodjecali, and (c) Birni Lafia.

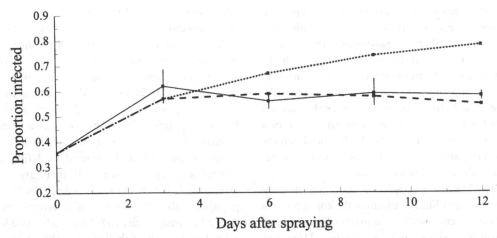

Figure 20.2. Changes in disease incidence following application of the mycopesticide for the observed data (mean for the three sites ± SE), the basic model output and the model output corrected for grasshopper mortality estimated from the cage incubation samples (see text for further details). —■—, observed; ···■···, model; – –■– –, model corrected for mortality.

Model simulations

In Fig. 20.2 we compare the levels of infection predicted by the model using the estimate of r, with the levels of infection observed in the grasshopper samples collected from the field after spraying. The model shows a reasonable fit with the field data, although it slightly underestimates the extent of residual infection over the first 3 days and then overestimates it subsequently. However, the two distributions are not significantly different when compared using a Kolmogorov–Smirnov test ($p = 0.329$), although this test has low power when there are few points in the distributions.

This overestimation in later observations could be due to the fact that since only live insects are sampled by sweep netting, cumulative disease incidence in the field is increasingly underestimated as insects treated by the spray application begin to die (either from the disease itself or through interactions with other mortality factors such as predators). Another factor could be immigration of uninfected insects into the sprayed plots and emigration of treated insects out of the plots. However, given that data were collected from the central areas of 8 ha plots and that the estimated mean movement rate of *H. daganensis* is

11 m/day (M. B. Thomas, unpublished results), dispersal is not expected to have had a significant effect during the first 12 days.

The overestimation due to the expected effects of mortality in the field is corrected by adjusting the model output using the cumulative mortality data from the incubated insects sampled after spraying on day 0, the assumption being that mortality in these treated populations held in the laboratory is representative of the mortality of treated populations in the field. These data give average mortalities for the three sites of 0% on days 0 and 3, 12% on day 6, 22% on day 9, and 30% on day 12. Multiplying the predicted levels of infection for each day by the proportion of the population expected to be alive on these days (e.g., 1×0.36 for day 3, 0.88×0.57 for day 6, etc) greatly improves the fit and the significance level of the prediction ($p = 0.82$), revealing that the model provides a good estimate of the effects of the spray residue on overall disease levels averaged across the three sites (Fig. 20.2). In contrast to other species such as *Zonocerus variegatus* (L.) (see Thomas *et al.*, 1996a; Langewald *et al.*, 1997), this also shows that mortality rates in the field correspond closely to mortality in the cage samples suggesting little net delay in disease incubation in the field. It should be noted that including mortality is not an adjustment of the model *per se* (i.e., the model does predict very well the total number of grasshoppers that become infected from direct and residual contact with the spray) but is simply an adjustment of the output to bring it in line with what is actually measurable in the field (i.e., if grasshoppers are dying then they are not available for sampling and total cumulative infection cannot be measured). It should also be noted that the data used to parameterize the basic model are independent of those used in validation.

Having obtained a satisfactory estimate of *r*, we can now use eqn (20.2) to examine the consequences of residual infection on total mortality by the end of the season for a single spray application. Moreover, by using different parameter values for *r* we can also speculate about the consequences of different residual infection profiles. This is reasonable given that a wide range of infection profiles have been reported in a number of studies conducted at different sites (Jenkins & Thomas, 1996; Thomas *et al.*, 1996a; Kooyman *et al.*, 1997). The reasons for these differences are not clear, although differences in application rate, vegetation density, rainfall, and cloud cover between years and sites are all likely to play a role. What is important is that the infectivity of the spray residue is variable and examining the relative impacts of different infection profiles is worthwhile with a view to potential manipulation. In Fig. 20.3(a), therefore, we examine the total mortality from a spray application in relation to direct spray contact rate, for a range of spray residues with different half-lives but with the same initial risk of infection of 0.156. In Fig. 20.3(b) we do a similar analysis but vary the initial infectivity of the spray residue and keep half-life constant at 6.8 days. These figures reveal that even small increases in half-life or initial infectivity of a spray residue can contribute significant additional mortality by the end of the season. For the highest values of $T_{\frac{1}{2}}$ and *P*, this additional mortality results in excellent overall mortality for even very low initial spray contact rates. This relative contribution of the spray residue declines as spray hit rate increases; if the spray contact rate is 100% there is nothing to be gained through residual infection. For the measured spray residue, which in this range

of parameter values has intermediate persistence and initial infectivity, an initial spray contact rate of 36% actually results in more than 80% mortality; a considerable increase over a spray with no residual activity and a major contribution to overall control.

Using eqn (20.2) we can also examine the long-term consequences of residual infection on grasshopper population dynamics. To do this we need an estimate of the finite rate of increase, *F*, of the grasshopper population and we need to define a spray control strategy. Unfortunately there are few estimates of *F* for Sahelian grasshoppers. However, population dynamic studies of *Zonocerus variegatus* (L.), a key pest species in subtropical Africa but also extending into the Sahelian zone (Steedman, 1990) have suggested a range of values up to $F = 10$ (Chapman *et al.*, 1979, 1986) so an intermediate value of $F = 5$ was selected for the model. Similarly, there are no good estimates of economic threshold densities or spray action thresholds. For the sake of the model we assume the population is sprayed whenever it exceeds a threshold density of 10 grasshoppers per m². This is an arbitrary figure representing intermediate grasshopper densities, but note that in the current case the threshold merely acts as linear scale factor, and altering it will not alter our general conclusions. We assume that spraying occurs no more than once a year.

In Fig. 20.4 we present contour plots of spray frequency as a function of direct spray contact rate and either half-life of the residue or initial risk of infection. The contours represent the number of times the pathogen is sprayed per year (averaged over many generations) in response to the grasshopper density exceeding the spray threshold. Spray frequencies higher than 1 represent parameter combinations where the pest population always exceeds the spray threshold, i.e., spraying would be required more than once a season to effect control. This region is represented by the shaded area in the figures.

Figure 20.4(a) clearly shows that for sprays with very low half-lives, high spray contact rates are necessary for effective control. Under these conditions, small additional increases in spray contact rate result in relatively large reductions in spray frequency. For low to intermediate spray contact rates, control is only

(a)

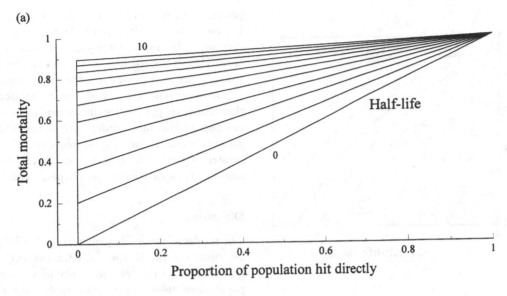

(b)

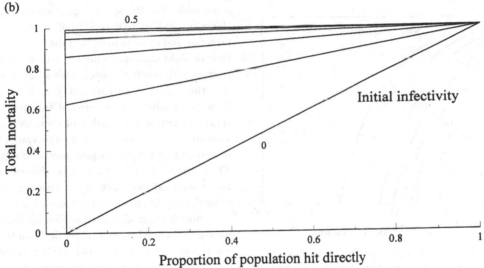

Figure 20.3. (a) The effect of persistence of a spray residue on total mortality after 90 days with respect to direct initial spray contact. The different lines show increases in the half-life of the spray residue from 0 to 10 days (shown in increments of 1 day) with the initial risk of infection set at $P = 0.156$. (b) As for (a) except the different lines show increases in the initial risk of infection of the spray residue from 0 to 0.5 (shown in increments of 0.1) with the half-life set at $T_{1/2} = 6.8$ days.

(a)

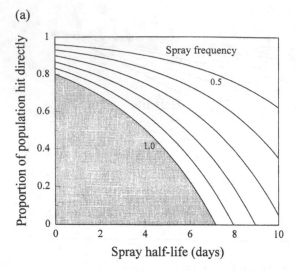

(b)

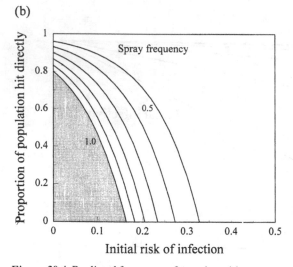

Figure 20.4. Predicted frequency of spraying with a *Metarhizium flavoviride*-based mycopesticide as a function of the direct contact rate of the spray and either (a) half-life of the spray residue with initial risk of infection set at $P = 0.156$, or (b) initial risk of infection with half-life set at $T_{1/2} = 6.8$ days. Spraying occurs when the grasshopper density is above a spray action threshold of $10/m^2$. The different lines show increases in spray frequency (average number per spray year) from 0.5 to 1.0 (shown in increments of 0.1). The shaded region indicates parameter combinations where spray frequency exceeds once per year. Mean finite rate of increase of the grasshopper population is $F = 5$.

possible with sprays that have long residual activity. For sprays that have little residual infection, repeated spraying within a season is required. A similar pattern is revealed in Fig. 20.4(b) for variation in the initial risk of infection. However, the spray frequency appears more sensitive to changes in initial infectivity than changes in persistence since a small proportional increase in the initial infectivity has a relatively greater effect on reducing spray frequency (and hence greater impact on long-term pest control) than an equivalent proportional increase in persistence.

Discussion

The results of this study demonstrate that infection *via* contact with the spray residue can contribute substantially to the total mortality of grasshopper populations following a spray application of *M. flavoviride*. For the spray trial conducted in north Benin in 1994, more than 50% of the predicted total infection and more than 40% of the observable infection in field-collected samples was attributable to contact with residual spores. This study also reveals that the extent of this mortality is determined by both the persistence and initial infectivity of the spray deposit; altering either of these by even small amounts can have a large effect on mortality within a season and on long-term pest population dynamics. These results show the value of combining empirical and theoretical techniques in interpreting and evaluating the impacts of spray applications.

Although the models used to identify these results were extrapolated from a single field study, they do not themselves rely on further field data since they only aim to identify what consequences different levels of persistence should have. The principal assumption in these models is that the pattern of infection from a spray residue follows a negative exponential. This is a biologically reasonable model given that several field-based studies have identified this to be the case (e.g., Jenkins & Thomas, 1996; Thomas *et al.*; 1996a; Kooyman *et al.*; 1997). These studies also reveal that the level of residual infection is variable and that both persistence and initial infectivity may be manipulated. These are important factors with respect to the economics of mycopesticide use and implementation of a control program. For example, in systems where direct contact between spray and insect is limited (such as the rice grasshopper system described here where a combination of high, dense grass vegetation together with the behavioral escape response of the host prevented high initial contact rates), contact with residual spores is essential for effective control. Hence, maximizing residual infectivity through increases in persistence and/or initial infectivity is likely to have a significant impact on the cost:benefit ratio of a control program. In contrast, under conditions where direct contact between spray and insect is not limiting, or where prolonged activity of the spray residue is considered undesirable (potentially in some conservation areas, for example), efforts to maximize efficiency of the spray application will be of greatest benefit. This could include increasing the initial infectivity of the residue but not the persistence. Thus, understanding the mechanisms that influence the pattern of residual infectivity is an important consideration in the development of optimum spray strategies.

A model with horizontal transmission

Population dynamic model

Having examined the effect of the spray residue, we now go on to examine the contribution of horizontal transmission of the pathogen to overall mortality following a spray application. The basic model developed to do this is illustrated in Fig. 20.5. The first thing to note is that in order to examine just the effects of horizontal transmission, the effects of direct contact with the spray and residual pick-up are combined into a single mortality figure. Following application the grasshoppers contacted by the pathogen incubate the disease for a certain time and then die. On death they enter a pool of non–infectious cadavers, which they have fixed probability per unit time of leaving, to enter the pool of infectious cadavers. These infectious individuals, which in the field represent sporulating cadavers, then have a fixed probability per unit time of leaving this stage, after which they play no further part in the dynamics. It is possible to object to this simple

formulation of the model since in reality each cadaver goes through a build up and then a decline in individual infectivity (see below). However, rigorous derivation of a model using this latter approach yields exactly the same model as the one just described (see Thomas *et al.*, 1995).

Mathematically the basic model looks like this:

$$\frac{dH}{dt} = -\beta A_1(t)H(t) \tag{20.4}$$

$$\frac{dA_1}{dt} = c[A_0(t) - A_1(t)] \tag{20.5}$$

$$\frac{dA_0}{dt} = \begin{cases} c\,[\beta A_1(t-\tau)H(t-\tau) - A_0(t)] & t > \tau \\ \quad -cA_0(t) & \text{otherwise} \end{cases} \tag{20.6}$$

H is the population of healthy grasshoppers, A_1 is proportional to the population of infectious cadavers and A_0 is proportional to the population of pre-infectious cadavers. τ is the delay between infection and death, β is the transmission coefficient from infectious cadavers to healthy hosts and c is a *per capita* rate of movement from one cadaver stage to the next. The model requires an initial pulse of infection in order to get started (typically a spray event). In model terms this instantaneously reduces H by the number of grasshoppers hit by spraying (say SH_0), and τ days later instantaneously increases A_0 by cSH_0. This model of the disease dynamics typically applies only for part of the year, since transmission can only occur during the wet season. As in the residue model, we therefore caricature the dry season host dynamics by multiplying the grasshoppers left alive at the end of the previous season by a 'Finite Rate of Increase' term, F, to obtain the expected host density at the start of the next season. The between-season changes in host population dynamics also include an immigration term, R, which represents the mean number of grasshoppers that arrive in an area independent of density last year due to redistribution between feeding and oviposition sites. Furthermore, although we do not include it in the initial model simulations outlined below, the model can also allow for carry-over of the pathogen between wet seasons by scaling infectivity of the cadaver pool by a constant proportion representing pathogen survival (see Thomas *et*

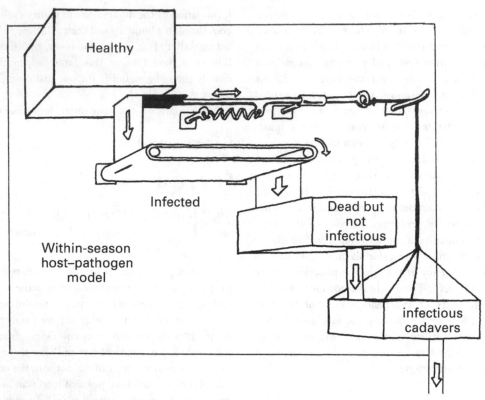

Figure 20.5. Diagrammatic representation of the basic host–pathogen model describing horizontal transmission of the pathogen. The top container represents the population of healthy hosts. Opening the door in the bottom of this container allows healthy hosts to become infected. The period of infection is represented by the conveyor belt. At the end of this belt hosts die and enter the pre-infectious cadaver bin, from which they pour out into the infectious cadaver bin. The contents of this bin controls the rate of infection of healthy hosts.

al., 1996b). Within a season, this basic form of the model omits all sources of mortality other than the pathogen: a conservative assumption if we are asking questions about how effective control is likely to be. The model also assumes a proportional mixing form for the transmission process, a standard assumption in many host–pathogen models (see Chapter 17).

Infectivity of cadavers

Once again, we use *H. daganensis* as a study organism. The infectivity of *H. daganensis* cadavers was estimated using a similar bioassay technique to that described above. However, instead of placing cages over a spray residue, populations of uninfected

hosts were exposed in field cages to bioassay for changes in infectivity through time of infected cadavers placed on the soil surface. Full details of these experiments are given in Thomas *et al.* (1995). For our current purposes it is sufficient to note that the transmission coefficient, β, and the transition rate for movement of cadavers between infectious stages, c, were estimated as 0.4 and 0.11, respectively.

Model simulations

As with the spray residue, we use the model to examine the total mortality of grasshopper populations when sprayed with a biopesticide with different

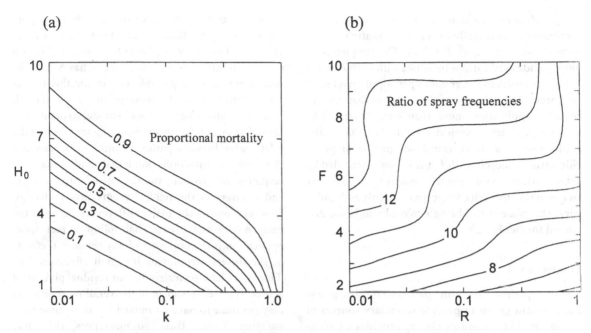

Figure 20.6. (a) The expected proportion of *Hieroglyphus daganensis* killed 90 days after spraying with *Metarhizium flavoviride*, as predicted by the model given in the text and using the experimentally determined parameters, c and β. The proportion killed is shown as a function of the proportion hit directly by spraying, k, and the initial grasshopper density H_0. (b) Predicted frequency of spraying with a non-persistent chemical pesticide divided by frequency of spraying with *M. flavoviride*, as a function of F, the mean finite rate of increase, and R, the mean number of grasshoppers that arrive in an area independent of the grasshopper density last year. Spraying is assumed to occur when the density is above a spray action threshold of $10/m^2$.

initial spray contact rates (Fig. 20.6(a)). This figure illustrates two important points. First, the bio-pesticide acts in a density-dependent manner. This results in high density populations suffering proportionately greater mortality for any given spray contact rate. Second, the pathogen is extremely efficient and even very low spray contact rates result in high mortality by the end of the season. This suggests the possibility for developing novel control strategies based on low-level pathogen applications.

The long-term consequences of this additional mortality (Fig. 20.6(b)) are measured by the relative advantage of the biopesticide over a chemical assuming that the chemical pesticide and biopesticide have the same initial spray contact rate of 50% and that contact with either spray is lethal (i.e., a simplifying assumption ignoring sublethal effects of either spray). The relative advantage is measured as the ratio of the predicted frequency of spraying with a non-persistent chemical pesticide to the frequency of spraying with a *Metarhizium*-based biopesticide for our spray action threshold of $10/m^2$. This ratio is examined as a function of F, the mean finite rate of increase, and R, the mean number of grasshoppers that arrive in an area independent of the grasshopper density last year. As stated previously, there are few estimates of F for Sahelian grasshoppers so $F = 10$ was selected as an upper limit. Similarly, no accurate estimates of R are available so a range of 0 to $1/m^2$ was selected to represent localized movement of *H. daganensis*, which exists in relatively discrete populations in specific habitats (M. B. Thomas, unpublished data and Steedman, 1990).

For these ranges of values, the biopesticide is between 6 and 13 times more efficient than the non-persistent chemical pesticide. The spray frequency of the chemical pesticide is insensitive to R but increases proportionally with F, necessitating several sprays a year at higher F values. The frequency of biopesticide applications increases with both F and R but always reduces the grasshopper population below the threshold by the end of the season so that it never requires application more than once a year. The contour plots are complicated by the fact that whilst application of a conventional non-persistent pesticide varies smoothly with F, the system controlled by a biopesticide tends towards multiyear cycles, the frequencies of which change quite sharply as F and R vary: this issue is still being explored and is not discussed further here.

Discussion

These experimental results demonstrate that grasshoppers can go on to provide secondary sources of pathogen. The secondary cycling provides a biological mechanism for enhancing the persistence of the biopesticide. The models demonstrate that this increases the efficacy of the biopesticide, especially when host densities are high, since under these conditions an epidemic is promoted, killing a high proportion of hosts. Interestingly, however, for the parameter ranges determined empirically for this model, the pathogen fails to control the grasshoppers in a sustainable manner. Although not shown here, whatever proportion of the pathogen we allow to persist from one year to the next, either the host or the pathogen is effectively driven to extinction. Since host extinction is most unlikely in practice, what this really means is that sustained 'classical' biological control is unlikely to succeed using this pathogen, at least under these model assumptions. Nonetheless, the results of this study do highlight the possibility of exploiting the biological properties of relatively ineffective indigenous pathogens to develop biopesticides that act in a density-dependent manner. This possibility has rarely been considered in the development or strategic use of entomopathogens in biological control. Unlike the residual infection example, the predictions from the simple horizontal transmission model have yet to be fully validated in the field. A number of studies have suggested

secondary cycling of the pathogen following spray applications (e.g., Baker *et al.*, 1994; Langewald *et al.*, 1997; Thomas *et al.*, 1997) but a major effect on overall mortality as indicated here has yet to be demonstrated. One possible reason for this is that some of the basic model assumptions may be flawed. In order to identify any errors in model structure and formulation, studies on a range of factors such as the relationship between pathogen and host density and the rate of infection, sublethal and behavioral responses to infection, and the fate of infected hosts and cadavers in the field, are currently underway. However, one of the major differences between the basic model and events that follow a real spray application is the omission of any sequential effects of the different routes of infection after spraying. That is, we have seen already that residual pick-up of spores can itself result in high overall mortality and may continue to have an impact for some time after spraying. Under these circumstances, the spray application does more than just provide a small pulse of infection. How this influences secondary cycling and which of these routes of infection is the most important in terms of practical control is unclear. This is examined in the following section.

The combined effects of residual infection and horizontal transmission on the impact of a spray application

The basic residual pick-up and horizontal transmission models can be combined easily to give a single model, for which the within-year dynamics are governed by the following equations:

$$\frac{\mathrm{d}H}{\mathrm{d}t} = -(\beta A_1(t) + Pe^{-at})H(t) \tag{20.7}$$

$$\frac{\mathrm{d}A_1}{\mathrm{d}t} = c[A_0(t) - A_1(t)] \tag{20.8}$$

$$\frac{\mathrm{d}A_0}{\mathrm{d}t} = \begin{cases} c[(\beta A_1(t-\tau) + Pe^{-a(t-\tau)})H(t-\tau) \\ \quad - A_0(t)] - cA_0(t) & \begin{array}{l} t > \tau \\ \text{otherwise} \end{array} \end{cases} \tag{20.9}$$

and between-year dynamics are unchanged, from the previous models. In Fig. 20.7 we show a number of

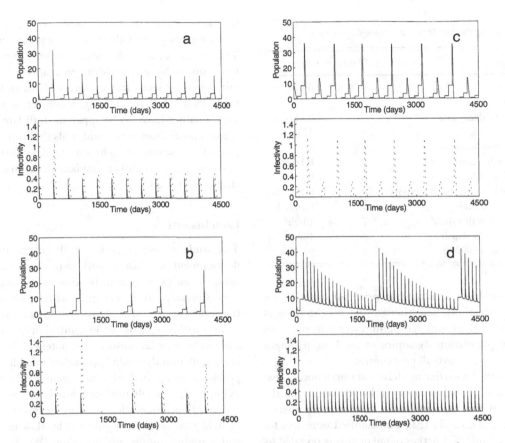

Figure 20.7. Dynamics of host and infectivity under different scenarios, illustrating the impact of different model components. Each pair of panels shows host population in the upper panel and infectivity due to spray residue (continuous lines) and secondary cycling (broken lines), in the lower panel. Only dynamics within the wet season are plotted, so the time axes are cumulative days within wet seasons.
(a) These panels are for a model including annual immigration, secondary cycling and residual pick-up.
(b) These panels are without annual immigration. (c) These panels leave out residual pick-up.
(d) These panels leave out secondary cycling.

outputs from the combined model that are based on our simple strategy of spraying once at the start of a season if grasshopper densities exceed 10/m². The parameters used are shown in Table 20.1. The overall conclusion from these outputs is that residual pick-up appears to kill more grasshoppers because it acts first while numbers are still high, but secondary cycling clears up those remaining. Thus, although residual pick-up produces rapid results, it is secondary cycling that will drastically reduce the host population next year (at least if a large enough area is treated) and which, in turn, contributes to better

overall control reducing spray frequencies and restricting population peaks. This result is important in any context in which the choice of pathogen strain or formulation involves a trade-off between direct and residual kill rates, and the potential for secondary cycling. For example, a trade-off between virulence and pathogen reproduction (i.e., spore production) has been noted for a number of fungal isolates tested in the LUBILOSA programme (LUBILOSA, unpublished data). Similar relationships have also been identified for certain viruses (Hails, 1997). Thus, although high virulence may be a desirable

Table 20.1. *Parameters used in model example*

Quantity	Symbol	Value
Infectivity m²/day	β	0.4
Infectivity decay rate	c	0.11/day
Time to death	τ	12 days
Host rate of increase	F	5
Host immigration	R	0.1/m²
Spray initial hit rate	S	0.15
Spray initial residual infectivity	P	0.156/day
Spray decay rate	a	0.102/day
Spray action threshold	T	10/m²
Initial healthy density	H	10/m²
Length of season	d	90 days

trait (virulence is often the principal criterion for isolate selection), selecting isolates on the basis of this factor alone may have unforeseen consequences for the population dynamics of the host–pathogen interaction and overall pest control.

As with the earlier models, it is worth noting that the general conclusion above hinges largely on the values of the experimentally estimated parameters (Table 20.1), rather than unquantified estimates for the host biology. Further detail on this is revealed by examining model outputs under slightly different scenarios. Figure 20.7(a) shows a basic run using the baseline parameter values set out in Table 20.1. Figure 20.7(b) shows the same model but with annual immigration (R) omitted. Figure 20.7(c) shows the model with residual pick-up omitted, and Fig. 20.7(d) shows the model with secondary cycling omitted.

It can be seen that omission of an annual immigration term leads to less predictable dynamics, although such a model is only appropriate if the area being sprayed is large enough such that untreated source populations for local immigration are removed. More interestingly, the omission of the high impact residual pick-up reveals that although spraying is required more often, it is still not necessary every year. Finally, the run in which secondary cycling is omitted emphasizes the importance of this mechanism; although residual pick-up causes high

mortality, spraying is required very frequently (almost every year) if there is no secondary cycling. Thus, it can be seen that under realistic conditions when direct hit and residual pick-up have a high initial impact, secondary cycling still has an important, albeit subtle role to play. Moreover, secondary cycling is not likely to be apparent until late in the season and so short-term field trials that run for just part of the season (which is a standard feature of nearly all trials to date) are not best suited to demonstrating its effects.

Conclusions

The study of insect–pathogen dynamics and the development of insect pathogens as biocontrol agents is an exciting and, in some ways, neglected area of research. In this chapter, we have focused on the development and application of insect–pathogen models within a particular biocontrol program. Our aim has been to demonstrate the potential for utilizing population dynamic approaches to address real problems in applied pest management. Here, for example, the models provide a link between laboratory and field observations and aid the interpretation of field trial results complicated by slow mortality and variable routes of infection. We have not attempted to address insect–pathogen interactions from a broad theoretical standpoint and instead refer readers to Godfray & Briggs (Chapter 17), in which there are some excellent studies on more fundamental aspects of insect–pathogen dynamics. That said, the work presented here does provide some novel insights relevant to the development of insect pathogens as biocontrol agents in general. Because of the biological nature of the active ingredient, a biopesticide may have a density-dependent component to its action, which could have significant consequences for the economics of biopesticide use. This is seen in the final model where although in absolute terms, secondary cycling of the pathogen appears to have little impact, its effect on reducing the frequency of spray applications is most pronounced. Related to this then, in order to fully evaluate the efficacy of a biopesticide, we need to understand that like any living control agent, the effectiveness of a pathogen depends not just on its capacity to kill pests but also its capacity to reproduce on pests and thereby continue and compound its killing action. The interaction between these functional and numerical components of biopesticide activity can be subtle and may not be apparent from short-term field studies. This may sound intuitively obvious but in a recent study examining the future potential of biopesticides, Waage (1997) concluded that much of the current development of biopesticides by the multinational pest control industry was aimed at exploiting the functional rather than the numerical responses of living natural enemies. This holds not only for products such as *Bt* and entomophilic nematodes but also viruses and fungi, since commercial development favors a traditional chemical pesticide model of quick kill, low persistence and frequent application. Waage (1997) also points out that new opportunities to increase speed of kill *via* genetic engineering will likely contribute to this further. What we identify here is that there is a need for an increased appreciation of both the similarities and differences between biopesticides and chemical pesticides, and that taking a more ecological approach could reveal opportunities for utilizing biological properties of biopesticides that fall outside the chemical model.

References

Amatobi, C. L., Apeji, S. A. & Oyidi, O. (1988) Effects of farming practices on populations of two grasshopper pests (*Kraussaria angulifer* Krauss and *Oedaleus senegalensis* Krauss (Orthoptera: Acrididae)) in northern Nigeria. *Tropical Pest Management* **34**: 173–9.

Baker, G. L., Milner, R. J., Lutton, G. G. & Watson, D. (1994) Preliminary field trial on control of *Phaulacridium vittatum* (Sjöstedt) (Orthoptera: Acrididae) populations with *Metarhizium flavoviride* Gams and Rozsypal (Deuteromycotina: Hyphomycetes). *Journal of the Australian Entomological Society* **33**: 190–2.

Bateman, R. P., Carey, M., Moore, D. & Prior, C. (1993) Oil formulations of entomopathogenic fungi infect desert locusts at low humidities. *Annals of Applied Biology* **122**: 145–52.

Bidochka, M. J. & Khachatourians, G. G. (1991) Microbial and protozoan pathogens of grasshoppers and locusts as potential biocontrol agents. *Biocontrol Science and Technology* **1**: 243–59.

Brader, L. (1988) Control of grasshoppers and migratory locusts. *BCPC Crop Protection Conference* 4A-1: 283–8.

Chapman, R. F., Page, W. W. & Cook, A. G. (1979) A study of population changes in the grasshopper, *Zonocerus variegatus*, in southern Nigeria. *Journal of Animal Ecology* **48**: 247–70.

Chapman, R. F., Page, W. W. & McCaffery, A. R. (1986) Bionomics of the variegated grasshopper (*Zonocerus variegatus*) in west and central Africa. *Annual Review of Entomology* **30**: 479–505.

Cunningham, G. L. (1992) APHIS: grasshopper integrated pest management in the United States – a cooperative project with emphasis on biological control. In *Biological Control of Locusts and Grasshoppers*, ed. C. J. Lomer & C. Prior, pp. 21–5. Wallingford: CAB International.

Douro-Kpindou, O. K., Godonou, I., Houssou, A., Lomer, C. J. & Shah, P. A. (1995) Control of *Zonocerus variegatus* with ULV formulation of *Metarhizium flavoviride* conidia. *Biocontrol Science and Technology* **5**: 131–9.

Hails, R. S. (1997) The ecology of baculoviruses: towards the design of viral pest control strategies. *BCPC Symposium Proceedings* **68**: 53–62.

Jenkins, K. (1994) Fungus wraps up locusts and grasshoppers. *Rural Research* **164**: 4–7.

Jenkins, N. E. & Thomas, M. B. (1996) Effect of formulation and application method on the efficacy of aerial and submerged conidia of *Metarhizium flavoviride* for locust and grasshopper control. *Pesticide Science* **46**: 299–306.

Joffe, S. R. (1995) *Desert Locust Management – A Time for Change*. World Bank Discussion Papers no. 284. Washington, DC: The World Bank.

Johnson, D. L., Goettel, M. S., Bradley, C., ven der Paauw, H. & Maiga, B. (1992) Field trials with the entomopathogenic fungus *Beauveria bassiana* against grasshoppers in Mali. In *Biological Control of Locusts and Grasshoppers*, ed. C. J. Lomer & C. Prior, pp. 296–310. Wallingford: CAB International.

Kooyman, C., Bateman, R. P., Langewald, J., Lomer, C. J., Ouambama, Z. & Thomas, M. B. (1997) Operational-scale application of entomopathogenic fungi for control of Sahelian grasshoppers. *Proceedings of the Royal Society of London, Series B* **264**: 541–6.

Langewald, J., Thomas, M. B., Lomer, C. J. & Douro-Kpindou, O. K. (1997) Use of *Metarhizium flavoviride* for control of *Zonocerus variegatus*: a model relating mortality in caged field samples with disease development in the field. *Entomologia Experimentalis et Applicata* 82: 1–8.

Lobo-Lima, M. L., Brito, J. M. & Henry, J. E. (1992) Biological control of grasshoppers on Cape Verde. In *Biological Control of Locusts and Grasshoppers*, ed. C. J. Lomer & C. Prior, pp. 287–95. Wallingford: CAB International.

Lomer, C. J. & Prior, C. (eds) (1992) *Biological Control of Locusts and Grasshoppers*. Wallingford: CAB International.

Lomer, C. J., Bateman, R. P., Godonou, I., Kpindou, D., Shah, P. A., Paraiso, A. & Prior, C. (1993) Field infection of *Zonocerus variegatus* following application of an oil-based formulation of *Metarhizium flavoviride* conidia. *Biocontrol Science and Technology* 3: 337–46.

Lomer, C. J., Thomas, M. B., Godonou, I., Shah, P., Douro-Kpindou, O.-K. & Langewald, J. (1997) Control of grasshoppers, particularly *Hieroglyphus daganensis*, in northern Benin using *Metarhizium flavoviride*. In *Microbial Control of Grasshoppers and Locusts*, ed. M. S. Goettel & D. L. Johnson, pp. 301–11. Memoirs of the Entomological Society of Canada, no. 171: Entomological Society of Canada.

Office of Technical Assessment (1990) *A Plague of Locusts – Special Report*. OTA-F-450. Washington, DC: US Government Printing Office.

Prior, C. & Greathead, D. J. (1989) Biological control of locusts: the potential for exploitation of pathogens. *FAO Plant Protection Bulletin* 3: 231–8.

Prior, C. & Streett, D. A. (1997) Strategies for the use of entomopathogenic agents in the biological control of locusts and grasshoppers. In *Microbial Control of Grasshoppers and Locusts*, ed. M. S. Goettel & D. L. Johnson, pp. 5–25. Memoirs of the Entomological Society of Canada, no. 171: Entomological Society of Canada.

Prior, C. Lomer, C. J., Herren, H., Paraiso, A., Kooyman, C. & Smit, J. J. (1992) The IIBC/IITA/DFPV collaborative research programme on the biological control of locusts and grasshoppers. In *Biological Control of Locusts and Grasshoppers*, ed. C. J. Lomer & C. Prior, pp. 8–20. Wallingford: CAB International.

Sekizawa, J., Eto, M., Miyamoto, J. & Matsuo, M. (1992) Fenitrothion. *Environmental Health Criteria 133*. Geneva: World Health Organization.

Steedman, A. (ed.) (1990). *Locust Handbook*, 3rd edn. Chatham: Natural Resources Institute.

Streett, D. A. & McGuire, M. R. (1990) Pathogenic diseases of grasshoppers. In *Biology of Grasshoppers*, ed. R. F. Chapman & A. Joern, pp. 483–516. New York: Wiley Interscience.

Thomas, M. B., Wood, S. N. & Lomer, C. J. (1995) Biological control of locusts and grasshoppers using a fungal pathogen: the importance of secondary cycling. *Proceedings of the Royal Society of London, Series B* 259: 265–70.

Thomas, M. B., Langewald, J. & Wood, S. N. (1996a). Evaluating the effects of a biopesticide on populations of the variegated grasshopper, *Zonocerus variegatus*. *Journal of Applied Ecology* 33: 1509 -16.

Thomas, M. B., Gbongboui, C. & Lomer, C. J. (1996b). Between-season survival of the grasshopper pathogen *Metarhizium flavoviride* in the Sahel. *Biocontrol Science and Technology* 6: 569–73.

Thomas, M. B., Wood, S. N., Langewald, J. & Lomer, C. J. (1997) Persistence of biopesticides and consequences for biological control of grasshoppers and locusts. *Pesticide Science* 49: 56–7.

Waage, J. K. (1997) Biopesticides at the crossroads: IPM products or chemical clones? *BCPC Symposium Proceedings* 68: 11–20.

Zimmermann, G., Zelazny, B., Kleespies, R. & Welling, M. (1994) Biological control of African locusts by entomopathogenic microorganisms. In *New Trends in Locust Control*, ed. S. Krall & H. Wilps, pp. 127–38. Eschborn, Germany: GTZ.

21

Dose–response relationships in biocontrol of plant disease and their use to define pathogen refuge size

KENNETH B. JOHNSON

Introduction

Biological control of plant disease is concerned with the effect of antagonistic microbes on the ability of a pathogen to survive and to infect its host (Cook & Baker, 1983; Handelsman & Stabb, 1996). Typically, in practical applications, high doses of antagonistic microbes are introduced inundatively or augmentatively to soil, seed or foliar environments. Once established in soil or on plant surfaces, populations of the microbes then suppress pathogen populations by one of several mechanisms including production of antibiotics, competition for essential resources, parasitism of pathogen propagules, or by induction of antimicrobial resistance responses in host tissues (Handelsman & Stabb, 1996).

Like biological control of insect pests or weeds, biocontrol of plant pathogens *via* inundative or augmentative introductions of antagonistic microbes is not usually 100% effective. Nonetheless, microbial antagonists can provide consistent suppression of disease and, consequently, some of these microbes are beginning to be used successfully in commercial agriculture. The use of these microbial products has been initiated without a strong theoretical foundation for understanding the relative effectiveness of individual antagonist strains. Recently, I hypothesized that quantitative concepts developed to describe host–pathogen interactions provided a valuable resource that could be applicable to interactions between antagonist and pathogen populations (Johnson, 1994). Evaluation of this hypothesis led to the extension of a pathogen dose–disease response relationship to account for the effect of the initial population size (dose) of an antagonistic microbe on the incidence of disease. The extended relationship describes disease suppression by inundatively introduced antagonists, but requires the inclusion of a refuge parameter (asymptote) that defines the maximum proportion of the pathogen population potentially rendered ineffective by the antagonist introduction. Consideration and measurement of pathogen refuge size provides a new framework for examining how mechanism and environment affect biological control, and for comparison of the relative efficacy of individual antagonist strains.

Dose–response model for biological control of plant disease

Pathogen dose–disease response relationships are commonly described by the expression:

$$y = 1 - e^{-kx} \tag{21.1}$$

where x is the inoculum density of a randomly distributed pathogen, y is the proportion of plant units that become diseased (infected), and k is a constant that represents the efficiency at which pathogen propagules infect plant units (Baker, 1978; Gregory, 1948; Van der Plank, 1975). Equation (21.1) is non-linear because, by chance, some plant units are infected more than once before others are infected for the first time. Gregory's (1948) multiple infection transformation $(-\ln(1-y) = kx)$ linearizes eqn (21.1).

Analogously, an equation that defines the effect of an antagonistic microbe on a pathogen population can be expressed as:

$$x_i / x = 1 - e^{-cz} \tag{21.2}$$

where z is antagonist density, c represents the efficiency of an antagonist propagule, and x_i / x is the proportion of effective pathogen units that become *ineffective* as a result of the antagonist introduction.

[385]

The variable, x_i/x, is intended to be inclusive of the various mechanisms by which inoculum of a pathogen can be rendered ineffective (Cook & Baker, 1983; Handelsman & Stabb, 1996).

An important modification to eqn (21.2) is the addition of an asymptote:

$$x_i/x = A[1 - e(-cz)] \qquad (21.3)$$

where the asymptote, A, represents the maximum proportion of effective pathogen units that can be rendered ineffective by antagonist introduction. The concept of an asymptote or 'refuge', which defines a proportion of the population not susceptible to influence or attack, has been observed previously in plant–pathogen interactions (Neher & Campbell, 1992), and was recently hypothesized to be an important factor influencing the impact of parasitoids introduced to control insect pests (Hawkins *et al.*, 1993; see also Chapter 4).

Because of small size, direct observation and differentiation of effective and ineffective pathogen propagules is not straightforward in most plant pathosystems. However, the magnitude of A and its effect on antagonist efficacy can be estimated by using the proportion of plant units diseased as a measure of the effectiveness of pathogen propagules. This is expressed mathematically by combining eqns (21.1) and (21.3):

$$y = 1 - e\{-kx[(1 - A) + Ae(-cz)]\} \qquad (21.4)$$

where the part of the equation to the right of x is the proportion of pathogen units that *remain* effective (i.e., $1 - x_i/x$).

In eqn (21.4), reductions in disease are dependent functionally on the density of the antagonist, and the prediction is made that disease control will be significant over a wide range of pathogen densities as long as the density of the antagonist and the value of the asymptote, A, remain high (Fig. 21.1(a)). However, as the value of A declines (and the refuge size for the pathogen increases), eqn (21.4) predicts that the amount of disease suppression obtained by an antagonistic microbe is more heavily influenced by the inoculum density of the pathogen (Figs 21.1(b) and (c)). Limits to disease suppression imposed by values of A that are less than 1.0 cannot be compensated for

by increasing antagonist dose (z) or the efficiency parameter, c. In contrast, reductions in c (e.g., 20%) can be compensated for by increasing the dose by the same proportion.

Values for the parameters A and c are most easily estimated from antagonist dose/plant disease response data by applying non-linear regression procedures to a transformed version of eqn (21.4):

$$-\ln(1 - y) = kx[(1 - A) + Ae(-cz)] \qquad (21.5)$$

Reasonable first estimates of kx (the product of pathogen efficiency × pathogen density, which at $z = 0$ is the y-intercept of eqn (21.5) and A are obtained as described by Johnson (1994). With these estimates, observed data can then be subjected to Gompertz-type transformation ($-\ln(-\ln[(1 - y)/e(-kx(1 - A))])$)) and regressed linearly on initial antagonist density (z) to provide a first estimate of c (Johnson, 1994). Initial estimates of A, c, and kx are then used in non-linear regression analysis to provide an iterative fit of eqns 21.4 or 21.5, and final estimates and standard errors of A and c. Rejection of the null hypothesis, H_0: $A = 1.0$, is essential in order to define a pathogen refuge.

Evidence for plant pathogen refuges in biological control

The first data to which eqn (21.5) was compared came from a study by Mandeel & Baker (1991). In this study, severity of Fusarium wilt of cucumber was evaluated in soils pre-amended with one of two non-pathogenic isolates of the fungus, *Fusarium oxysporum*, which are antagonistic to the cucumber pathogen, *Fusarium oxysporum* f. sp. *cucumerinum*. Parameter estimates for eqn (21.5) resulted in similar values of c for both antagonist isolates (i.e., isolates C5 and C14) but different values of the refuge parameter, A (Fig. 21.2(a); Johnson, 1994). The interpretation was that *F. oxysporum* isolate C14 has an ability to render ineffective a higher proportion of *F. o. cucumerinum* propagules than does isolate C5. In other words, the pathogen refuge was smaller when isolate C14 was introduced compared to when isolate C5 was introduced. To explore this result, an attempt was made to model the same data by varying c and

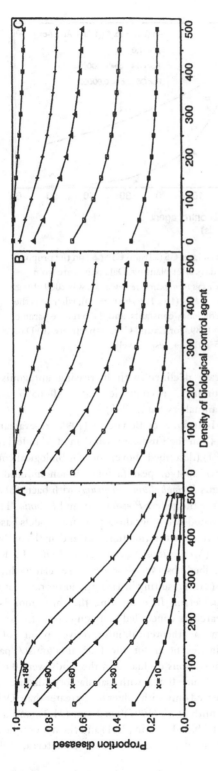

Figure 21.1. Proportion of disease (y) in response to antagonist density (z) at five densities of a pathogen (x). The curves are described by: $y = 1 - \exp(-kx((1 - A) + A\exp(-cz)))$, with values of the efficiency constants, k and c, assumed to be 0.035 and 0.007, respectively. The value of the refuge parameter, A, was either (a) 1.0, (b) 0.8, or (c) 0.6. ■, $x = 10$; □, $x = 30$; ▲, $x = 60$; +, $x = 90$; ⋈, $x = 180$.

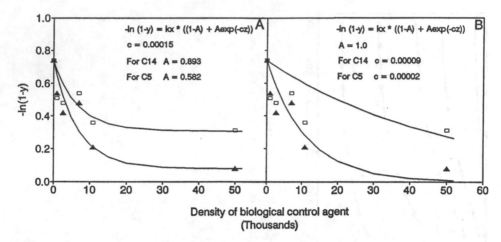

Density of biological control agent (Thousands)

Figure 21.2. Effect of non-pathogenic *Fusarium oxysporum* isolates C5 ($\square$) and C14 ($\blacktriangle$) on the proportion of cucumber plants diseased, y, with Fusarium wilt 55 days after planting. Data points are from Mandeel & Baker (1991); density of the pathogen, *Fusarium oxysporum cucumerinum* (x), was 2000 cfu/g. Curves were fit to the data based on the equation: $-\ln(1-y) = kx \cdot ((1-A) + A\exp(-cz))$, where z is the density of either isolate C5 or C14, k ($= 0.00037$) and c are efficiency constants, and A is a refuge parameter. (a) Data described with $c = 0.00015$ and $A = 0.582$ or $A = 0.893$ for C5 and C14, respectively. (b) Data described with $A = 1.0$ and $c = 0.00002$ or $c = 0.00009$ for C5 and C14, respectively.

assuming A had a value of 1.0 (no refuge). The dose–response relationship for antagonist isolate C14 could be approximated by a model in which only c was considered (Fig. 21.2(b)) but this conclusion could not be extended to the data for isolate C5. Thus, defining the difference between C5 and C14 in terms of their effect on a pathogen's refuge size was apparently superior to differentiating these antagonists solely on the basis of an efficiency constant.

The hypothesis that the efficacy of inundatively applied antagonistic microbes can be summarized with two parameters (A and c) has led to several additional efforts to evaluate concepts contained within eqns (21.4) and (21.5). In a growth chamber study, Raaijmakers *et al.* (1995) obtained responses similar to those shown in Fig. 21.2(a) for the biological control of Fusarium wilt of radish (caused by *Fusarium oxysporum* f.sp. *raphani*). Suppression of Fusarium wilt was provided by a strain of *Pseudomonas putida* or by a strain of *P. fluorescens*, for which the stated mechanisms of biocontrol for these bacteria were siderophore-mediated iron competition, and the induction of systemic-acquired host resistance, respectively. For each strain, incidence of

disease declined with increasing antagonist dose, becoming asymptotic at 40–65% control (Raaijmakers *et al.*, 1995).

Montesinos & Bonaterra (1996) re-evaluated eqn (21.4) on the Fusarium wilt data of Mandeel & Baker (1991) (described above), and for biological control of brown spot of pear (a foliar lesion induced by the fungus *Stemphylium vesicarum*) with bacterial antagonists in the genera *Pseudomonas* and *Erwinia*. They also evaluated two other dose–response models based on a hyperbolic saturation function, and on the probit function (Montesinos & Bonaterra, 1996). Each model described observed dose–response relationships, with eqn (21.4) resulting in a slightly lower range of relative model error. In comparing the three models, these researchers concluded that eqn (21.4) also was useful from a theoretical/interpretive point of view. Estimates of A for the four antagonist/ pathogen combinations evaluated in the study were 0.65, 0.93, 0.95, and 0.97, with satisfactory disease control observed only with values of A greater than 0.93.

Smith *et al.* (1997) used eqn (21.4), and a hybrid model based on eqn (21.1) and the hyperbolic saturation model (Montesinos & Bonaterra, 1996), to develop an analytical approach by which they could assess the relative susceptibility of different tomato genotypes to damping-off by the fungus, *Pythium torulosum*, in combination with an assessment of the ability of the same tomato genotypes to support populations of bacterial antagonists of *P. torulosum*. Numerous dose–response experiments led to the conclusion that antagonist dose–disease response modeling was useful for measuring the individual contributions of host resistance and antagonistic microbes on disease suppression. Host genotypes that were poor supporters of bacterial antagonists were characterized as having small values of *A*. Comparing the hybrid model to eqn (21.4), Smith *et al.* concluded that the hybrid model provided a better overall fit of the data, but that eqn (21.4) provided the most precise estimates of pathogen refuge size. Estimates of *A* ranged from 0.0 to 0.94 for individual host genotype/antagonist strain combinations; effective biological control was observed only for those combinations with values of *A* in the range of 0.87 to 0.94.

Many other studies have examined the effect of dose of an antagonistic microbe on the development of plant disease (Adams & Fravel, 1990; Bull *et al.*, 1991; Mandeel & Baker, 1991; Vanneste *et al.*, 1992; Fukui *et al.*, 1994a; Korsten *et al.*, 1995; Schisler *et al.*, 1997). Nearly all of these studies, however, assumed that disease suppression was a linear function of the logarithm of the applied antagonist density, for which extrapolation to higher densities results in greater disease suppression. For several of these studies, back-transformation of the initial antagonist density to an arithmetic scale revealed that the relationship between the density and the degree of disease suppression became asymptotic (Johnson, 1994). In each case, these asymptotes appeared to define a level of disease suppression that could not be exceeded with increased doses of the antagonist.

Significance of a refuge concept for the biocontrol of plant disease

Equation (21.4) implies that microbes antagonistic to plant pathogens are governed by two parameters, one that accounts for the relative efficiency of the microbe propagule in rendering pathogen propagules ineffective, and one that defines the proportion of the pathogen propagules that are potentially susceptible to the antagonist. In contrast to older dose–response models used commonly in the evaluation of biocontrol of plant pathogens (Bull *et al.*, 1991; Fukui *et al.*, 1994a; Korsten *et al.*, 1995; Schisler *et al.*, 1997), eqn (21.4) directly addresses the question as to whether higher antagonist doses can compensate for deficiencies in disease suppression. Equation (21.4), for example, predicts that increased antagonist density (z) can compensate for decreased efficiency of the antagonist (c); however, if eqn (21.4) also contains a large refuge (small value of *A*), increased density of the antagonist microbe will cause only negligible increases in disease control. Determining that pathogen refuges exist and that they define a maximum level of disease suppression is valuable for understanding the limitations that apply to biological control systems. This is particularly important because pathogen refuges, while defined in experiments testing large doses of antagonist inoculum, affect the efficiency of disease suppression over the full range of antagonist densities (Fig. 21.2(a)).

Acceptance of the refuge concept could change expectations when implementing biological control strategies. For example, realization that dose–response curves become asymptotic may lead to more efficient use of antagonist inoculum with the recognition that most benefits are obtained with relatively low doses and that increasing dose above a certain point will not appreciably increase disease control. Equation (21.4) also outlines a meaningful approach to the problem of antagonist evaluation. Because *A* is proportional, values of this parameter can be compared among strains and pathosystems. Empirical 'rules of thumb' between an antagonist's value of *A* and the likelihood of economic disease control may result. Future efforts also may find that strategies designed to enhance biocontrol, such as the use of antagonist mixtures, also may involve changing pathogen refuge size. For instance, in research concerned with biocontrol of postharvest rots of apple, Janisiewicz (1996) has shown that mixtures of antagonistic microbes that were selected to broaden the nutritional niche occupied by the control organisms can suppress infection by fruit rot fungi more effectively than either strain alone.

Factors affecting pathogen refuge size

Modeling antagonist dose–plant disease response relationships can lead to the recognition of a pathogen refuge within a specific system, but the models themselves reveal little as to why the refuge exists. Among plant pathosystems, it is likely that several factors affect refuge size. These factors include non-uniform or incompletely overlapping distributions of the pathogen and antagonist populations (Adams, 1990), differential rates of response to environmental stimuli among pathogen and antagonist populations (e.g., temperature, water potential: Cook & Baker, 1983; Paulitz, 1990), and differential abilities among pathogens and antagonists to compete for nutritional resources (Loper & Ishimaru, 1991; Wilson & Lindow, 1994). Furthermore, fungal propagules or bacterial cells are commonly located in aggregates (Adams, 1990) or assemblages (Foster *et al.*, 1983; Fukui *et al.*, 1994b) that may not be equally accessible or susceptible to an antagonistic microbe. Pathogen populations also are heterogeneous genetically, and thus, individual propagules may vary in their ability to compete for limited resources (Loper, 1990), or in their sensitivity to antibiotics (Sule & Kado, 1980; Mazzola *et al.*, 1994; Handelsman & Stabb, 1996).

As an example, we have recently conducted dose–response experiments to examine antibiotic production by an antagonistic microbe for its effect on pathogen refuge size. These experiments measured suppression of crown gall of tomato, caused by the bacterium *Agrobacterium tumefaciens*, by two strains of the antagonistic microbe, *Agrobacterium radiobacter*. Strain K84 produces the antibiotic, agrocin (Kerr & Htay, 1974), whereas the other strain, K84*agr⁻*, does not. Biocontrol of a strain of *A. tumefaciens* resistant to agrocin (Cooksey & Moore, 1982) also was included as a treatment in the experiment. Results in Fig. 21.3 show that the effect of antagonist dose on crown gall incidence was

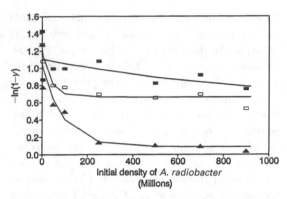

Figure 21.3. Incidence of crown gall of tomato as affected by the dose of *Agrobacterium radiobacter* applied prior to inoculation with the pathogen, *A. tumefaciens*. Data points represent incidence of disease (corrected for multiple infection) resulting from: (i) treatment with an agrocin-producing strain of *A. radiobacter* followed by inoculation with an agrocin-sensitive strain of the pathogen (▲); (ii) treatment with an agrocin-deficient strain of *A. radiobacter* followed by inoculation with an agrocin-sensitive strain of the pathogen (□); and (iii) treatment with an agrocin-producing strain of *A. radiobacter* followed by inoculation with an agrocin-resistant strain of the pathogen (■).

described by eqn (21.5), and that agrocin production by *A. radiobacter* (or resistance to agrocin in the pathogen) contributed significantly to the estimated size of the pathogen refuge (J. A. DiLeone & K. B. Johnson, unpublished results). The value of A estimated for the combination of the agrocin-producing antagonist and agrocin-sensitive pathogen was 0.92. Values of A for the agrocin-deficient antagonist and agrocin-sensitive pathogen, and for the agrocin-producing antagonist and agrocin-resistant pathogen were 0.46 and 0.47, respectively.

Summary

Models that describe or predict the effects of introduced antagonists on plant disease suppression are essential for the optimization of biological control in agriculture. A dose–response model developed for plant pathogen–plant host interactions was extended to the relationship between an antagonist and its effect on pathogen and plant populations. The derived model describes observed antagonist dose–plant disease response relationships, but requires inclusion of a refuge parameter that defines the maximum proportion of a pathogen population potentially rendered ineffective by an antagonist introduction. The size of this refuge greatly influences the amount of disease suppression obtained by the use of a biological agent. The refuge concept for inundative biocontrol of plant disease extends and reaffirms the premise that plant pathogen populations are heterogeneous, and that suppression of the disease they cause faces certain limitations owing to this heterogeneity. Practically, this approach and its associated equations become important to biological control because clear methods are established for evaluating and mathematically describing these limitations. The model also provides a framework for evaluating hypotheses concerned with how the mechanism of biological control and environment relate to refuge size, and for empirical comparison of the relative efficacy of individual antagonist strains. The refuge concept inherent within the model may serve as a bridge to the theory of biological control of other organisms.

References

Adams, P. B. (1990) The potential of mycoparasites for biological control of plant diseases. *Annual Review of Phytopathology* 28: 59–72.

Adams, P. B. & Fravel, D. R. (1990) Economical biological control of Sclerotinia lettuce drop by *Sporidesmium sclerotivorum*. *Phytopathology* 80: 1120–4.

Baker, R. (1978) Inoculum potential. In *Plant Disease: An Advanced Treatise, Volume II, How Disease Develops in Populations*, ed. J. G. Horsfall & E. B. Cowling, pp. 137–57. New York: Academic Press.

Bull, C. T., Weller, D. M. & Thomashow, L. S. (1991) Relationship between root colonization and suppression of *Gaeumannomyces graminis* var. *tritici* by *Pseudomonas fluorescens* strain 2–79. *Phytopathology* 81: 954–9.

Cook, R. J. & Baker, K. F. (1983) *The Nature and Practice of Biological Control of Plant Pathogens*. St. Paul, MN: The American Phytopathological Society.

Cooksey, D. A. & Moore, L. W. (1982) High frequency spontaneous mutations to agrocin 84 resistance in *Agrobacterium tumefaciens* and *A. rhizogenes*. *Physiological Plant Pathology* 20: 129–35.

Foster, R. C., Rovira, A. D. & Cock, T. W. (1983) *Ultrastructure of the Root–Soil Interface*. St. Paul, MN: The American Phytopathological Society.

Fukui, R., Schroth, M. N., Hendson, M., Hancock, J. G. & Firestone, M. K. (1994a) Growth patterns and metabolic activity of pseudomonads in sugar beet spermospheres: relationship to pericarp colonization by *Pythium ultimum*. *Phytopathology* 84: 1331–8.

Fukui, R., Poinar, E. I., Bauer, P. H., Schroth, M. N., Hendson, M., Wang, X.-L. & Hancock, J. G. (1994b) Spatial colonization patterns and interaction of bacteria on inoculated sugar beet seed. *Phytopathology* 84: 1338–45.

Gregory, P. H. (1948) The multiple infection transformation. *Annals of Applied Biology* 35: 412–17.

Handelsman, J. & Stabb, E. (1996) Biocontrol of soilborne plant pathogens. *Plant Cell* 8: 1855–69.

Hawkins, B. A., Thomas, M. B. & Hochberg, M. E. (1993) Refuge theory and biological control. *Science* 262: 1429–32.

Johnson, K. B. (1994) Dose-response relationships and inundative biological control. *Phytopathology* 84: 780–4.

Janisiewicz, W. (1996) Ecological diversity, niche overlap, and coexistence of antagonists used in developing mixtures for biocontrol of postharvest diseases of apples. *Phytopathology* 86: 473–9.

Kerr, A. & Htay, K. (1974) Biological control of crown gall through bacteriocin production. *Physiological Plant Pathology* 4: 37–44.

Korsten, L., de Jager, E. S., De Villiers, E. E., Lourens, A., Kotzé, J. M. & Wehner, F. C. (1995) Evaluation of bacterial epiphytes isolated from avocado leaf and fruit surfaces for biocontrol of avocado postharvest diseases. *Plant Disease* 79: 1149–56.

Loper, J. E. (1990) Molecular and biochemical basis for activities of biological control agents: the role of siderophores. In *New Directions in Biological Control*, ed. R. Baker & P. E. Dunn, pp. 735–47. New York: Alan R. Liss.

Loper, J. E. & Ishamuru, C. A. (1991) Factors influencing siderophore-mediated biocontrol activity of rhizosphere *Pseudomonas* spp. In *The Rhizosphere and Plant Growth*, ed. D. L. Keister & P. B. Cregan, pp. 253–61. Amsterdam: Kluwer Academic Publishers.

Mandeel, Q. & Baker, R. (1991) Mechanisms involved in biological control of Fusarium wilt of cucumber with strains of nonpathogenic *Fusarium oxysporum*. *Phytopathology* 81: 462–9.

Mazzola, M., Fujimoto, D. K. & Cook, R. J. (1994) Differential sensitivity of *Gaeumannomyces graminis* populations to antibiotics produced by biocontrol fluorescent pseudomonads. *Phytopathology* 84: 1091. (Abstract).

Montesinos, E. & Bonaterra, A. (1996) Dose-response models in biological control of plant pathogens: an empirical verification. *Phytopathology* 86: 464–72.

Neher, D. A. & Campbell, C. L. (1992) Underestimation of disease progress rates with the logistic, mono-molecular, and Gompertz models when maximum disease is less than 100 percent. *Phytopathology* 82: 811–14.

Paulitz, T. C. (1990) Biochemical and ecological aspects of competition in biological control. In *New Directions in Biological Control*, ed. R. Baker & P. E. Dunn, pp. 713–24. New York: Alan R. Liss.

Raaijmakers, J. M., Leeman, M., van Oorschot, M. M. P., van der Sluis, I., Schippers, B. & Bakker, P. A. H. M. (1995) Dose-response relationships in biological control of Fusarium wilt of radish by *Pseudomonas* spp. *Phytopathology* 85: 1075–81.

Schisler, D. A., Slininger, P. J. & Bothast, R. J. (1997) Effect of antagonist cell concentration and two-strain mixtures on biological control of Fusarium dry rot of potatoes. *Phytopathology* 87: 177–83.

Smith, K. P., Handelsman, J. & Goodman, R. M. (1997) Modeling dose-response relationships in biological control: partitioning host responses to the pathogen and biocontrol agent. *Phytopathology* 87:720–9.

Sule, S. & Kado, C. I. (1980) Agrocin resistance in virulent derivatives of *Agrobacterium tumefacians* harboring the pTi plasmid. *Physiological Plant Pathology* 17: 347–56.

Van der Plank, J. E. (1975) *Principles of Plant Infection*. New York: Academic Press.

Vanneste, J. L., Yu, J. & Beer, S. V. (1992) Role of antibiotic production by *Erwinia herbicola* Eh252 in biological control of *Erwinia amylovora*. *Journal of Bacteriology* 174: 2785–96.

Wilson, M. E. & Lindow, S. E. (1994) Ecological similarity and coexistence of epiphytic ice-nucleating (ice$^+$) *Pseudomonas syringae* strains and a non-ice-nucleating (ice$^-$) biological control agent. *Applied and Environmental Microbiology* 60: 3128–37.

Index